ADVANCES IN

Applied Microbiology

VOLUME 15

CONTRIBUTORS TO THIS VOLUME

George B. Boder

Charles L. Cooney

L. E. Erickson

L. T. Fan

John C. Godfrey

Ladislav J. Haňka

R. H. Haskins

W. F. Hink

Irving S. Johnson

I. C. Kao

David W. Levine

Vedpal S. Malik

Carlos J. Muller

Kenneth E. Price

Joseph L. Sardinas

P. S. Shah

Irwin W. Sizer

K. L. Smiley

G. W. Strandberg

A. Dinsmoor Webb

ADVANCES IN

Applied Microbiology

Edited by D. PERLMAN

School of Pharmacy
The University of Wisconsin
Madison, Wisconsin

VOLUME 15

 1972

ACADEMIC PRESS, New York and London

COPYRIGHT © 1972, BY ACADEMIC PRESS, INC.
ALL RIGHTS RESERVED.
NO PART OF THIS PUBLICATION MAY BE REPRODUCED OR
TRANSMITTED IN ANY FORM OR BY ANY MEANS, ELECTRONIC
OR MECHANICAL, INCLUDING PHOTOCOPY, RECORDING, OR ANY
INFORMATION STORAGE AND RETRIEVAL SYSTEM, WITHOUT
PERMISSION IN WRITING FROM THE PUBLISHER.

ACADEMIC PRESS, INC.
111 Fifth Avenue, New York, New York 10003

United Kingdom Edition published by
ACADEMIC PRESS, INC. (LONDON) LTD.
24/28 Oval Road, London NW1

LIBRARY OF CONGRESS CATALOG CARD NUMBER: 59-13823

PRINTED IN THE UNITED STATES OF AMERICA

CONTENTS

LIST OF CONTRIBUTORS.. ix

Medical Applications of Microbial Enzymes

IRWIN W. SIZER

I.	Introduction	1
II.	Enzymes Active in the Digestive System...	2
III.	Enzymes Active on the Blood Coagulation System	4
IV.	Use of Microbial Enzymes in the Treatment of Infection, Inflammation, and Burns..	5
V.	Use of Microbial Enzymes in the Treatment of Cancer	6
VI.	Microbial Enzymes Active on Connective Tissue	8
VII.	Medical Applications of Bacterial Enzymes *in Vitro*...........................	10
	References	11

Immobilized Enzymes

K. L. SMILEY AND G. W. STRANDBERG

I.	Introduction	13
II.	Carriers...	14
III.	Properties of Bound Enzymes	26
IV.	Methods of Application..	30
V.	Potential Applications..	32
VI.	Current Status and Future Prospects	35
	References	35

Microbial Rennets

JOSEPH L. SARDINAS

I.	Introduction	39
II.	Rennets: General Considerations...	40
III.	Higher Plant Rennets	45
IV.	Animal Rennets	46
V.	Microbial Rennets..	48
VI.	General Comments	65
	References	66

Volatile Aroma Components of Wines and Other Fermented Beverages

A. Dinsmoor Webb and Carlos J. Muller

I.	Introduction	75
II.	List of Volatile Aroma Compounds	77
III.	Biosynthetic Pathways and Interrelationships among Aroma Compounds	131
IV.	Summary	142
	References	142

Correlative Microbiological Assays

Ladislav J. Haňka

I.	Introduction	147
II.	Materials and Methods	148
III.	Results	149
IV.	Discussion	154
V.	Summary	156
	References	156

Insect Tissue Culture

W. F. Hink

I.	Introduction	157
II.	Equipment and Sources of Supplies	158
III.	Media Formulation	160
IV.	Nutrition and Metabolism	166
V.	Primary Cultures	169
VI.	Cell Lines	180
VII.	Virus Studies	190
VIII.	Protozoa Studies	202
IX.	Rickettsia Studies	202
X.	Interactions of Biologically Active Materials with Insect Cells and Tissues	203
XI.	Conclusion	207
	References	209

Metabolites from Animal and Plant Cell Culture

Irving S. Johnson and George B. Boder

I.	Introduction	215
II.	Macromolecules	216
III.	Hormones	219
IV.	Alkaloids, Glycosides, Steroidal Glycosides	220
V.	Lipids, Sterols, and Steroids	221

VI.	Miscellaneous Metabolites	223
VII.	Future Applications	225
	References	227

Structure-Activity Relationships in Coumermycins

John C. Godfrey and Kenneth E. Price

I.	Introduction	231
II.	The Natural Coumermycins	232
III.	Semisynthetic Coumermycins	250
	References	294

Chloramphenicol

Vedpal S. Malik

I.	Introduction	297
II.	Chloramphenicol as a Growth Inhibitor	299
III.	Effect on Morphology	300
IV.	Mode of Action	300
V.	Structure-Activity Relationships	303
VI.	Differential Effects on Proteins and Protein Synthesis	305
VII.	Effect on RNA Metabolism	307
VIII.	Effect on Energy-Producing Systems	308
IX.	Effect on Cell Wall Biosynthesis	310
X.	Effect on Biosynthesis of Aromatic Compounds	311
XI.	Effect on Antibiotic Biosynthesis	312
XII.	Resistance to Chloramphenicol	312
XIII.	Chloramphenicol as a Mutagen	318
XIV.	Chloramphenicol as a Therapeutic Agent	318
XV.	Chloramphenicol as a Secondary Metabolite	319
XVI.	Biosynthesis of Chloramphenicol	321
XVII.	Catabolism of Chloramphenicol by the Producing Organism	323
XVIII.	Effect of Chloramphenicol on the Producing Organism	323
XIX.	Regulation of Chloramphenicol Biosynthesis	328
XX.	Effect of p-Nitrophenylserinol on Chloramphenicol Biosynthesis	330
	References	331

Microbial Utilization of Methanol

Charles L. Cooney and David W. Levine

I.	Introduction	337
II.	Aerobic Utilization of Methanol	338
III.	Anaerobic Utilization of Methanol	350
IV.	Single-Cell Protein from Methanol	355
V.	Fermentation Products from Methanol	362
VI.	Summary and Conclusions	362
	References	363

Modeling of Growth Processes with Two Liquid Phases: A Review of Drop Phenomena, Mixing, and Growth

P. S. Shah, L. T. Fan, I. C. Kao, and L. E. Erickson

I.	Introduction	367
II.	Dispersion and Coalescence Phenomena of Liquid Drops	368
III.	Mathematical Modeling	402
	Nomenclature	409
	References	410

Microbiology and Fermentations in the Prairie Regional Laboratory of the National Research Council of Canada 1946–1971

R. H. Haskins

I.	Historical	415
II.	Microbial Studies	419
III.	Higher Plant Studies and Plant Cell Culture	433
IV.	In Retrospect—and the Future	434
	References	435

Author Index	437
Subject Index	461
Contents of Previous Volumes	463

LIST OF CONTRIBUTORS

Numbers in parentheses indicate the pages on which the authors' contributions begin.

GEORGE B. BODER, *The Lilly Research Laboratories, Eli Lilly and Company, Indianapolis, Indiana* (215)

CHARLES L. COONEY, *Department of Nutrition and Food Science, Massachusetts Institute of Technology, Cambridge, Massachusetts* (337)

L. E. ERICKSON, *Department of Chemical Engineering, Kansas State University, Manhattan, Kansas* (367)

L. T. FAN, *Department of Chemical Engineering, Kansas State University, Manhattan, Kansas* (367)

JOHN C. GODFREY, *Bristol Laboratories, Division of Bristol-Myers Company, Syracuse, New York* (231)

LADISLAV J. HAŇKA, *Research Laboratories, The Upjohn Company, Kalamazoo, Michigan* (147)

R. H. HASKINS, *National Research Council of Canada, Prairie Regional Laboratory, Saskatoon, Saskatchewan, Canada* (415)

W. F. HINK, *Department of Entomology, The Ohio State University, Columbus, Ohio* (157)

IRVING S. JOHNSON, *The Lilly Research Laboratories, Eli Lilly and Company, Indianapolis, Indiana* (215)

I. C. KAO, *Department of Chemical Engineering, Kansas State University, Manhattan, Kansas* (367)

DAVID W. LEVINE, *Department of Nutrition and Food Science, Massachusetts Institute of Technology, Cambridge, Massachusetts* (337)

VEDPAL S. MALIK, *Department of Nutrition and Food Science, Massachusetts Institute of Technology, Cambridge, Massachusetts* (297)

CARLOS J. MULLER, *Department of Chemistry, Louisiana Tech University, Ruston, Louisiana* (75)

KENNETH E. PRICE, *Bristol Laboratories, Division of Bristol-Myers Company, Syracuse, New York* (231)

JOSEPH L. SARDINAS, *Central Research Division, Pfizer Inc., Groton, Connecticut* (39)

P. S. SHAH*, *Department of Chemical Engineering, Kansas State University, Manhattan, Kansas* (367)

IRWIN W. SIZER, *Department of Biology, Massachusetts Institute of Technology, Cambridge, Massachusetts* (1)

K. L. SMILEY, *Northern Regional Research Laboratory, Agricultural Research Service, United States Department of Agriculture, Peoria, Illinois* (13)

G. W. STRANDBERG, *Northern Regional Research Laboratory, Agricultural Research Service, United States Department of Agriculture, Peoria, Illinois* (13)

A. DINSMOOR WEBB, *Department of Viticulture and Enology, University of California, Davis, California* (75)

*Present address: Argonne National Laboratories, Argonne, Illinois.

Medical Applications of Microbial Enzymes

IRWIN W. SIZER

Department of Biology,
Massachusetts Institute of Technology,
Cambridge, Massachusetts

I.	Introduction	1
II.	Enzymes Active in the Digestive System	2
III.	Enzymes Active on the Blood Coagulation System	4
IV.	Use of Microbial Enzymes in the Treatment of Infection, Inflammation, and Burns	5
V.	Use of Microbial Enzymes in the Treatment of Cancer	6
VI.	Microbial Enzymes Active on Connective Tissue	8
VII.	Medical Applications of Bacterial Enzymes *in Vitro*	10
	References	11

I. Introduction

The modification of foods by microbial enzymes prior to ingestion was practiced in ancient times and occurred long before primitive man had any knowledge of what an enzyme was or how it operated. Early examples of enzyme application were in cheese manufacture, breadmaking, bating of hides, and a host of fermentation processes involved in the brewing of beer, wine, spirits, and vinegar. Microbial enzymes as digestive aids were slow to be scientifically applied but despite this fact were sometimes effective as incidental contaminants of food. Yeast has been used medically not only as a source of vitamins but also to combat constipation and to stimulate normal digestion by the action of yeast proteases and amylases.

Since animal and plant enzymes as well as microbial enzymes are widely used in medicine, one must ask how microbial enzymes differ, if at all, from the rest. In general, there is a remarkable similarity between microbial enzymes and those of higher organisms. Enzyme parameters, such as pH optimum, activation energy, temperature sensitivity, inhibition, substrate specificity and affinity, are likely to be quite comparable but by no means identical properties of enzymes in all living organisms (Gutfriend, 1965). At the molecular level the same is also true. The amino acid sequence of a particular enzyme bears considerable resemblance in different organisms although the further apart two species are phylogenetically the greater the number of amino acid substitutions which have occurred in the primordial peptide chain (Byers *et al.*, 1970). In the case of enzymes which exist in a polymeric form, the active form may not always be identical in the microbial and animal species. For example, alcohol dehydrogenase of yeast has a molecular weight of 153,000 and contains four molecules of NAD (nicotinamide-adenine dinucleotide) and four atoms of zinc, while that of horse liver has a molecular weight of 73,000 and contains two residues each of NAD and zinc (Drum and Vallee, 1970). Although the differences between microbial enzymes and those of higher forms may seem small, the variations may well be significant enough to make it critically important in medical applications to use one enzyme in preference to another. Since most enzymes which are used medically are not crystalline, the impurities in the preparation cannot be ignored since they may interfere with the catalyzed reaction. Bacterial enzymes in particular may contain pyrogens or toxins which may have serious adverse effects on the patient. Hence, microbial enzymes often must be extensively purified before they can be used; e.g., collagenase from *Clostridium welchii* is too dangerous for human applications until all the toxin has been removed by successive purification steps (Mandl, 1961). Since all enzymes are proteins, they behave as antigens when injected and eventually stimulate the production of antibodies which, in addition to causing the usual allergic reactions, may combine with the enzyme and inactivate it. As might be predicted, a patient who has in this way become allergic or refractory to an enzyme from one organism may still react beneficially to the comparable enzyme from another species (Sizer, 1964).

A convenient way to review the medical applications of microbial enzymes is to consider them with reference to systems of the body or, in some cases where this is impractical, to deal with individual diseases. In the case of certain enzymes which exert beneficial activity on several body systems, mention will be made in several places in this review.

II. Enzymes Active in the Digestive System

Enzymes taken orally may be useful because of their activity in the digestive tract or because of their action on some other part of the body after their absorption from the small intestine. In either case they may need to be protected from hydrolysis by pepsin or pancreatic proteases through the use of enteric coatings or by binding the enzymes to insoluble particles such as plastics or porous glass (Weetall, 1969).

Microbial enzymes which supplement the action of amylase in saliva or pepsin in gastric juice are not very effective digestive aids because of the short time available for action in the mouth and because most are inert in the highly acidic stomach (pH about 2.0). Of some interest is the use of enzyme preparations from *Aspergillus oryzae* and *A. niger* in the form of a toothpaste by scientists in several different pharmaceutical companies. These fungal enzymes produced a significantly greater retardation of calculus (tartar) formation and the accumulation of soft accretions on teeth as compared with controls. In addition, tobacco staining was also lessened by the enzyme preparation. Along similar lines, Fitzgerald *et al.* (1968) used a culture filtrate of *Penicillium funiculosum* containing the enzyme dextranase. When caries (induced by *Streptococcus* infections) of hamster teeth were treated with dextranase (dispersed in food or drinking water), coronal plaque deposits on the molar teeth decreased and the progression of the disease was retarded. It is suggested that the enzyme brings about this beneficial effect by degrading the microbially produced extracellular polysaccharides of the dextran type which are present in the plaque matrix.

Enzymatic digestive aids effective in the small intestine have been extensively used for a long period of time. They are most often fungal in origin and preparations from *Aspergillus oryzae* and *A. niger* are the most commonly used, because of their high content of amylase and protease. Certain foods which contain indigestible cellulose fibers such as cucumbers, cabbage, and radishes may produce dyspepsia, flatulence, and other digestive disturbances. This condition may be relieved by cellulase of *Aspergillus oryzae* or *Tricoderma viride* but the latter has been approved for human use only in Japan (Cayle, 1962). Lipase from *Aspergillus oryzae* or *Candida lipolytica* taken orally may be useful in patients who have fatty stools, but careful limitation of fat in the diet is also indicated (Underkofler *et al.*, 1958). Of special interest is the possibility of treatment of genetic disease in which a single enzyme is lacking in the body. Administration of the missing enzyme by an oral or intravenous route may prove useful, and considerable progress in this area can be expected in the near future. An example is seen in the treatment of children who are unable to digest lactose in milk because of a deficiency in lactase (β-D-galactosidase). Addition of a crude lactase preparation from *Aspergillus niger* to the diet results in the digestion of lactose and the absorption of the glucose and galactose from the intestine (Cayle, 1962).

As in the case of enzymes which can be added to foods, the FDA (Food and Drug Administration) severely limits the enzymes on the GRAS (Generally Recognized As Safe) list which can be taken orally to those from *Saccharomyces cerevisiae* and *S. fragilis*, *Aspergillus niger* and *A. oryzae*, and *Bacillus subtilis*.

III. Enzymes Active on the Blood Coagulation System

The major problem in the cardiovascular system is the formation of blood clots in the arteries and veins and smaller blood vessels of the body. Thromoembolism can be serious wherever it occurs in the body, but can be quickly fatal if it occurs in the coronary artery of the heart or branches of the carotid artery of the brain. Although the blood coagulation system is most complicated and involves many enzyme-catalyzed steps, major attention has focused upon the last step: the conversion of soluble fibrinogen to insoluble fibrin, the major component of the clot. Enmeshed in the fibrin network of the clot are blood cells, especially erythrocytes, white blood cells, platelets, and plasma proteins. Microbial enzymes have been employed both for blocking the coagulation system and for dissolving the clot after it has been produced in an insoluble form. Many of the enzymes injected into the blood to bring about the dissolution of clots exert their action by acting as proteolytic enzymes to split a single arginyl-valine bond of plasminogen (profibrinolysin) thereby converting it to plasmin (fibrinolysin). The plasmin is responsible for digesting the fibrin in the clot (Sherry, 1968) thereby effecting its dissolution. Once digestion of a thrombus (a clot on the wall of a blood vessel) begins it may become dislodged and carried to another resting place to become a dangerous embolus if it clogs a critical blood vessel. Hence, it is crucial that a circulating blood clot be quickly and completely digested.

Although fibrinolysin can be used directly for the lysis of blood clots (Chazov, 1966), typically enzymes have been used which convert the normally occurring blood zymogen (profibrinolysin) to fibrinolysin. Most extensively employed for this purpose is streptokinase from a β-hemolytic *Streptococcus*. Until recently this preparation of streptokinase has been very crude (Varidase) and is contaminated by streptodornase (deoxyribonuclease) in large amounts. A relatively pure streptokinase has been prepared by Heimberger *et al.*, of Behringwerke, A. G. (Editorial, *Chem. & Eng. News*, 1971) who found that the enzyme molecule is fairly small (mol. wt. 47,000) consisting of a single polypeptide chain stabilized by disulfide bonds. The rate of conversion of plasminogen to plasmin by streptokinase is accelerated by *o*-thymoctic acid (Editorial, *Chem. & Eng. News*, 1971). Partially purified streptokinase may be administered buccally, intradermally, intramuscularly, parenterally, or intravenously—the latter is the method of choice when rapid action is required.

Despite the fact that millions of dollars have been spent on research and thousands of patients have been treated with streptokinase, clinical experience with this enzyme varies tremendously (Hume,

1970). On the negative side impurities in the poorly standardized preparations can cause trouble; action of streptokinase is not confined to fibrin (fibrinogen and other plasma proteins may also be hydrolyzed); some patients are sensitive to this bacterial preparation while others become immunized against streptokinase so that it is no longer effective. Highly purified (pyrogen-free) and carefully standardized streptokinase can, however, produce a satisfactory plasma thrombolytic state that is usually well tolerated except for impaired hemostasis. In a typical patient a high level of streptokinase in plasma can incite lysis of experimental and naturally occurring thromboemboli. The enzyme has found use in the treatment of a great many types of thromboembolic problems in several different organs of the body. For example, in a recent extensive study using streptokinase in acute myocardial infarction the mortality from this condition was halved.

Because of the difficulties obtained with a pyrogen-containing crude bacterial enzyme and because human blood contains some antistreptokinase, many workers have recently turned to the use of urokinase, a human enzyme obtained from urine (Sherry, 1968) with some initial success in converting plasminogen to plasmin. In view of the fact that thrombosis is the primary health hazard of white adults it is essential that research continue on thrombolytic enzymes with emphasis on both streptokinase and urokinase.

Because of the significance of the problem a variety of other microbial enzymes are being explored for possible use in thromboembolic diseases. Pisano *et al.* (1963) have reported on the interesting fibrinolytic properties of three different fungi of the genus *Cephalosporium*. Similarly other workers have obtained lysis of blood clots in cancer patients using a filtrate of a culture of *Aspergillus oryzae* B-1273. This enzyme preparation was only slightly toxic and was nonantigenic. Doubtless other useful microbial fibrinases will be found in the near future (Sakakihara *et al.*, 1970).

IV. Use of Microbial Enzymes in the Treatment of Infection, Inflammation, and Burns

In most types of infection and trauma there is an accumulation of pus, microorganisms, blood cells, serum, clots, and other kinds of debris. In view of this situation, it is not surprising to learn that enzymes, especially proteases, have been widely used clinically to treat such conditions. Although enzymes are most successful when applied locally, they have also been administered by other routes. In the case of enzymes which have been used as antiinflammatory agents they are often taken orally. Despite the fact that typical enzymes are inactivated in the digestive tract and would not be expected to be absorbed into the blood stream, many clinicians nonetheless have reported good results in treating inflammation by administering enzymes buccally. Although plant and animal enzymes seem to be preferred for the oral route, some use has also been made of microbial enzymes, especially streptokinase-streptodornase (Varidase) and *Aspergillus* enzymes (Bioprase, Nargase) (Underkofler *et al.*, 1958; Sameshima, 1966). Bruises, edema, sprains, and infectious areas of the body are often improved after such treatment. These same enzymes have also been applied locally for cleaning up dirty wounds, ulcers and necrotic tissue, infections, and for the debridement of third degree burns. When injected systemically the streptokinase-streptodornase preparation can relieve pain in traumatized areas (e.g., "black and blue" areas of the skin) by dissolving the blood clot. This enzyme preparation has also been used in the treatment of corneal herpes, facial edema, urinary and genital infections, acute otitis, and a variety of infectious conditions. While proteases are most commonly used for these conditions, other types of enzymes have also been employed in special conditions. For example, hyaluronidase, which hydrolyzes mucopolysaccharides, has been used as a "spreader" to assist local anesthetics in penetrating tissues. Similarly lysozyme from *Micrococcus lysodeikticus*, which attacks glycoproteins of bacterial walls, has proven valuable in the treatment of eye infections and of gastrointestinal disease. It is suggested that lysozyme may have some antiviral as well as antibacterial activity (Oldham, 1967).

Microbial enzymes used to treat infection are commonly administered in conjunction with other agents such as antibiotics. A very interesting medical application of an enzyme is its administration to those patients who are highly sensitive to penicillin or who inadvertently have received an overdose of it. Administration of penicillinase brings quick relief to such individuals (Jarvin and Berridge, 1969).

V. Use of Microbial Enzymes in the Treatment of Cancer

Differences between malignant and normal cells are quite subtle and are usually associated with biochemical factors related to the very rapid rate of growth of tumor cells. With this in mind streptodornase (Varidase) has been applied to Ehrlich tumor cells to hydrolyze their DNA (Nuzhina *et al.*, 1970). Lysozyme has also been used in the treatment of some types of cancer (Oldham, 1967). In situations where necrosis is involved in association with the tumor the proteolytic enzymes discussed in the previous section may find limited application in "cleaning up" the area. On the whole, however, these digestive enzymes have not been highly successful in causing the tumor to disappear.

An exciting breakthrough in the enzymatic treatment of cancer has resulted from the discovery of a metabolic difference between certain tumor and host cells. The tumor cell requires an extracellular source of L-asparagine and in its absence protein synthesis fails to occur in the cancer. L-Asparaginase hydrolyzes L-asparagine to aspartic acid and ammonia and is an enzyme found in many bacteria, but is often lacking in mammals (Wade and Rutter, 1970). When asparaginase from *Escherichia coli* is injected over a period of days into mice suffering from certain types of leukemia, marked improvement and often apparent cures occurred. The regression of lymphoma tumors in mice was not always permanent, however, and indeed the malignant cells often became resistant to the asparaginase. Riley virus tumors in mice, however, were completely cured by this enzyme (Wade and Rutter, 1970).

Patients suffering from acute lymphoblastic leukemia and treated with *E. coli* asparaginase showed up to 60% remission which lasted for 1 to 8 months. Relapse usually occurred despite continued treatment (Page and Alvarez, 1970). Undesirable side effects in some patients included nausea and vomiting, anorexia, and a change in indices of coagulation. Since asparaginase is a foreign protein it acts as an antigen with the result that repeated administration may lead to allergic reactions and production of antibodies which may inactivate the asparaginase. Despite these many disadvantages asparaginase has proven useful in the treatment of several kinds of leukemia, especially when used in conjunction with other anticancer drugs, in particular predisone and vincristine (Scott, 1968; Beard, 1970).

Because of the problems with *E. coli* asparaginase, other sources have been examined in the hopes of finding an enzyme with a low molecular weight (150,000), high affinity for substrate, and low toxicity. A pathogenic bacterium, *Erwinia caratovorum*, which produces soft rot in many vegetables, contains asparaginase many times more active than the *E. coli* enzyme (Wade and Rutter, 1970). Large-scale production of the *Erwinia* enzyme has made possible successful trials in curing lymphosarcoma in dogs: The purified enzyme is now undergoing extensive clinical trial against leukemia in humans.

An important modification in the method of application of asparaginase is to microencapsulate it before intravenous injection (Chang, 1971). Spherical ultrathin polymer membranes were prepared by polymerizing 1.6 hexamethylenediamine in the presence of a stabilized enzyme suspension. The semipermeable microcapsules containing the enzyme did not leak but permitted rapid diffusion of asparagine and end products into and out of the capsule. By using these microcapsules it is possible to avoid the hypersensitivity and immunological reactions normally produced by *E. coli* asparaginase. In addition, the encapsulated enzyme remained active in the blood for a much longer time. In comparison with the nonencapsulated enzyme the encapsulated form was much more effective in suppressing the growth of implanted mouse lymphosarcoma (Chang, 1971).

The possibilities of using microencapsulated enzymes for other medical applications are most intriguing, since such preparations will have the advantage of low toxicity, high stability, prolonged and controlled enzyme activity, and no foreign body reaction. Various techniques using a variety of plastics (especially nylon) are now undergoing development. Closely related are techniques for attaching enzyme molecules by chemical bonds to polymer molecules. Especially useful for this purpose is a porous type of glass granule being developed by Weetall and his associates (1970). As soon as enzymes which are either encapsulated or bonded to polymers become generally available they will be tried clinically for the treatment of a variety of medical problems.

Other biochemical differences between tumor and normal cells in addition to the ability to synthesize asparagine exist and are being actively explored. It appears that some myeloid leukemic cells cannot synthesize sufficient serine. Hence, the enzyme serine dehydratase might be of clinical value in treatment of this leukemia (Wade and Rutter, 1970). Similarly carboxypeptidase may inhibit the growth of leukemia cells by causing a depletion of cellular folic acid (Bretino *et. al.*, 1971).

VI. Microbial Enzymes Active on Connective Tissue

Since collagen is the major component of connective tissue, one could have predicted that there might be medical applications of an enzyme which cleaves the peptide bonds of collagen. Native collagen is highly resistant, however, to enzymatic attack and hence it required intensive research before an effective collagenase became available for medical use. By far the best source is *Clostridium histolyticum* of which several strains are now available which produce collagenase in high yield with minimum toxicity. Because clostridial toxin is so lethal, however, it is necessary to carefully purify the enzyme before it can be used on humans. The medical applications of collagenase, which will be published as a monograph in 1972, were discussed in a recent symposium. The various reports of the symposium have been summarized in a preliminary manner by Mandl (1970). Collagenase has proven to be a useful enzyme for growing cells in tissue culture since the enzyme does not damage cell membranes and hence can be used to disperse cells. Human skin fragments dispersed by bacterial collagenase could then be used as a multiple graft system to speed up the healing of third degree burns. The separated skin cells applied to the debrided burn site become foci for skin regeneration. In tissue culture collagenase can also be used to prevent the accumulation of connective tissue fibers in a human carcinoma as it is transferred through successive generations of rats, thereby rendering the tumor indefinitely transplantable.

In tooth transplantation Shulman et al. as reported by Mandl (1970) have treated the tooth allographs with bacterial collagenase prior to transplantation. As a result of the dissolution of collagen fibers in the periodontal ligament early rejection of the tooth was prevented in rhesus monkeys. In addition, the enzyme treatment reduced inflammation after transplantation, increased ankylosis, and resulted in prolonged survival of the tooth graft.

Collagenase may turn out to be a useful enzyme in the treatment of patients suffering from low back pain due to a "slipped disk." In this condition cartilage of the intervertebral disk protrudes and exerts pressure on the spinal nerve roots. Sterile collagenase injected into the nucleus pulposa between the vertebrae of dogs resulted in the dissolving of the cartilage without damage to surrounding tissue (Sussman and Mann, 1969). Similarly action by collagenase on the nucleus pulposa of cadavers was brought about with dissolution of fibrocartilage without action on hyaline cartilage or bone collagen. It should be noted that injection of chymopapain into the nucleus pulposa of patients suffering from "slipped disk" resulted in an attack on the chondromucoprotein and rapid relief of the prolapsed disk syndrome (Stern, 1969).

Cryoprostatectomy has been developed recently as a procedure for surgically removing the prostate after freezing it with the cryoprobe. In dogs the residual slough was subsequently dissolved by injection of collagenase into the prostate prior to inserting the cryoprobe. The collagenase prevented the residual slough from clogging the urethra yet did not damage vital tissues. Collagenase proved much more effective than other enzymes and produced total dissolution of the slough in 18 hours (Mandl, 1970).

Second and third degree burns have been treated with bacterial collagenase in a large number of patients prior to skin grafting. The enzyme appears to be equally useful in the treatment of human dermal ulcers. Beneficial effects on burns and ulcers can occur in as short a time as 3 days. Overall satisfactory results were obtained in 80% of the patients in 14 days. Healing of the exposed underlying tissue progresses well with no keloid formation or contracture (Mandl, 1970). Action of other proteases on burns and ulcers has been discussed in a previous section of this paper.

VII. Medical Applications of Bacterial Enzymes *in Vitro*

Because of their high specificity and rapid action enzymes are especially useful for quantitative chemical assay in the clinical laboratory. Measurement of common and exotic components of blood and urine using microbial enzymes in the assay procedure is yielding to automation in the hospital laboratory. In addition, the measurement of normal and abnormal enzymes in blood or amniotic fluid has become a useful diagnostic tool. Details of the use of microbial enzymes in the clinical laboratory have been presented in a number of monographs on the subject (Abderhalden, 1961; Dioguardi, 1961; Cruickshank, 1965; Netter, 1966).

A very interesting application of microbial enzymes is their use in the biosynthesis of medically important compounds. Microbial enzymes have proven useful in the production of corticosteroids from inert precursors. In this regard enzymes can bring about steroid hydroxylation, hydrogenation and dehydrogenation, epoxidation and cleavage of side chains. The steroid raw material diosgenin (from yams) is converted chemically to progesterone which is then oxidized by a fungal enzyme to cortisone (Oldham, 1967). It is also possible to convert by the use of microbial enzymes certain antibiotics to derivatives which possess interesting antibiotic and medical properties. For example, penicillin can be hydrolyzed to 6-amino-penicillanic acid and phenylacetic acid by penicillin acylase of *E. coli* or *Streptococcus lavendulae*. The penicillanic acid can in turn be used for the synthesis of potent nonallergenic penicillin derivatives (Jarvin and Berridge, 1969). It seems highly likely that there will develop in the future an increase in the use of enzymes for bringing about specific steps in the synthesis of complex chemical compounds which exhibit high pharmacological activity.

Acknowledgments

It is a pleasure to acknowledge the bibliographic assistance of Dr. Eugene R. L. Gaughran and Dr. Tibor Sipos of the Johnson and Johnson Research Center, of Dr. Edith Martin of Ethicon, Inc., and of Dr. Alfred Kupferberg of the Ortho Research Foundation.

References

Abderhalden, R. (1961). "Clinical Enzymology." Van Nostrand-Reinhold, Princeton, New Jersey.
Beard, M. E. J. (1970). *Brit. Med. J.* **1**, 191.
Bretino, J. R., O'Brien, P., and McCullough, J. L. (1971). *Science* **172**, 161.
Byers, U. S., Lambeth, D., Lardy, H. A., and Margoliash, E. (1970). *Fed. Proc., Fed. Amer. Soc. Exp. Biol.* **30**, 1286 (abstr.).
Cayle, T. (1962). *Wallerstein Lab. Commun.* **25**, 349.
Chang, T. M. S. (1971). *Nature (London)* **229**, 117.
Chazov, E. I. (1966). *Z. Gesamte Inn. Med. Ihre Grenzgeb.* **20**, 727.
Cruickshank, R. (1965). "Medical Microbiology," 11th ed. Williams & Wilkins, Baltimore, Maryland.
Dioguardi, N., ed. (1961). "European Symposium on Medical Enzymology." Academic Press, New York.
Drum, D. E., and Vallee, B. L. (1970). *Biochemistry* **9**, 4078.
Editorial. (1971). *Chem. & Eng. News* **49**, 18.
Fitzgerald, R. J., Keyes, P. H., Stoudt, T. H., and Spinell, D. M. (1968). *J. Amer. Dent. Ass.* **76**, 301.
Gutfriend, H. (1965). "An Introduction to the Study of Enzymes." Blackwell, Oxford.
Hume, D. (1970). *Arch. Surg. (Chicago)* **101**, 653.
Jarvin, B., and Berridge, N. J. (1969). *Chem. Ind. (London)* p. 1721.
Mandl, I. (1961). *Advan. Enzymol.* **23**, 164.
Mandl, I. (1970). *Science* **169**, 1234.
Netter, E. (1966). "Medical Microbiology," 5th ed. Davis, Philadelphia, Pennsylvania.
Nuzhina, A. M., et al. (1970). *Vop. Onkol.* **16**, 99.
Oldham, S. (1967). *Mfg. Chem. Aerosol News* August, 47.
Page, I. H., and Alvarez, W. C. (1970). *Mod. Med.* **12**, 28.
Pisano, M. S., Oleniacz, W. S., Mason, R. T., Fleischman, A. I., Vaccaro, S. E., and Catalano, G. R. (1963). *Appl. Microbiol.* **11**, 111.
Sakakihara, Y., Biery, D. N., Polishhook, R. D., Wagner, D. E., Baum, S., and Nusbaum, M. (1970). *Surg., Gynecol. Obstet.* **130**, 821.
Sameshima, H. (1966). *Acta Urol., Jap.* **12**, 1143.
Scott, R. B. (1968). *Proc. Roy. Soc. Med.* **61**, 471.
Sherry, S. (1968). *Ann. Intern. Med.* **69**, 415.
Sizer, I. W. (1964). *Advan. Appl. Microbiol.* **6**, 207.
Stern, I. J. (1969). *Clin. Orthop. Related Res.* **67**, 42.
Sussman, B. J., and Mann, M. (1969). *J. Neurosurg.* **31**, 628.
Underkofler, L. A., Barton, R. R., and Bennert, S. S. (1958). *Appl. Microbiol.* **6**, 212.
Wade, H. E., and Rutter, D. A. (1970). *Sci. J.* **6**, 62.
Weetall, H. H. (1969). *Science* **166**, 615.
Weetall, H. H. (1970). *Biochim. Biophys. Acta* **212**, 1.

Immobilized Enzymes

K. L. SMILEY AND G. W. STRANDBERG

*Northern Regional Research Laboratory, Agricultural Research Service,
United States Department of Agriculture, Peoria, Illinois*

I.	Introduction	13
II.	Carriers	14
	A. Cellulose and Cellulose Derivatives	14
	B. Polysaccharides other than Cellulose	15
	C. Organic Polymers	16
	D. Inorganic Supports	18
	E. Cross-Linking	19
III.	Properties of Bound Enzymes	26
	A. Stability	26
	B. Kinetic Patterns	28
IV.	Methods of Application	30
	A. Stirred Tank Reactors	30
	B. Columns	31
	C. Other Methods	31
V.	Potential Applications	32
	A. Industrial Uses	32
	B. Medical Uses	34
	C. Other Uses	34
VI.	Current Status and Future Prospects	35
	References	35

I. Introduction

Immobilized enzymes constitute a new class of heterogeneous catalysts with a high degree of specificity. They have potential application in medicine, food and industrial processing, waste treatment, and as research tools.

Water-insoluble enzymes are easily removable reagents that provide precise control over enzymatic reactions. In general, they are more stable under operating conditions than their soluble counterparts. Because they are readily recovered, they can be reused many times. Their insoluble nature makes it possible also to use them in continuous reactors.

There are four general methods of immobilizing enzymes: (1) They can be adsorbed on such materials as ion exchangers and clays. (2) They can be entrapped in gel lattices where they are physically restrained. (3) Enzymes can be covalently linked to an insoluble carrier through functional groups of the protein not involved in reactive sites. (4) Bifunctional reagents can cross-link the enzyme protein and cause insolubilization.

This review will concern itself with immobilization techniques, properties of immobilized enzymes, and possible applications of this new technology.

II. Carriers

A. CELLULOSE AND CELLULOSE DERIVATIVES

Cellulose has received much attention as a support for insolubilization of enzymes. The hydrazide of carboxymethyl cellulose (CM-cellulose) has been widely used as a support material. Micheel and Ewers (1949) and Mitz and Summaria (1961) used the Curtius azide method to attach enzymes to CM-cellulose. The method involves forming the methyl ester of CM-cellulose followed by reaction with hydrazine to form cellulose-hydrazide. This is then diazotized in the normal manner to form the diazonium salt that will couple with proteins, mainly with phenolic residues of tyrosine and, to a lesser extent, with either the indole nucleus of tryptophan or the imidazole nucleus of histidine.

Another successful method described by Kay and Lilly (1970) involves reacting 2-amino-4,6-dichloro-s-triazine with diethylaminoethyl cellulose (DEAE-cellulose). The product is then reacted with the enzyme, and coupling takes place at the position occupied by the remaining chlorine atom.

Barker et al. (1968a) attached α-amylase to microcrystalline cellulose by forming the 2-hydroxy-3-(p-nitrophenoxy) propyl ethers of cellulose from 1,2-epoxy-3-(p-nitrophenoxy) propane. The aryl nitro group was reduced to the aryl amine, which was diazotized by HNO_2 in N HCl. The diazonium salt was then reacted with the enzyme at a pH of 7.7.

Weliky et al. (1969) attached horseradish peroxidase to CM-cellulose by reacting a mixture of CM-cellulose, N,N'-dicyclohexylcarbodiimide, and the enzyme at 4°C for several days. The carbodiimide reacts directly with the carboxyl groups of CM-cellulose, and the method is recommended for enzymes labile to alkaline solutions.

Lynn and Falb (1969) attached several enzymes to aminoethyl cellulose (AE-cellulose) by cross-linking the cellulose derivative to the enzyme via glutaraldehyde. First a Schiff base is formed between the AE-cellulose and excess glutaraldehyde. Excess glutaraldehyde is required to minimize cross-linking between the amino groups of AE-cellulose. This reaction is followed by Schiff base formation with the second aldehydic group of glutaraldehyde and NH_2-protein.

An analogous procedure was described by Balcom et al. (1971) in which glutaraldehyde was used to cross-link catalase to DEAE-cellulose.

Anion exchange celluloses can also bind enzymes by virtue of their charge properties. Mitz (1956) showed that catalase adsorbed to DEAE-cellulose was active in the combined form. Complexing must be done with low ionic-strength buffers since addition of high ion concentrations will elute the enzyme from the carrier. Several enzymes have been found to be active in this form. Many workers have avoided this method of attachment to ion-exchange celluloses because they fear that the enzyme would be gradually eluted from the matrix. However, Mitz (1956), Tosa et al. (1966), Suzuki et al. (1966), Bachler et al. (1970), and Smiley (1971) found no detectable leaching from DEAE-cellulose during long periods of using the adsorbed enzymes. Nikolaev and Mardashev (1961) had comparable results with asparaginase ionically bound to CM-cellulose. Ionic binding is advantageous because the procedure is simple to perform and economical.

B. Polysaccharides other than Cellulose

1. Dialdehyde Starch

Goldstein et al. (1970) describe the immobilization of several enzymes on 90% oxidized dialdehyde starch. The polymer is reacted with methylenedianiline to form Schiff-base adducts to dialdehyde starch. The degree of cross-linking is minimized by addition of four to five parts by weight of methylenedianiline to one part of dialdehyde starch. Approximately 13% of the methylenedianiline remains available for diazotizing and coupling to an enzyme. Binding capacities are reportedly from 10 to 30 mg of protein/100 mg of carrier with approximately one-half of the bound enzyme showing activity.

2. Starch Anthranilates

Mehltretter and Weakley (1972) attached papain to cyanoethylated strach anthranilates via diazotization. The insoluble derivative showed appreciable activity against casein. This carrier is not readily degraded by microbial action, and the enzyme-starch complex showed good stability to storage at 40°C.

3. Dextran

Cross-linked dextran, known commercially as Sephadex,[1] has been

---

[1] The mention of firm names or trade products does not imply that they are endorsed or recommended by the Department of Agriculture over other firms or similar products not mentioned.

used extensively as a support for insolubilization of enzymes. Axén and Porath (1966) coupled several enzymes to p-amino-phenoxyhydroxypropyl ethers of Sephadex G-200. The amino group on the swollen polymer was activated by converting it to an isothiocyano group with thiophosgene. The isothiocyano-Sephadex was then coupled to the enzymes.

O'Neill et al. (1971a) attached chymotrypsin to soluble dextran by reaction of the dextran with 2-amino-4,6-dichloro-s-triazine followed by coupling to chymotrypsin. Although this derivative is soluble, the same procedure used with Sephadex would yield an insoluble derivative.

Axén et al. (1967a) and Axén and Ernback (1971) coupled enzymes to Sephadex by using cyanogen bromide to activate the polymer. A reactive imidocarbonate results that reacts with NH_2-protein under mild alkaline conditions to form an N-substituted imidocarbonate which, in turn, hydrolyzes to an isourea or N-substituted carbamate. Other polyhydroxy polymers, such as cellulose or agarose, can substitute for Sephadex.

Tosa et al. (1967) found that amino acid acylase could be tightly bound to DEAE-Sephadex by adsorption. Columns of amino acid acylase-DEAE-Sephadex complex were used to form L-amino acids from DL-acylamino acids. The columns were operated for several days without any evidence of the enzyme leaching from the columns.

4. Agarose

Agarose has served as a support in the immobilization of enzymes (Porath et al., 1967; Kato and Anfinsen, 1969; Axén and Ernback, 1971). This material has a relatively open structure and is desirable for enzymatic reactions involving large molecules, such as proteases and proteins (Axén et al., 1967b).

C. Organic Polymers

Silman et al. (1966) have prepared water-soluble enzyme derivatives by reacting an enzyme with diazotized-p-aminophenylalanine-lecuine copolymer. Wagner et al. (1968) chose a commercially available copolymer of L-alanine and L-glutamic acid to insolubilize chymotrypsin. A condensing agent, N-ethyl-5-phenylisoxazolium-3'-sulfonate, formed amide bonds between the carboxyl functions of the polymer and ϵ-amino groups of the enzyme.

Levin et al. (1964) pioneered the use of maleic-anhydride-ethylene copolymer as a matrix for attaching enzymes. The polymer is available commercially and ϵ-amino groups of lysine react directly with the anhydride groups in aqueous solutions.

Manecke (1962) immobilized several enzymes by reaction of a nitrated copolypmer of methacrylic acid-m-fluoroanalide and methacrylic acid. An active, insoluble, alcohol dehydrogenase was formed in this manner. This example is probably the first to show that an apoenzyme was active in insoluble form if the coenzyme was supplied with the substrate.

Polyaminostyrene beads have been used by Grubhofer and Schleith (1953) and Filippusson and Hornby (1970) as a carrier for enzymes. The polymer was activated by diazotization of the arylamine group followed by coupling to the enzyme.

Brown et al. (1968b) coupled apyrase to maleic-anhydride-ethylene copolymer, polyaminostyrene, and also polyvinylamine. Activity was measured by liberation of inorganic phosphorus from ATP (adenosine triphosphate). Only the first two polymers showed significant activity.

Polyurethane was the support chosen by Sato (1971). The amine groups of the polymer were coupled to a diazonium salt of an arylamino carboxylic acid, such as hippuric acid. The acid group was then coupled to a protein, such as serum albumin, via N,N'-diclohexylcarbodiimide. Finally, the enzyme was coupled to the serum albumin-polyurethane diazo hippurate disc in the presence of more N,N'-dicyclohexylcarbodiimide. When serum albumin was interspersed between the carrier and the protein, much more enzyme was coupled, and it was more stable under operating conditions.

Rothfus and Kennel (1970) adsorbed wheat β-amylase to wheat glutenin and obtained an active insoluble β-amylase. Enzymatic activity of the insoluble complex was quite low, and the adsorption seemed to be selective for one type of β-amylase.

Sundaram and Hornby (1970) and Hornby and Filippusson (1970) attached enzymes to the interior of nylon tubes. When the substrates were pumped through the tubes, then products emerged at the discharge end.

L-Asparaginase has been coupled by Weetall (1970a) to a dacron vascular prosthesis.

Enzymes have been extensively trapped in polyacrylamide gels and to a lesser extent in starch gels. Bernfeld and Wan (1963), Mosbach and Mosbach (1966), and Hicks and Updike (1966) described methods to immobilize enzymes in this gel matrix. Guilbault and Das (1970) entrapped cholinesterase and urease in starch gels, as well as in polyacrylamide gels. In principle, the method involves polymerizing acrylamide in the presence of an enzyme to form a rigid gel, which is made particulate by extrusion through a fine screen. Little or no leaching of enzymes from the gel can be noticed over extended periods.

D. INORGANIC SUPPORTS

1. Glass

Porous glass is frequently a support in the insolubilization of enzymes. Weetall (1969a) showed that alkaline phosphatase could be attached to glass by first silanizing the glass with 3-aminopropyltriethoxysilane. The resulting alkylamino glass was converted to an arylamino group by reaction with *p*-nitrobenzoylchloride followed by reduction of the nitro group to an amine. Diazotization of the amine enabled the derivatized glass to couple readily with the enzyme. Several other enzymes have been coupled to porous glass. Weetall and Baum (1970) coupled L-amino acid oxidase to glass; Royer and Green (1971) attached pronase to porous glass while Line *et al.* (1971) attached pepsin to porous glass by using a carbodiimide on the acid side to protect the pepsin from denaturing. Glutaraldehyde was selected by Robinson *et al.* (1971) to attach several enzymes to aminoalkyl glass.

Work by Messing (1969, 1970a,b) shows that proteins can be tenaciously adsorbed by glass. He found that papain and ribonuclease when adsorbed on porous glass showed enzymatic activity over long periods of time. Unfortunately, the specific activity of the enzyme-glass complexes was very low.

Barker *et al.* (1971) reported on the activation of glass surfaces as well as cellulose and nylon with titanium salts or other transition metals after which enzymes could be chelated to the support by simple contact. The specific activity of the immobilized enzyme chelates was quite high.

2. Other Carriers

Messing and Weetall (1970) and Weetall and Hersh (1970) successfully immobilized several enzymes on inorganic carriers in a manner similar to that used for glass. Amorphous siliceous materials used were silica gel and colloidal silica. Crystalline silicates used were bentonite and wollastonite. Enzymes could be attached to metal oxides, such as NiO, Al_2O_3, and hydroxyapatite.

Monson and Durand (1971) also attached invertase to bentonite by use of cyanuric chloride.

Such materials as silica gel, charcoal, kaolinite, and clays have been adsorbants for various enzymes. Although active preparations have been produced, this approach has not been as successful as entrapment or covalent binding.

E. Cross-linking

Insolubilization of enzymes can also be accomplished by cross-linking the enzymatic protein with a bifunctional reagent, such as glutaraldehyde or bisdiazobenzidine-2,2'-disulfonic acid. Quiocho and Richards (1964) treated crystals of carboxypeptidase A with glutaraldehyde. The cross-linked crystals were insoluble in M NaCl but showed significant carboxypeptidase activity. Habeeb (1967) and Glassmeyer and Ogle (1971) cross-linked trypsin with glutaraldehyde. Cross-linking at pH 7.0 or higher was enhanced if $(NH_4)_2SO_4$ was added (Habeeb, 1967). Ogata et al. (1968) used glutaraldehyde to cross-link subtilisin. They found as much as 30–50% saturated $(NH_4)_2SO_4$ was necessary for a high degree of cross-linking although acetone could substitute for the salt. Jansen and Olson (1969) insolubilized papain with glutaraldehyde. A protein precipitant was not required, but the pH of the reaction mixture was important. Invertase was coprecipitated with tannic acid by Negoro (1970) to produce an active, insoluble enzyme complex that was in some respects superior to the soluble enzyme. When Schejter and Bar-Eli (1970) cross-linked catalase with glutaraldehyde, they obtained a water-insoluble product that had about 10% the activity of native catalase.

The cross-linking technique can also be used to prepare enzymatically active membranes. Goldman et al. (1968) adsorbed papain on collodion membranes. They used bisdiazobenzidine-2,2'-disulfonic acid to cross-link the adsorbed enzyme molecules on the membrane. Failure to cross-link the enzyme resulted in rapid desorption of papain on incubation with the substrate. Broun et al. (1969) adsorbed glucose oxidase on cellophane membranes and cross-linked the protein with glutaraldehyde.

Goldman and Lenhoff (1971) and Goldman et al. (1971) adsorbed glucose-6-phosphate dehydrogenase and alkaline phosphatase, respectively, on collodion without benefit of cross-linking. The dehydrogenase was, however, selectively desorbed by either NADP (nicotinamide-adenine dinucleotide phosphate) or glucose-6-phosphate or a combination of these substrates. On the other hand, alkaline phosphatase desorbed only slowly from the membrane.

Table I shows the carriers, other than polyacrylamide gel, that have been used as supports for immobilization of enzymes. Enzymes that have been immobilized by entrapment in polyacrylamide gel are listed in Table II. A table listing specific enzymes that have been immobilized can be found in a review by Melrose (1971).

TABLE I
CARRIERS USED FOR IMMOBILIZATION OF ENZYMES

Carrier	Enzyme bound	Binding agent	Reference
Microcrystalline cellulose	α-Amylase	Diazotized 3-(p-aminophenoxy)-2-hydroxypropyl ethers	Barker et al. (1968a)
	β-Amylase	Diazotized 3-(p-aminophenoxy)-2-hydroxypropyl ethers	Barker et al. (1968b)
	γ-Amylase (glucoamylase)	Diazotized 3-(p-aminophenoxy)-2-hydroxypropyl ethers	Barker et al. (1968b)
Cellulose sheets	Pyruvate kinase	2,4,6-Trichloro-s-triazine	Kay et al. (1968)
	β-Galactosidase	2,4,6-Trichloro-s-triazine	Kay et al. (1968)
	Chymotrypsin	2,4,6-Trichloro-s-triazine	Kay et al. (1968)
Cellulose fiber	Catalase	Cyanuric chloride	Surinov and Manoilov (1966)
DEAE-Cellulose sheets	Lactic dehydrogenase	2,4,6-Trichloro-s-triazine	Kay et al. (1968)
	β-Galactosidase	2,4,6-Trichloro-s-triazine	Kay et al. (1968)
	Creatine kinase	2,4,6-Trichloro-s-triazine	Kay et al. (1968)
DEAE-Cellulose fibers	Amino acid acylase	Ionic	Tosa et al. (1966)
	Catalase	Ionic	Mitz (1956)
	Catalase	Glutaraldehyde	Balcom et al. (1971)
	Chymotrypsin	2-NH_2-4,6-dichloro-s-triazine	Kay and Lilly (1970)
	Glucoamylase	Ionic	Bachler et al. (1970)
	Glucoamylase	2-NH_2-4,6-dichloro-s-triazine	O'Neill et al. (1971b)
	Invertase	Ionic	Suzuki et al. (1966)
CM-cellulose	Asparaginase	Ionic	Nikolaev and Mardashev (1961)
	Asparaginase	Diazotized CM-cellulose hydrazide	Brown et al. (1966)
	Adenosine triphosphatase	2-NH_2-4,6-dichloro-s-triazine	Kay and Lilly (1970)
	Chymotrypsin	Diazotized CM-cellulose hydrazide	Mitz and Summaria (1961)

Carrier	Enzyme	Method	Reference
	Glucoamylase	Diazotized CM-cellulose hydrazide	Maeda and Suzuki (1970)
	Horseradish peroxidase	Dicyclohexylcarbodiimide	Weliky et al. (1969)
	Bromelain	Diazotized CM-cellulose hydrazide	Crock et al. (1970)
	Ficin	Diazotized CM-cellulose hydrazide	Crock et al. (1970)
	Papain	Diazotized CM-cellulose hydrazide	Crock et al. (1970)
	Trypsin	Diazotized CM-cellulose hydrazide	Crock et al. (1970)
	Apyrase	Diazotized CM-cellulose hydrazide	Whittam et al. (1968)
	Ribonuclease	Diazotized CM-cellulose hydrazide	Lilly et al. (1965)
Aminoethyl cellulose	Aldolase	Glutaraldehyde	Lynn and Falb (1969)
	Glycerol-3-phosphate dehydrogenase	Glutaraldehyde	Lynn and Falb (1969)
	Fructose-1,6-diphosphatase	Glutaraldehyde	Lynn and Falb (1969)
	Rennin	Glutaraldehyde	Green and Crutchfield (1969)
	Trypsin	Glutaraldehyde	Habeeb (1967)
p-Aminobenzyl cellulose	Chymotrypsin	Diazonium salt	Mitz and Summaria (1961); Lilly et al. (1965)
	Ribonuclease	Diazonium salt	Lilly et al. (1965)
	Papain	Diazonium salt	Goldstein et al. (1970)
m-Aminobenzyloxy methyl ether of cellulose	Catalase	Diazonium salt	Surinov and Manoilov (1966)
	Chymotrypsin	Diazonium salt	Surinov and Manoilov (1966)
	Ribonuclease	Diazonium salt	Surinov and Manoilov (1966)
Starch gel	Cholinesterase	Entrapment	Guilbault and Das (1970)
	Urease	Entrapment	Guilbault and Das (1970)
Dialdehyde starch	Papain	Methylenedianaline	Goldstein et al. (1970)
	Trypsin	Methylenedianaline	Goldstein et al. (1970)
	Subtilopeptidase A	Methylenedianaline	Goldstein et al. (1970)

(Continued)

TABLE I (*Continued*)

Carrier	Enzyme bound	Binding agent	Reference
Dacron	L-Asparaginase	Diazotization of an arylamine Dacron	Weetall (1970a)
Acrylamide-acrylic acid copolymer	Hexokinase and glucose-6-phosphate dehydrogenase	1-Cyclohexyl-3-(2-morpholinoethyl)carbodiimide metho-*p*-toluenesulfonate	Mosbach and Mattiasson (1970)
Polyacrylamide beads	Acid phosphatase	Glutaraldehyde	Weston and Avrameas (1971)
	Glucose oxidase	Glutaraldehyde	Weston and Avrameas (1971)
	Trypsin	Glutaraldehyde	Weston and Avrameas (1971)
	Chymotrypsin	Glutaraldehyde	Weston and Avrameas (1971)
Porous glass (aryl amine derivative)	Acylase	Diazonium salt	Weetall (1969b)
	Invertase	Diazonium salt	Weetall and Mason (1971)
	Papain	Diazonium salt or isothiocyanate derivative	Weetall (1969c)
	Pronase	Diazonium salt	Royer and Green (1971)
	Trypsin	Diazonium salt or isothiocyanate derivative	Weetall (1969c)
	Glucose oxidase	Diazonium salt or isothiocyanate derivative	Messing and Weetall (1970)
	Ficin	Diazonium salt or isothiocyanate derivative	Messing and Weetall (1970)

Carrier	Enzyme	Method	Reference
Porous glass (alkylamine derivative)	Urease	Diazonium salt or isothiocyanate derivative	Messing and Weetall (1970)
	Peroxidase	Diazonium salt	Messing and Weetall (1970)
	L-Amino acid oxidase	Diazonium salt	Weetall and Baum (1970)
	Alkaline phosphatase	Diazonium salt	Weetall (1969a); Messing and Weetall (1970)
	Steroid esterase	Diazonium salt	Grove et al. (1971)
	Deoxyribonuclease I	Diazonium salt	Neurath and Weetall (1970)
	Pepsin	1-Cyclohexyl-3-(2-morpholino-ethyl)carbodiimide	Line et al. (1971)
	Chymotrypsin	Glutaraldehyde	Robinson et al. (1971)
	β-Galactosidase	Glutaraldehyde	Robinson et al. (1971)
Porous glass (underivatized)	Papain	Adsorption	Messing (1970a)
	Ribonuclease	Adsorption	Messing (1970b)
Colloidal silica	Alkaline protease	Isothiocyanoalkylsilane derivative	Messing and Weetall (1970)
	Glucose oxidase	Isothiocyanoalkylsilane derivative	Messing and Weetall (1970)
Alumina	Glucose oxidase	Isothiocyanoalkylsilane derivative	Messing and Weetall (1970)
Hydroxyapatite	Glucose oxidase	Isothiocyanoalkylsilane derivative	Messing and Weetall (1970)
Nickel oxide	Peroxidase	Isothiocyanoalkylsilane derivative	Messing and Weetall (1970)
	Glucose oxidase	Isothiocyanoalkylsilane derivative	Weetall and Hersh (1970)

TABLE II
Enzymes Immobilized by Entrapment in Polyacrylamide Gel

Enzyme	Reference
Alcohol dehydrogenase	Wieland et al. (1966)
Aldolase	Bernfeld and Wan (1963)
Aldolase	Brown et al. (1968a)
α-Amylase	Bernfeld and Wan (1963)
β-Amylase	Bernfeld and Wan (1963)
Cholinesterase	Guilbault and Das (1970)
Chymotrypsin	Bernfeld and Wan (1963)
Citrate synthase	Mosbach (1970)
Glucose isomerase	Strandberg and Smiley (1971)
Glucose oxidase	Hicks and Updike (1966)
Glucose-6-phosphate dehydrogenase	Mosbach and Mattiasson (1970)
Hexokinase	Mosbach and Mattiasson (1970)
Lactic dehydrogenase	Hicks and Updike (1966)
Orsellinic acid decarboxylase	Mosbach and Mosbach (1966)
Papain	Bernfeld and Wan (1963)
Phosphofructokinase	Brown et al. (1968a)
Phosphoglucose isomerase	Brown et al. (1968a)
Phosphoglycerate mutase	Bernfeld et al. (1969)
Ribonuclease	Bernfeld and Wan (1963)
Steroid dehydrogenase	Mosbach and Larsson (1970)
Trypsin	Bernfeld and Wan (1963)
Trypsin	Mosbach and Mosbach (1966)
Urease	Guilbault and Das (1970)

III. Properties of Bound Enzymes

The responses of several insolublized enzymes to parameters such as pH, temperature, substrate concentration, inhibitors, and the like, frequently differ from those of the native enzymes. The differences have been attributed to the physical and chemical properties of the carriers that alter the microenvironment around the bound enzyme. There are general patterns of response with particular types of enzymes and carriers, but each enzyme-carrier complex can have its own peculiar characteristics.

A. Stability

The stability of insolubilized enzymes to adverse conditions is often greater than the native form. However, the same enzyme when bound to different carriers can have either more or less stability than the native enzyme. For instance, cross-linked papain, adsorbed on a collodion membrane, has less thermal stability (Goldman et al., 1968) while a papain-porous glass complex has greater thermal stability (Weetall, 1969c). It is not certain whether differences in thermal stability are related to the type of carrier or to binding procedure. Polyacrylamide entrapment, which does not involve chemical attachment, should not increase heat resistance, but Bernfeld and Bieber (1969) found an unexplained exception with rabbit muscle enolase.

In some cases there is increased resistance of insolubilized enzymes to inhibition by metal ions and other inhibitors, but generally there appear to be no significant alterations in resistance or sensitivity to inhibitory agents.

Storage stability of insolubilized enzyme under wet or dry conditions is often greater than that of the native enzyme although stability varies with the enzyme and with the carrier. Weetall (1970b) studied several enzymes insolubilized on both inorganic and organic carriers. Enzymes covalently coupled to inorganic carriers had greater longevity in storage than those bound to organic carriers. Weetall also found that enzymes coupled to inorganic carriers by sulfonamide linkage were not so stable under the same conditions as azo-linked enzymes. These results are surprising because one would expect that a hydrophobic support would tend to unfold the tertiary structure of the protein with loss of enzymatic activity. On the other hand, hydrophilic polymers could be expected to protect the tertiary structure and thus prolong storage life.

The benefits of increased stability are realized most during the actual use of insolubilized enzymes for repeated or continuous substrate conversion. It is not necessary that increased stability be provided as some soluble enzymes withstand reaction conditions for long periods. It is required that the enzyme-carrier complex resist degradation. Insolubilized enzymes remain tightly bound to their carriers during storage and use. Losses in activity are due primarily to enzyme inactivation rather than to disruption of the enzyme-carrier complex.

There are numerous reports of long-term usage of insolubilized enzymes. For example, Tosa et al. (1967) worked with columns of aminoacylase bound to different DEAE-Sephadex beads for 20 days. Depending on the form of DEAE-Sephadex, from 40 to 65% of the activity remained. At the Northern Laboratory, glucoamylase ionicly bound to DEAE-cellulose remained active in a continuous stirred reactor for 25 days with only a 15% loss in activity (Smiley, 1971). We have also coupled glucose isomerase to porous glass beads (Strandberg and Smiley, 1972). In a batch-type system, the free enzyme was inactivated in 3 to 4 hr while the enzyme-glass complex retained 50% of its activity after 24 hr. However, a column of the glass-bound enzyme had one-half the original activity after 12 to 14 days; after it ran for 46 days at 60°C, 11% of the activity remained. Increased stability in the batch reaction is attributed to the covalent attachment of the enzyme to the beads. The great increase in stability of the enzyme column results partly from attachment but also because there is a continuous change of the reaction mixture. Glucose isomerase entrapped in polyacrylamide had the same sensitivity to batch reaction conditions as the free enzyme, but when used as a column, it too was active after several days of operation (Strandberg and Smiley, 1971).

The stabilization of enzymes by insolubilization extends to a protection from proteolytic digestion (Tosa et al., 1969a; Negoro, 1970; Glassmeyer and Ogle, 1971) and for proteases, protection from self-digestion (Crock et al., 1970).

B. Kinetic Patterns

1. pH

Alterations that occur in pH profiles of enzymes insolubilized on charged carriers are believed due to a difference in electrostatic field around the bound enzyme caused by the inherent charge of the carrier. Polyanionic carriers cause an alakaline shift in the pH profile of the bound enzyme as compared to the native enzyme. A more acidic profile occurs with polycationic carriers. Also, there is a narrowing of the pH profile near the optimum in both cases. The pH shift can be overcome by increasing the ionic strength of the medium. The theoretical considerations of these phenomena have been treated extensively by Goldstein (1970).

An unexpected response of polyacrylamide-entrapped phosphoglycerate mutase to pH was reported by Bernfeld et al. (1969). The acid shift of one pH unit was not explainable by a charge on the carrier. Polyacrylamide-entrapped rabbit muscle enolase (Bernfeld and Bieber, 1969) showed no pH shift, but the profile was broader than with the soluble enzyme. Perhaps these findings can be explained by alterations in the environment within the gel matrix. Goldman et al. (1968) demonstrated that the pH shift, which occurred with papain, adsorbed in a collodion membrane, and then crosslinked to itself, was due to a lowering of the local pH within the membrane as a result of acid production during the enzymatic hydrolysis of substrate.

2. K_m

An increase or decrease in the apparent K_m of enzymes insolublized on polyanionic or polycationic carriers is attributed primarily to charge-charge interactions between carrier and substrate. If the carrier has a charge opposite that of the substrate, the apparent K_m decreases. When the charges are alike, the apparent K_m increases. Little or no change is found with neutral carriers. Several workers have proposed mathematical expressions of the charge-charge interactions that occur (Hornby et al., 1968; Goldstein and Katchalski, 1968; Wharton et al., 1968).

Lilly et al. (1966) determined that the apparent K_m for ficin-carboxymethyl cellulose in packed beds depended on flow rate of the substrate solution through the column. They explained this determination by considering the enzyme-carrier particle as being surrounded by a diffusion layer whose thickness is inversely related to the rate of flow of solution past the particle. The diffusion of substrate through this layer is inversely related to its thickness. Therefore, the concentration of substrate at the enzyme surface will be lower than that in the bulk solution at low flow rates and the apparent K_m will be increased. Hornby et al. (1968) proposed mathematical expressions to include the effects of both charge-charge interactions and diffusion limitation on the apparent K_m.

Wilson et al. (1968a) pointed out that other factors are likely involved in altering the kinetics of insolubilized enzymes. It is not known to what extent substrate binding sites are affected either physically or chemically by attaching enzymes to solid supports. Active sites on the enzyme may be directly involved in binding to the carrier, and the tertiary structure of the enzyme molecule may be changed. Generally, there is a drastic reduction in enzyme activity which accompanies binding, indicating that these changes do occur.

3. Substrate Specificity

There is little information available about alterations in substrate specificity for bound enzymes except for the proteases. Insoluble proteases have higher activity toward low-molecular-weight substrates than toward those of high molecular weight (Bar-Eli and Katchalski, 1963; Cresswell and Sanderson, 1970). It is thought that there is stearic hindrance by the carrier resulting in restricted access to catalytic sites on the enzyme. Likely, action of some protein inhibitors would be affected similarly. Alterations in substrate binding sites or the tertiary structure could also affect the affinity of an enzyme for different substrates, as well as required cofactors.

Because the substrate and products would need to penetrate the pores of the gel matrix, polyacrylamide entrapment should be limited to enzymes with low-molecular-weight substrates. Bernfeld and Wan (1963) demonstrated that entrapment of proteases and amylases is feasible although activity is limited to that portion of the enzyme trapped at or near the surface of the gel.

IV. Methods of Application

Insoluble enzymes can be used for repeated or continuous substrate conversion, and most applications take advantage of this feature. The factors that affect insoluble-enzyme activity, as discussed in Section III, must be considered in developing methods of applying enzyme-carrier complexes. Also critical to any method is the physical and chemical nature of the carrier. The carrier must not affect or be affected by the reaction mixture or other material with which it will be in contact. Particularly important in this respect are carriers needed for medically significant enzymes since the enzyme-carrier complex may be in contact with body fluids and tissues.

Columns have been the most frequent means of working with insoluble enzymes. Batch or continuously stirred reactors, membranes, and a few specific techniques are among other methods reported. Technical aspects of design and function are not discussed except to indicate where particular problems might be encountered in using a method.

A. STIRRED TANK REACTORS

Stirred tank reactors are operated as batch or continuous feed systems. The insoluble enzyme is removed from a batch reaction by some suitable means (filtration or centrifugation) and then used on the next batch of substrate. The enzyme-carrier complex is physically retained in a continuous stirred tank reactor by a filter while the substrate, maintained at a constant level, is continuously replaced.

Two physical factors of stirred tank reactors that affect substrate conversion are stirring speed and flow rate. Naturally, the flow rate is governed by the capacity of the insoluble enzyme to convert a given amount of substrate during the holdup time in the reactor. Also, Lilly and Sharp (1968) showed that the reaction rate in a stirred tank of CM-cellulose-chymotrypsin depended upon the agitation rate, a condition indicating a diffusion-limiting effect. The effect of diffusion limitation on insoluble enzymes was described here earlier.

The agitation required in a stirred tank reactor can be detrimental to the carrier because of shearing. Physical disruption of the carrier does not affect the enzyme carrier bond or activity, but handling of the complex might be more difficult. One variation in reactor design is to bind the enzyme directly to the agitator. We have envisioned a rotating drum with the enzyme attached to it to achieve such a technique. Weetall and Havewala (1972) placed glass beads in rigid screen baskets attached to a stirring motor. The rotating baskets provided agitation for the tank but protected the beads from breaking. Weetall and Hersh (1970) described the binding of enzymes to NiO screens, which can be dipped or swirled in the reaction tank.

B. Columns

Columns or packed bed reactors have been used for single- and multiple-step continuous substrate conversions and for analytical purposes. Handling an enzyme column is similar in most respects to column chromatography. Flow rates depend physically upon the type of carrier, bed volume, and viscosity of the substrate solution. The rate of conversion of substrate also determines flow rate.

O'Neill et al. (1971b) compared the efficiency of a column and continuously stirred tank reactor using amyloglucosidase covalently linked to DEAE-cellulose. With maltose as a substrate, under conditions which favored column operation, the column exhibited lower efficiency than expected. O'Neill et al. attributed low efficiency to an inability of the substrate solution to permeate completely the column bed at low flow rates (and low pressure drop). Apparently there was a lack of enzyme-substrate contact. When the choice exists, it would be best to examine both continuously stirred tank reactors and column reactors.

Enzymes have been attached to porous cellulose sheets and used as enzyme filters. Wilson et al. (1968b) suggested that at low flow rates the substrate solution passes through only a portion of the pores of the cellulose sheets they used. Increased flow rates would evenly distribute the flow and would increase reaction capacity of the sheets, as they observed with pyruvate kinase attached to DEAE-cellulose.

C. Other Methods

A discussion of other methods of applying insoluble enzymes tends to lead more into their potential uses since it is through the need for specific applications that innovative developments occur.

Columns and stirred reactors seem more applicable to industrial uses with carriers and reactor design modified to overcome specific problems. As an example of carrier modification, Weetall and Havewala (1972) described a metal-coated porous glass that is more resistant to chemical degradation than regular porous glass.

Analytical enzymes have usually been applied in column form. One example of an enzyme electrode is that reported by Updike and Hicks (1967) in which a gel-entrapped enzyme is coated on a specific ion electrode. Test strips consisting of the enzyme bound to paper containing sensitive dyes are described (Weetall and Weliky, 1966).

There has been a great deal of study on methods for using insoluble enzymes to treat diseases and metabolic disorders and for artificial kidneys. Semipermeable microcapsules of nylon and collodion (Chang, 1964, 1969; Chang and Poznansky, 1968) and tubes of nylon (Sundaram and Hornby, 1970) and of Dacron (Weetall, 1970a) are potential methods of applying medical enzymes (see Section V).

Lately there have been reports of immobilized enzyme reactors where the soluble enzyme is retained within the reactor by means of an ultrafiltration membrane (Butterworth *et al.*, 1970; Ghose and Kostick, 1970; Marshall and Whelan, 1971). This system is applicable to high-molecular-weight substrates, such as starch, proteins, and lipids. However, there is no operational stability afforded to the enzyme. O'Neill *et al.* (1971a) added to the usability of this type of system by attaching the enzyme to a high-molecular-weight, but soluble, dextran. The enzyme has more stability and the porosity of the ultrafiltration membrane can be varied to recover a wider variety of products.

V. Potential Applications

A. Industrial Uses

A major reason why more enzymes have not achieved commercial use is because of their high cost of preparation. Overall costs of producing an enzyme might be reduced if it could be reused or used in a continuous process. Enzyme insolubilization offers such a possibility. Even if the cost of the enzyme is not prohibitive, the fact that the insoluble form can be used for long periods may justify using an insoluble enzyme.

Although our knowledge of binding enzymes to insoluble carriers has expanded greatly in the past few years, more work is needed on improved carriers and binding techniques and also on methods of handling insoluble enzymes in diverse fields of application.

Only two industrial insolubilized enzyme processes are in operation. In Japan, columns of DEAE-cellulose-aminoacylase are used for the continuous resolution of a mixture of DL-amino acids to the L-form (Tosa *et al.*, 1969b). Insoluble penicillin amidase reportedly serves in England and in the United States for splitting the side chain from penicillins to form 6-amino penicillanic acid as a starting point for the production of synthetic penicillins (Self *et al.*, 1969).

Table III lists several other insoluble-enzyme systems that are being investigated. References are omitted from the table either because the enzymes are referred to in the text or because no specific reports on their use are available.

TABLE III
PROPOSED INDUSTRIAL USES OF INSOLUBILIZED ENZYMES

Enzyme	Carrier	Use
Amino acid acylase	DEAE-Cellulose	Resolution of DL-amino acids
Penicillin amidase	Triazinyl cellulose	Penicillin modification
Glucoamylase	DEAE-Cellulose, glass	Conversion of starch to dextrose
	s-Triazinyl cellulose	Glucose syrup manufacture
Glucose isomerase	Polyacrylamide, glass	Conversion of glucose to fructose
Steroid-modifying enzymes	Polyacrylamide, glass	Steroid modification
Invertase	DEAE-Cellulose	Hydrolysis of sucrose
Glycolytic enzymes	Polyacrylamide	Production of metabolic intermediates
α-Amylase	Glass	Treatment of white water during paper manufacturing
		Glucose syrup manufacture
α-Galactosidase	Glass	Removal of flatulence factor from soybean products
β-Galactosidase	Cellulose, glass	Hydrolysis of lactose in milk or whey
Proteases	Several	Clarification of beverages
Trypsin	Glass	Increasing shelf-life of milk
Catalase	Glass	Increasing shelf-life of milk
Pepsin	Glass	Cheesemaking

Glucoamylase is an example of a low-cost enzyme that has considerable industrial use. Smiley (1972) showed that glucoamylase immobilized on either DEAE-cellulose or porous glass converted up to 60 times more starch to glucose than did soluble glucoamylase. Improvements in support materials should further widen this difference. Increase in dextrose production per unit of enzyme coupled with advantages of continuous operation may make the immobilized glucoamylase process industrially attractive.

Insoluble glucose-isomerase in column form is being studied on a pilot-plant scale at the Northern Laboratory and by others.

Immobilized α-amylase could aid paper manufacturers in disposing of excess "white water." This waste stream contains colloidal starch which holds cellulose fibers and other ingredients of the paper in suspension so that they ultimately are discharged into lakes and streams. Treating white water with α-amylase readily breaks the starch colloid allowing the fibers and other solids to settle where they can be recovered and recycled. Use of soluble amylase is not economically feasible because of the large volumes of white water involved. Similarly, immobilized α-amylase and glucoamylase could improve the process for manufacturing glucose syrups from starch. Because insoluble amylases can readily be removed from the substrate solution at any desired point, precise control of the reaction is possible. Wilson and Lilly (1969) demonstrated the feasibility of this approach.

Insoluble enzymes amenable to multistep processes are exemplified by the work of Wilson et al. (1968b), Mattiasson and Mosbach (1971), and Mosbach and Mattiasson (1970) with glycolytic enzymes and by Mosbach and Larsson (1970) with steroid-modifying systems.

B. Medical Uses

Insoluble enzymes offer great hope for the treatment of metabolic disorders, such as phenylketonuria, and there is active work in this area. The primary advantage is in eliminating the repeated injection of the soluble enzyme into the bloodstream where immune responses can occur. Of course, it is essential that the enzyme-carrier bond be rigid and that the carrier be nonreactive with tissues and body fluids. Several methods of applying these enzymes are being investigated. Chang (1969) has proposed the use of enzymes entrapped in semipermeable microcapsules of nylon or collodion. Similarly, such enzymes as urease, encapsulated or attached to other carriers, are being developed for use in artificial kidney devices. L-Asparaginase has been attached to a Dacron vascular prosthesis and implanted in a dog (Weetall, 1970a).

C. Other Uses

A few laboratories have employed insolubilized enzymes as analytical devices for analytical and clinical use in the form of columns, probes, and test strips. The difficulty in preparing and using these devices as compared with other methods of analysis has limited their acceptance. Few insoluble enzymes are commercially available and these are primarily proteases. Other laboratory uses of insoluble enzymes are for affinity chromatography, structural studies of proteins and nucleic acids, and attempts at producing the artificial *in vivo* conditions of cellular membrane-bound enzymes.

In a *Chemical and Engineering News* report (1972), Humphrey and Pye of the University of Pennsylvania proposed that enzymes like phenoloxidase be incorporated in columns for air pollution control. Treatment of waste effluents and monitoring of air and water pollution are included in this general area.

VI. Current Status and Future Prospects

The current interest in immobilized enzymes stems from work started about 15 years ago in Europe and Israel. At present there are a large number of laboratories involved in immobilized enzyme studies and systems are getting quite sophisticated. For example, Mattiasson and Mosbach (1971) reported the kinetic behavior of a matrix-bound three-enzyme system consisting of β-galactosidase, hexokinase, and glucose-6-phosphate dehydrogenase. Efficiency of the immobilized enzyme system, before reaching steady state, was higher than for the corresponding soluble enzyme system. Likewise, the efficiency of the three-enzyme system was sufficiently greater than the two-enzyme system consisting of hexokinase-glucose-6-phosphate dehydrogenase-coupled system to indicate a cumulative effect.

Also, the bioengineering community has begun to show great interest in immobilized enzymes, and data on fixed enzyme reactors are starting to appear in the literature.

An interesting study that is a corollary of immobilized enzymes is the work of Weibel *et al.* (1971) in which they showed that coenzymes could be insolubilized on glass beads and function in an alcohol-dehydrogenase system.

In the future, increased use of immobilized enzymes can be expected in the fields of biochemical research and analysis. They should find a place in medical applications and in drug modifications. They have promising potential as heterogeneous catalysts in a variety of industrial processes, and they should find application in treatment of waste streams from many industrial operations.

REFERENCES

Axén, R., and Ernback, S. (1971). *Eur. J. Biochem.* **18,** 351.
Axén, R., and Porath, J. (1966). *Nature (London)* **210,** 367.
Axén, R., Porath, J., and Ernback, S. (1967a). *Nature (London)* **214,** 1302.
Axén, R., Porath, J., and Ernback, S. (1967b). *Nature (London)* **215,** 1491.
Bachler, M. J., Strandberg, G. W., and Smiley, K. L. (1970). *Biotechnol. Bioeng.* **12,** 85.
Balcom, J., Foulkes, P., Olson, N. F., and Richardson, T. (1971). *Process Biochem.* **6,** 42.
Bar-Eli, A., and Katchalski, E. (1960). *Nature (London)* **188,** 856.
Bar-Eli, A., and Katchalski, E. (1963). *J. Biol. Chem.* **238,** 1690.
Barker, S. A., Somers, P. J., and Epton, R. (1968a). *Carbohyd. Res.* **8,** 491.
Barker, S. A., Somers, P. J., and Epton, R. (1968b). *Carbohyd. Res.* **9,** 257.
Barker, S. A., Emery, A. N., and Novais, J. M. (1971). *Process Biochem.* **6,** 11.
Bernfeld, P., and Bieber, R. E. (1969). *Arch. Biochem. Biophys.* **131,** 587.
Bernfeld, P., and Wan, J. (1963). *Science* **142,** 678.
Bernfeld, P., Bieber, R. E., and Watson, D. M. (1969). *Biochim. Biophys. Acta* **191,** 570.
Broun, G., Selegny, E., Avrameas, S., and Thomas, D. (1969). *Biochim. Biophys. Acta* **185,** 260.

Brown, H. D., Chattopadhyay, S. K., and Patel, A. (1966). *Biochem. Biophys. Res. Commun.* **25**, 304.
Brown, H. D., Patel, A. B., and Chattopadhyay, S. K. (1968a). *J. Chromatogr.* **35**, 103.
Brown, H. D., Patel, A. B., Chattopadhyay, S. K., and Pennington, S. N. (1968b). *Enzymologia* **35**, 233.
Butterworth, T. A., Wang, D. I. C., and Sinskey, A. J. (1970). *Biotechnol. Bioeng.* **12**, 615.
Chang, T. M. S. (1964). *Science* **146**, 524.
Chang, T. M. S. (1969). *Sci. Tools* **16**, 33.
Chang, T. M. S., and Poznansky, M. J. (1968). *Nature (London)* **218**, 243.
Cresswell, P., and Sanderson, A. R. (1970). *Biochem. J.* **119**, 447.
Crock, E. M., Brocklehurst, K., and Wharton, C. W. (1970). *In* "Methods in Enzymology" (G. Perlman and L. Lorand, eds.), Vol. 19, pp. 963–978. Academic Press, New York.
Filippusson, H., and Hornby, W. E. (1970). *Biochem. J.* **120**, 215.
Ghose, T. K., and Kostick, J. A. (1970). *Biotechnol. Bioeng.* **12**, 921.
Glassmeyer, C. K., and Ogle, J. D. (1971). *Biochemistry* **10**, 786.
Goldman, R., and Lenhoff, H. M. (1971). *Biochim. Biophys. Acta* **242**, 514.
Goldman, R., Kedam, O., Silman, J. H., Caplan, S. R., and Katchalski, E. (1968). *Biochemistry* **7**, 486.
Goldman, R., Kedam, O., and Katchalski, E. (1971). *Biochemistry* **10**, 165.
Goldstein, L. (1970). *In* "Methods in Enzymology" (G. Perlman and L. Lorand, eds.), Vol. 19, pp. 935–962. Academic Press, New York.
Goldstein, L., and Katchalski, E. (1968). *Z. Anal. Chem.* **243**, 375.
Goldstein, L., Pecht, M., Blumberg, S., Atlas, D., and Levin, Y. (1970). *Biochemistry* **9**, 2322.
Green, M. L., and Crutchfield, G. (1969). *Biochem. J.* **115**, 183.
Grove, M. J., Strandberg, G. W., and Smiley, K. L. (1971). *Biotechnol. Bioeng.* **13**, 709.
Grubhofer, N., and Schleith, L. (1953). *Naturwissenschaften* **40**, 508.
Guilbault, G. G., and Das, J. (1970). *Anal. Biochem.* **33**, 341.
Gutman, M., and Rimon, A. (1964). *Can. J. Biochem.* **42**, 1339.
Habeeb, A. F. S. A. (1967). *Arch. Biochem. Biophys.* **119**, 264.
Hicks, G. P., and Updike, S. J. (1966). *Anal. Chem.* **38**, 726.
Hornby, W. E., and Filippusson, H. (1970). *Biochim. Biophys. Acta* **220**, 343.
Hornby, W. E., Lilly, M. D., and Crook, E. M. (1968). *Biochem. J.* **107**, 669.
Humphrey, A. E., and Pye, E. K. (1972). *Chem. & Eng. News* Jan. 3, p. 25.
Jansen, E. F., and Olson, A. C. (1969). *Arch. Biochem. Biophys.* **129**, 221.
Katchalski, E. (1962). *In* "Polyamino Acids Polypeptides and Proteins" (M. Stahmann, ed.), p. 283. Univ. of Wisconsin Press, Madison.
Kato, J., and Anfinsen, C. B. (1969). *J. Biol. Chem.* **244**, 5849.
Kay, G., and Lilly, M. D. (1970). *Biochim. Biophys. Acta* **198**, 276.
Kay, G., Lilly, M. D., Sharp, A. K., and Wilson, R. J. H. (1968). *Nature (London)* **217**, 641.
Levin, Y., Pecht, M., Goldstein, L., and Katchalski, E. (1964). *Biochemistry* **3**, 1905.
Lilly, M. D., and Sharp, A. K. (1968). *Chem. Eng. (London)* **215**, CE12.
Lilly, M. D., Money, C., Hornby, W., and Crook, E. M. (1965). *Biochem. J.* **95**, 45.
Lilly, M. D., Hornby, W. E., and Crook, E. M. (1966). *Biochem. J.* **100**, 718.
Line, W. F., Kwong, A., and Weetall, H. H. (1971). *Biochim. Biophys. Acta* **242**, 194.
Lynn, J., and Falb, R. D. (1969). *Abstr. Pap., 158th Meet., Amer. Chem. Soc.* BIOL 298.
Maeda, H., and Suzuki, H. (1970). *J. Agr. Chem. Soc. Jap.* **44**, 547.
Manecke, G. (1962). *Pure Appl. Chem.* **4**, 507.
Marshall, J. J., and Whelan, W. J. (1971). *Chem. Ind. (London)* p. 701.
Mattiasson, B., and Mosbach, K. (1971). *Biochim. Biophys. Acta* **235**, 253.

Mehltretter, C. L., and Weakley, F. B. (1972). *Biotechnol. Bioeng.* **14**, 281.
Melrose, G. J. H. (1971). *Rev. Pure Appl. Chem.* **21**, 83.
Messing, R. A. (1969). *J. Amer. Chem. Soc.* **91**, 2370.
Messing, R. A. (1970a). *Enzymologia* **38**, 39.
Messing, R. A. (1970b). *Enzymologia* **38**, 370.
Messing, R. A., and Weetall, H. H. (1970). U.S. Pat. No. 3,519,538.
Micheel, F., and Ewers, J. (1949). *Makromol. Chem.* **3**, 200.
Mitz, M. A. (1956). *Science* **123**, 1076.
Mitz, M. A., and Summaria, L. J. (1961). *Nature (London)* **189**, 576.
Monson, P., and Durand, G. (1971). *FEBS Lett.* **16**, 39.
Mosbach, K. (1970). *Acta Chem. Scand.* **24**, 2084.
Mosbach, K., and Larsson, P. O. (1970). *Biotechnol. Bioeng.* **12**, 19.
Mosbach, K., and Mattiasson, B. (1970). *Acta Chem. Scand.* **24**, 2093.
Mosbach, K., and Mosbach, R. (1966). *Acta Chem. Scand.* **20**, 2807.
Negoro, H. (1970). *J. Ferment. Technol.* **48**, 689.
Neurath, A. R., and Weetall, H. H. (1970). *FEBS Lett.* **8**, 253.
Nikolaev, A. Ya., and Mardashev, S. R. (1961). *Biokhimiya* **26**, 641.
Ogata, K., Otteson, M., and Svendsen, J. (1968). *Biochim. Biophys. Acta* **159**, 403.
O'Neill, S. P., Wykes, J. R., Dunnill, P., and Lilly, M. D. (1971a). *Biotechnol. Bioeng.* **13**, 319.
O'Neill, S. P., Dunnill, P., and Lilly, M. D. (1971b). *Biotechnol. Bioeng.* **13**, 337.
Porath, J., Axén, R., and Ernback, S. (1967). *Nature (London)* **215**, 1491.
Quiocho, F. A., and Richards, F. M. (1964). *Proc. Nat. Acad. Sci. U.S.* **52**, 833.
Riesel, E., and Katchalski, E. (1964). *J. Biol. Chem.* **239**, 1521.
Rimon, S., Stupp, Y., and Rimon, A. (1966). *Can. J. Biochem.* **44**, 415.
Robinson, P. J., Dunnill, P., and Lilly, M. D. (1971). *Biochim. Biophys. Acta* **242**, 659.
Rothfus, J. A., and Kennel, S. J. (1970). *Cereal Chem.* **47**, 140.
Royer, G. P., and Green, G. M. (1971). *Biochem. Biophys. Res. Commun.* **44**, 426.
Sato, T. R. (1971). U.S. Pat. No. 3,574,062.
Schejter, A., and Bar-Eli, A. (1970). *Arch. Biochem. Biophys.* **136**, 325.
Self, D. A., Kay, G., Lilly, M. D., and Dunnill, P. (1969). *Biotechnol. Bioeng.* **11**, 337.
Silman, I. H., Wellner, D., and Katchalski, E. (1963). *Isr. J. Chem.* **1**, 65.
Silman, I. H., Albu-Weissenberg, M., and Katchalski, E. (1966). *Biopolymers* **4**, 441.
Smiley, K. L. (1971). *Biotechnol. Bioeng.* **13**, 309.
Smiley, K. L. (1972). *Proc. Int. Symp. Conversion Manufacture Foodstuffs Micro-Organisms. 1971* (in press).
Strandberg, G. W., and Smiley, K. L. (1971). *Appl. Microbiol.* **21**, 588.
Strandberg, G. W., and Smiley, K. L. (1972). *Biotechnol. Bioeng.* **14**, 509.
Sundaram, P. V., and Hornby, W. E. (1970). *FEBS Lett.* **10**, 325.
Surinov, B. P., and Manoilov, S. E. (1966). *Biokhimiya* **31**, 387.
Suzuki, H., Ozawa, Y., and Maeda, H. (1966). *Agr. Biol. Chem.* **30**, 807.
Tosa, T., Mori, T., Fuse, N., and Chibata, I. (1966). *Enzymologia* **31**, 244.
Tosa, T., Mori, T., Fuse, N., and Chibata, I. (1967). *Biotechnol. Bioeng.* **9**, 603.
Tosa, T., Mori, T., Fuse, N., and Chibata, I. (1969a). *Agr. Biol. Chem.* **33**, 1047.
Tosa, T., Mori, T., and Chibata, I. (1969b). *Agr. Biol. Chem.* **33**, 1053.
Updike, S. J., and Hicks, G. P. (1967). *Nature (London)* **214**, 986.
Wagner, T., Hsu, C. J., and Kellehev, G. (1968). *Biochem. J.* **108**, 892.
Weetall, H. H. (1969a). *Nature (London)* **223**, 959.
Weetall, H. H. (1969b). *Abstr. Pap., 158th Meet. Amer. Chem. Soc.* BIOL 153.
Weetall, H. H. (1969c). *Science* **166**, 615.

Weetall, H. H. (1970a). *J. Biomed. Mater. Res.* **4**, 597.
Weetall, H. H. (1970b). *Biochim. Biophys. Acta* **212**, 1.
Weetall, H. H., and Baum, G. (1970). *Biotechnol. Bioeng.* **12**, 399.
Weetall, H. H., and Havewala, N. B. (1972). *Biotechnol. Bioeng.* (in press).
Weetall, H. H., and Hersh, L. S. (1970). *Biochim. Biophys. Acta* **206**, 54.
Weetall, H. H., and Mason, R. D. (1971). *Bacteriol. Proc.* A44.
Weetall, H. H., and Weliky, N. (1966). *Anal. Biochem.* **14**, 160.
Weibel, M. K., Weetall, H. H., and Bright, H. J. (1971). *Biochem. Biophys. Res. Commun.* **44**, 347.
Weliky, N., Brown, F. S., and Dale, E. C. (1969). *Arch. Biochem. Biophys.* **131**, 1.
Weston, P. D., and Avrameas, S. (1971). *Biochem. Biophys. Res. Commun.* **45**, 1574.
Wharton, C. W., Crock, E. M., and Brocklehurst, K. (1968). *Eur. J. Biochem.* **6**, 572.
Whittam, R., Edwards, B. A., and Wheeler, K. P. (1968). *Biochem. J.* **107**, 3.
Wieland, T., Determann, H., and Bunning, K. (1966). *Z. Naturforsch. B* **21**, 1003.
Wilson, R. J. H., and Lilly, M. D. (1969). *Biotechnol. Bioeng.* **11**, 349.
Wilson, R. J. H., Kay, G., and Lilly, M. D. (1968a). *Biochem. J.* **108**, 845.
Wilson, R. J. H., Kay, G., and Lilly, M. D. (1968b). *Biochem. J.* **109**, 137.

Microbial Rennets

JOSEPH L. SARDINAS

*Central Research Division, Pfizer Inc.,
Groton, Connecticut*

I.	Introduction	39
II.	Rennets: General Considerations	40
	A. Definitions	40
	B. Process of Coagulation	40
	C. Assays	41
	D. Criteria for Rennets	44
III.	Higher Plant Rennets	45
IV.	Animal Rennets	46
	A. Pepsin	46
	B. Calf Rennin	47
V.	Microbial Rennets	48
	A. General Survey	48
	B. Bacterial Rennets	49
	C. Fungal Rennets	54
	D. Commercialized Fungal Rennets	59
VI.	General Comments	65
	References	66

I. Introduction

One of man's highly esteemed and nourishing foods is cheese. The nutritious merit of this savory comestible is attributable to the concentrated protein and fat from milk, as well as to attendant vitamins, minerals, and other nutriments. Though all cheese can be categorized into one of eighteen basic varieties, there are perhaps over a thousand different, tasty cheeses easily distinguished by sophisticated turophile palates (Brown, 1966; Sanders, 1953).

Primordial cheese probably derived from the natural process of milk souring. In this process, a train of events is initiated via the fermentation of lactose to lactic acid by indigenous microorganisms. Lactic acid accumulates to lower the pH sufficiently to coagulate casein. The origin of use of vegetable or animal extracts to permit more practical and directed curdling of milk is lost in antiquity, though there are engaging legends to account for it (Brown, 1966; Petersen, 1939; Sanders, 1953; Wilson and Reinbold, 1965). Microbial rennets, on the other hand, are relatively recent innovations. The first recorded references to potential microbial rennets are by Conn in 1892 and Gorini, also in 1892.

By far, the most common rennet enzyme in commercial use throughout the world is derived from the abomasum, or fourth stomach, of unweaned calves. It is referred to generally as rennet extract. The amount added to a given quantity of milk for coagulation can vary substantially with the type of cheese to be produced. In the United States, for example, 2 to 4 ounces of rennet are added to 1000 pounds of milk to produce about 100 pounds of Cheddar cheese, while as little as 1/30 of an ounce may be used for cottage cheese.

In 1970, over 2.2 billion pounds of cheese were consumed in the United States (U.S. Department of Agriculture, 1971). Cheese consumption outside the United States is substantially higher. The milk coagulant market, therefore, is clearly a large-volume enzyme market.

Calf rennet remains the industry standard against which other coagulants are measured. However, declining calf slaughter has resulted in a worldwide shortage of stomach raw materials for rennet production. Price of the enzyme in the United States has increased about threefold during the past decade. Added to all this is the fact that very substantial numbers of people in the world are opposed to the use of certain animal secretions in cheese production on grounds or religion, morality, or diet. All these factors contribute to the worldwide search for animal rennet substitutes of microbial or plant origin.

II. Rennets: General Considerations

A. DEFINITIONS

In the cheese industry, the term calf rennet generally refers to an enzyme extract which is obtained from the fourth stomach of calves and is used to coagulate milk for cheese production. The pure, milk-clotting enzyme present in the crude rennet preparation is known as rennin, or chymosin (note that *renin* is an enzyme derived from the kidney and is associated with hypertension). Often, calf rennet is referred to loosely as animal rennet. At infrequent times, the term chymase appears in the literature as a synonym for rennet. In more general usage, however, any milk-clotting enzyme preparation yielding a relatively stable curd is designated a rennet (or rennin, if pure). These enzymes most often are identified by some term associated with their source.

B. PROCESS OF COAGULATION

Cheese is produced throughout the world from the milks of a large variety of animals, e.g., buffaloes, camels, goats, llama, reindeer, sheep, yak, zebu, etc. (Sardinas, 1969). Of course, the preponderant proportion of cheese produced on a commercial scale derives from cow's milk. In practice, rennet is added to prepared milk to cause coagulation of casein and consequent curd formation. The curd then is treated further to yield a cheese whose type is determined by the nature of the processing employed.

The mechanism of coagulation and curd formation is quite an intricate process which is not completely understood. There have been extensive investigations and hypotheses offered to explain the phenomenon (Berridge, 1954; Dyachenko and Slavyanova, 1962; Garnier *et al.*, 1968; R. J. Hill and Wake, 1969; Lindqvist, 1963a,b; Reed, 1966; Wheelock and Knight, 1969). In general, there are two distinct phases during coagulation which can be resolved into enzymatic and nonenzymatic reactions. Resolution of the two stages can be achieved by maintaining milk at either: (a) a low temperature, (b) a relatively high pH, or (c) very low levels of calcium. Though milk will not curdle upon the addition of rennin under any of these conditions, an enzymatic reaction does occur, nonetheless. If the treated milk containing added rennin is restored to its normal state, coagulation will ensue immediately. The current theory holds that k-casein behaves as a protective colloid, stabilizing the entire casein micelle (Waugh and Hipple, 1956). Rennin enzymes initially hydrolyze an especially labile bond of k-casein to a glycopeptide and para-k-casein. The latter protein, along with other fractions of casein, precipitate in the presence of calcium ions. Some workers in the field refer to the cheese ripening period, during which rennin continues to act on the various casein fractions, as the third stage (El-Negoumy, 1970; Lindqvist, 1963b). Recently, Tuszynski (1971) reported the resolution of casein clotting *per se* into three separate phases, namely, enzymatic, flocculation, and gelification. There are excellent updated reviews of coagulation mechanism and curd formation by Beeby *et al.* (1971), MacKinlay and Wake (1971), and Waugh (1971).

C. Assays

1. Milk Clotting

There is no universally accepted standard assay procedure for determining the milk-clotting potency of rennets. In their review, Babbar *et al.* (1965) described more than a half dozen different methods. Among the more common procedures are those of Balls and Hoover (1937), Berridge (1952a,b), Ernstrom (1958), Kunitz (1935), and Soxhlet (1877). In his excellent review, Foltmann (1970) describes in detail his modified version of Berridge's technique. There is considerable variation between assay techniques. For example, the obvious substrate, milk, may be whole or skim, raw or pasteurized, reconstituted dry whole or nonfat milk solids powder, etc. If the last is used, the method of drying the milk (spray or roller) constitutes still another variation. Of course, the constituents of milk also may vary with locale or season or both (Bernatonis and Mickiene, 1970). Kruger et al. (1968) argue the merits of assorted types of milk substrate available for use.

Milk-clotting assays also vary in the use (or nonuse) of buffers and cofactors (e.g., calcium), as well as in pH, temperature, mixing techniques, end-point determinations, and concentrations of ingredients. In comparing assays, it is found that the range of some parameters is quite broad. For example, temperature may range from about 30 to 45°C and pH from 5.0 to 6.7. There also is great diversity in mixing techniques as well as in end-point determinations. The latter encompasses subjective and objective methods. Quite commonly, the milk-clotting end-point is fixed as the instant discrete particles (macro- or micro-curds) are visually detected in a film of milk substrate (Babbar et al., 1965; Berridge, 1961; Davis and MacDonald, 1952; Reed, 1966). More reproducible, and possibly sharper, endpoints are possible by the use of: (a) viscometric techniques which measure viscosity increase at coagulation (Babbar et al., 1965; Richardson et al., 1971; Scott Blair and Oosthuizen, 1962; Vamos Vigyazo et al., 1969a), (b) recorded changes (increase) in sound velocity at the moment of milk curdling (Everson and Winder, 1968), and (c) automatic "thromboelastographic" measurement of clot formation (deMan and Batra, 1964; Tarodo de la Fuente et al., 1969).

In the assays of animal rennet, the rate of milk clotting is inversely proportional to the concentration of enzyme. In the assays of animal rennet substitutes, clotting time also should fall within the range in which it is proportional to the reciprocal of the enzyme concentration. Since there is such a spread of modified milk-clotting assays, reported enzyme activities are understood to be only relative. This makes it quite difficult, if not impossible, to compare potencies. In these assays, animal rennet or rennin generally is employed as a standard control. It is a common experience with enzyme assays to find that results obtained in the laboratory do not necessarily translate into predicted performance in the field (Collier, 1970). In the particular case of potential rennet substitutes, there are a multitude of possible causes. For example, the assay milk and cheese manufacturer's milk may vary in pH, calcium content, temperature, etc. Unless the activity of the rennet substitute parallels animal rennet's exactly, under all conditions, the rate of coagulation and curd formation may vary considerably. It is prudent for an enzyme producer to tailor his assay to conditions which approximate, as closely as possible, those encountered in the field (Collier, 1970).

2. *Protease*

The lack of a universally accepted standard assay procedure also applies to protease assays. The International Union of Biochemistry has attempted to define a standard unit of enzyme activity, but parameters of pH, substrate, additives, etc. obviously cannot be specified uniformly for all enzymes (Collier, 1970). It is well known that many, if not most, proteolytic enzymes clot milk. In fact, the coagulation of milk can be used as an assay of proteolytic activity (Gorini and Lanzavecchia, 1954; Ilany-Feigenbaum and Netzer, 1969; Kunitz, 1935). There is contrasting evidence that the milk-clotting and proteolytic activities of an enzyme may be inseparable (Gorini and Lanzavecchia, 1954), or separable (Skelton, 1971). An important consideration in the evaluation of a rennet is the ratio of milk-clotting potency to proteolytic strength. Animal rennin possesses one of the highest, if not *the* highest, ratio and is the standard for potential substitutes.

As noted above, proteolytic assays vary widely. Substrates may be casein, B chain of insulin, hemoglobin, synthetic peptides (R. D. Hill, 1969), etc. As expected, the usual parameters: pH, temperature, reaction times, etc., also range broadly. In a typical quantitative assay, the enzyme and substrate are allowed to react and the undigested substrate is precipitated with trichloroacetic acid. The clear filtrate, or supernatant, is recovered and the absorbance of liberated aromatic amino acids measured at 280 nm. The extent of absorbance is an index of enzyme activity. There are many published descriptions of proteolytic assay methods commonly used on rennets (e.g., Anson, 1938; Arima *et al.*, 1970; Behnke, 1967; Ernstrom, 1958; Fox, 1969; Hagihara *et al.*, 1958; McDonald and Chen, 1965; Mickelsen and Fish, 1970; Whitaker, 1970). There are two short, but excellent reviews on proteolytic enzymes. One covers the subject in general (Spickett, 1971). The other reviews microbial proteases more specifically (Keay, 1971).

3. *Lipase*

A contaminating component which may accompany rennets is the enzyme lipase. This enzyme usually is present only in minor quantities. However, potential rennet substitutes should be tested, since excessive lipase activity may lead to undesirable flavor (e.g., rancidity) development in cheese (Hammer and Babel, 1957; Somkuti *et al.*,

1969a). On the other hand, certain levels of lipase activity may advantageously contribute to cheese flavor (Oi et al., 1969) or even may help accelerate the curing of cheese (Richardson et al., 1967). Specific procedures for lipase assays can be found in the references cited immediately above and in Bier (1955). Additional information on lipases is available from Aldridge (1961), Groves (1971), Reed (1966), and other standard references.

D. Criteria for Rennets

In order for a microbial (or higher plant) rennet enzyme preparation to substitute satisfactorily for animal rennet, it must meet certain criteria. One of the first requirements is that the preparation must effectively coagulate milk without excessively hydrolyzing the resulting curd during, or at, maturation. Furthermore, any contaminating enzymes should manifest only nominal activity. Under certain circumstances, however, such enzymes may favorably contribute to the overall quality of specific cheeses (e.g., lipases in certain Italian cheese varieties). The rennet substitute must be nontoxic and devoid of antibiotic activity, as well as free of pathogens. In fact, there should be few microorganisms, and gas formers, minimally. Preferably, the rennet substitute should be readily water soluble and possess acceptable color and odor. It also should exhibit reasonable shelf life.

Ultimately, the overall quality of cheese produced, particularly as determined on an industrial scale, will decide the practical value of any animal rennet substitute. A paramount consideration for the cheese manufacturer is the yield of curd from milk. Clearly, there should be no significant loss of protein or fat during curdling. During ripening, there must be proper body development. This implies good body consistency and texture with no "openness," seams, reticulations, crystalline deposits, "graininess," "mealiness," "curdiness," etc.

Undoubtedly, one of the most rigorous demands made of an animal rennet substitute relates to the organoleptic quality of the cheese produced. Though quite a number of flavor defects develop in cheese (rancidity, sourness, etc.), by far the most common organoleptic failure experienced with substitutes is bitterness of the cheese. The cause of bitterness in cheese is recognized to result from the accumulation of particular peptides which impart this universally distasteful flavor (Carr et al., 1956; Czulak, 1959; Gordon and Speck, 1965a,b; Harwalkar and Elliott, 1965). These bitter peptides can be released from casein by diverse proteinases, including those of microorganisms, as well as by chemical hydrolysis (Carr et al., 1956; Harwalkar and Elliott, 1971; Harwalkar and Seitz, 1971; Matoba et al., 1970; Sullivan and Jago, 1970). Not unexpectedly, hydrolysis of the peptides eliminates the bitterness (Edwards, 1969; Fujimaki *et al.*, 1970; Sato *et al.*, 1969; Stadhouders, 1962; Tokita, 1969; Toolens, 1968; Zvyagintsev *et al.*, 1971a). In 1970, Sullivan and Jago suggested that spontaneous cyclization of an exposed N-terminal glutamine group confers resistance to hydrolysis of bitter peptides by ordinary aminopeptidases. These researchers further theorized that, in addition to this terminal group, a high proline distribution on the bitter peptide chain inhibits most endopeptidase and carboxypeptidase activities. In accord with this hypothesis, "nonbitter" bacterial starter cultures possess enzymes capable of attacking the cyclized group.

The actual mechanism of bitter flavor development involves an interplay of the acidity of the milk, the strains of starter cultures used, and the concentration of salt (sodium chloride) in the milk (Lawrence and Gilles, 1969). Some starter cultures can hydrolyze bitter peptides regardless of the pH, while others cannot (Czulak and Shimmin, 1961; Jago, 1962). It is well known that sodium chloride can inhibit development of bitter flavor in Cheddar cheese. Fox and Walley in 1971 reported that high levels of sodium chloride (5 to 10%) inhibited the proteolysis of β-casein without affecting the hydrolysis of α_s-casein. They also found that rennin produces bitter flavor from β-casein but not from α_s-casein. It is their conclusion that sodium chloride may control development of bitter flavor in Cheddar cheese by inhibiting the hydrolysis of β-casein. The fact that bitterness develops only from the hydrolysis of β-casein, argues against the theory of Sullivan and Jago (see above) since the glutamic acid content of α_s- and β-casein is about equal.

There are thorough reviews by Schormuller (1968) and Margalith and Schwartz (1970) which provide additional detailed information on the subjects of flavors and off-flavors.

III. Higher Plant Rennets

Many proteolytic enzymes of vegetable origin are known to be milk coagulants. These enzymes seemingly are ubiquitous in the diverse plant structures. They are present in buds, flowers, fruits, latex, leaves, roots, sap, and seeds. Reviews by Babbar *et al.* (1965), Dewane (1960), Greenberg (1955), Sardinas (1969), and Veringa (1961) contain references to the study of potential rennets from plant species representing more than two dozen different plant genera. Some of these commonly known plants include: ash gourd, fig, milkweed, papaya, pineapple, prickly artichoke, pumpkin, and soybean.

Most plant proteinases exhibit rather potent proteolytic activity. Unfortunately, the ratio of milk-clotting activity to proteolytic activity is much too low for cheese-making purposes. However, a few of these enzymes are reported to produce satisfactory cheese (type of cheese often unspecified). For example, a proteolytic enzyme from *Cucurbita pepo* (pumpkin) was reported by Berkowitz-Hundert and Leibowitz (1963) and Berkowitz-Hundert et al. (1964, 1965) to produce satisfactory cheese. Ilany-Feigenbaum and Netzer (1969) also reported production of suitable Cheddar-type cheese with a variety of animal and vegetable enzymes whose proteolytic potency was partially inactivated by an assortment of treatments.

At present there is no commercially available rennet derived from a higher plant. However, there is no explicit reason why a satisfactory plant rennet should not ultimately become available, even if only as a partial replacement for calf rennet.

IV. Animal Rennets

It was noted earlier that animal rennet is obtained from the fourth stomach of unweaned calves. If any calves have been fed something other than milk, then some pepsin will be present in the rennet extract (Berridge, 1954). Other animals also may be a source of rennets, e.g., buffalo, chicken, goat, rabbit, sheep, and swine (Alais et al., 1962; Dewane, 1960). Just recently, a patent was issued in Israel for the recovery of a rennet from the proventriculus tissue of chickens for the manufacture of cheese (Yeda Research and Development Co., Ltd., 1971). It is probable that there are other animals with rennets, but studies are lacking. Apparently, humans, dogs, and possibly young pigs, do not secrete a rennin (Berridge, 1961).

There are several animal proteases, in addition to rennin, which coagulate milk. These are chymotrypsin, gastricsin, pepsin, and trypsin. Since chymotrypsin and trypsin yield frankly poor cheese, studies have focused on the more promising pepsin (Alais et al., 1962; Dewane, 1960; Ilany-Feigenbaum and Netzer, 1969; Reed, 1966). Gastricsin is an enzyme different from, but similar to, pepsin. There are no reports in the literature on the use of gastricsin in cheese making.

A. Pepsin

Pepsin is an acid protease with a pH optimum of about 2.0, an isoelectric point below pH 1.08, and a molecular weight of about 34,000 (Laskowski, 1961; Tang, 1970; Tang et al., 1959). The enzyme loses its activity rapidly in neutral or alkaline solutions, though at acid pH (below 6.0) it is relatively stable (Tang, 1970). The amino acid content of porcine pepsin is as follows: ala_{16}, arg_2, asp_{40}, $\frac{1}{2} cys_6$, glu_{26}, gly_{34}, his_1, $ileu_{23}$, leu_{28}, lys_1, met_4, phe_{14}, pro_{16}, ser_{43}, thr_{25}, try_6, tyr_{16}, and val_{20} (Ryle, 1970; Tang, 1970). Though pepsin resembles calf rennin in some respects (at one time it was thought to be identical to rennin), its use as a 100% replacement for animal rennet in cheese making is limited (Emmons et al., 1971). Fox (1969) has found that bovine pepsin is a rennet superior to porcine pepsin (properties of bovine pepsin have been reviewed by Kassel and Meitner, 1970). Nonetheless, as a rennet, pepsin suffers the following shortcomings: (a) setting time for curd formation is extended; (b) the curd is not as firm as animal rennin's; (c) there is a loss of fat in the whey, (d) the cheese flavor generally is bland; and (e) above pH 6.5, pepsin activity falls off so rapidly that its use is limited in the production of sweet cheese (e.g., Swiss) and some Italian varieties. However, a mixture of pepsin and rennet is used commercially (Babel, 1967; Chapman and Burnett, 1968; Melachouris and Tuckey, 1964). The percentage of pepsin mixed with animal rennet can vary, but most frequently it is a 1:1 mixture. In general, the 1:1 mixture produces quite satisfactory cheese, including Cheddar. However, animal rennet still produces superior cheese. Emmons et al. (1971) reported a series of studies with animal rennet, pepsin, and a 1:1 mixture of pepsin:animal rennet. They concluded that the 1:1 mixture produces essentially high quality Cheddar cheese, though slower rates of coagulation, some loss of fat in whey, and somewhat poorer texture may be encountered. Defects with the 1:1 mixture are not experienced uniformly, and in any case, are not extreme (Emmons et al., 1971). The production of satisfactory cheese coupled with the low cost of pepsin makes the 1:1 mixture a very attractive coagulant to the cheese manufacturer, and accounts for its commercial success. Babel indicated in 1967 that 75% of the cheese produced in the United States was manufactured with the 1:1 mixture. This increased to 80% by 1970 (Davis, 1971).

B. Calf Rennin

The zymogen, prorennin, is secreted in the abomasa of milk-fed calves. It is the inactive precursor of the milk-clotting enzyme, rennin. Conversion of prorennin to rennin occurs at pH's below 5.0, optimally at pH 2.0 (Foltmann, 1970). Rennin has been crystallized by a number of investigators (Berridge, 1945, 1961). Bunn et al. (1971) reported detailed crystallographic studies of the enzyme.

Rennin is a rather strong acid protease inasmuch as it hydrolyzes hemoglobin to about the same extent as pepsin and trypsin (Reed,

1966). It also hydrolyzes α-casein to a much greater extent than it does β-casein (Ledford et al., 1968). In solution, it is reported to be most stable at pH 5.3 to 6.3 (Foltmann, 1970). Interestingly, relatively good stability is reexpressed specifically at pH 2.0 (Mickelsen and Ernstrom, 1967). The enzyme has an isoelectric point of about pH 4.5 and a molecular weight of 30,000 to 40,000 (Foltmann, 1970; Reed, 1966). The optimum pH for proteolysis is about 3.5, though for hemoglobin it is specifically 3.7 (Berridge, 1961; Foltmann, 1970). The amino acid composition is as follows: ala_{13}, arg_5, asp_{30}, $\frac{1}{2}cys_6$, glu_{29}, gly_{25}, his_4, $ileu_{15}$, leu_{19}, lys_8, met_7, phe_{14}, pro_{12}, ser_{27}, thr_{18}, try_4, tyr_{15}, and val_{21} (Foltmann, 1970).

Crystalline rennin is not a homogeneous material (Ernstrom, 1958). It has been separated chromatographically into three fractions which manifest relatively minor differences in their properties (Foltmann, 1970). Oruntaeva and Seitov (1971) also separated lamb rennin into three fractions. Some of the differences noted between the fractions relate to: amino acid composition, milk-clotting potencies, and ratio of milk-clotting activity to proteolytic activity.

There is no doubt that rennin and pepsin (especially bovine pepsin) resemble each other, but only in a general fashion. For example, both enzymes: (a) derive from zymogens, (b) are gastric acid proteases, (c) act similarly on the peptide bond of k-casein, (d) have about the same molecular weights, (e) have similar proteolytic specificities on the B chain of oxidized insulin, and (f) have high ratios of milk-clotting activity to proteolytic activity, though rennin's is clearly higher (Fox, 1969).

V. Microbial Rennets

It already has been noted that most, if not all, proteolytic enzymes can coagulate milk. It is not surprising, therefore, that microorganisms represent a large, inviting reservoir of potential animal rennet substitutes. However, the preponderant number of microbial rennets is much too proteolytic for cheese making. Nonetheless, there are a few which are commercially available as animal rennet substitutes.

A. General Survey

The earliest scientific reports on milk-clotting enzymes of microbial origin are by Conn (1892) and Gorini (1892). In 1921, Takamine and Takamine were granted a patent in the United States for the use of rennets produced by species of *Aspergillus, Mucor,* and *Penicillium,* as well as by *Eurotium oryzae.* In his introductory review, Wahlin

(1928) refers to a number of bacteria described up to that time to produce rennet-like enzymes. This included *Ascobacillus citreus*, *Bacillus amylobacter*, *B. indicus*, and *Proteus mirabilis*. It was not until much later, however, that more concerted efforts were directed to finding practical, microbiologically derived, animal rennet substitutes. There are numerous reviews, e.g., Babbar *et al.* (1965), Dewane (1960), Food and Agriculture Organization of the United Nations (1968), Iwasaki *et al.* (1968), Naudts (1969), Sardinas (1969), Veringa (1961), and Zvyagintsev *et al.* (1971b), which list a host of bacteria and fungi elaborating enzymes that catalyze the coagulation of milk. There also are reports in the literature broadly describing the results of screening large numbers of microorganisms for rennets. Tendler and Burkholder (1961) canvassed 950 thermophilic actinomycetes. Two genera were represented, namely, 129 isolates of *Streptomyces* and 821 *Thermoactinomyces*. Only 4 of the *Streptomyces* species clotted milk, while 408 of the *Thermoactinomyces* evidenced rennet activity. Srinivasan *et al.* (1964) screened 230 fungi and 43 bacteria. Many of these organisms produced enzymes capable of curdling milk and, not unexpectedly, also exhibited variable proteolytic activity. No specific cheese trials were described nor were any of the organisms identified. Schulz *et al.* (1967) investigated approximately 500 microorganisms, most of which were unidentified aerobic spore formers. They reported that about 8% of the organisms produced coagulants suitable for cheese making. Camembert cheese was prepared with some of the enzymes, but the quality of the cheese was not indicated. More recently, large-scale screenings were conducted by Arima *et al.* (1967) and Sardinas (1968). Arima and his colleagues tested 800 bacteria and fungi. Sardinas evaluated 381 bacteria and 540 fungi. Prins and Nielsen (1970) reported an extensive screening of thermophilic organisms. There are many other reports in the literature of unidentified organisms producing milk-clotting enzymes (e.g., Babbar and Sampathkumar, 1970; Conn, 1892; Su and Chen, 1971), or of cheese trials conducted with rennets derived from unidentified microorganisms (e.g., Masek *et al.*, 1970).

B. Bacterial Rennets

A number of bacteria reported to produce milk-clotting enzymes are listed in Table I. The genera in the table represent less than 7% of all genera listed in the seventh edition of Bergey's Manual of Determinative Bacteriology (Breed *et al.*, 1957). Considering the ubiquity of proteolytic enzymes among bacterial species, this appears to be a poor representation. However, investigators do not necessarily bother to identify all organisms in their screening programs, particularly producers of potent proteases. In addition, most screening programs are geared to the study of aerobic mesophiles or thermophiles. It is more difficult, possibly less rewarding, and unattractive from a manufacturing point of view to screen pathogens, psychrophiles, anaerobes, phototrophs, etc.

TABLE I
BACTERIA REPORTED TO PRODUCE MILK COAGULANTS

Bacteria	References
Alcaligenes sp.	Annibaldi (1962a,b)
Ascobacillus citreus	Wahlin (1928)
Bacillus amylobacter	Wahlin (1928)
B. brevis	Shimwell and Evans (1944); A. Singh et al. (1967)
B. calidolactis	Hussang and Hammer (1928)
B. cereus	Melachouris and Tuckey (1968); Srinivasan et al. (1962a,b)
B. coagulans	Babbar et al. (1965)
B. firmus	Arima et al. (1967)
B. fusiformis	Shimwell and Evans (1944)
B. indicus	Wahlin (1928)
B. licheniformis	Damodaran et al. (1955)
B. megatherium	A. Singh et al. (1967)
B. mesentericus	Barkan et al. (1965); Dimitrov et al. (1970); Emanuiloff (1956); Oosthuizen (1962); Shimwell and Evans (1944)
B. mycoides	Godo Shusei Co. (1968)
B. polymyxa	Imai et al. (1970); Irie et al. (1966)
B. sphericus	Arima et al. (1967)
B. subtilis	Murray and Kendall (1969); Murray and Prince (1970); Puhan (1969); Shimwell and Evans (1944); A. Singh et al. (1967); Srinivasan et al. (1962a,b); Tipograf et al. (1966)
B. thermoproteolyticus	Iwasaki et al. (1968)
Corynebacterium hoagii	Arima et al. (1967)
Escherichia sp.	Annibaldi (1962a,b)
Lactobacillus helveticus	Arima et al. (1967)
Proteus mirabilis	Wahlin (1928)
Pseudomonas aeruginosa	Iwasaki et al. (1968)
Pseudomonas fluorescens	Peterson and Gunderson (1960)
Pseudomonas myxogenes	Babbar et al. (1965); Tsugo and Yamauchi (1959)
Pseudomonas schuylkilliensis	Arima et al. (1967)
Serratia marcescens	Gorini (1892, 1930, 1931); Wahlin (1928)
Staphylococcus quadrigeminus	Loeb (1902)

(Continued)

TABLE I *(Continued)*

Bacteria	References
Staphylococcus spp.	Christensen (1960)
Streptococcus faecalis-lactis	Annibaldi (1962a,b)
Streptococcus faecalis var. *liquefaciens*	Somkuti and Babel (1966)
Streptococcus liquefaciens	Babbar *et al.* (1965)
Streptococcus zymogenes	Babbar *et al.* (1965)
Streptomyces albus	Arima *et al.* (1967)
Streptomyces ehimensis	Arima *et al.* (1967)
Streptomyces griseochromogenus	Arima *et al.* (1967)
Streptomyces griseus	Babbar *et al.* (1965); Dewane (1960)
Streptomyces hachijoensis	Arima *et al.* (1967)
Streptomyces rimosus	Arima *et al.* (1967)
Streptomyces rubescens	Arima *et al.* (1967)
Thermoactinomyces spp.	Tendler and Burkholder (1961)
Vibrio cholerae	Wahlin (1928)

1. Serratia marcescens

One of the first identified bacteria reported to produce a milk-curdling enzyme was *Bacillus prodigiosus,* or *Serratia marcescens* as it is now classified (Gorini, 1892). Conn in 1892 also described rennet-producing bacteria, but the organisms were not identified. In 1928, Wahlin undertook a study of the enzyme of *S. marcescens.* Both Gorini and Wahlin found the enzyme to be thermostable and to have a temperature optimum for milk-clotting activity of 40 to 50°C. Wahlin learned that rennin and the bacterial enzyme were influenced in a similar manner by calcium ion. On the other hand, oxalate inhibited the microbial enzyme to a lesser extent than it inhibited animal rennet. The rennet enzyme of *S. marcescens* coagulated heated milk more readily than did animal rennet. There are no reports of cheese production with this enzyme (Veringa, 1961).

2. Bacillus polymyxa

Other bacterial rennets, especially those derived from the genus *Bacillus,* have received due consideration. For example, *Bacillus polymyxa* is reported to produce a valuable enzyme. In fact, patents have been issued for the use of this enzyme as an animal rennet substitute (Godo Shusei Co., 1968, 1969; Imai *et al.,* 1970). The enzyme appears to be essentially a thermolabile alkaline protease. Its activity and stability range between pH 4.0 and 9.0. In the examples cited in the patents, cheese can be produced from whole milk, or a mixture of whole milk and reconstituted nonfat milk solids (4:1). Calcium salts are added to milk for potentiation of the enzyme. There is a small, but distinct, loss in curd yield. The examples in the patents indicate that cheese (type not specified) aged 4 months, had no bitter flavor and appeared to be equal to control cheeses. Kikuchi and Toyoda (1969) prepared Gouda cheese with the enzyme and found the whey to be turbid. They reported the development of bitter flavor in the cheese.

3. Bacillus mesentericus

The milk-clotting enzyme of *Bacillus mesentericus* has been studied by a number of investigators (Barkan *et al.*, 1965; Dimitrov *et al.*, 1970; Emanuiloff, 1956; Oosthuizen, 1962; Orosin *et al.*, 1970; Shimwell and Evans, 1944; Tipograf *et al.*, 1966). There is a possibility that the rennet of *B. mesentericus* is accompanied by more potent proteases (Orosin *et al.*, 1970). In the production of "Dutch-type" cheese, the enzyme, used at full strength, yielded an atypical flavor which was often bitter. The body proved to be hard and crumbly. Used as a 1:1 mixture with calf rennet, however, the prepared cheese had a satisfactory consistency and a flavor comparable to the animal rennet control (Barkan *et al.*, 1965).

4. Bacillus cereus

A bacterial rennet of some promise is that of *Bacillus cereus*. Choudhery and Mikolajcik (1970, 1971), Melachouris and Tuckey (1968), and Srinivasan *et al.* (1962a,b), among others, have studied the enzyme. Miles Laboratories, Inc., has obtained extensive patent coverage on this coagulant (Sardinas, 1969). In the patents, the enzyme is stated to be destroyed within 20 minutes at 70°C and to maintain its milk curdling potency down to pH 4.8. Melachouris and Tuckey (1968) found the enzyme to be destroyed in 3 minutes at 65°C and to be less sensitive to pH changes in substrate than calf rennet. They also reported the enzyme expressed maximum rennet activity at 75 to 80°C, in contrast to 40 to 45°C for animal rennet. Srinivasan *et al.* (1962a,b) described results of the production of Cheddar cheese with this bacterial enzyme. The cheese was free of bitter flavor even after 6 to 8 months of ripening, though the texture was noted to be "hard and acid." The enzyme does not appear to be as active as animal rennet during the ripening period inasmuch as there is delayed maturation. A. Singh *et al.* (1967) also produced Cheddar cheese with the enzyme and reported it to be free of bitterness. Melachouris and Tuckey (1968) detected greater casein proteolysis with the microbial enzyme than with animal rennet. The specific proteolytic action of the bacterial rennet on α-casein was somewhat similar to that obtained with calf rennin, though the former hydrolyzed β-casein to a greater degree. Choudhery and Mikolajcik (1970, 1971) also observed a distinctly greater degradation of β-casein than α-casein by the rennet of *B. cereus*. In 1969, Naudts reviewed some favorable results of cheese trials with the coagulent of *B. cereus*.

5. Bacillus subtilis

The milk coagulant of *Bacillus subtilis* is one of the more extensively studied bacterial rennets. In 1944, Shimwell and Evans were granted a patent for the production of this enzyme for cheese making. Since then, other patents have been issued (e.g., Murray and Kendall, 1969; Murray and Prince, 1970) as a result of interest in the enzyme by John Labatt, Ltd., of Ontario, Canada. *Bacillus subtilis* is known to produce three proteolytic enzymes, namely, an acid, a neutral, and an alkaline protease (Murray and Kendall, 1969; Puhan, 1969). The ratios of the three enzymes can be made to vary according to the strain used and the conditions of culture. Puhan (1969) indicated the neutral protease to be present in his preparations in the largest amount, the acid protease in the lowest amount, and the alkaline protease intermediate in quantity. On the other hand, Murray and Prince (1970) excluded almost entirely elaboration of the alkaline protease by using an asporulant mutant grown under their specified conditions. If alkaline protease is present, it can be removed by ion-exchange chromatography, inhibited by potato inhibitor, or thermally inactivated (Murray and Kendall, 1969). According to Murray and Kendall, milk-clotting activity of the microbial rennet is dependent on at least one monovalent metal ion (e.g., potassium) and one divalent metal ion (e.g., magnesium). Puhan (1966, 1969) studied the production and properties of *B. subtilis* proteases. He found neutral protease to be stable at pH 6.5 to 10.0 and inhibited by ethylenediaminetetraacetate. It released 38% more nonprotein nitrogen than animal rennet up to the point of clotting milk. Alkaline protease proved to be stable in the range of pH 5.0 to 10.0 and inhibited by potato inhibitor. The enzyme manifested esterase activity. It also released three times more nonprotein nitrogen than animal rennet under the same conditions that were used for testing the neutral protease. Puhan further reported that neutral protease digested α- and β-caseins nonspecifically and rapidly. Tsuru *et al.* (1967) also studied the proteolytic specificities of neutral proteases. Characteristics of *B. subtilis* neutral protease have been reviewed by Yasunobu and McConn (1970).

In 1966, Tipograf et al. claimed to have adsorbed onto kieselguhr up to 94% of the rennet produced by *B. subtilis*. The enzyme can be eluted with 0.2 M phosphate-citrate buffer at pH 6.0. Godo Shusei Co. (1968) describe recovery of practically all milk-clotting activity from broth by precipitation with ammonium sulfate at 0.3 to 0.8 of saturation. Production of good, bitter-free Cheddar cheese with the *B. subtilis* rennet is claimed by the aforementioned patentees (i.e., Murray and Kendall, 1969; Murray and Prince, 1970; Shimwell and Evans, 1944), as well as by A. Singh et al. (1967), and Srinivasan et al. (1962a,b). Naudts (1969) also reviews briefly some other favorable cheese trials conducted with the rennet of *B. subtilis*.

At present there is no bacterial coagulant available on a large commercial scale to substitute for animal rennet. In spite of some seemingly satisfactory cheese trials, some difficulties must prevail. However, the promise of these enzymes remains. They may find particular use for given cheeses, or as partial substitutes in mixtures with other enzymes, such as pepsin, animal rennet, or even some other microbial rennets. Also, there still are legions of bacteria whose potential is yet to be explored.

C. Fungal Rennets

A rather large number of fungi have been reported to produce milk coagulants (Table II). The tabulation is not offered as an exhaustive documentation. For example, some fungi have been determined to produce quite weak milk coagulants (Osman et al., 1969a). The 38 genera listed in Table II represent less than 1% of all known mycota genera [according to Ainsworth (1961) there are some 4300 fungal genera]. Of course, most fungal coagulants are much too proteolytic for use as rennets. Nonetheless, some have undergone more than a cursory evaluation because of their promising properties. In fact, the rennets of 3 fungal species (*Endothia parasitica, Mucor miehei*, and *M. pusillus*) have proved sufficiently suitable for large-scale commercialization.

1. Entomophthorales

Oringer (1960) and Whitehill et al. (1960) were issued patents for the manufacture of milk-curdling enzymes produced by organisms from 3 genera of the Order Entomophthorales, namely, *Basidiobolus, Conidiobolus*, and *Entomophthora*. The enzyme produced by these organisms was described to possess a proteolytic activity ranging from pH 4.0 to 11.0, with an optimum at pH 9.0. The stability range also was from 4.0 to 11.0. The isoelectric point was determined to be at

10.2, and the molecular weight estimated at about 30,000. It was stated that the enzyme coagulated milk in a manner similar to that of animal rennet. Furthermore, it possesses properties superior to animal rennet in that higher yields and firmer cheese curds were obtained.

2. *Aspergillus candidus*

In 1965, Veselov et al. described a rennet elaborated by *Aspergillus candidus*. The enzyme displayed greatest milk-curdling activity at pH 5.6 to 5.87. It was more stable than animal rennet at a pH exceeding 6.9 and was inactivated at 60°C. According to Veselov et al., the enzyme within 24 hr yielded a cheese with a slight bitter aftertaste. Studies by Chebotarev et al. (1969a,b) and Dimitrov et al. (1969) on the production and evaluation of this microbial enzyme indicated that a lower yield of cheese resulted, and a greater proteolytic activity was manifested within 1 month of coagulation.

3. *Byssochlamys fulva*

Knight (1966) published his investigation of 39 molds (representing 14 genera) for rennin-like enzymes. In his preliminary screen he found 8 species of fungi (and one mutant) which produced good curds without defects (see Table II). He considered *Byssochlamys fulva* to be the best organism for further studies. Reps et al. (1969) studied the enzyme and found the optimum temperature to be 64 to 66°C (vs. 39 to 41°C for animal rennet); the enzyme was inactivated at 70°C. They also determined that activity response of the microbial rennet to pH changes, though somewhat similar to animal rennet, was actually less sensitive. In general, the proteolytic activity was considered to be insignificantly different from animal rennet and manifested itself only during the enzymatic phase of milk coagulation.

4. *Rhizopus chinensis*

Fukumoto et al. (1967) purified and crystallized the protease of *Rhizopus chinensis*. Tsuru et al. (1969) published further on the substrate specificity of this potent enzyme. This acid protease is most active at pH 2.9 to 3.3, and stable over the range of pH 2.8 to 6.5. It is inactivated by ferric ions. The enzyme has a modicum of thermal stability since it withstands a temperature of 60°C for 15 minutes. In fact, it retains 85% of its activity after 15 minutes exposure to 65°C. However, the enzyme is completely inactivated at 75°C. The interesting aspect of this enzyme is the relatively high ratio of milk-clotting activity to proteolytic activity. In their assays, Fukumoto and his colleagues determined the ratio of *R. chinensis* rennet to be 2.91. This

TABLE II
Fungi Reported to Produce Milk Coagulants

Fungi	References
Absidia lichtheimii	Arima *et al.* (1967)
Armillaria mellea	Kawai and Mukai (1970)
Ascochyta visa	Arima and Iwasaki (1964); Grimberg (1965)
Aspergillus candidus	Chebotarev *et al.* (1969a,b); Veselov *et al.* (1965)
Aspergillus flavus	Knight (1966); Paleva (1969)
Aspergillus glaucus	Sannabadthi *et al.* (1970)
Aspergillus nidulans	Sannabadthi *et al.* (1970)
Aspergillus niger	Osman *et al.* (1969a,b)
Aspergillus oryzae	Paleva (1969); Tsugo and Yamauchi (1959)
Aspergillus parasiticus	Paleva (1969); Paleva and Popova (1965)
Aspergillus terricola	Dyachenko and Slavyanova (1962)
Aspergillus usameii	Fox (1968)
Basidiobolus ranarum	Oringer (1960); Whitehill *et al.* (1960)
Byssochlamys fulva	Knight (1966); Reps *et al.* (1969, 1970)
Chaetomium brasilliense	Arima *et al.* (1967)
Chaetomium globosum	Knight (1966)
Cladosporium herbarum	Sannabadthi *et al.* (1970)
Colletotrichum atramentarium	Arima and Iwasaki (1964); Arima *et al.* (1967)
Colletotrichum lindenmuthianum	Arima *et al.* (1967)
Conidiobolus brefeldianus	Oringer (1960); Whitehill *et al.* (1960)
Conidiobolus villosus	Oringer (1960); Whitehill *et al.* (1960)
Coprinus macrorhizus	Kawai and Mukai (1970)
Coriolus consors	Kawai and Mukai (1970); Mukai and Kawai (1970a)
Coriolus hirsutus	Kawai and Mukai (1970); Mukai and Kawai (1970a)
Coriolus versicolor	Shichiji and Saeki (1970)
Daedaleopsis styracina	Kawai and Mukai (1970)
Elfvingia applanata	Kawai and Mukai (1970)
Endothia parasitica	Sardinas (1965, 1966, 1968)
Entomophthora apiculata	Oringer (1960); Whitehill *et al.* (1960)
Entomophthora coronata	Oringer (1960); Whitehill *et al.* (1960)
Epicoccum purpurascens	Osman *et al.* (1969a,b)
Eurotium oryzae	Takamine and Takamine (1921)
Flammulina velutipes	Kawai and Mukai (1970)
Fomes fomentarius	Kawai and Mukai (1970)
Fomitopsis annosa	Kawai and Mukai (1970); Mukai and Kawai (1970b)
Fomitopsis castanea	Kawai and Mukai (1970)
Fomitopsis cytisina	Kawai and Mukai (1970)
Fomitopsis pinicola	Kawai and Mukai (1970); Mukai and Kawai (1970b)
Fomitopsis rosea	Kawai and Mukai (1970); Mukai and Kawai (1970b)

(Continued)

TABLE II *(Continued)*

Fungi	References
Fusarium moniliforme	Knight (1966)
Gliocladium roseum	Knight (1966)
Gloeophyllum saepiarium	Kawai and Mukai (1970)
Irpex lacteus	Kawai and Mukai (1970)
Lampteromyces japonicus	Kawai and Mukai (1970)
Lenzites betulina	Kawai and Mukai (1970)
Lenzites saepiaria	Mukai and Kawai (1971)
Monascus anka	Arima and Iwasaki (1964)
Mucor hiemalis	Knight (1966)
Mucor manchuricus	Arima *et al.* (1967)
Mucor miehei	Aunstrup (1968a,b)
Mucor pusillus	Arima *et al.* (1967)
Mucor rouxii	Dewane (1960)
Mucor spinescens	Arima *et al.* (1967)
Panellus stipticus	Kawai and Mukai (1970)
Panus rudis	Dewane (1960)
Penicillium chrysogenum	Knight (1966)
Rhizopus achlamydosporus	Arima *et al.* (1967)
R. batatae	Arima *et al.* (1967)
R. candidus	Arima *et al.* (1967)
R. chinensis	Arima *et al.* (1967); Fukumoto *et al.* (1967)
R. chinniary [sic]	Arima *et al.* (1967)
R. chungkuoensis	Arima *et al.* (1967)
R. delemar	Arima *et al.* (1967)
R. japonicus	Arima *et al.* (1967)
R. niveus	Arima *et al.* (1967)
R. nigricans	Arima *et al.* (1967)
R. nodosus	Arima *et al.* (1967)
R. pseudokiensis	Arima *et al.* (1967)
R. oligosporus	Wang *et al.* (1969)
R. oryzae	Arima *et al.* (1967)
R. pekae	Arima *et al.* (1967)
R. salebrosus	Arima *et al.* (1967)
R. thermosus	Arima *et al.* (1967)
Schizophyllum commune	Kawai and Mukai (1970)
Sclerotium oryzae-sativa	Arima and Iwasaki (1964)
Scopulariopsis brevicaulis	K. Singh and Vezina (1971)
Syncephalastrum racemosum	Sannabadthi *et al.* (1970)
Trametes sanguinea	Kawai and Mukai (1970)

was comparable to that of pepsin in their assay (2.94). Partially purified animal rennin exhibited a ratio of 16.00. In contrast, the neutral protease of *Bacillus subtilis* var. *amylosaccharitius* had a ratio of 1.15, *Trametes sanguinea* a ratio of 1.09, and *Aspergillus saitoi*, 0.35. No information relating to cheese trials with the coagulant of

R. chinensis as a partial or complete replacement for animal rennet was uncovered at the time of this writing.

5. *Rhizopus niveus*

During their screening tests for a rennet of microbiological origin, Arima et al. (1967) noted that *Rhizopus niveus* produced a milk coagulant. In 1970, Fukumoto was awarded a patent for use of the enzyme as a rennet. In the patent, the enzyme is described as an acid protease which makes good quality Cheddar-type cheese if the enzyme is added to milk while it is at pH 5.5 to 6.4 and the temperature is raised above 40°C. Calcium chloride is added at a concentration of 0.05% (w/w). It is stated in the patent that, under these conditions, the ratio of milk-clotting activity to proteolytic activity increases "enormously," so that loss of soluble nitrogen decreases to the same level as the loss experienced with animal rennet. The Cheddar-type cheese was ripened for 3 months.

6. *Rhizopus oligosporus*

Another milk coagulant derived from a species of *Rhizopus* was reported by Wang and colleagues in 1969. The organism, *R. oligosporus*, produced a proteolytic enzyme complex. Their studies showed that the enzyme was stable at 40°C, but inactivated quickly at 60°C. Its stability was within the range of pH 3.0 to 6.0. The ratio of milk-clotting activity to proteolytic activity was lower than animal rennet's. In 1970, Wang and Hesseltine published the results of further studies with the enzyme. They successfully fractionated the proteolytic enzyme complex into five distinct components which they labeled A through E. All fractions responded similarly to inhibitors and exhibited the same pH stabilities. However, they had different pH and temperature optima. The five fractions could be divided into two groups (A, B, C, and D, E) on the basis of the ratios of milk-clotting activity to proteolytic activity. The first group (A, B, C) had a better ratio than the second group (D, E). Fractions A and B were successfully crystallized. There is no information relating to actual cheese production with any of the enzymes.

7. *Basidiomycetes*

Kawai and Mukai (1970) undertook a screening program limited to the evaluation of 44 strains of Basidiomycetes for potential rennet producers. They found a number of genera which elaborated milk coagulants (see Table II). The most promising of these were species of *Coriolus, Fomitopsis,* and *Irpex*. A number of patents have been issued for the use of the rennets secreted by these organisms (Mukai and Kawai, 1970a,b, 1971; Shichiji and Saeki, 1970). In actual cheese trials, the rennet of *Irpex lacteus* appeared to be the best. The rennet of *Fomitopsis pinicola* yielded a Cheddar cheese which developed a slight bitter taste after 5 months' ripening. Cheddar cheese made with the rennet of *I. lacteus*, on the other hand, was of good quality. Nakanishi and Itoh (1970) purified and partially characterized the rennet of *F. pinicola*. The enzyme was observed to be twice as strong as animal rennet at 35°C. Milk-clotting activity could be increased greatly by the addition of calcium, or by decreasing the pH from 6.8 to 6.0. Optimum temperature for curdling of milk was determined to be 48°C. At 55°C, only 22% of the activity was lost. Kawai (1970) studied the acid proteases of the 3 genera of Basidiomycetes noted above. Their optimum pH for hydrolysis of casein is 2.5. Enzymatic proteolysis of milk at pH 6.0 continued to increase with time to a greater extent than that of animal rennet. The coagulant of *I. lacetus* proved to be the least sensitive to pH and calcium variations in milk. The optimum temperature of the enzyme is 55 to 60°C. Interestingly, the rennet of *I. lacteus* evidences only 20% of its optimum activity at 30°C. The stability of *I. lacteus* and *F. pinicola* rennets extends from pH 3.0 to 5.0; for *Coriolus consors*, it is only pH 3.0 to 4.0. The ratio of milk-clotting activity to proteolytic activity of the rennet of *I. lacteus* is more than twice as great as that of the other two Basidiomycete coagulants. In fact, it is about equivalent to the ratio exhibited by the rennet of *Mucor pusillus*.

D. Commercialized Fungal Rennets

One of the first microbial rennets to be commercialized on a large scale (in 1967) is the rennet enzyme of *Endothia parasitica*. It is sold by Pfizer Inc., New York, under the name *Sure-Curd*. Internationally, it also is sold as *Suparen*. Another microbial rennet, which appeared on the market at about the same time as Pfizer's product, is derived from *Mucor pusillus* Lindt. Though it is sold by a number of companies, e.g., Dairyland Food Laboratories, Incorporated, Wisconsin, the sole patentee is Meito Sangyo K. K. of Japan. The product is sold under a number of names, e.g., *Novadel* and *Emporase*. Most recently, a third microbial rennet has appeared on the scene. It is derived from strains of *M. miehei*. There are several companies associated with enzymes from these strains. They are: Miles Laboratories, Incorporated, Illinois; Novo Industri, A/S, Copenhagen; and Wallerstein Company, Division of Baxter Laboratories, Incorporated, Illinois. The products are named: *Rennilase* (Novo) and *Fromase* (Wallerstein). Miles Laboratories does not yet have its product on the market.

1. Endothia parasitica Rennet

The rennet of *Endothia parasitica* was first described by Sardinas in 1965. His subsequent reports (1967, 1968) elaborated on characteristics of the enzyme. In the original evaluation, it was learned that *E. parasitica* was a unique species of the genus. None of 6 other species of *Endothia* synthesized significant levels of rennet, while all 8 available strains of *E. parasitica* collected from diverse sources produced the enzyme. A study of the genetics and nutritional essentials of *E. parasitica* was conducted by Puhalla and Anagnostakis (1971). The rennet enzyme of this fungal organism was crystallized and characterized (Sardinas, 1968). It is a thermolabile acid protease, destroyed within 5 minutes at 60°C. It is stable in a pH range of about 4.0 to 5.5, with the optimum at 4.5. Its isoelectric point is at pH 5.5, though Hagemeyer *et al.* (1968) reported an isoelectric point of less than pH 4.6. The molecular weight has been estimated to be in the range of 34,000 to 39,000. Hagemeyer *et al.* reported a figure of 37,500. The pH optima on acid-denatured hemoglobin and casein have been determined by Larson and Whitaker (1970a) to be 2.0 and 2.5, respectively. They also found that over the pH range of 5.1 to 6.5, the milk-clotting activity was much less sensitive to the pH of the substrate than was that of animal rennet. In fact, the rennet of *E. parasitica* also is less sensitive than animal rennet to calcium ion concentration variations in milk (Alais and Novak, 1968; Pedersen, 1970). This is of some importance, since pH and calcium concentrations in milk vary with season and geography. Larson and Whitaker (1970a) also found that the rennet of *E. parasitica* displayed broader specificity than either animal rennet or pepsin on the oxidized B chain of insulin. They reported (1970b) that maximum stability of the enzyme was at pH 3.8 to 4.5, and that at 50°C for 30 minutes, only 30% of the activity was lost. Rotini and Sequi (1970), and Sequi and Rotini (1970) also studied the stability and purification of the enzyme.

In 1970, Whitaker reviewed much of the technical data relating to the rennet of *Endothia parasitica*. His review contains the following amino acid analysis of this rennin: ala_{29}, arg_2, asp_{26}, $½ cys_2$, glu_{15}, gly_{38}, his_3, $ileu_{17}$, leu_{19}, lys_{12}, met_1, phe_{13}, pro_{13}, ser_{44}, thr_{50}, try_3, tyr_{19}, and val_{22}. This analysis is at variance with Sardinas' findings (1968). The latter did not find any lysine, methionine, or tryptophan in his crystalline preparation on two separate assays. Single crystals of *E. parasitica* rennin have been grown and studied by X-ray diffraction (Moews and Bunn, 1970).

The ratio of milk-clotting activity to proteolytic activity of the *Endothia parasitica* rennin is among the higher of microbial rennets.

Alais and Novak (1968, 1970) studied the enzyme and found its essential enzymological and technological properties in regard to cheese making, to be similar to animal rennet. The mode of milk coagulation is quite comparable to animal rennet's. Many types of cheeses have been successfully prepared with this rennet. In fact, some cheeses are produced with qualities judged to be superior to those prepared with animal rennet. Some cheeses processed satisfactorily with the *E. parasitica* rennet include: Brie, Camembert, Cheddar, Colby, Emmentaler, Gruyère, Italian varieties, Limburger, Monterey, Munster, Swiss, etc. (Bolliger and Schilt, 1969; Puhan and Steffen, 1967; Ramet and Schluter, 1970; Ramet *et al.*, 1969; Shovers and Bavisotto, 1967). Though good Cheddar cheese can be made with this fungal rennet, organoleptic failures are experienced. Therefore, it is not recommended at present as a 100% replacement for animal rennet in the preparation of long-hold Cheddar. Morris and McKenzie (1970) have produced good Cheddar cheese using a 1:1 mixture of *E. parasitica* and animal rennets. There are many other reports on the use of this fungal rennet in cheese production. For example, Maubois and Mocquot (1969) observed a 1% loss in yield with the rennet in the production of Camembert-type cheese. On the other hand, Ramet *et al.* (1969) noted slightly higher yields with the same type cheese. There is no doubt that the use in certain cheese of this rennet, no less than other microbial rennets, may require some minor adjustments in cheese-making technique in order to be optimally employed (Kikuchi *et al.*, 1968b; Resmini *et al.*, 1971). A feature that emerges from many of the studies cited above is the accelerated flavor and body development during cheese maturation. Commercial value of this feature still remains to be proved.

2. *Mucor pusillus* Rennet

A French patent was issued in 1963 to Meito Sangyo K. K. for the production and use in cheese making of the rennet elaborated by *Mucor pusillus* Lindt. Subsequently, a number of reports were published concerning the screening techniques used, as well as characteristics of the crude enzyme (Arima *et al.*, 1967; Iwasaki *et al.*, 1967a,b,c). In 1968, Arima *et al.* described the successful crystallization of the enzyme. Yu *et al.* (1969) investigated some of the properties. Detailed information on the properties of *M. pusillus* coagulant is provided by the fine review of Arima *et al.* (1970). Huang (1970) also studied the enzyme in some detail. The enzyme is an acid protease possessing a somewhat greater thermostability than animal rennet. The less purified material digested milk optimally at pH 3.5, though crystalline material optimally digested k-casein at pH 4.5 and hemoglobin at pH 4.0. The isoelectric point is in the neighborhood of pH 3.5 to 3.8. Stability of the enzyme is in the range of pH 4.0 to 6.0, peaking at pH 5.0. Milk is curdled most actively at pH 5.5. The molecular weight was estimated to be in the range of 29,000 to 30,600. Amino acid composition was as follows: $ala_{16\text{-}17}$, arg_4, asp_{44}, ½ cys_2, glu_{20}, gly_{34}, $his_{1\text{-}2}$, $ileu_{12}$, leu_{15}, $lys_{11\text{-}12}$, met_3, phe_{19}, pro_{14}, ser_{22}, thr_{21}, $try_{2\text{-}3}$, tyr_{13}, and val_{24}.

This enzyme has one of the higher ratios of milk-clotting activity to proteolytic activity exhibited by microbial rennets. Its activity is more resistant to pH changes than calf rennet, though the proteolytic activity is less specific. Some metal ions inhibit its activity (e.g., ferric, mercuric, zinc, etc.). Milk-clotting activity is potentiated to a larger degree by calcium ion concentration than is animal rennet's. Yu et al. (1971) have related histidine residues in the enzyme to the active site. Somkuti and Babel (1967, 1968a,b, 1969), as well as Somkuti et al. (1969b), studied the production and purification of the M. pusillus enzyme. They reported the presence of ample lipase produced under their fermentation conditions and with their strains, namely, NRRL 2543 and PCC 410. Richardson et al. (1967) also found the lipase activity and considered it an advantage in some cheeses for acceleration of curing. A polypeptide antibiotic active against grampositive bacteria also was produced by Somkuti's strains (Somkuti and Walter, 1970). In 1971, Trop and Pinsky reported that addition of Mucor pusillus rennet to animal rennet caused an increase in milk-clotting activity to a greater extent than could be expected on the basis of the sum of their separate activities. This synergistic effect suggested to them that each enzyme coagulated milk via different mechanisms.

Satisfactory cheeses have been reported to be made with Mucor pusillus rennet, e.g., Brick, Butter, Camembert, Cheddar, Cottage, Edam, Gouda, Italian varieties, and Tilsit (Arima et al., 1970; Richardson et al., 1967; Robertson and Gilles, 1969; Schulz et al., 1967). There are other numerous reports on the use of M. pusillus rennet in cheese manufacture. Some note development of bitterness and body defects in cheese, e.g., Babel and Somkuti (1968), Kikuchi and Toyoda (1969), and Kikuchi et al. (1968a,b). On the other hand, Kikuchi et al. (1968c) and Pedersen (1969) indicate the production of acceptable cheese by modifying somewhat the cheese-making procedure. There also is a profusion of reports comparing the rennets of both Endothia parasitica and M. pusillus with animal rennet and pepsin in cheese making. Of interest is the finding that fungal rennets hydrolyze β-casein to a greater extent than do animal rennet and pepsin (Ledford et al., 1968; Mickelsen and Fish, 1970). It is beyond the scope of this paper to attempt an evaluation of all reports on cheese-making trials. Suffice it to say that the data tend to give mixed results. Often enough, cheese trials are conducted with one batch test for each rennet, in small volume and under a confusing array of varied conditions. A few references are listed for the interested reader: Edelsten et al. (1969), Hamdy and Edelsten (1970), Kiss (1969), Kyla-Siurola and Antila (1970), Praprotnik (1968), and Vamos Vigyazo et al. (1969a). Naudts (1969) and Zvyagintsev et al. (1971b) also review results of cheese-making trials with the rennets produced by E. parasitica and M. pusillus.

3. Mucor miehei Rennet

A second *Mucor* species, *M. miehei*, which also produces a rennet, has been found. In fact, a number of strains, claimed to be distinct from each other, have been reported, and patents for some of these have been issued. The strains are: *M. miehei* Cooney et Emerson CBS 370.65 (Aunstrup, 1968a,b – Novo Industri A/S), *M. miehei* NRRL 3420 (Charles et al., 1970 – Miles Laboratories, Inc), *M. miehei* ATCC 16457 (Feldman, 1968, 1969 – Baxter Laboratories, Inc.), and *M. miehei* NRRL 3169 (Baxter Laboratories, Inc., 1970). This number of strains can lead to a confusing state of affairs, since it is not always clear which strain is being evaluated in reported cheese trials. The situation may become even more involved if still other strains of *M. miehei* are discovered to produce useful rennets. It is certain that the genus *Mucor* will be systematically scrutinized (e.g., Veselov et al., 1968).

Ottensen and Rickert (1970a,b) purified and characterized the rennet produced by *Mucor miehei* Cooney et Emerson CBS 370.65. It is an acid protease with an optimum pH activity at 4.5 on denatured hemoglobin, and at pH 4.0 on the B chain of oxidized insulin. Rickert (1970) also studied the activity of the protease on the latter substrate and found it to resemble the specificities of animal rennet and pepsin. Ottensen and Rickert observed that, in broth, the enzyme remained stable at a pH of 4.5 and a temperature of 4°C for at least 8 months. Even at 38°C, it retained greater than 90% of its activity in the range of pH 3.0 to 6.0 during 8 days of incubation. Its isoelectric point was determined to be at pH 4.2. The molecular weight was estimated to be about 38,000. Its amino acid composition was found to be as follows (calculated on the basis of one residue of histidine): ala_{15}, arg_4, asp_{26}, $½cys_2$, glu_{14}, gly_{19}, his_1, $ileu_{11}$, leu_{12}, lys_5, met_4, phe_{12}, pro_{10}, ser_{21}, thr$_{17}$, try$_2$, tyr$_{10}$, and val$_{15}$. Approximately 6% of the purified material was composed of hexosamine and neutral hexoses. The ratio of milk-clotting activity to proteolytic activity was approximately equivalent to the ratio for *M. pusillus* rennet.

In 1971, Sternberg crystallized the principal rennet (representing 51% of total activity) of *Mucor miehei* NRRL 3420. Not unexpectedly, this enzyme also is an acid protease. It is not metal dependent, nor does it possess serine or SH active groups. Its molecular weight is in the range of 34,000 to 39,000. The amino acid composition is as follows: ala$_{22}$, arg$_4$, asp$_{42}$, ½ cys$_4$, glu$_{15}$, gly$_{24}$, his$_2$, ileu$_{11}$, leu$_{14}$, lys$_8$, met$_5$, orn$_1$, phe$_{14}$, pro$_{11}$, ser$_{25}$, thr$_{18}$, try$_3$, tyr$_{13}$, and val$_{16}$. The presence of ornithine (orn) in this enzyme, if confirmed, is a rather novel finding. Urea-denatured hemoglobin is maximally proteolyzed in a range of pH 4.1 to 4.4, acid-denatured hemoglobin, at pH 3.5. The enzyme proved to be stable over the wide range of pH 2.0 to 6.0 at a temperature of 40°C for 24 hr. Calcium ions (6.8×10^{-3} M calcium chloride) protected the activity somewhat. Though there are some resemblances to the rennet of *M. pusillus*, the differences between the two enzymes are preponderant. Charles *et al.* (1970) claim the ratio of milk-clotting activity to proteolytic activity exhibited by *M. miehei* NRRL 3420 to be the highest among microbial rennets.

Edelsten and Jensen (1970) succeeded in fractionating the rennet of *M. miehei* into three main, milk-coagulating fractions labeled: A, B, and C. Fractions A and C were most active at 18°C, while component B was inactive at 18°C and most active at 35°C. In cheese trials, Behnke and Siewert (1969) observed a delayed reaction with the *M. miehei* rennet in the primary phase of coagulation. They speculated that it might be due to a mode of action by the rennet of *M. miehei* different from that of calf rennet. On the other hand, the tertiary phase (ripening stage) was accelerated. The Edam and Camembert cheeses they produced developed some bitterness after 49 and 21 days, respectively. All in all, Behnke and Siewert did not consider the organoleptic defect significant. Limburger and Tollenser cheeses proved to be quite suitable. Hamdy (1970) studied the production of Domiati cheese. He indicated that the cheese was organoleptically satisfactory. However, a slightly higher proteolysis was noted with the cheese made with the *M. miehei* rennet than with the cheese prepared with animal rennet. Prins and Nielsen (1970) refer to the production of good Cheddar cheese with the rennet of *M. miehei*.

Thomasow *et al.* (1971) also observed acceleration of curing in Edam, Tilsit, and Butter cheeses. No bitterness developed with these cheeses. They observed that a rise in temperature affected the

M. miehei rennet activity more than animal rennet, causing increased protein hydrolysis. A comparison of the factors affecting coagulation strength of the three commercially available microbial rennets (i.e., from *E. parasitica, M. pusillus,* and *M. miehei*) was made by Hamdy and Edelsten (1970). They reported that at somewhat over 0.6% concentration of sodium chloride in milk, coagulation by animal rennet was prolonged. It required greater than 1.5% sodium chloride to increase the coagulation time of the three microbial rennets. Conversely, calcium chloride reduced clotting time for all rennets, though *M. pusillus* proved most sensitive. Optimum temperature for coagulation was 42°C, with the three fungal rennets being more sensitive than animal rennet in the range of 37 to 42°C. The rennet of *M. miehei* evidenced the greatest sensitivity to temperature.

VI. General Comments

There is no question but that the commercially available fungal rennets described in Section V, D can and do make good cheese. Indeed, they even possess the potential for producing cheeses with qualities superior to those prepared with animal rennet. Of course, defects do appear in some cheeses, particularly the long-hold varieties. However, this must be accepted as a challenge rather than as an immutable failing. After all, animal rennet itself, peerless though it may be, is not an incontestably ideal milk coagulant. In addition to the drawbacks noted in the first section of this paper, other shortcomings may be cited. For example, the enzymatic composition of commercially available animal rennet is generally inconstant. The pepsin content has been found in some extracts to exceed 25% of the total milk-clotting activity. Not surprisingly, an analysis of a 1:1 mixture of animal rennet:pepsin revealed as much as 75% pepsin to be present (Pedersen, 1970). Animal rennet also manifests an undesired sensitivity to pH variations in milk. Furthermore, it also must be acknowledged that a certain proportion of poor cheeses does result from time to time with animal rennet.

The fungal rennets might perform better in some cheeses if the cheese-manufacturing procedure were to be modified somewhat to accommodate the characteristics of the enzymes. For example, appropriate starter cultures might need to be selected, as well as suitable pH and temperature profiles. Such variations also might permit small but advantageous reduction in the amount of fungal rennet added to milk for coagulation.

The future of microbial rennets in the cheese manufacturing industry will be decided in part by the innovative and enterprising technologists in the field, as well as by the overall economics involved. One can easily envision possibilities for the use of insolubilized rennets in large columns through which cooled milk is percolated, effecting thereby the primary phase of coagulation. Subsequent heating of the reacted milk would induce the secondary phase of coagulation. Efforts at insolubilization of rennin, unsuccessful though thay have been to date, already have been made (Green and Crutchfield, 1969). On the other hand, a measure of success has been achieved with insolubilized pepsin in the continuous coagulation of reconstituted nonfat milk solids (Ferrier et al., 1971). It is only a matter of time before success is attained and a commercially feasible procedure is realized. Microbial enzyme recovery techniques also are being constantly improved to yield more desirable rennets (Moelker and Matthijsen, 1971; Organon, 1970; Schleich, 1971). There is every reason to expect that the future bodes well for microbial rennets. Their increasing use at present already represents a challenge to the primacy of rennet extract.

ACKNOWLEDGMENT

The author wishes to express grateful appreciation to his colleagues, especially Dr. John B. Routien, and Messrs. J. W. Davisson and John Shovers, for their valuable suggestions and criticisms during preparation of this review.

REFERENCES

Ainsworth, G. C. (1961). "Ainsworth and Bisby's Dictionary of Fungi." Commonw. Mycol. Inst., Kew, Surrey, England.
Alais, C., and Novak, G. (1968). *Lait* **48**, 393–418.
Alais, C., and Novak, G. (1970). *Int. Dairy Congr., Proc., 18th, 1970* Vol. IE, p 979.
Alais, C., Dutheil, H., and Bose, J. (1962). *Int. Dairy Congr., Proc., 16th, 1962* Vol. B, pp. 643–654.
Aldridge, W. N. (1961). *In* "Biochemist's Handbook" (C. Long, ed.), pp. 280–281. Van Nostrand-Reinhold, Princeton, New Jersey.
Annibaldi, S. (1962a). *Int. Dairy Congr., Proc., 16th, 1962* Vol. B, pp. 381–384.
Annibaldi, S. (1962b). *Int. Dairy Congr., Proc., 16th, 1962* Vol. B, pp. 545–556.
Anson, M. L. (1938). *J. Gen. Physiol.* **22**, 79–89.
Arima, K., and Iwasaki, S. (1964). U.S. Pat. No. 3,151,039.
Arima, K., Iwasaki, S., and Tamura, G. (1967). *Agr. Biol. Chem.* **31**, 540–545.
Arima, K., Yu, J., Iwasaki, S., and Tamura, G. (1968). *Appl. Microbiol.* **16**, 1727–1733.
Arima, K., Yu, J., and Iwasaki, S. (1970). *In* "Methods in Enzymology" (G. Perlman and L. Lorand, eds.), Vol. 19, pp. 446–459. Academic Press, New York.
Aunstrup, K. (1968a). Brit. Pat. No. 1,108,287.
Aunstrup, K. (1968b). Fr. Pat. No. 1,521,368.
Babbar, I. J., and Sampathkumar, B. (1970). Indian Pat. No. 118,023; *Chem. Abstr.* **74**, 30739t (1971).
Babbar, I. J., Srinivasan, R. A., Chakravorty, S. C., and Dudani, A. T. (1965). *Indian J. Dairy Sci.* **18**, 89–95.

Babel, F. J. (1967). *Dairy Ind.* **32**, 901–904.
Babel, F. J., and Somkuti, G. A. (1968). *J. Dairy Sci.* **51**, 937 (M59).
Balls, A. K., and Hoover, S. R. (1937). *J. Biol. Chem.* **121**, 737–745.
Barkan, S. M., Ramaranova, O. Kh., and Yulius, A. A. (1965). *Dairy Sci. Abstr.* **27**, 152.
Baxter Laboratories, Inc. (1970). Brit. Pat. No. 1,207,892.
Beeby, R., Hill, R. D., and Snow, N. S. (1971). *In* "Milk Proteins: Chemistry and Molecular Biology" (H. A. McKenzie, ed.), Vol. 2, pp. 422–427. Academic Press, New York.
Behnke, U. (1967). *Milchwissenschaft* **22**, 563–569.
Behnke, U., and Siewert, R. (1969). *FBM Milch Stand.* **11**, 66–72.
Berkowitz-Hundert, R., and Leibowitz, J. (1963). *Enzymologia* **25**, 257–260.
Berkowitz-Hundert, R., Leibowitz, J., and Ilany-Feigenbaum, J. (1964). *Enzymologia* **27**, 332–342.
Berkowitz-Hundert, R., Ilany-Feigenbaum, J., and Leibowitz, J. (1965). *Enzymologia* **29**, 98–100.
Bernatonis, J., and Mickiene, N. (1970). *Chem. Abstr.* **75**, 18871t.
Berridge, N. J. (1945). *Biochem. J.* **39**, 179–186.
Berridge, N. J. (1952a). *Analyst* **77**, 57–62.
Berridge, N. J. (1952b). *J. Dairy Res.* **19**, 328–329.
Berridge, N. J. (1954). *Advan. Enzymol. Relat. Subj. Biochem.* **16**, 423–448.
Berridge, N. J. (1961). *In* "Biochemist's Handbook" (C. Long, ed.), pp. 299–301. Van Nostrand-Reinhold, Princeton, New Jersey.
Bier, M. (1955). *In* "Methods in Enzymology" (S. P. Colowick and N. O. Kaplan, eds.), Vol. 1, pp. 627–651. Academic Press, New York.
Bolliger, O., and Schilt, P. (1969). *Schweiz. Milchztg.* **95**, 1029–1034.
Breed, R. S., Murray, E. G. D., and Smith, N. R. (1957). "Bergey's Manual of Determinative Bacteriology," 7th ed. Williams & Wilkins, Baltimore, Maryland.
Brown, R. C. (1966). "The Complete Book of Cheese," Crown (Gramercy), New York.
Bunn, C. W., Moews, P. C., and Baumber, M. E. (1971). *Proc. Roy. Soc., Ser. B* **178**, 245–258.
Carr, J. W., Loughheed, T. C., and Baker, B. E. (1956). *J. Sci. Food Agr.* **7**, 629–637.
Chapman, H. R., and Burnett, J. (1968). *Dairy Ind.* **33**, 308–311.
Charles, R. L., Gertzman, D. P., and Malachouris, N. (1970). U.S. Pat. No. 3,549,390.
Chebotarev, A. I., Durova, Zh. I., and Petina, T. A. (1969a). *Prikl. Biokhim. Mikrobiol.* **5**, 451–454.
Chebotarev, A. I., Durova, Zh. I., and Petina, T. A. (1969b). *Moloch. Prom.* **30**, 10–11.
Choudhery, A. K., and Mikolajcik, E. M. (1970). *J. Dairy Sci.* **53**, 363–366.
Choudhery, A. K., and Mikolajcik, E. M. (1971). *J. Dairy Sci.* **54**, 321–325.
Christensen, P. A. (1960). *Acta Pathol. Microbiol. Scand.* **49**, 95–97.
Collier, B. (1970). *Process Biochem.* **5**, 39–42.
Conn, H. W. (1892). *Centralbl. Bacteriol.* **12**, 223–227.
Czulak, J. (1959). *Aust. J. Dairy Technol.* **14**, 177–179.
Czulak, J., and Shimmin, P. D. (1961). *Aust. J. Dairy Technol.* **16**, 96–98.
Damodaran, M., Govindarajan, V. S., and Subramanian, S. S. (1955). *Biochim. Biophys. Acta* **17**, 99–110.
Davis, J. G. (1971). *Dairy Ind.* **36**, 135–141.
Davis, J. G., and MacDonald, F. J. (1952). "Richmond's Dairy Chemistry," 5th ed. Griffin, London.
deMan, J. M., and Batra, S. C. (1964). *Dairy Ind.* **29**, 32–33.
Dewane, R. A. (1960). "Rennet Substitutes." Coronet, Milwaukee, Wisconsin.

Dimitrov, D., Veselov, I. Ya., Petina, T. A., and Martirosova, L. A. (1969). *Prikl. Biokhim. Mikrobiol.* **5**, 47–51.

Dimitrov, D., Veselov, I. Ya., Petina, T. A., and Tipograf, D. Ya. (1970). *Prikl. Biokhim. Mikrobiol.* **6**, 173–177.

Dyachenko, P. F., and Slavyanova, V. V. (1962). *Int. Dairy Congr., Proc., 16th, 1962* Vol. 2, pp. 349–352.

Edelsten, D., and Jensen, J. S. (1970). *Int. Dairy Congr., Proc., 18th, 1970* Vol. IE, p. 280.

Edelsten, D., Hamdy, A., and El Kousy, L. (1969). *Kgl. Vet.- Landbohoejsk. Arsskr.* pp. 201–212.

Edwards, J. L. (1969). Ph.D. Thesis, Cornell University, Ithaca, New York.

El-Negoumy, A. M. (1970). *J. Dairy Res.* **37**, 437–444.

Emanuiloff, I. (1956). *C. R. Acad. Bulg. Sci.* **9**, 71–74.

Emmons, D. B., Petrasovits, A., Gillan, R. H., and Bain, J. M. (1971). *J. Can. Inst. Food Technol.* **4**, 31–37.

Ernstrom, C. A. (1958). *J. Dairy Sci.* **41**, 1663–1670.

Everson, T. C., and Winder, W. C. (1968). *J. Dairy Sci.* **51**, 940.

Feldman, L. I. (1968). Ger. Pat. No. 1,962,575.

Feldman, L. I. (1969). Fr. Pat. No. 1,556,473.

Ferrier, L. K., Richardson, T., and Olson, N. F. (1971). *J. Dairy Sci.* **54**, 762.

Foltmann, B. (1970). *In* "Methods in Enzymology" (G. Perlman and L. Lorand, eds.), Vol. 19, pp. 421–436. Academic Press, New York.

Food and Agriculture Organization of the United Nations (1968). "Ad Hoc Consultation on World Shortage of Rennet in Cheesemaking" Rep. No. AN 1968/3. FAO, Rome.

Fox, P. F. (1968). *Ir. J. Agr. Res.* **7**, 251–260.

Fox, P. F. (1969). *J. Dairy Res.* **36**, 427–434.

Fox, P. F., and Walley, B. F. (1971). *J. Dairy Res.* **38**, 165–170.

Fujimaki, M., Yamashita, M., Arai, S., and Kato, H. (1970). *Agr. Biol. Chem.* **34**, 483–484.

Fukumoto, J. (1970). Jap. Pat. No. 70 39020.

Fukumoto, J., Tsuru, D., and Yamamoto, T. (1967). *Agr. Biol. Chem.* **31**, 710–717.

Garnier, J., Mocquot, G., Ribadeau-Dumas, B., and Maubois, J. L. (1968). *Ann. Nutr. Aliment.* **22**, 13495–552.

Godo Shusei Co. (1968). Fr. Pat. No. 1,540,317.

Godo Shusei Co. (1969). Fr. Pat. No. 1,569,837.

Gordon, D. F., Jr., and Speck, M. L. (1965a). *J. Dairy Sci.* **48**, 499–500.

Gordon, D. F., Jr., and Speck, M. L. (1965b). *Appl. Microbiol.* **13**, 537–542.

Gorini, C. (1892). *Riv. Ing. Sanit.* **3**, 527–530.

Gorini, C. (1930). *J. Bacteriol.* **20**, 297–298.

Gorini, C. (1931). *Milchwirt. Forsch.* **12**, 199–200.

Gorini, L., and Lanzavecchia, G. (1954). *Biochim. Biophys. Acta* **14**, 407–414.

Green, M. L., and Crutchfield, G. (1969). *Biochem. J.* **115**, 183–190.

Greenberg, D. M. (1955). *In* "Methods in Enzymology" (S. P. Colowick and N. O. Kaplan, eds.), Vol. 2, pp. 54–64. Academic Press, New York.

Grimberg, M. V. (1965). *Fette, Seifen, Anstrichm.* **67**, 271–273.

Groves, M. L. (1971). *In* "Milk Proteins: Chemistry and Molecular Biology" (H. A. McKenzie, ed.), Vol. 2, pp. 367–418. Academic Press, New York.

Hagemeyer, K., Fawwal, I., and Whitaker, J. R. (1968). *J. Dairy Sci.* **51**, 1916–1922.

Hagihara, B., Matsubara, H., Nakai, M., and Okunuki, K. (1958). *J. Biochem. (Tokyo)* **45**, 185–194.

Hamdy, A. (1970). *Int. Dairy Congr., Proc., 18th, 1970* Vol. IE, p. 350.
Hamdy, A., and Edelsten, D. (1970). *Milchwissenschaft* **25**, 450–453.
Hammer, B. W., and Babel, F. J. (1957). "Dairy Bacteriology," 4th ed. Wiley, New York.
Harwalkar, V. R., and Elliott, J. A. (1965). *Pap., 60th Annu. Meet., Amer. Dairy Sci. Ass., Lexington, Kentucky* Pap. No. N92.
Harwalkar, V. R., and Elliott, J. A. (1971). *J. Dairy Sci.* **54**, 8–11.
Harwalkar, V. R., and Seitz, E. W. (1971). *J. Dairy Sci.* **54**, 12–14.
Hill, R. D. (1969). *J. Dairy Res.* **36**, 409–415.
Hill, R. J., and Wake, R. G. (1969). *Nature (London)* **221**, 635.
Huang, C. (1970). Ph.D. Thesis, Purdue University, Ann Arbor, Michigan.
Hussang, R. V., and Hammer, B. W. (1928). *J. Bacteriol.* **15**, 179–188.
Ilany-Feigenbaum, J., and Netzer, A. (1969). *J. Dairy Sci.* **52**, 43–46.
Imai, T., Irie, Y., Matsudo, C., and Kanazawa, Y. (1970). Can. Pat. No. 852,711.
Irie, S., Kanazawa, Y., and Imai, N. (1966). *Annu. Meet. Agr. Chem. Soc. Jap.* p. 136.
Iwasaki, S., Tamura, G., and Arima, K. (1967a). *Agr. Biol. Chem.* **31**, 546–551.
Iwasaki, S., Yasui, T., Tamura, G., and Arima, K. (1967b). *Agr. Biol. Chem.* **31**, 1421–1426.
Iwasaki, S., Yasui, T., Tamura, G., and Arima, K. (1967c). *Agr. Biol. Chem.* **31**, 1427–1433.
Iwasaki, S., Yu, J., Tamura, G., and Arima, K. (1968). Paper presented at the 3rd International Fermentation Symposium (Unpublished).
Jago, G. R. (1962). *Aust. J. Dairy Technol.* **17**, 83–85.
Kassell, B., and Meitner, P. A. (1970). *In* "Methods in Enzymology" (G. Perlman and L. Lorand, eds.), Vol. 19, pp. 337–347. Academic Press, New York.
Kawai, M. (1970). *Agr. Biol. Chem.* **34**, 164–169.
Kawai, M., and Mukai, N. (1970). *Agr. Biol. Chem.* **34**, 159–163.
Keay, L. (1971). *Process Biochem.* **6**, 17–21.
Kikuchi, T., and Toyoda, S. (1969). *Yukijirushi Nyngyo Gijutsu Kenkyujo Hokoku* **71**, 11–24.
Kikuchi, T., Takafuji, S., Toyoda, S., and Suzuki, Y. (1968a). *Yukijirushi Nyngyo Gijutsu Kenkyujo Hokoku* **70**, 1–6.
Kikuchi, T., Takafuji, S., Toyoda, S., and Sukegawa, K. (1968b). *Yukijirushi Nyngyo Gijutsu Kenkyujo Hokoku* **70**, 7–12.
Kikuchi, T., Toyoda, S., Ahiko, K., and Suzuki, Y. (1968c). *Yukijirushi Nyngyo Gijutsu Kenkyujo Hokoku* **70**, 13–22.
Kiss, E. (1969). *Elelmiszertudomany* **3**, 83–88.
Knight, S. G. (1966). *Can. J. Microbiol.* **12**, 420–422.
Kruger, W., Krenkel, H., and Gust, H.-P. (1968). *Nahrung* **12**, 157–162.
Kunitz, M. (1935). *J. Gen. Physiol.* **18**, 459–466.
Kyla-Siurola, A. L., and Antila, V. (1970). *Int. Dairy Congr., Proc., 18th, 1970* Vol. IE, p. 283.
Larson, M. K., and Whitaker, J. R. (1970a). *J. Dairy Sci.* **53**, 253–261.
Larson, M. K., and Whitaker, J. R. (1970b). *J. Dairy Sci.* **53**, 262–269.
Laskowski, M. (1961). *In* "Biochemist's Handbook" (C. Long, ed.), pp. 298–299. Van Nostrand-Reinhold, Princeton, New Jersey.
Lawrence, R. C., and Gilles, J. (1969). *N. Z. J. Dairy Technol.* **4**, 189–196.
Ledford, R. A., Chen, J. H., and Nath, K. R. (1968). *J. Dairy Sci.* **51**, 792–794.
Lindqvist, B. (1963a). *Dairy Sci. Abstr.* **25**, 257–264.
Lindqvist, B. (1963b). *Dairy Sci. Abstr.* **25**, 299–305.
Loeb, A. (1902). *Centralbl. Bacteriol.* **32**, 471–477.
McDonald, C. E., and Chen, L. L. (1965). *Anal. Biochem.* **10**, 175–177.

MacKinlay, A. G., and Wake R. G. (1971). *In* "Milk Proteins: Chemistry and Molecular Biology" (H. A. McKenzie, ed.), Vol. 2, pp. 198–212. Academic Press, New York.
Margalith, P., and Schwartz, Y. (1970). *Advan. Appl. Microbiol.* **12**, 64–68.
Masek, J., Havlova, J., and Tepley, M. (1970). *Prum. Potravin.* **21**, 386–389.
Matoba, T., Hayashi, R., and Hata, T. (1970). *Agr. Biol. Chem.* **34**, 1235–1243.
Maubois, J. L., and Mocquot, G. (1969). *Lait* **49**, 497–506.
Meito Sangyo, K. K. (1963). Fr. Pat. No. 1,320,231.
Melachouris, N. P., and Tuckey, S. L. (1964). *J. Dairy Sci.* **47**, 1–7.
Melachouris, N. P., and Tuckey, S. L. (1968). *J. Dairy Sci.* **51**, 650–655.
Mickelsen, R., and Ernstrom, C. A. (1967). *J. Dairy Sci.* **50**, 645–650.
Mickelsen, R., and Fish, N. L. (1970). *J. Dairy Sci.* **53**, 704–710.
Moelker, H. C. T., and Matthijsen, R. (1971). U.S. Pat. No. 3,591,388.
Moews, P. C., and Bunn, C. W. (1970). *J. Mol. Biol.* **54**, 395–397.
Morris, T. A., and McKenzie, I. J. (1970). *Int. Dairy Congr., Proc., 18th, 1970* Vol. IE, p. 293.
Mukai, N., and Kawai, M. (1970a). Jap. Pat. No. 70 17,149.
Mukai, N., and Kawai, M. (1970b). Jap. Pat. No. 70 17,588.
Mukai, N., and Kawai, M. (1971). U.S. Pat. No. 3,607,655.
Murray, E. D., and Kendall, M. S. (1969). U.S. Pat. 3,482,997.
Murray, E. D., and Prince, M. P. (1970). U.S. Pat. No. 3,507,750.
Nakanishi, T., and Itoh, M. (1970). *Rakuno Kagaku No Kenkyu* **17**, A94–A101.
Naudts, I. M. (1969). *53rd Int. Dairy Fed. Sess.* I-DOC-51 (1969 revision).
Oi. S., Yamazaki, O., Sawada, A., and Satomura, Y. (1969). *Agr. Biol. Chem.* **33**, 729–738.
Oosthuizen, J. C. (1962). *J. Dairy Res.* **29**, 297–305.
Organon, N. V. (1970). Fr. Pat. No. 1,592,965.
Oringer, K. (1960). U.S. Pat. No. 2,927,060.
Orosin, B., Veselov, I. Ya., Smirnova, T. A., and Ivanova, T. V. (1970). *Prikl. Biokhim. Mikrobiol.* **6**, 660–665.
Oruntaeva, K. B., and Seitov, Z. S. (1971). *Biokhimiya* **36**, 18–21.
Osman, H. G., Abdel-Fattah, A. F., Abdel-Samie, M., and Mabrouk, S. S. (1969a). *J. Gen. Microbiol.* **59**, 125–129.
Osman, H. G., Abdel-Fattah, A. F., and Mabrouk, S. S. (1969b). *J. Gen. Microbiol.* **59**, 131–135.
Ottensen, M., and Rickert, W. (1970a). *C. R. Trav. Lab. Carlsberg* **37**, 301–325.
Ottensen, M., and Rickert, W. (1970b). *In* "Methods in Enzymology" (G. Perlman and L. Lorand, eds.), Vol. 19, pp. 459–460. Academic Press, New York.
Paleva, N. S. (1969). *Mikrobiologiya* **38**, 1002–1005.
Paleva, N. S., and Popova, N. V. (1965). *Ferment. Spirt. Prom.* **31**, 6–8.
Pedersen, A. H. (1969). *Chem. Abstr.* **71**, 59750p.
Pedersen, A. H. (1970). 179th Rep. Dairy Ind., Dan. Gov. Res. Inst.
Petersen, W. E. (1939). "Dairy Science," Lippincott, Philadelphia, Pennsylvania.
Peterson, A. C., and Gunderson, M. F. (1960). *J. Appl. Microbiol.* **8**, 98–104.
Praprotnik, V. (1968). *Deut. Molk.-Ztg.* **89**, 2058–2060.
Prins, J., and Nielsen, T. K. (1970). *Process Biochem.* **5**, 34–35.
Puhalla, J. E., and Anagnostakis, S. L. (1971). *Phytopathology* **61**, 169–173.
Puhan, Z. (1966). *Int. Dairy Congr., Proc., 17th, 1966* Vol. 4, pp. 199–204.
Puhan, Z. (1969). *J. Dairy Sci.* **52**, 1372–1378.
Puhan, Z., and Steffen, C. (1967). *Schweiz. Milchztg.* **93**, 937–951.
Ramet, J. P., and Schluter, A. S. (1970). *Deut. Molk.-Ztg.* **91**, 1822–1829.
Ramet, J. P., Alais, C., and Weber, F. (1969). *Lait* **49**, 40–52.

Reed, G. (1966). "Enzymes in Food Processing." Academic Press, New York.
Reps, A., Poznanski, S., and Kowalska, W. (1969). *Bull. Acad. Pol. Sci., Ser. Sci. Biol.* **17**, 535–541.
Reps, A., Poznanski, S., and Kowalska, W. (1970). *Milchwissenschaft* **25**, 146–150.
Resmini, P., Volonterio, G., Saracchi, S., and Bozzolati, M. (1971). *Ind. Agr.* **9**, 6–18.
Richardson, G. H., Nelson, J. H., Lubnow, R. E., and Schwarberg, R. L. (1967). *J. Dairy Sci.* **50**, 1066–1072.
Richardson, G. H., Gandhi, N. R., Divatia, M. A., and Ernstrom, C. A. (1971). *J. Dairy Sci.* **54**, 182–186.
Rickert, W. (1970). *C. R. Trav. Lab. Carlsberg* **38**, 1–17.
Robertson, P. S., and Gilles, J. (1969). *N. Z. J. Dairy Technol.* **4**, 128–132.
Rotini, O. T., and Sequi, P. (1970). *Agrochimica* **16**, 386–391.
Ryle, A. P. (1970). *In* "Methods in Enzymology" (G. Perlman and L. Lorand, eds.), Vol. 19, pp. 316–336. Academic Press, New York.
Sanders, G. P. (1953). Cheese Varieties. *U.S., Dep. Agr., Agr. Handb.* **54**.
Sannabadthi, S. S., Srinivasan, R. A., and Laxminarayana, H. (1970). *Int. Dairy Congr., Proc., 18th, 1970* Vol. IE, p. 278.
Sardinas, J. L. (1965). Fr. Pat. No. 1,401,474.
Sardinas, J. L. (1966). U.S. Pat. No. 3,275,453.
Sardinas, J. L. (1967). *Abstr. Pap., 154th Meet., Amer. Chem. Soc.* Pap No. Q29.
Sardinas, J. L. (1968). *Appl. Microbiol.* **16**, 248–255.
Sardinas, J. L. (1969). *Process Biochem.* **4**, 13–16, 21.
Sato, Y., Sekiguchi, Y., Chiba, Y., and Ikai, M. (1969). *Agr. Biol. Chem.* **33**, A21.
Schleich, H. (1971). U.S. Pat. No. 3,616,233.
Schormuller, J. (1968). *Advan. Food Res.* **16**, 231–334.
Schulz, M. E., Voss, E., Sell, H., and Mrowetz, G. (1967). *Milchwissenschaft* **22**, 139–144.
Scott Blair, G. W., and Oosthuizen, J. C. (1962). *J. Dairy Res.* **29**, 37–46.
Sequi, P., and Rotini, O. T. (1970). *Agrochimica* **16**, 379–385.
Shichiji, S., and Saeki, Y. (1970). Jap. Pat. No. 70 33,754.
Shimwell, J. L., and Evans, J. E. (1944). Brit. Pat. No. 565,788.
Shovers, J., and Bavisotto, V. S. (1967). *J. Dairy Sci.* **50**, 942–943.
Singh, A., Kuila, R. K., Dutta, S. M., Babbar, I. J., Srinivasan, R. A., and Dudani, A. T. (1967). *J. Dairy Sci.* **50**, 1886–1890.
Singh, K., and Vezina, C. (1971). *Can. J. Microbiol.* **17**, 1029–1042.
Skelton, G. S. (1971). *Enzymologia* **40**, 170–172.
Somkuti, G. A., and Babel, F. J. (1966). *J. Dairy Sci.* **49**, 700.
Somkuti, G. A., and Babel, F. J. (1967). *Appl. Microbiol.* **15**, 1309–1312.
Somkuti, G. A., and Babel, F. J. (1968a). *Appl. Microbiol.* **16**, 617–619.
Somkuti, G. A., and Babel, F. J. (1968b). *J. Bacteriol.* **95**, 1407–1414.
Somkuti, G. A., and Babel, F. J. (1969). *J. Dairy Sci.* **52**, 535–536.
Somkuti, G. A., and Walter, M. M. (1970). *Proc. Soc. Exp. Biol. Med.* **133**, 780–785.
Somkuti, G. A., Babel, F. J., and Somkuti, A. C. (1969a). *Appl. Microbiol.* **17**, 606–610.
Somkuti, G. A., Bable, F. J., and Somkuti, A. C. (1969b). *J. Dairy Sci.* **52**, 1104–1106.
Soxhlet, F. (1877). *Milch-ztg.* **6**, 495–501.
Spickett, R. G. W. (1971). *Chem. Ind. (London)* No. 3, pp. 83–94.
Srinivasan, R. A., Anantharamaiah, S. N., Ananthakrishnan, C. P., and Iya, K. K. (1962a). *Int. Dairy Congr., Proc., 16th, 1962* Vol. B, pp. 401–409.
Srinivasan, R. A., Anantharamaiah, S. N., Keshavamurthy, N., Ananthakrishnan, C. P., and Iya, K. K. (1962b). *Int. Dairy Congr., Proc., 16th, 1962* Vol. B, pp. 506–512.

Srinivasan, R. A., Iyengar, M. K. K., Babbar, I. J., Chakravorty, S. C., Dudani, A. T., and Iya, K. K. (1964). *Appl. Microbiol.* **12**, 475–478.
Stadhouders, J. (1962). *Int. Dairy Congr., Proc., 16th, 1962* Vol. B, pp. 353–361.
Sternberg, M. Z. (1971). *J. Dairy Sci.* **54**, 159–167.
Su, Y., and Chen, W.-P. (1971). *Chem. Abstr.* **74**, 139715h.
Sullivan, J. J., and Jago, G. R. (1970). *Aust. J. Dairy Technol.* **25**, 111.
Takamine, J., and Takamine, J., Jr. (1921). U.S. Pat. No. 1,391,219.
Tang, J. (1970). *In* "Methods in Enzymology" (G. Perlman and L. Lorand, eds.), Vol. 19, pp. 406–421. Academic Press, New York.
Tang, J., Wolf, S., Caputto, R., and Trucco, R. E. (1959). *J. Biol. Chem.* **234**, 1174–1178.
Tarodo de la Fuente, B., Alais, C., and Frentz, R. (1969). *Lait* **49**, 400–416.
Tendler, M. D., and Burkholder, P. R. (1961). *Appl. Microbiol.* **9**, 394–399.
Thomasow, J., Mrowetz, G., and Schmanke, E. (1971). *Kiel. Milchwirt, Forschungsber.* **23**, 57–68.
Tipograf, D. Ya., Veselov, A. I., Nyong, L. V., and Mosichev, M. S. (1966). *Prikl. Biokhim. Mikrobiol.* **2**, 45–50.
Tokita, F. (1969). *Chem. Abstr.* **70**, 86351v.
Toolens, H. P. (1968). Fr. Pat. No. 1,566,447.
Trop, M., and Pinsky, A. (1971). *J. Dairy Sci.* **54**, 5–7.
Tsugo, T., and Yamauchi, K. (1959). *Int. Dairy Congr., Proc., 15th, 1959* Vol. 2, pp. 636–642.
Tsuru, D., Kira, H., Yamamoto, T., and Fukumoto, J. (1967). *Agr. Biol. Chem.* **31**, 718–723.
Tsuru, D., Hattori, A., Tsuji, H., Yamamoto, T., and Fukumoto, J. (1969). *Agr. Biol. Chem.* **33**, 1419–1426.
Tuszynski, W. B. (1971). *J. Dairy Res.* **38**, 115–125.
U.S. Department of Agriculture. (1971). "Dairy Situation," DS-334, p. 12. Econ. Res. Serv., Washington, D.C.
Vamos Vigyazo, L., Pozsar Hajnal, K., Gajzago, I., and Hegedus, V. E. (1969a). *Elelmiszertudomany* **3**, 13–23.
Vamos Vigyazo, L., Pozsar Hajnal, K., and Davis Szekeres, A. (1969b). *Elelmiszertudomany* **3**, 43–54.
Veringa, H. A. (1961). *Dairy Sci. Abstr.* **23**, 197–200.
Veselov, I. Ya., Tipograf, D. Ya., and Petina, T. A. (1965). *Priklbiokhim. Mikrobiol.* **1**, 52–56.
Veselov, I. Ya., Mosichev, M. S., and Tipograf, D. Ya. (1968). *Mikrobiologiya* **37**, 616–619.
Wahlin, J. G. (1928). *J. Bacteriol.* **16**, 355–373.
Wang, H. L., and Hesseltine, C. W. (1970). *Arch. Biochem. Biophys.* **140**, 459–463.
Wang, H. L., Ruttle, D., and Hesseltine, C. W. (1969). *Can. J. Microbiol.* **15**, 99–104.
Waugh, D. F. (1971). *In* "Milk Proteins: Chemistry and Molecular Biology" (H. A. McKenzie, ed.), Vol. 2, pp. 75–79. Academic Press, New York.
Waugh, D. F., and Hippel, P. H. (1956). *J. Amer. Chem. Soc.* **78**, 4576–4582.
Wheelock, J. V., and Knight, D. J. (1969). *J. Dairy Res.* **36**, 183–190.
Whitaker, J. R. (1970). *In* "Methods in Enzymology" (G. Perlman and L. Lorand, eds.), Vol. 19, pp. 436–445. Academic Press, New York.
Whitehill, A. R., Ablondi, F. B., Mowat, J. H., and Krupka, G. (1960). U.S. Pat. No. 2,936,265.
Wilson, H. L., and Reinbold, G. W. (1965). "American Cheese Varieties," Vol. 2. Pfizer Inc., New York.

Yasunobu, K. T., and McConn, J. (1970). *In* "Methods in Enzymology" (G. Perlman and L. Lorand, eds.), Vol. 19, pp. 560–575. Academic Press, New York.
Yeda Research and Development Co., Ltd. (1971). Isra. Pat. No. 30,520.
Yu, J., Tamura, G., and Arima, K. (1969). *Biochim. Biophys. Acta* **171,** 138–144.
Yu, J., Tamura, G., and Arima, K. (1971). *Agr. Biol. Chem.* **35,** 1194–1199.
Zvyagintsev, V. I., Gudkov, A. V., Tolkachev, A. N., and Buzov, I. P. (1971a). *Moloch. Prom.* **32,** 19–22.
Zvyagintsev, V. I., Sergeeva, E. G., and Gudkov, A. V. (1971b). *Prikl. Biokhim. Mikrobiol.* **1,** 259–271.

Volatile Aroma Components of Wines and Other Fermented Beverages

A. Dinsmoor Webb

*Department of Viticulture and Enology,
University of California, Davis, California*

AND

Carlos J. Muller

Department of Chemistry, Louisiana Tech University, Ruston, Louisiana

I.	Introduction	75
II.	List of Volatile Aroma Compounds	77
III.	Biosynthetic Pathways and Interrelationships among Aroma Compounds	131
	A. General Considerations	131
	B. Higher Alcohols	133
	C. Fatty Acids	134
	D. Ethers, Aldehydes, and Acetals	138
	E. Phenolics	139
	F. Nitrogen-Containing Compounds	139
	G. Lactones	139
	H. Sulfur-Containing Compounds	141
IV.	Summary	142
	References	142

I. Introduction

Alcoholic beverages appeal to man for many reasons. Among these, of course, are the physiological effects, the appeal to the eye because of the clarity and the colors, the pleasant tastes, and especially the aroma and bouquet which are appreciated through the sense of smell. Chemists have been subjecting wines, beers, and distilled spirits to analysis since the earliest days of these chemical procedures. The major components in most alcoholic beverages were isolated and identified at a fairly early date. The compounds primarily responsible for aroma and bouquet however, have presented a much more difficult problem—primarily because they are present in such very low concentrations. Studies of the aroma and bouquet compounds in alcoholic beverages, thus, can effectively be divided into two groups—pre and post gas chromatographic studies.

The chemical and technical literature contains numerous lists of aroma and bouquet components of fermented beverages. Among these may be mentioned the review of the organic constituents of wines by Amerine (1954), that of Webb (1967), the lists of volatile components of food by Weurman and Van Straten (1969), and a list by Kahn (1969). Many other shorter lists have been published and are cited in the publications mentioned above. Any list of compounds can, of course, only reflect those substances reported up to the time of publication, but many of the lists are incomplete because of an inadequate literature search. In addition to the incompleteness certain ones of the more lengthy publications were difficult to use because the compounds were listed according to functional groups. This is, of course, a logical system for monofunctional compounds but many of the substances encountered in aroma and bouquet mixtures are multifunctional. Further, and certainly a more serious criticism, is the fact that certain ones of the lists report compounds as being present on the basis of insufficient or inadequate data, and there is at least one case of the listing of reference or standard compounds, which had been added to the sample to permit quantitation, as rigorously identified components of the substance under investigation. In some cases of original reports, not enough data is presented to substantiate the claimed presence of the compound. This review will attempt to list all of the volatile aroma and bouquet compounds for which adequate data have been published. It is recognized that in some cases the decision is difficult, and in these cases it will be the intention to err on the side of conservatism. In cases of repeated identifications of compounds, all of the separate instances are not necessarily tabulated.

A fairly wide range of alcoholic beverages has been studied with respect to volatile aroma and bouquet compounds. These beverages all have the one factor in common that they contain the products of the activity of the yeast enzymes on a sugar substrate. They differ widely, of course, in that there are other substances present besides the sugars in the base material before fermentation and they are further modified by cellaring or processing operations. A very distinct group is those beverages produced by distillation of the wine or beer. We shall attempt in this review to include all of the volatile aroma and bouquet substances listed as present in the various types of wines, beers, and spirits.

Compounds are listed by empirical formula, following the *Chemical Abstracts* system of alphabetical arrangement of the symbols for the elements with the exception that C for carbon and H for hydrogen take precedence over the other symbols. Compounds having the same empirical formula are arranged alphabetically according to the names used in the *Chemical Abstracts* subject indexes.

Following the section on tabulation of aroma compounds, we discuss briefly the possible mode of biosynthesis of some of the more interesting of the compounds found in the aroma complex of alcoholic beverages.

II. List of Volatile Aroma Compounds

CH_2O
Formaldehyde
Identified by Hrdlicka *et al.* (1968) in beer by means of thin-layer chromatography.

CH_2O_2
Formic acid
Identified in bread pre-ferment by Hunter *et al.* (1961) by gas chromatography of free acid and of ethyl ester. Found in Sauvignon blanc wines by Chaudhary *et al.* (1968), in sherries by Webb *et al.* (1964b), and by Van Wyk *et al.* (1967a) in White Riesling wine, all by gas chromatography. The last authors prepared the hexyl ester and verified its identity by comparison of infrared spectra.

CH_4O
Methanol
Identified in passion fruit wine by Muller *et al.* (1964), and in *Vitis rotundifolia* grapes by Kepner *et al.* (1956) by gas and silica column chromatography respectively. In each case the 3,5-dinitrobenzoate derivative was prepared (m. 105–106°, no depression on admixture with known methanol derivative). Found in beer by Ahrenst-Larsen and Hansen (1964), and in sherry by Preobrazhenskii *et al.* (1969) by gas chromatographic techniques. Proved present in whiskey by Kahn *et al.* (1969) using gas chromatography–mass spectrometric procedures.

CH_4S
Methanethiol
Identified by Dellweg *et al.* (1969) in brandies at head-space levels less than 1 µg/liter using gas chromatography and a flame color detector.

CH_5N
Methylamine
Identified by Drawert (1965) in wines by paper and gas chromatography, and by Puputti and Suomalainen (1969) in wines, using gas chromatography.

$C_2H_2O_2$
Glyoxal

Found by Palamand *et al.* (1970a) in beer by thin-layer chromatography of the bis-2,4-dinitrophenylhydrazone.

$C_2H_2O_4$
Oxalic acid

Identified in beer by Marinelli *et al.* (1968) by gas chromatography of the methyl ester and by infrared analysis, and found by Chaudhary *et al.* (1968) in wine from botrytis-infected grapes by gas chromatography of the free acid and the methyl ester.

C_2H_4O
Acetaldehyde

Identified in numerous wine types and in beers and spirits by different workers by preparation of the 2,4-dinitrophenylhydrazone and its separation by paper, column, or thin-layer chromatography. Shown present in bourbon by Kahn *et al.* (1969) using gas chromatography–mass spectrometry.

$C_2H_4O_2$
Acetic acid

Identified in fusel oil from brandy by Webb *et al.* (1952) by ion exchange, preparation of the *p*-phenylphenacyl derivative (m. 108–110°, no depression), in bread pre-ferment by Hunter *et al.* (1961) using gas chromatography of the free acid and the ethyl ester, and by Kahn *et al.* (1969) using gas chromatography and mass spectrometry on a sample of bourbon whiskey. Found by Marinelli *et al.* (1968) in beer. Acetic acid has been identified as a component of nearly all fermented media by many different researchers using many different techniques.

C_2H_6O
Ethanol

Present in all fermented beverages.

C_2H_6S
Ethanethiol

Identified by Dellweg *et al.* (1969) in several different spirits using gas chromatography and a flame color detector.

Sulfide, dimethyl

Identified by Ahrenst-Larsen and Hansen (1964) in beer, and by

Dellweg et al. (1969) in several different spirits by gas chromatographic techniques. Found in rum by Liebich et al. (1970) using gas chromatography–mass spectrometry.

$C_2H_6S_2$
Disulfide, bis (methyl)
Identified by Dellweg et al. (1969) in several different spirits by gas chromatography and flame color detector.

C_2H_7N
Dimethylamine
Identified in two white wines and one red wine by Puputti and Suomalainen (1969) by gas chromatography.
Ethylamine
Identified in wines by Drawert (1965) using paper and gas chromatography, and by Puputti and Suomalainen (1969) in three different wines by gas chromatography and thin-layer chromatography.

C_2H_7NO
Ethanolamine
Identified by Puputti and Suomalainen (1969) in red Burgundy wine by thin-layer chromatography.

C_3H_4O
Acrolein (propenal)
Identified by Hrdlicka et al. (1968) in beer by thin-layer chromatography, and by Kahn et al. (1969) in bourbon whiskey by gas chromatography–mass spectrometry.

$C_3H_4O_2$
Pyruvaldehyde (methylglyoxal)
Identified in beer by Hrdlicka et al. (1968) by thin-layer chromatography, and by Palamand et al. (1970b) using gas chromatography–mass spectrometry and by preparation of the 2,4-dinitrophenylhydrazone.

$C_3H_4O_3$
Pyruvic acid (2-oxopropionic acid)
Identified in bread pre-ferment by Cole et al. (1966) by paper chromatography of the 2,4-dinitrophenylhydrazone and by chemical means, by Harrison and Collins (1968) in beer by capillary gas chromatography with electron capture detector of the methyl ester, and by Marinelli et al. (1968) in beer by gas chromatography and infrared analysis of the methyl ester.

$C_3H_4O_4$
3-Hydroxy-2-oxopropionic acid (3-hydroxypyruvic acid)
Identified in bread pre-ferment by Cole *et al.* (1966) by paper chromatography of the 2,4-dinitrophenylhydrazone and by chemical means.
Malonic acid
Identified in beer by Marinelli *et al.* (1968) by gas chromatography and infrared analysis of the methyl ester.

C_3H_6O
Acetone (2-propanone)
Identified by Hennig and Villforth (1942) in wine by preparation of derivatives, by Drawert and Rapp (1968) in wines, by gas chromatography on several different columns, and by Ahrenst-Larsen and Hansen (1964) in beer, by gas chromatography.
Allyl alcohol (2-propen-1-ol)
Identified by Takayama and Mizuuchi (1966) in industrial alcohol by gas chromatography, infrared, and preparation of derivatives. Also found by Kahn *et al.* (1969) in bourbon whiskey by gas chromatography–mass spectrometry.
Propionaldehyde
Identified by Almashi (1965) in wines by paper chromatography of the 2,4-dinitrophenylhydrazone, and by Kahn *et al.* (1969) in bourbon whiskey by gas chromatography–mass spectrometry. Also found in rum by Liebich *et al.* (1970) using gas chromatography–mass spectrometry.

$C_3H_6O_2$
Ethyl formate
Identified in beer and several wine types by gas chromatography by various workers. Found in bourbon by Kahn *et al.* (1969) using gas chromatography–mass spectrometry.
Methyl acetate
Found in beer by Kepner *et al.* (1963) using gas chromatography.
Propionic acid
Identified by Van Wyk *et al.* (1967a) in White Riesling wine by gas chromatography of the free acid and the methyl and hexyl esters. Infrared spectra of the esters agreed with knowns. Found in bread pre-ferment by Hunter *et al.* (1961) by gas chromatography of the free acid and the ethyl ester.

$C_3H_6O_3$
Lactic acid
Identified in White Riesling wine by Van Wyk *et al.* (1967a) by gas

chromatography of the ethyl ester, by Kepner *et al.* (1968) in sherries by gas chromatography of methyl esters, and in Cabernet Sauvignon and Ruby Cabernet wines by Kepner *et al.* (1969) by gas chromatography of the methyl esters. Found in beer by Marinelli *et al.* (1968) using gas chromatography of the methyl ester. In each case infrared spectra confirmed the identification.

C_3H_8O
1-Propanol

Identified in grape brandy fusel oil by Webb *et al.* (1952) by distillation analysis and preparation of the 3,5-dinitrobenzoate ester. Propanol had b_{760} 97.3–97.7°, n_D^{25} 1.3839; derivative m. 72.0–72.5° (no depression on mixing with known ester). 1-Propanol has been found in nearly all fermented beverages by many workers using several different analytical techniques.

2-Propanol (isopropyl alcohol)

Identified in wines by Drawert and Rapp (1968) using gas chromatography on several different columns, in beer by Drawert and Tressl (1969) by gas chromatography, functional group separations, and spectroscopy, and in sherries by Preobrazhenskii *et al.* (1969) by gas chromatography.

C_3H_8S
Ethylmethylsulfide

Identified by Kahn *et al.* (1969) in bourbon whiskey, and by Liebich *et al.* (1970) in rum, by gas chromatography–mass spectrometry.

C_3H_9N
Propylamine

Identified in wines by Drawert (1965) by paper chromatography and gas chromatography.

Isopropylamine

Found in two white wines and one red wine by Puputti and Suomalainen (1969) by gas chromatography, and wines by Drawert (1965), by paper and gas chromatography.

$C_4H_4O_4$
Fumaric acid (trans-1,4-butendioic acid)

Found in beer by Marinelli *et al.* (1968) by gas chromatography of the methyl ester. Confirmed by infrared spectra.

C_4H_6O
2-Butenal

Found in beer by Hrdlicka *et al.* (1968) by thin-layer chromatography. Whether *cis* or *trans* isomer not established.

$C_4H_6O_2$
2,3-Butanedione (biacetyl)
　　Identified in rotundifolia wine by Kepner and Webb (1956) by distillation analysis and preparation of the bis 2,4-dinitrophenylhydrazone (m. 317°, no depression on mixing with known derivative). 2,3-Butanedione has been found in many beers, wines, and spirits by several workers. Found in bread pre-ferment by D. E. Smith and Coffman (1960) by gas chromatography, infrared, and mass spectrometry.
4-Hydroxybutyric acid γ-lactone (2(3H)dihydrofuranone)
　　Found by Webb and Kepner (1962) in Australian flor sherry by gas chromatography, infrared spectra, and preparation of 4-hydroxybutyrate-phenylhydrazide (m. 82.5–93.0°, no depression on mixing with known). Found in bread pre-ferment by D. E. Smith and Coffman (1960) by gas chromatography, infrared and mass spectrometric techniques. Found in nearly all fermented and distilled beverages by many workers using several techniques.
Crotonic acid (2-butenoic acid)
　　Identified in bread pre-ferment by Hunter et al. (1961) by gas chromatography of free acid and ethyl ester.

$C_4H_6O_3$
2-Oxobutyric acid (α-ketobutyric acid)
　　Identified in beer by Harrison and Collins (1968) by means of capillary gas chromatography with an electron capture detector as the methyl ester.

$C_4H_6O_4$
Succinic acid
　　Identified in White Riesling wine by Van Wyk et al. (1967a), in flor sherries by Kepner et al. (1968), and in Cabernet Sauvignon and Ruby Cabernet wines by Kepner et al. (1969) by gas chromatography as the dimethyl ester and by determination that known and unknown methyl ester infrared spectra were identical. Found in beer by Marinelli et al. (1968) using gas chromatography and infrared of the methyl ester.

$C_4H_6O_5$
Malic acid (hydroxysuccinic acid)
　　Identified in White Riesling wine by Van Wyk et al. (1967a) by means of gas chromatography and infrared analysis of the ethyl ester. Found in beer by Marinelli et al. (1968) using gas chromatography and infrared of the methyl ester. One of the main acids of grapes and wines.

C_4H_8O
2-Butanone
Identified by Muller *et al.* (1964) in passion fruit wine by gas chromatography and by preparation of the 2,4-dinitrophenylhydrazone (m. 115.5–116.2°, no depression on mixing with known). Found in bourbon by Kahn *et al.* (1969) using gas chromatography–mass spectrometry.
Isobutyraldehyde (2-methylpropanal)
Identified by Kepner and Webb (1956) in rotundifolia wine by distillation analysis and preparation of the 2,4-dinitrophenylhydrazone (m. 178–183°, no depression on admixture with known). Isobutyraldehyde was also found in sherries by Webb *et al.* (1964b), in tokay wine by Almashi (1965), and in bourbon by Kahn *et al.* (1969). It was found in Jamaica rum by Maarse and ten Noever de Brauw (1966) and in rum by Liebich *et al.* (1970), both using gas chromatography–mass spectrometry.

$C_4H_8O_2$
3-Hydroxy-2-butanone (acetoin)
Identified in bread pre-ferments by D. E. Smith and Coffman (1960) using gas chromatography, infrared, and mass spectrometry. Found in beers by Ahrenst-Larsen and Hansen (1964) and Hrdlicka *et al.* (1968) using gas chromatography and thin-layer chromatography respectively. Found in sherries by Webb *et al.* (1967b) and in wines by Ronkainen *et al.* (1970) using gas chromatography.
Butyric acid
Extracted from fusel oil by use of ion-exchange resin by Webb *et al.* (1952). *p*-Phenylphenacyl ester prepared, column chromatographed, and crystallized (m. 79.7–80.8°, no depression on admixture with known). Found in bread pre-ferments by Hunter *et al.* (1961), in several other varieties of wines by other workers, and in bourbon by Kahn *et al.* (1969) using gas chromatography–mass spectrometry.
Ethyl acetate
Identified in rotundifolia wine by Kepner *et al.* (1956) by distillation analysis, saponification of the fraction, and preparation of the 3,5-dinitrobenzoate of the alcohol and the *p*-phenylazophenacyl derivative of the acid. Found in sherry by Webb and Kepner (1962), in beer by Ahrenst-Larsen and Hansen (1964), and in several other wines by other workers using gas chromatographic techniques. Confirmed as a bourbon component by Kahn *et al.* (1969) using gas chromatography–mass spectrometry.
Isobutyric acid (2-methylpropionic acid)
Identified in bread pre-ferment by Hunter *et al.* (1961) using gas

chromatography of the free acid and the ethyl ester. Confirmed as a component of sherries (Webb et al., 1964b), Cabernet Sauvignon (Webb et al., 1964a), White Riesling (Van Wyk et al., 1967b), and Sauvignon blanc (Chaudhary et al., 1968) by means of gas chromatographic techniques. Found in bourbon by Kahn et al. (1969) using gas chromatography–mass spectrometry.

Isopropyl formate

Reported a component of flor sherries by Preobrazhenskii et al. (1969) using gas chromatographic techniques.

C_4H_9N
Pyrrolidine (tetrahydropyrrole)

Identified in wines by Drawert (1965) by gas and paper chromatography, and in beers by Slaughter (1970) using thin-layer chromatography.

C_4H_9NO
Acetamide, N-ethyl

Identified in White Riesling wine by Van Wyk et al. (1967b) using gas chromatography for isolation and infrared spectra for proof of identity.

$C_4H_{10}O$
1-Butanol

Confirmed in grape brandy fusel oil by Webb et al. (1952) by distillation analysis (b_{760} 118.1–118.2°; n_D^{25} 1.3972; acetylation equivalent 74.8; 3,5-dinitrobenzoate ester m. 62.8–63.0° with no depression on admixture with known butyl 3,5-dinitrobenzoate). Identified in many other wines, beers, and spirits by other workers using gas chromatographic techniques.

2-Butanol

Identified by Webb et al. (1952) as a minor component of grape brandy fusel oil by distillation analysis (b_{760} 99.6–99.8°; n_D^{25} 1.3947; acetylation equivalent 78.6; observed rotation, 2 dm., −10.0°; 3,5-dinitrobenzoate m. 89.0–90.5°). Found in other fusel oils, beers, wines, and bourbon by other workers.

2-Methyl-1-propanol (isobutyl alcohol)

Confirmed as a major component of fusel oils by Webb et al. (1952) by distillation analysis (b_{760} 107.3–107.5°; n_D^{25} 1.3936; acetylation equivalent 75.0; 3,5-dinitrobenzoate m. 84.5–85.0°, no depression on admixture with known isobutyl 3,5-dinitrobenzoate). Found in many other fusel oils, and in wines, beers, and spirits by other workers using several techniques. Mode of production from amino acids and other precursors investigated by Yamada et al. (1963).

2-Methyl-2-propanol

Found in juice and wine of Grenache grapes by Stevens et al. (1969) using gas chromatography–mass spectrometry.

$C_4H_{10}OS$
3-(Methylthio)propanol

Identified in Carbernet Sauvignon and Ruby Cabernet wines by Muller et al. (1971) using gas chromatography, infrared, and mass spectrometry. Infrared and mass spectra given.

$C_4H_{10}O_2$
(−) 2,3-Butanediol

Identified in bread pre-ferment by D. E. Smith and Coffman (1960) by gas chromatography, infrared, mass spectrometry, and preparation of di-p-nitrobenzoate, m. 143°, alpha −52°. Found in White Riesling wine by Van Wyk et al. (1967b) using gas chromatrography and infrared techniques.

(±) 2,3-Butanediol

Identified in bread pre-ferment by Smith and Coffman (1960) by gas chromatography, infrared, mass spectrometry, and preparation of the di-p-nitrobenzoate, m. 192°, alpha 0°. Found in White Riesling wine by Van Wyk et al. (1967b) by gas chromatography and infrared techniques.

$C_4H_{10}S$
Sulfide, diethyl

Identified by Dellweg et al. (1969) in several different spirits by gas chromatography with a flame color detector.

$C_4H_{10}S_2$
Disulfide, bis (ethyl)

Identified by Dellweg et al. (1969) in several different spirits by gas chromatography with a flame color detector.

$C_4H_{11}N$
Butylamine

Identified in wines by Drawert (1965) by gas and paper chromatography, and by Puputti and Suomalainen (1969) in wines by gas chromatography.

Isobutylamine

Found in two white wines and one red wine by Puputti and Suomalainen (1969) using gas and thin-layer chromatography, and in wines by Drawert (1965) by gas and paper chromatography.

$C_4H_{12}N_2$
Putrescine (1,4-diaminobutane)

Found in red Burgundy wine by Puputti and Suomalainen (1969) using thin-layer chromatography.

$C_5H_4O_2$
2-Furaldehyde

Found in sherries by Webb *et al.* (1964b) by gas chromatography, in beer by Ahrenst-Larsen and Hansen (1964) by gas chromatography, and in tokay wine by Almashi (1965) by paper chromatography of the 2,4-dinitrophenylhydrazone. Identified in bourbon by Kahn *et al.* (1969), and in Japanese and Scotch whiskies by Nishimura and Masuda (1971) by gas chromatographic–mass spectrometric techniques.

$C_5H_4O_3$
2-Furoic acid

Identified by Kepner *et al.* (1968) in sherry, and by Kepner *et al.* (1969) in Cabernet Sauvignon and Ruby Cabernet wines by gas chromatography and infrared techniques. Found by Liebich *et al.* (1970) in rum using gas chromatography–mass spectrometry.

3-Furoic acid

Identified by Liebich *et al.* (1970) in rum by gas chromatography–mass spectrometry.

$C_5H_6N_2$
2-Methylpyrazine

Identified in rum by Liebich *et al.* (1970) by gas chromatography–mass spectrometry.

C_5H_6O
2-Methylfuran

Identified in rum by Liebich *et al.* (1970) using gas chromatography–mass spectrometry.

$C_5H_6O_2$
Furfuryl alcohol

Found in beer by Strating and Venema (1961) using gas chromatography and infrared, in Sauvignon blanc wines by Chaudhary *et al.* (1968) and in sherries by Preobrazhenskii *et al.* (1969) by gas chromatography. Identified in Japanese and Scotch whiskies by Nishimura and Masuda (1971) using gas chromatography–mass spectrometry.

$C_5H_6O_5$
2-Oxoglutaric acid (α-ketoglutaric acid)
Found by Cole *et al.* (1966) in bread pre-ferment by paper chromatography of the 2,4-dinitrophenylhydrazone, and by chemical procedures.

C_5H_8O
Cyclopentanone
Identified by Yamaguchi *et al.* (1966) in industrial alcohol by gas chromatography, infrared, and preparation of the 2,4-dinitrophenylhydrazone.
2-Pentenal
Found in cognac by Schaefer and Timms (1970) by thin-layer chromatography of the 2,4-dinitrophenylhydrazone.
3-Penten-2-one
Identified in rum by Liebich *et al.* (1970) by gas chromatography–mass spectrometry.

$C_5H_8O_2$
2-Methyl-3-oxo-tetrahydrofuran
Identified in rums by Maarse and ten Noever de Brauw (1966) and Liebich *et al.* (1970) each using gas chromatography–mass spectrometry.
2,3-Pentanedione
Found in whiskies by Suomalainen and Nykänen (1970) by means of thin-layer chromatography of the 2,4-dinitrophenylhydrazone.

$C_5H_8O_3$
2-Oxoisovaleric acid (2-oxo-3-methylbutanoic acid)
Identified in bread pre-ferment by Cole *et al.* (1966) by paper chromatography of the 2,4-dinitrophenylhydrazone, and by Harrison and Collins (1968) in beer by gas chromatography with electron capture detector of the methyl ester.

$C_5H_8O_3S$
4-(Methylthio)-2-oxobutyric acid
Identified in bread pre-ferment by Cole *et al.* (1966) by paper chro-

matography of the 2,4-dinitrophenylhydrazone, and by chemical techniques.

$C_5H_8O_4$
Glutaric acid
Found by Kepner et al. (1968) in sherry and by Kepner et al. (1969) in Cabernet Sauvignon and Ruby Cabernet wines by gas chromatographic and infrared analysis of the dimethyl ester of the extracted acid.

$C_5H_9N_3$
Histamine
Identified in red Burgundy wine by Puputti and Suomalainen (1969) by thin-layer chromatography, and by Ough (1971) in very small amounts in several California wines by a chemical flourescence method.

$C_5H_{10}O$
Isovaleraldehyde
Identified in tokay wine by Almashi (1965) by paper chromatography of 2,4-dinitrophenylhydrazone derivative, and in bourbon by Kahn et al. (1969) and in rum by Liebich et al. (1970) by means of gas chromatography–mass spectrometry.
2-Methylbutyraldehyde
Found in rum by Liebich et al. (1970) by gas chromatography–mass spectrometry.
2-Pentanone
Identified in bourbon whiskey by Kahn et al. (1969) using gas chromatography–mass spectrometry.

$C_5H_{10}O_2$
3-Ethoxypropionaldehyde
Identified in bourbon whiskey by Kahn et al. (1969) using gas chromatography–mass spectrometry.
Ethyl propionate
Identified in beer by Ahrenst-Larsen and Hansen (1964) and in Sauvignon blanc wine by Chaudhary et al. (1968) by gas chromatographic techniques. Found in bourbon whiskey by Kahn et al. (1969) using gas chromatography–mass spectrometry.
Isopropyl acetate
Reported present in sherry by Preobrazhenskii et al. (1969) on the basis of gas chromatography.
2-Methylbutyric acid
Identified in White Riesling wine by Van Wyk et al. (1967a) by gas

chromatography of the free acid and the methyl and hexyl esters, and by Kahn et al. (1969) in bourbon whiskey by gas chromatography–mass spectrometry.

3-Methylbutyric acid (isovaleric acid)

Identified in fusel oil by Kepner and Webb (1961), in bread pre-ferment by Hunter et al. (1961), in sherries by Webb et al. (1964b), and in Sauvignon blanc wine by Chaudhary et al. (1968), all using gas chromatography. Van Wyk et al. (1967a) used gas chromatographic and infrared techniques to identify isovaleric acid (as the hexyl ester) in White Riesling wine, and Kahn et al. (1969) found it in bourbon whiskey using gas chromatography–mass spectrometry.

1-Propyl acetate

Identified in White Riesling wine by Van Wyk et al. (1967b) and in sherry by Preobrazhenskii et al. (1969) by gas chromatography. Found in bourbon by Kahn et al. (1969) using gas chromatography–mass spectrometry.

Valeric acid (1-pentanoic acid)

Found in bread pre-ferment by Hunter et al. (1961), in sherries by Webb et al. (1964b), and in Sauvignon blanc wine by Chaudhary et al. (1968) using gas chromatography. Using gas chromatography–mass spectrometry, Kahn et al. (1969) identified valeric acid in bourbon whiskey.

$C_5H_{10}O_3$

3-Ethoxypropionic acid

Identified in rum by Liebich et al. (1970) using gas chromatography–mass spectrometry.

Ethyl hydracrylate (ethyl 3-hydroxypropionate)

Identified by Webb et al. (1967b) in sherries, and by Webb et al. (1969) in Cabernet Sauvignon and Ruby Cabernet wines using gas chromatographic and infrared techniques.

Ethyl lactate

Identified by Sihto et al. (1962) in two Finnish berry wines, by Webb and Kepner (1962) in a flor sherry, by Ahrenst-Larsen and Hansen (1964) in beer, and by Webb et al. (1964a) in a Bordeaux red by gas chromatographic procedures. Using gas chromatography and infrared techniques, Webb et al. (1967b) found ethyl lactate in Spanish film sherries, Van Wyk et al. (1967b) identified it in White Riesling wine, and Webb et al. (1969) showed its presence in Cabernet Sauvignon and Ruby Cabernet wines. It was found in bourbon whiskey by Kahn et al. (1969) using gas chromatography–mass spectrometry. Ethyl lactate was found in bread pre-ferment by D. E.

Smith and Coffman (1960) by gas chromatography, infrared, and mass spectrometry.

2-Hydroxyisovaleric acid (2-hydroxy-3-methylbutyric acid)

Identified by Kepner et al. (1968) in Spanish and California film sherries by gas chromatographic and infrared techniques.

1,3-Propanediol monoacetate

Found by Van Wyk et al. (1967b) in White Riesling wine by gas chromatographic and infrared techniques, and by D. E. Smith and Coffman (1960) in bread pre-ferment using gas chromatography, infrared, and mass spectrometry.

$C_5H_{10}S$

Prenyl mercaptan (3-Methyl-2-butene-1-thiol)

Identified as the sunlight flavor compound in beer by Obata and Tanaka (1965) by crystallization as the 2,4-dinitrophenyl derivative, m. 78–79°. Identity verified by paper chromatography and infrared.

$C_5H_{12}O$

Isopentyl alcohol (3-methyl-1-butanol, isoamyl alcohol)

Presence in fusel oils confirmed by means of distillation analysis and preparation of derivatives by Webb et al. (1952), by Ikeda et al. (1956) (b_{740} 131.6°; n_D^{25} 1.4047; 3,5-dinitrobenzoate, m. 63.0–64.2°, no depression on admixture with known), and by Hirose et al. (1962). Gas chromatography was used by Kepner and Webb (1961), and by Sihto et al. (1962) to demonstrate the presence of isopentyl alcohol in several other fusel oils. Isopentyl alcohol has been shown a constituent of Australian flor sherry by Webb and Kepner (1962), of different types of California sherry by Webb et al. (1964b), of passion fruit wine by Muller et al. (1964), of beers by Ahrenst-Larsen and Hansen (1964), of Sauvignon blanc wine by Chaudhary et al. (1968), and of Russian sherries by Preobrazhenskii et al. (1969) all using gas chromatographic techniques. Using gas chromatography and infrared, Webb et al. (1967b) showed that isopentyl alcohol was present in Spanish sherries, and Van Wyk et al. (1967b) found it in White Riesling wine. Gas chromatography–mass spectrometry was used by Kahn et al (1969) to demonstrate the presence of isopentyl alcohol in bourbon, and by Nishimura and Masuda (1971) to find it in Japanese and Scotch whiskies.

2-Methyl-1-butanol

Identified by distillation analysis and preparation of derivatives by Webb et al. (1952), by Ikeda et al. (1956) (b_{740} 128.5–129.0°; n_D^{25} 1.4081; d^{25} 0.8141; observed rotation, 2 dm. −8.70°; 3,5-dinitrobenzoate m. 81.8–83.0°, no depression on admixture with known), by Terry et al.

(1960) (b_{760} 128.5°; n_D^{25} 1.4082; d_D^{25} 0.8161; $[\alpha]_D^{25}$ -9.50 ± 0.012), and by Hirose et al. (1962). Using simple gas chromatographic procedures, Kepner and Webb (1961) and Sihto et al. (1962) found 2-methylbutanol in several different fusel oils, while it was found by Webb and Kepner (1962) in an Australian flor sherry, by Muller et al. (1964) in passion fruit wine, and by Webb et al. (1964b) in three types of California sherry. Van Wyk et al. (1967b) and Webb et al. (1967b) using gas chromatography and infrared techniques found 2-methylbutanol in White Riesling and Spanish flor sherries, respectively. The compound was found in bourbon by Kahn et al. (1969) using gas chromatography–mass spectrometry.

2-Methyl-2-butanol

Identified in rum by Liebich et al. (1970) by gas chromatography–mass spectrometry.

1-Pentanol (pentyl alcohol, amyl alcohol)

Identified in grape brandy fusel oil by distillation anaylsis and preparation of derivatives by Webb et al. (1952); (3,5-dinitrobenzoate-1-naphthylamine addition compound m. 80–81°, no depression on admixture of known), and by Hirose et al. (1962). Found in beer by Ahrenst-Larsen and Hansen (1964), and in Russian sherries by Preobraznenskii et al. (1969) using simple gas chromatographic techniques, and Scotch whiskey by Nishimura and Masuda (1971) using gas chromatography–mass spectrometry.

2-Pentanol

Found in wines by Drawert and Rapp (1968) by gas chromatography on several different columns.

3-Pentanol

Found in rum by Liebich et al. (1970) by gas chromatography–mass spectrometry.

$C_5H_{12}O_2$

Diethoxymethane

Identified in bourbon whiskey by Kahn et al. (1969) by means of gas chromatography–mass spectrometry, and in Jamaica rum by Maarse and ten Noever de Brauw (1966) by the same technique.

1-Ethoxy-1-methoxyethane

Identified in rum by Liebich et al. (1970) by gas chromatography–mass spectrometry.

$C_5H_{13}N$

Pentylamine (amyl amine)

Identified in wines by Drawert (1965) and by Puputti and Suomalainen (1969) by gas and paper chromatographic techniques.

Isopentylamine (1-amino-3-methylbutane)
Found in wines by Drawert (1965) and by Puputti and Suomalainen (1969) by gas and paper chromatographic techniques.

C_6H_6
Benzene
Identified in bourbon whiskey by Kahn *et al.* (1969) using gas chromatography–mass spectrometry.

C_6H_6O
Phenol
Identified in Sauvignon blanc wines by Chaudhary *et al.* (1968) by means of simple gas chromatography, and in Japanese and Scotch whiskies by Nishimura and Masuda (1971) using gas chromatography–mass spectrometry.

$C_6H_6O_2$
2-Furyl methyl ketone (2-acetylfuran)
Identified in Japanese and Scotch whiskies by Nishimura and Masuda (1971) by means of gas chromatography–mass spectrometry.
3-Furyl methyl ketone (3-acetylfuran)
Identified in Japanese and Scotch whiskies by Nishimura and Masuda (1971) using gas chromatography–mass spectrometry.
5-Methyl-2-furaldehyde
Identified in bourbon by Kahn *et al.* (1969), and in Japanese and Scotch whiskies by Nishimura and Masuda (1971) using gas chromatography–mass spectrometry. Found in rum by Lieblich *et al.* (1970) by the same technique.

C_6H_7N
2-Picoline (2-methylpyridine)
Identified in Japanese and Scotch whiskies by Nishimura and Masuda (1971) by means of gas chromatography–mass spectrometry.
4-Picoline (4-methylpyridine)
Identified in Japanese and Scotch whiskies by Nishimura and Masuda (1971) by using gas chromatography–mass spectrometry.

$C_6H_8N_2$
2,3-Dimethylpyrazine
Found in Scotch whiskey by Nishimura and Masuda (1971) using gas chromatography–mass spectrometry.

2,5-Dimethylpyrazine
Found in Japanese and Scotch whiskies by Nishimura and Masuda (1971) using gas chromatography–mass spectrometry, and in rum by Liebich *et al.* (1970) using the same method.

2,6-Dimethylpyrazine
Identified in rum by Liebich *et al.* (1970) by gas chromatography–mass spectrometry.

2-Ethylpyrazine
Identified in Scotch whiskey by Nishimura and Masuda (1971) by means of gas chromatography–mass spectrometry.

$C_6H_8O_3$
5-Acetyldihydro-2(3H)-furanone   (4-acetyl-4-hydroxybutanoic   acid γ-lactone)
Identified in flor sherries by Augustyn *et al.* (1971) by gas chromatography, mass spectrometry, and infrared. Infrared and mass spectra given.

$C_6H_8O_6$
1,2,3-Propanetricarboxylic acid (tricarballylic acid)
Found by Kepner *et al.* (1968) in Spanish sherries using gas chromatographic and infrared techniques.

$C_6H_8O_7$
Citric acid
Identified in beer by Marinelli *et al.* (1968) by gas chromatography of the methyl ester and by infrared techniques.

$C_6H_{10}O$
2-Hexenal
Identified by Kepner and Webb (1956) in rotundifolia wine by distillation analysis and preparation of the 2,4-dinitrophenylhydrazone (m. 141–143°, no depression on admixture with known), by Hrdlicka *et al.* (1968) in beer by thin-layer chromatography, and by Drawert and Rapp (1968) in wines by gas chromatography on several columns.

$C_6H_{10}O_2$
2-Hexenoic acid
Identified by Clarke *et al.* (1962) in beer by means of gas chromatography of the free acid and of the methyl ester.

3-Hexenoic acid
Identified in beer by Clarke *et al.* (1962) by gas chromatography of the free acid and of the methyl ester.

$C_6H_{10}O_3$

2,3-Dihydroxy-3,3-dimethylbutyric acid γ-lactone (pantolactone)

Identified in Californian and Spanish flor sherries by Webb et al. (1967b) by gas chromatographic and infrared techniques.

Ethyl acetoacetate (ethyl 3-oxobutyrate)

Found in sherry by Preobrazhenskii et al. (1969) by gas chromatography.

Ethyl 4-oxobutyrate

Identified by mass spectrometry in Ruby Cabernet wine by Muller et al. (1972a).

5-(1-Hydroxyethyl)dihydro-2(3H)furanone(4,5-dihydroxyhexanoic acid γ-lactone)

Two isomers identified in California and Spanish flor sherries by Muller et al. (1969). The lower gas chromatographic retention time isomer was (−) 4R:5R or 4S:5S and had $[\alpha]_{578}^{23} -31°$ and a low intensity of odor. The other isomer was (+) 4R:5S or 4S:5R, had $[\alpha]_{578}^{23} +5°$, and a winelike odor. Infrared, NMR (nuclear magnetic resonance), and mass spectra are given.

3-Methyl-2-oxopentanoic acid

Identified in bread pre-ferment by Cole et al. (1966) by paper chromatography of the 2,4-dinitrophenylhydrazone and by chemical methods. Found in beer by Harrison and Collins (1968) by gas chromatography with electron capture detector of the methyl ester.

4-Methyl-2-oxopentanoic acid (α-keto isocaproic acid)

Identified in bread pre-ferment by Cole et al. (1966) by paper chromatography of the 2,4-dinitrophenylhydrazone, and by chemical methods. Found in beer by Harrison and Collins (1968) by gas chromatography with electron capture detector of the methyl ester.

$C_6H_{10}O_4$

Ethyl acid succinate

Identified in Bordeaux Cabernet Sauvignon wine by gas chromatography and infrared analysis of the acid and the methyl ester by Webb et al. (1964a). Infrared spectrum of methyl ethyl succinate given. Found in sherries by Webb et al. (1964b), in White Riesling wine by Van Wyk et al. (1967a), and in Sauvignon blanc by Chaudhary et al. (1968) using gas chromatography and infrared.

Diethyl oxalate

Found by Chaudhary et al. (1968) in Sauvignon blanc wines using gas chromatography on several columns.

5-Hydroxy-4-oxohexanoic acid

Found by Hirabayashi and Harada (1969) in shake culture of yeast growing on ethanol; identified by paper chromatography, infrared, and NMR.

$C_6H_{10}O_5$
Ethyl acid malate
Identified by Webb *et al.* (1967a) in Californian flor sherry. Wine isomer has free carboxyl group adjacent to the carbinol group. Infrared and mass spectra of both isomers of methyl ethyl malate given. Ethyl acid malate found in White Riesling wine by Van Wyk *et al.* (1967a) using gas chromatography and infrared techniques.

$C_6H_{10}O_6$
Ethyl acid tartrate
Identified by Webb *et al.* (1967a) in Californian flor sherry using ion exchange and gas chromatography together with infrared analysis (spectrum of methyl ethyl tartrate given).

$C_6H_{10}S$
Diallyl sulfide
Found by Dellweg *et al.* (1969) in pomace brandy using gas chromatography with a flame color detector.

$C_6H_{12}O$
Hexanal
Found by Kepner and Webb (1956) by distillation analysis and preparation of derivative in rotundifolia wine (2,4-dinitrophenylhydrazone m. 105–105.8°, no depression on admixture of known).
2-Hexanone (butyl methyl ketone)
Found in beer by Hrdlicka *et al.* (1968) by thin-layer chromatography.
cis-3-Hexen-1-ol
Identified in wine from Pedro grapes by Wagener and Wagener (1968), and in cider by Williams and Tucknott (1971) using gas chromatography and infrared techniques.

$C_6H_{12}O_2$
Butyl acetate
Found by Ahrenst-Larsen and Hansen (1964) in beer, and by Preobrazhenskii *et al.* (1969) in Russian sherry by simple gas chromatographic techniques.
Ethyl butyrate
Identified in Sauvignon blanc wines by Chaudhary *et al.* (1968) by

gas chromatography on several columns, and by Kahn et al. (1969) in bourbon by gas chromatography–mass spectrometry.

Ethyl isobutyrate

Identified in Australian flor sherry by Webb and Kepner (1962), in passion fruit wine by Muller et al. (1964), and in Sauvignon blanc wines by Chaudhary et al. (1968) all using gas chromatographic techniques. Found in bourbon whiskey by Kahn et al. (1969) by gas chromatography–mass spectrometry.

4-Ethoxy-2-butanone

Identified in rum by Leibich et al. (1970) by gas chromatography–mass spectrometry.

Hexanoic acid (caproic acid)

Found in bread pre-ferment by Hunter et al. (1961), in muscat raisin fusel oil by Kepner and Webb (1961), in a Bordeaux Cabernet Sauvignon wine by Webb et al. (1964a), in three types of sherry by Webb et al. (1964b), in White Riesling wine by Van Wyk et al. (1967a), in Sauvignon blanc wines by Chaudhary et al. (1968), and in Russian sherry by Preobrazhenskii et al. (1969) all using gas chromatographic techniques. Identified by gas chromatographic and infrared techniques in Spanish sherries by Kepner et al. (1968), and in Cabernet Sauvignon and Ruby Cabernet wines by Kepner et al. (1969). Kahn et al. (1969) found hexanoic acid in bourbon by gas chromatographic–mass spectrometric techniques.

Isobutyl acetate

Identified by gas chromatographic techniques in three sherry types by Webb et al. (1964b), in a Bordeaux Cabernet Sauvignon wine by Webb et al. (1964a), in White Riesling wine by Van Wyk et al. (1967b), and in Russian sherry by Preobrazhenskii et al. (1969). Kahn et al. (1969) found isobutyl acetate in bourbon whiskey using gas chromatography–mass spectrometry.

Isopentyl formate (isoamyl formate)

Found by Liebich et al. (1970) in rum using gas chromatography–mass spectrometry, and by Williams and Tucknott (1971) in cider by gas chromatography, infrared, and mass spectrometry.

4-Methylvaleric acid (isocaproic acid)

Found in bread pre-ferment by Hunter et al. (1961) using gas chromatographic techniques.

2,4,5-Trimethyl-1,3-dioxolane (2,3-butanediol acetaldehyde acetal)

Found in Californian and Spanish sherries by Webb et al. (1967b) using gas chromatographic and infrared techniques.

$C_6H_{12}O_3$
(−) 2,3-Butanediol monoacetate
Found by Van Wyk et al. (1967b) in White Riesling wine using gas chromatography and infrared procedures (spectrum given).

Ethyl 3-hydroxybutyrate
Identified in White Riesling wine by Van Wyk et al. (1967b), in Spanish flor sherries by Webb et al. (1967b), and in Cabernet Sauvignon and Ruby Cabernet wines by Webb et al. (1969) using gas chromatographic and infrared techniques.

Ethyl 4-hydroxybutryate
Identified by gas chromatographic and infrared techniques in Californian and Spanish flor sherries by Webb et al. (1967b).

2-Hydroxyhexanoic acid
Identified by Kepner et al. (1968) in Californian and Spanish flor sherries, and in Cabernet Sauvignon and Ruby Cabernet wines by Kepner et al. (1969) using gas chromatography and infrared techniques.

2-Hydroxy-4-methylvaleric acid (2-hydroxyisocaproic acid)
Identified by Van Wyk et al. (1967a) in White Riesling wine by gas chromatography and infrared techniques.

$C_6H_{14}O$
1-Hexanol (hexyl alcohol)
Identified by Webb et al. (1952) in fusel oil from grape brandy by distillation analysis and preparation of derivative (b_{49} 89.2–89.7°; n_D^{25} 1.4147; 3,5-dinitrobenzoate, m. 56–58°, no depression on admixture with known). Also found in fusel oils by Ikeda et al. (1956) (b_{740} 158.1–158.5°), by Kepner and Webb (1961), and by Hirose et al. (1962), the latter two using gas chromatographic procedures. Hexanol has been found in rotundifolia wine by Kepner and Webb (1956), in Australian flor sherry by Webb and Kepner (1962), in passion fruit by Muller et al. (1964), in Cabernet Sauvignon by Webb et al. (1964a), in beer by Ahrenst-Larsen and Hansen (1964), in three types of Californian sherry by Webb et al. (1964b), in White Riesling wine by Van Wyk et al. (1967b), in Sauvignon blanc wines by Chaudhary et al. (1968), in Russian sherries by Preobrazhenskii et al. (1969), and in Cabernet Sauvignon and Ruby Cabernet wines by Webb et al. (1969), all using gas chromatographic and in some cases infrared techniques. Kahn et al. (1969) found the compound in bourbon whiskey by means of gas chromatography–mass spectrometry.

2-Hexanol
Identified in wines by Drawert and Rapp (1968) by gas chromatography on several different columns.
3-Methyl-1-pentanol
Identified in White Riesling wine by Van Wyk et al. (1967b), and in Cabernet Sauvignon and Ruby Cabernet wines by Webb et al. (1969) by gas chromatographic and infrared techniques. Found in bourbon whiskey by Kahn et al. (1969) by gas chromatography–mass spectrometry.
4-Methyl-1-pentanol (isohexyl alcohol)
Identified in White Riesling wine by Van Wyk et al. (1967b) using gas chromatographic and infrared techniques.

$C_6H_{14}O_2$
Acetaldehyde diethyl acetal (1,1-diethoxyethane)
Identified in sulfite spirits by Sihto et al. (1962) by gas chromatographic techniques. Found in passion fruit wine by Muller et al. (1964), in Cabernet Sauvignon wine by Webb et al. (1964a), in three types of Californian sherries by Webb et al. (1964b), in White Riesling wine by Van Wyk et al. (1967b), in Sauvignon blanc wine by Chaudhary et al. (1968), and in Russian sherries by Preobrazhenskii et al. (1969), all by gas chromatographic methods. Kahn et al. (1969) found acetal in bourbon using gas chromatographic–mass spectrometric methods.

$C_6H_{14}S$
Butyl ethyl sulfide
Identified by Dellweg et al. (1969) in a sample of fruit brandy by gas chromatography with flame color detector.
Diisopropyl sulfide
Identified in pomace brandy and German wine brandy by Dellweg et al. (1969) using gas chromatography with a flame color detector.

$C_6H_{14}S_2$
Disulfide, bis (isopropyl) (diisopropyldisulfide)
Found in a sample of fruit brandy by Dellweg et al. (1969) using gas chromatography with a flame color detector.

$C_6H_{15}N$
Hexylamine
Identified in two white wines and one red wine by Puputti and Suomalainen (1969) by gas and thin-layer chromatography.

C_7H_6O
Benzaldehyde
Identified by Hirose *et al.* (1962) in corn fusel oil by chromatographic procedures and preparation of derivatives, and in bourbon by Kahn *et al.* (1969) by gas chromatography–mass spectrometry. Found in wines by Drawert and Rapp (1968) by gas chromatography on several columns.

$C_7H_6O_2$
Benzoic acid
Found in Sauvignon blanc wines by Chaudhary *et al.* (1968) using gas chromatography on several different columns, and by Kepner *et al.* (1968) in Californian and Spanish flor sherries by means of gas chromatography and infrared methods. Identified in beer by Marinelli *et al.* (1968) using gas chromatography of the methyl ester and infrared.

$C_7H_6O_3$
4-Hydroxybenzoic acid
Identified in cognac by Schaefer and Timms (1970) by paper chromatographic methods.
Salicylic acid (o-hydroxybenzoic acid)
Identified in Sauvignon blanc wines by Chaudhary *et al.* (1968), and in Cabernet Sauvignon and Ruby Cabernet wines by Kepner *et al.* (1969) by use of gas chromatographic and infrared techniques.

$C_7H_6O_4$
Protocatechuic acid (3,4-dihydroxybenzoic acid)
Found in cognac by Schaefer and Timms (1970) by paper chromatographic techniques.

$C_7H_6O_5$
Gallic acid (3,4,5-trihydroxybenzoic acid)
Identified in cognac by Schaefer and Timms (1970) by paper chromatographic procedures.

C_7H_8
Toluene
Identified in bourbon whiskey by Kahn *et al.* (1969) using gas chromatography–mass spectrometry.

C_7H_8O
Benzyl alcohol
Identified by Hirose *et al.* (1962) in corn fusel oil by chromatographic procedures and the preparation of derivatives. Found in Californian and Spanish flor sherries by Webb *et al.* (1967b), in Sauvignon blanc wines by Chaudhary *et al.* (1968), and in Cabernet Sauvignon and Ruby Cabernet wines by Webb *et al.* (1969) by means of gas chromatographic and infrared techniques. Also identified in grape and berry wines fermented by flor yeasts by Suomalainen and Nykänen (1966), and in beer by Drawert and Tressl (1969) by gas chromatographic procedures.

m-*Cresol (3-methylphenol)*
Found in Sauvignon blanc wine, made from grapes infected with botrytis, by Chaudhary *et al.* (1968) using gas chromatographic and infrared techniques.

o-*Cresol (2-methylphenol)*
Found in Japanese and Scotch whiskies by Nishimura and Masuda (1971) by gas chromatography–mass spectrometry.

p-*Cresol (4-methylphenol)*
Identified in Japanese and Scotch whiskies by Nishimura and Masuda (1971) using gas chromatography–mass spectrometry.

$C_7H_8O_2$
o-*Methoxyphenol (guaiacol)*
Identified in Japanese and Scotch whiskies by Nishimura and Masuda (1971) and in rum by Liebich *et al.* (1970) both using gas chromatography-mass spectrometry.

Methyl 5-methyl-2-furyl ketone (5-methyl-2-acetylfuran)
Identified in Scotch whisky by Nishimura and Masuda (1971) using gas chromatography-mass spectrometry.

$C_7H_8O_3$
Ethyl 2-furoate (ethyl 2-furancarboxylate)
Identified in rum by Liebich *et al.* (1970) by gas chromatography–mass spectrometry.

C_7H_9N
2,5-Lutidine (2,5-dimethylpyridine)
Identified in Japanese whisky by Nishimura and Masuda (1971) using gas chromatography–mass spectrometry.

2,6-Lutidine (2,6-dimethylpyridine)
Identified in Japanese whisky by Nishimura and Masuda (1971) by means of gas chromatography–mass spectrometry.
3,5-Lutidine (3,5-dimethylpyridine)
Identified in Japanese whisky by Nishimura and Masuda (1971) using gas chromatography–mass spectrometry.
2-Ethylpyridine
Identified by Nishimura and Masuda (1971) in Japanese and Scotch whiskies by means of gas chromatography–mass spectrometry.
4-Ethylpyridine
Identified in Scotch whisky by Nishimura and Masuda (1971) using gas chromatography–mass spectrometry.

$C_7H_9NO_2$
Methyl anthranilate
Identified in labrusca grapes by Power and Chesnut (1921) by chemical methods.

$C_7H_{10}N_2$
2-Ethyl-3-methylpyrazine
Identified in rum by Liebich et al. (1970) by gas chromatography–mass spectrometry.
2-Ethyl-5-methylpyrazine
Identified by Nishimura and Masuda (1971) in Japanese and Scotch whiskies using gas chromatography–mass spectrometry.
2-Ethyl-6-methylpyrazine
Identified in rum by Liebich et al. (1970) by gas chromatography–mass spectrometry.
Trimethylpyrazine
Found in Japanese and Scotch whiskies by Nishimura and Masuda (1971) by means of gas chromatography–mass spectrometry.

$C_7H_{10}O_4$
4-Carboethoxy-4-hydroxybutyric acid γ-lactone (5-carboethoxydihydro-2(3H)-furanone)
Identified in Californian and Spanish flor sherries by Webb et al. (1967b), and in Cabernet Sauvignon and Ruby Cabernet wines by Webb et al. (1969) using gas chromatographic and infrared techniques.

$C_7H_{11}NO_3$
5-Oxoproline, ethyl ester (ethyl pyroglutamate)
Identified in Californian and Spanish flor sherries by Webb et al. (1967b) using gas chromatography, infrared, and NMR spectra.

$C_7H_{13}NO_3$
N-acetylalanine, ethyl ester
Found in Californian and Spanish flor sherries by Webb et al. (1967b) using gas chromatographic and infrared techniques.

$C_7H_{14}O$
2-Heptanone (methyl pentyl ketone)
Identified in beer by Hrdlicka et al. (1968) by thin-layer chromatography of the 2,4-dinitrophenylhydrazone.
1-Hepten-3-ol
Found in beer by Drawert and Tressl (1969) by gas chromatography, functional group analysis, and spectrometry.

$C_7H_{14}O_2$
Acrolein diethyl acetal (1,1-diethoxy-2-propene)
Identified in bourbon whiskey by Kahn et al. (1969) using gas chromatography–mass spectrometry.
2,4-Dimethyl-5-ethyl-1,3-dioxolane
Found in Grenache grape juice and wine by Stevens et al. (1969) by gas chromatography–mass spectrometry.
Ethyl isovalerate (ethyl 3-methylbutyrate)
Found in passion fruit wine by Muller et al. (1964), in Sauvignon blanc wines by Chaudhary et al. (1968), and in Russian sherries by Preobrazhenskii et al. (1969), all using simple gas chromatographic procedures.
Ethyl valerate
Identified in bourbon whiskey by Kahn et al. (1969), and in Russian sherries by Preobrazhenskii et al. (1969) by means of gas chromatography–mass spectrometry and simple gas chromatography respectively.
4-Ethoxy-2-pentanone
Identified in rum by Liebich et al. (1970) by gas chromatography–mass spectrometry.
Heptanoic acid
Found in bread pre-ferment by Hunter et al. (1961) by gas chromatography of the free acid and the ethyl ester, in Californian sherries by Webb et al. (1964b) by gas chromatography of the free acid and the methyl ester, in Sauvignon blanc wines by Chaudhary et al. (1968) by gas chromatography of free acids and methyl esters, and in bourbon whiskey by Kahn et al. (1969) by gas chromatography–mass spectrometry.
Isopentyl acetate (3-methylbutyl acetate)
Using simple gas chromatographic techniques, isopentyl acetate was found in an Australian flor sherry by Webb and Kepner (1962), in

beers by Sihto et al. (1962), and by Ahrenst-Larsen and Hansen (1964), in three types of Californian sherries by Webb et al. (1964b), in a Bordeaux Cabernet Sauvignon wine by Webb et al. (1964a), in passion fruit wine by Muller et al. (1964), in Sauvignon blanc wines by Chaudhary et al. (1968), and in Russian sherries by Preobrazhenskii et al. (1969). It was found in White Riesling wine by Van Wyk et al. (1967b) using gas chromatography and infrared techniques. Kahn et al. (1969) found isopentyl acetate in bourbon by gas chromatography–mass spectrometry.

2-Methylbutyl acetate

Identified in White Riesling wine by Van Wyk et al. (1967b) by gas chromatographic and infrared techniques.

Pentyl acetate

Reported in Russian sherry by Preobrazhenskii et al. (1969) on the basis of simple gas chromatography.

$C_7H_{14}O_3$

1,1-Diethoxy-2-propanone (pyruvaldehyde diethyl acetal)

Found in rum by Liebich et al. (1970) by gas chromatography–mass spectrometry.

Ethyl 3-ethoxypropionate

Identified by Kahn et al. (1969) in bourbon whiskey by gas chromatography–mass spectrometry.

Ethyl 2-hydroxy-3-methylbutyrate

Identified in Californian and Spanish flor sherries by Webb et al. (1967b) using gas chromatographic and infrared techniques.

$C_7H_{15}NO$

N-isopentylacetamide

Isolated from a submerged-culture flor sherry by gas chromatographic techniques by Webb et al. (1966) and identified through infrared and mass spectra. Infrared spectrum given.

C_7H_{16}

Heptane

Identified by Kahn et al. (1969) in bourbon whiskey by gas chromatography–mass spectrometry.

$C_7H_{16}O$

1-Heptanol

Found in fusel oils by Hirose et al. (1962) by distillation analysis, gas chromatography, and infrared analysis. Using simple gas cromatography, Chaudhary et al. (1968) found heptanol in Sauvignon blanc wines, and Preobrazhenskii et al. (1969) found it in Russian sherries. It was identified by Kahn et al. (1969) in bourbon whiskey by means of gas chromatography–mass spectrometry.

2-Heptanol (amyl methyl carbinol)

Proved present in corn fusel oil by Hirose et al. (1962) using distillation analysis, gas chromatography, and infrared techniques. Found in sweet potato fusel oil by Taira (1963) by gas chromatography and infrared, in beer by Drawert and Tressl (1969), and in wines by Drawert and Rapp (1968) using gas chromatography on several columns.

$C_7H_{16}O_2$

Acetaldehyde ethyl propyl acetal (1-Ethoxy-1-propoxyethane)

Found by Kahn et al. (1969) in bourbon whiskey by gas chromatography–mass spectrometry, in Grenache grapes and wines by Stevens et al. (1969), and in rum by Liebich et al. (1970), both using gas chromatography–mass spectrometry.

Propionaldehyde diethyl acetal (1,1-Diethoxypropane)

Found in bourbon whiskey by Kahn et al. (1969) using gas chromatography–mass spectrometry, in Jamaica rum by Maarse and ten Noever de Brauw (1966), and in cognac by Schaefer and Timms (1970) both using gas chromatography–mass spectrometry.

C_8H_8

Styrene

Identified in bourbon whiskey by Kahn et al. (1969) by gas chromatography–mass spectrometry.

C_8H_8O

Acetophenone (methyl phenyl ketone)

Identified in corn fusel oil by Hirose et al. (1962) by distillation analysis, gas chromatography, preparation of derivative, and infrared.

$C_8H_8O_2$

Phenylacetic acid

Identified in Sauvignon blanc wines by Chaudhary et al. (1968), in sherries by Kepner et al. (1968), and in Cabernet Sauvignon and Ruby Cabernet wines by Kepner et al. (1969), all using gas chromatography and infrared.

2′-Hydroxyacetophenone (2-hydroxyphenyl methyl ketone)

Identified in Scotch whisky by Nishimura and Masuda (1971) using gas chromatography–mass spectrometry, and in rum by Liebich et al. (1970) using the same method.

$C_8H_8O_3$
Methyl salicylate
Reported a probable constituent of fusel oil by Webb et al. (1952) on the basis of distillation analysis, saponification of the ester, and preparation of solid derivatives. Found in rum by Liebich et al. (1970) using gas chromatography–mass spectrometry.
Vanillin
Found in bourbon whiskey by Kahn et al. (1969) using gas chromatography–mass spectrometry, by Hennig and Villforth (1942) in wine by preparation of a derivative, and by Baldwin et al. (1967) in spirits by paper chromatography and UV (ultraviolet) spectrometry.

$C_8H_8O_4$
Vanillic acid (4-hydroxy-3-methoxybenzoic acid)
Found by Kepner et al. (1968) in Californian and Spanish flor sherries by means of gas chromatography and infrared techniques. Using the same techniques, Kepner et al. (1969) identified vanillic acid in Cabernet Sauvignon and Ruby Cabernet wines. Identified by Otsuka and Imai (1964) in an extract of oak chips, by paper and thin-layer chromatography, and by Schaefer and Timms (1970) in cognac, by paper chromatography.

$C_8H_{10}O$
o-*Ethylphenol*
Identified in Japanese and Scotch whiskies by Nishimura and Masuda (1971) using gas chromatography–mass spectrometry.
p-*Ethylphenol*
Identified in Cabernet Sauvignon and Ruby Cabernet wines by Webb et al. (1969) using gas chromatographic and infrared techniques. Found a constituent of Japanese and Scotch whiskies by Nishimura and Masuda (1971) by means of gas chromatography–mass spectrometry, and identified in cider by Williams and Tucknott (1971) by gas chromatography, mass spectrometry, and infrared techniques.
Phenethyl alcohol (2-phenylethanol)
Identified in fusel oils by Ikeda et al. (1956), and by Hirose et al. (1962) using distillation analysis and derivative preparation. Ikeda reports 3,5-dinitrobenzoate ester m. 106–108° with no depression on admixture with known. Phenethyl alcohol was found in rotundifolia wine by Kepner et al. (1956) by distillation analysis and preparation of the 3,5-dinitrobenzoate, m. 106–107°. Sihto et al. (1962) found the compound in beers using gas chromatography, and it has been identified in Australian flor sherry by Webb and Kepner (1962), in three

Californian sherry types by Webb et al. (1964b), in passion fruit wine by Muller et al. (1964), in a Bordeaux Cabernet Sauvignon wine by Webb et al. (1964a), in White Riesling wine by Van Wyk et al. (1967b), in Spanish and Californian flor sherries by Webb et al. (1967b), in Sauvignon blanc wines by Chaudhary et al. (1968), in Cabernet Sauvignon and Ruby Cabernet wines by Webb et al. (1969), and in Russian sherries by Preobrazhenskii et al. (1969), all using gas chromatography and some infrared techniques. By means of gas chromatography-mass spectrometry Kahn et al. (1969) have found phenethyl alcohol in bourbon whiskey, and Nishimura and Masuda (1971) found it in Japanese and Scotch whiskies. Äyräpää (1965) has shown that phenethyl alcohol is formed both from amino acids and from sugars during fermentation.

2,6-Xylenol (2,6-dimethylphenol)

Identified in Scotch whisky by Nishimura and Masuda (1971) using gas chromatography-mass spectrometry.

$C_8H_{10}O_2$

2-Methoxy-4-methylphenol (4-methylguaiacol)

Found in Japanese and Scotch whiskies by Nishimura and Masuda (1971) using gas chromatography-mass spectrometry, and by Liebich et al. (1970) in rum by the same technique.

2-Methoxy-6-methylphenol (6-methylguaiacol)

Found by Nishimura and Masuda (1971) in Japanese whisky using gas chromatography-mass spectrometry.

Tyrosol (p-hydroxyphenethyl alcohol)

Found in wine and beer by Ehrlich (1917) using classical procedures. Presence confirmed in wine and beer by Nykänen et al. (1966) on separation by gas chromatography followed by determination of the melting point and the infrared spectrum.

$C_8H_{11}N$

5-Ethyl-2-methylpyridine

Found in Japanese whisky by Nishimura and Masuda (1971) using gas chromatography-mass spectrometry.

2-Isopropylpyridine

Found in Japanese and Scotch whiskies by Nishimura and Masuda (1971) using gas chromatography-mass spectrometry.

2-Phenethylamine

Identified in wines by Drawert (1965) by paper and gas chromatography.

$C_8H_{11}NO$
Tyramine
Identified in wines by Puputti and Suomalainen (1969) by thin-layer chromatography.

$C_8H_{12}N_2$
2,5-Diethylpyrazine
Identified in Japanese whisky by Nishimura and Masuda (1971) using gas chromatography–mass spectrometry.
2,5-Dimethyl-3-ethylpyrazine
Found in Japanese and Scotch whiskies by Nishimura and Masuda (1971) by means of gas chromatography–mass spectrometry, and in rum by Liebich et al. (1970) using the same method.
3,5-Dimethyl-2-ethylpyrazine
Identified in rum by Liebich et al. (1970) by gas chromatography–mass spectrometry.
Tetramethylpyrazine
Identified in Japanese and Scotch whiskies by Nishimura and Masuda (1971) using gas chromatography–mass spectrometry.

$C_8H_{14}O_2$
5-Hydroxyoctanoic acid δ-lactone (δ-octalactone)
Identified in rum by Liebich et al. (1970) using gas chromatography–mass spectrometry.

$C_8H_{14}O_4$
Diethyl succinate
Identified in Australian flor sherry by Webb and Kepner (1962) using gas chromatography, infrared, and the preparation of derivatives of the parts of the saponified ester (p-phenylazophenacyl succinate m. 219.5–221.5°, no depression on admixture with known). Using gas chromatography, and in the later researches, infrared, Webb et al. (1964b) found diethyl succinate in three Californian sherry types, Webb et al. (1964a) found it in a Bordeaux Cabernet Sauvignon wine, Van Wyk et al. (1967b) found it in White Riesling wines, Webb et al. (1967b) found it in Spanish flor sherries, Chaudhary et al. (1968) found it in Sauvignon blanc wines, and Webb et al. (1969) demonstrated its presence in Cabernet Sauvignon and Ruby Cabernet wines. Kahn et al. (1969) found diethyl succinate in bourbon whiskey by gas chromatography–mass spectrometry.

$C_8H_{14}O_5$
Diethyl malate

Found in Australian flor sherry by Webb and Kepner (1962) using gas chromatography and preparation of derivatives. Found in Spanish and Californian flor sherries by Webb *et al.* (1967b), in White Riesling wine by Van Wyk *et al.* (1967b), and in Cabernet Sauvignon and Ruby Cabernet wines by Webb *et al.* (1969) using gas chromatographic and infrared techniques.

$C_8H_{14}O_6$
Diethyl tartrate

Identified in Spanish and Californian flor sherries by Webb *et al.* (1967b) using gas chromatographic and infrared techniques. Also found in Cabernet Sauvignon and Ruby Cabernet wines by Webb *et al.* (1969) by the same procedures.

$C_8H_{16}O$
1-Octene-3-ol (matsutakeol)

Found by Hirose *et al.* (1962) in corn fusel oil by distillation analysis, gas chromatography, infrared, and preparation of a derivative, and identified by Drawert and Tressl (1969) in beer by gas chromatography, functional group analysis, and spectroscopy.

Hexyl methyl ketone (2-octanone)

Found in corn fusel oil by Hirose *et al.* (1962) by distillation analysis, gas chromatography, infrared, and preparation of the 2,4-dinitrophenylhydrazone.

$C_8H_{16}O_2$
Ethyl hexanoate (ethyl caproate)

Identified in fusel oils by Webb *et al.* (1952), by Kepner and Webb (1961), and by Hirose *et al.* (1962) using distillation analysis and gas chromatography. Found in beer by Ahrenst-Larsen and Hansen (1964) by gas chromatography. Found in Australian flor sherry by Webb and Kepner (1962), in three types of Californian sherries by Webb *et al.* (1964b), in passion fruit wine by Muller *et al.* (1964), in a Bordeaux Cabernet Sauvignon wine by Webb *et al.* (1964a), in White Riesling by Van Wyk *et al.* (1967b), in Sauvignon blanc wines by Chaudhary *et al.* (1968), and in Russian sherry by Preobrazhenskii *et al.* (1969), all using gas chromatography, and some, infrared techniques.

Hexyl acetate

Identified in fusel oils by Kepner and Webb (1961), and by Hirose *et al.* (1962) by gas chromatographic and infrared techniques. Found in Australian flor sherry by Webb and Kepner (1962), in three types of Californian sherries by Webb et al. (1964b), in a Bordeaux Cabernet Sauvignon by Webb et al. (1964a), in White Riesling wine by Van Wyk et al. (1967b), and in Sauvignon blanc wines by Chaudhary et al. (1968), all using gas chromatographic techniques.

Isobutyl isobutyrate

Identified by gas chromatographic techniques in Australian flor sherry by Webb and Kepner (1962), in passion fruit wine by Muller et al. (1964), and in Russian sherry by Preobrazhenskii et al. (1969).

Isooctanoic acid

Found in rum by Liebich et al. (1970) using gas chromatography–mass spectrometry.

Octanoic acid (caprylic acid)

Identified in bread pre-ferment by Hunter et al. (1961) using gas chromatography, and by using the same techniques, by Kepner and Webb (1961) in fusel oil. Found by Webb et al. (1964a) in Bordeaux Cabernet Sauvignon, by Webb et al. (1964b) in three types of Californian sherry, by Van Wyk et al. (1967a) in White Riesling wine, by Chaudhary et al. (1968) in Sauvignon blanc wines, and by Preobrazhenskii et al. (1969) in Russian sherry, all using gas chromatography. By means of gas chromatography and infrared techniques Kepner et al. (1968) found octanoic acid in Spanish and Californian flor sherries, and Kepner et al. (1969) found it in Cabernet Sauvignon and Ruby Cabernet wines as well. It was found in bourbon whiskey by Kahn et al. (1969) using gas chromatography–mass spectrometry.

$C_8H_{16}O_3$

Ethyl 2-hydroxy-4-methylpentanoate (ethyl 2-hydroxyisocaproate)

Identified by Webb et al. (1967b) in Spanish and Californian flor sherries, by Van Wyk et al. (1967b) in White Riesling wine, and by Webb et al. (1969) in Cabernet Sauvignon and Ruby Cabernet wines, all by gas chromatographic and infrared techniques.

3-Hydroxyoctanoic acid

Identified by Kepner et al. (1968) in Spanish and Californian flor sherries by gas chromatographic and infrared techniques.

Isopentyl lactate

Identified by Webb et al. (1964a) in a Bordeaux Cabernet Sauvignon wine, and by Webb et al. (1964b) in three types of Californian sherries by simple gas chromatographic techniques.

$C_8H_{18}O$

1-Octanol

Identified in fusel oils by Hirose et al. (1962) by distillation analysis, gas chromatography, infrared, and preparation of derivative. Found in three types of Californian sherries by Webb et al. (1964b), in Sauvignon blanc wines by Chaudhary et al. (1968), and in Russian sherry by Preobrazhenskii et al. (1969), by gas chromatographic techniques. Identified in bourbon whiskey by Kahn et al. (1969) using gas chromatography–mass spectrometry.

2-*Octanol*

Identified in Sauvignon blanc wines by Chaudhary et al. (1968) by gas chromatography, in beer by Drawert and Tressl (1969), and in wines by Drawert and Rapp (1968) by gas chromatography on several columns.

$C_8H_{18}O_2$
Acetaldehyde dipropyl acetal (1,1-dipropoxyethane)

Identified in Grenache juice and wine by Stevens et al. (1969), and in rum by Liebich et al. (1970) both by gas chromatography–mass spectrometry.

Acetaldehyde butyl ethyl acetal (1-butoxy-1-ethoxyethane)

Identified in Jamaica rum by Maarse and ten Noever de Brauw (1966) by gas chromatography–mass spectrometry.

Acetaldehyde ethyl isobutyl acetal (1-ethoxy-1-(2-methylpropoxy)-ethane)

Found in Jamaica rum by Maarse and ten Noever de Brauw (1966) by gas chromatography–mass spectrometry.

Butyraldehyde diethyl acetal (1,1-diethoxybutane)

Found in cognac by Schaefer and Timms (1970) and in rum by Liebich et al. (1970) both by gas chromatography–mass spectrometry. Also identified in Jamaica rum by Maarse and ten Noever de Brauw (1966) by the same procedures.

Isobutyraldehyde diethyl acetal (1,1-diethoxy-2-methylpropane)

Identified in cognac by Schaefer and Timms (1970), in rum by Liebich et al. (1970), and in bourbon whiskey by Kahn et al. (1969), all by gas chromatography–mass spectrometry.

C_9H_7N
Quinoline

Identified in Japanese and Scotch whiskies by Nishimura and Masuda (1971) by gas chromatography–mass spectrometry.

C_9H_8O
Cinnamaldehyde

Identified in wine by Hennig and Villforth (1942) by preparation of derivative.

$C_9H_8O_4$
p-*Hydroxyphenyl pyruvic acid*
Found in bread pre-ferment by Cole et al. (1966) by paper chromatography of the 2,4-dinitrophenylhydrazone and by chemical methods.

$C_9H_{10}O_2$
Ethyl benzoate
Identified in corn fusel oil by Hirose et al. (1962) using distillation analysis, gas chromatography, and infrared. Found in bourbon whiskey by Kahn et al. (1969) by means of gas chromatography–mass spectrometry.
2'-Hydroxy-5'-methylacetophenone (2-hydroxy-5-methylphenyl methyl ketone)
Found by Nishimura and Masuda (1971) in Scotch whisky by gas chromatography–mass spectrometry.
2-Phenethyl formate
Found in Grenache juice and wine by Stevens et al. (1969), and in cider by Williams and Tucknott (1971) both by gas chromatography–mass spectrometry.

$C_9H_{10}O_3$
3-Phenyllactic acid (2-hydroxy-3-phenylpropionic acid)
Found in Spanish and Californian flor sherries by Kepner et al. (1968) by gas chromatographic and infrared techniques. Identified by Kepner et al. (1969) in Cabernet Sauvignon and Ruby Cabernet wines by gas chromatography and infrared.

$C_9H_{10}O_4$
Syringaldehyde (3,5-dimethoxy-4-hydroxybenzaldehyde)
Found in spirits by Baldwin et al. (1967) using paper chromatography and UV techniques.

$C_9H_{10}O_5$
Syringic acid (3,5-dimethoxy-4-hydroxybenzoic acid)
Found in an alcoholic extract of oak chips by Otsuka and Imai (1964) and in cognac by Schaefer and Timms (1970) by paper chromatographic techniques.

$C_9H_{12}O$
o-*Isopropylphenol*
Identified in Japanese whisky by Nishimura and Masuda (1971) by means of gas chromatography–mass spectrometry.

$C_9H_{12}O_2$
2-(3-Furyl)-5-methyltetrahydrofuran
Found in corn fusel oil by Hirose *et al.* (1962) by distillation analysis, gas chromatography, and infrared.
2-Ethoxy-4-ethylphenol (p-*ethylguaiacol*)
Found by Nishimura and Masuda (1971) in Japanese and Scotch whiskies by means of gas chromatography–mass spectrometry, in rums by Liebich *et al.* (1970), and in cider by Williams and Tucknott (1971) by the same procedures.

$C_9H_{14}O$
trans-2-cis-6-*Nonadienal*
Found in beer by Visser and Lindsay (1971) using gas chromatography–mass spectrometry. Mass spectrum given.

$C_9H_{16}O$
trans-2-*Nonenal*
Found in beer by Visser and Lindsay (1971) using gas chromatography–mass spectrometry.

$C_9H_{16}O_2$
4-Hydroxy-3-methyloctanoic acid γ-lactone (5-butyl-4-methyldihydro-2(3H)furanone)
Found in whiskies by Suomalainen and Nykänen (1970) by gas chromatography, infrared, and mass spectrometry.
trans-*4-Hydroxy-3-methyloctanoic acid γ-lactone*
Identified in Japanese and Scotch whiskies by Nishimura and Masuda (1971) by gas chromatography–mass spectrometry.
cis-*4-Hydroxy-3-methyloctanoic acid γ-lactone*
Identified in Japanese and Scotch whiskies by Nishimura and Masuda (1971) by gas chromatography–mass spectrometry.
4-Hydroxynonanoic acid γ-lactone
Identified in bourbon whiskey by Kahn *et al.* (1969) using gas chromatography–mass spectrometry, and in Japanese and Scotch whiskies by Nishimura and Masuda (1971) by the same methods. Also found in rums by Liebich *et al.* (1970) using the same procedures.
5-Hydroxynonanoic acid Δ-lactone
Found in bourbon whiskey by Kahn *et al.* (1969) by means of gas chromatography–mass spectrometry.

$C_9H_{16}O_4$
Azelaic acid (nonanedioic acid)
Found in Spanish and Californian flor sherries by Kepner *et al.*

(1968), and in Cabernet Sauvignon and Ruby Cabernet wines by Kepner *et al.* (1969) using gas chromatography and infrared techniques.

$C_9H_{18}O$
Heptyl methyl ketone (2-nonanone)
Found by Hirose *et al.* (1962) in corn fusel oil by means of distillation analysis, gas chromatography, and infrared, and in cognac by Schaefer and Timms (1970) by gas chromatography–mass spectrometry.

$C_9H_{18}O_2$
Ethyl heptanoate (ethyl enanthate)
Found in fusel oils by Kepner and Webb (1961), and in corn fusel oil by Hirose *et al.* (1962) by distillation analysis and gas chromatography. Found in passion fruit wine by Muller *et al.* (1964), in Sauvignon blanc wines by Chaudhary *et al.* (1968), and in Russian sherry by Preobrazhenskii *et al.* (1969) by simple gas chromatography. Kahn *et al.* (1969) found bourbon whiskey to contain ethyl heptanoate by gas chromatography–mass spectrometry. Found in rums by Maarse and ten Noever de Brauw (1966) and by Liebich *et al.* (1970) using gas chromatography–mass spectrometry.
Heptyl acetate
Identified in corn fusel oil by Hirose *et al.* (1962) by distillation analysis, gas chromatography, and infrared.
Isobutyl valerate
Identified in Sauvignon blanc wines by Chaudhary *et al.* (1968) by gas chromatography on several different columns.
Methyl octanoate
Found in rum by Liebich *et al.* (1970) by gas chromatography–mass spectrometry.
Nonanoic acid (pelargonic acid)
Found in bread pre-ferment by Hunter *et al.* (1961) using gas chromatography. By the same methods nonanoic acid was identified in three types of Californian sherries by Webb *et al.* (1964b), in a Bordeaux Cabernet Sauvignon wine by Webb *et al.* (1964a), in White Riesling wine by Van Wyk *et al.* (1967a), and in Sauvignon blanc wines by Chaudhary *et al.* (1968). Kahn *et al.* (1969) found the acid in bourbon whiskey by gas chromatography–mass spectrometry.

$C_9H_{20}O$
1-Nonanol
Identified in corn and molasses fusel oils by Hirose *et al.* (1962) by distillation analysis, gas chromatography, and infrared.

2-Nonanol

Found in corn fusel oil by Hirose *et al.* (1962) using distillation analysis, gas chromatography, and infrared. Found in wines by Drawert and Rapp (1968), and in beers by Drawert and Tressl (1969) using gas chromatography on several different columns.

$C_9H_{20}O_2$

Acetaldehyde ethyl 2-methylbutyl acetal [1-ethoxy-1-(2-methylbutoxy)ethane]

Found in submerged-culture flor sherry by Galetto *et al.* (1966) by gas chromatography, preparation of derivatives, and infrared techniques. Infrared spectrum given. Found in rum by Maarse and ten Noever de Brauw (1966) by gas chromatography–mass spectrometry.

Acetaldehyde ethyl isopentyl acetal [1-ethoxy-1-(3-methylbutoxy)ethane]

Identified in submerged-culture flor sherry by Galetto *et al.* (1966) by gas chromatography, preparation of derivatives, and infrared techniques. Infrared spectra given. Found in rum by Maarse and ten Noever de Brauw (1966) by gas chromatography–mass spectrometry.

Isobutyraldehyde ethyl propyl acetal (1-ethoxy-1-propoxy-2-methylpropane)

Identified in rums by Liebich *et al.* (1970) by gas chromatography–mass spectrometry.

Isovaleraldehyde diethyl acetal (1,1-diethoxy-3-methylbutane)

Found in cognac by Schaefer and Timms (1970) by gas chromatography and gas chromatography–mass spectrometry.

2-Methylbutyraldehyde diethyl acetal (1,1-diethoxy-2-methylbutane)

Identified in Jamaica rum by Maarse and ten Noever de Brauw (1966), in cognac by Schaefer and Timms (1970), and in rums by Liebich *et al.* (1970) all using gas chromatography–mass spectrometry.

Propionaldehyde ethyl isobutyl acetal [1-ethoxy-1-(2-methylpropoxy)propane)

Identified in rums by Liebich *et al.* (1970) by gas chromatography–mass spectrometry.

$C_9H_{20}O_3$

1,1,3-Triethoxypropane

Identified in bourbon whiskey by Kahn *et al.* (1969) by gas chromatography–mass spectrometry.

$C_{10}H_8$
Naphthalene
Identified in corn fusel oil by Hirose et al. (1962) by distillation analysis, gas chromatography, and infrared.

$C_{10}H_8O_4$
Scopoletin (7-Hydroxy-6-methoxycoumarin)
Found in spirits by Baldwin et al. (1967) by paper chromatography and UV techniques.

$C_{10}H_9N$
2-Methylquinoline
Identified in Japanese and Scotch whiskies by Nishimura and Masuda (1971) by gas chromatography–mass spectrometry.
6-Methylquinoline
Identified in Japanese whiskey by Nishimura and Masuda (1971) by gas chromatography–mass spectrometry.

$C_{10}H_{10}O_3$
Coniferaldehyde (4-hydroxy-3-methoxycinnamaldehyde)
Identified in spirits by Baldwin et al. (1967) by paper chromatography and UV analysis.

$C_{10}H_{10}O_4$
Dimethyl phthalate
Found in White Riesling wine by Van Wyk et al. (1967b) using gas chromatography and infrared techniques.
Ferulic acid (4-hydroxy-3-methoxycinnamic acid)
Found in cognac by Schaefer and Timms (1970) by paper chromatrography.

$C_{10}H_{11}NO$
Tryptophol (2-indolylethanol)
Found in wine, beer, and distillery residue by Ehrlich (1917) using classical analytical techniques. Presence confirmed in wines and beers by Nykänen et al. (1966) by gas chromatography, melting point, and infrared.

$C_{10}H_{12}O_2$
4-Allyl-2-methoxyphenol (eugenol)
Found in Japanese and Scotch whiskies by Nishimura and Masuda (1971) by gas chromatography–mass spectrometry.

Ethyl phenylacetate

Found in corn fusel oil by Hirose *et al.* (1962) by distillation analysis, gas chromatography, and infrared.

Isoeugenol (2-methoxy-4-propenylphenol)

Identified by Liebich *et al.* (1970) in rums by gas chromatography–mass spectrometry.

2-Phenethyl acetate

Identified in fusel oils by Kepner and Webb (1961), and by Hirose *et al.* (1962) by means of distillation analysis and gas chromatography. Found by Webb and Kepner (1962) in Australian flor sherry, by Webb *et al.* (1964a) in a Bordeaux Cabernet Sauvignon, by Webb *et al.* (1964b) in three types of Californian sherries, by Van Wyk *et al.* (1967b) in White Riesling wine, by Chaudhary *et al.* (1968) in Sauvignon blanc wines, and by Preobrazhenskii *et al.* (1969) in Russian sherries, all by simple gas chromatography. Found in whiskey by Kayahara *et al.* (1964) by gas chromatography. Identified in Cabernet Sauvignon wines by Webb *et al.* (1969) using gas chromatographic and infrared techniques. Kahn *et al.* (1969) identified 2-phenethyl acetate in bourbon whiskey by gas chromatography–mass spectrometry.

$C_{10}H_{12}O_4$

Ethyl 4-hydroxy-3-methoxybenzoate

Found in rum by Liebich *et al.* (1970) by gas chromatography–mass spectrometry.

$C_{10}H_{13}NO$

N-(2-phenethyl)acetamide

Identified by Webb *et al.* (1966) in submerged culture sherry by gas chromatographic and infrared techniques.

$C_{10}H_{16}$

p-*Mentha-1,8-diene (limonene)*

Found by Hirose *et al.* (1962) in corn fusel oil by distillation analysis, gas chromatography, and infrared.

2-Pinene

Identified in bourbon whiskey by Kahn *et al.* (1969) using gas chromatography–mass spectrometry.

$C_{10}H_{16}O$

Camphor (2-bornanone)

Identified in corn fusel oil by Hirose *et al.* (1962) by distillation analysis, gas chromatography, and infrared.

$C_{10}H_{18}O$
Borneol (2-bornanol)
Identified by Hirose *et al.* (1962) in corn fusel oil by distillation analysis, gas chromatography, and infrared.
trans-*3,7-Dimethyl-2,6-octadien-1-ol (geraniol)*
Found in whisky by Nykänen and Suomalainen (1963) by gas chromatography, and in sweet potato fusel oil by Taira (1963) using distillation analysis, gas chromatography, and infrared.
3,7-Dimethyl-1,6-octadien-3-ol (linaloöl)
Found by Hirose *et al.* (1962) in corn fusel oil by distillation analysis, gas chromatography, and infrared, by Pisarnitskii (1966) in Cabernet and Riesling wines by thin-layer chromatography, and by Van Wyk *et al.* (1967b) in White Riesling wine by gas chromatography and infrared techniques. Identified by Wenzel and de Vries (1968) in muscat wines by gas chromatography, and ultraviolet and infrared spectroscopy. Infrared spectrum given. Found in beer by Drawert and Tressel (1969) by gas chromatography.
3,7-Dimethyl-2,6-octadien-1-ol (nerol)
Identified in beer by Drawert and Tressl (1969) by gas chromatography, functional group analysis, and spectroscopy.
p-*Menth-1-en-8-ol (α-terpineol)*
Identified as a flavor factor in cognac by Schaefer and Timms (1970) using gas chromatography–mass spectrometry, and in beer by Drawert and Tressl (1969) by gas chromatography, functional group analysis, and spectroscopy.

$C_{10}H_{18}O_2$
9-Decenoic acid
Identified by Van Wyk *et al.* (1967a) in White Riesling wine by gas chromatography and infrared, and by Kepner *et al.* (1969) in Cabernet Sauvignon and Ruby Cabernet wines by the same techniques.
4-Hydroxydecanoic acid γ-lactone [5-hexyldihydro-2(3H)furanone]
Identified in rum by Liebich *et al.* (1970) by gas chromatography–mass spectrometry.
5-Hydroxydecanoic acid δ-lactone (δ-decalactone)
Identified in rum by Liebich *et al.* (1970) by gas chromatography–mass spectrometry.

$C_{10}H_{20}O$
3,7-Dimethyl-6-octen-1-ol (citronellol)
Identified in corn fusel oil by Hirose *et al.* (1962) by distillation analysis, gas chromatography, and infrared.

$C_{10}H_{20}O_2$
Decanoic acid (capric acid)

Found by Hunter *et al.* (1961) in bread pre-ferment by gas chromatography, by Kepner and Webb (1961) in fusel oil using gas chromatography, by Webb *et al.* (1964a) in a Bordeaux Cabernet Sauvignon, by Van Wyk *et al.* (1967a) in White Riesling wine by gas chromatography and infrared techniques, by Chaudhary *et al.* (1968) in Sauvignon blanc wines by gas chromatography, by Kepner *et al.* (1969) in Cabernet Sauvignon and Ruby Cabernet wines by gas chromatography and infrared procedures, by Preobrazhenskii *et al.* (1969) using gas chromatography on Russian sherries, and by Kahn *et al.* (1969) in bourbon whiskey using gas chromatography–mass spectrometry.

Ethyl octanoate

Identified in fusel oils by Webb *et al.* (1952) using distillation analysis and preparation of derivatives, by Kepner and Webb (1961) in fusel oils by gas chromatography, and by Hirose *et al.* (1962) in fusel oils by distillation analysis, gas chromatography, and infrared techniques. Found in rum and whiskies by Sihto *et al.* (1962) using gas chromatography, and by Webb and Kepner (1962) in Australian flor sherry, by Muller *et al.* (1964) in passion fruit wine, by Webb *et al.* (1964a) in a Bordeaux Cabernet Sauvignon, by Webb *et al.* (1964b) in three types of Californian sherry, by Van Wyk *et al.*, (1967b) in White Riesling wine, by Chaudhary *et al.* (1968) in Sauvignon blanc, and by Webb *et al.* (1969) in Cabernet Sauvignon and Ruby Cabernet wines by gas chromatography and, in some cases, by infrared techniques. Found in bourbon whiskey by Kahn *et al.* (1969) using gas chromatography–mass spectrometry, and in rum by Liebich *et al.* (1970) by the same technique.

Hexyl butyrate

Identified in Sauvignon blanc wines by Chaudhary *et al.* (1968) by gas chromatography.

Hexyl isobutyrate

Found by Webb *et al.* (1964b) in three types of Californian sherries by gas chromatographic techniques, and by Chaudhary *et al.* (1968) in Sauvignon blanc wines by gas chromatography.

Isobutyl hexanoate

Identified by Webb and Kepner (1962) in Australian flor sherry by gas chromatographic and chemical techniques, by Webb *et al.* (1964b) in three types of Californian sherries by gas chromatography, and by Chaudhary *et al.* (1968) by gas chromatography in Sauvignon blanc wines.

Isopentyl isovalerate
Found by Webb and Kepner (1962) in Australian flor sherry by gas chromatographic and chemical procedures.

Octyl acetate
Found by Hirose et al. (1962) in corn fusel oil by distillation analysis, gas chromatography, and infrared.

$C_{10}H_{20}O_4$
Ethyl isopentyl succinate
Identified in wine by Hardy and Ramshaw (1970) by gas chromatography–mass spectrometry.

$C_{10}H_{22}O$
1-Decanol
Found by Hirose et al. (1962) in corn fusel oil by distillation analysis, gas chromatography, and infrared, and by Preobrazhenskii et al. (1969) by gas chromatography.

2-Decanol
Identified in wines by Drawert and Rapp (1968) using gas chromatography on several columns, and in beers by Drawert and Tressl (1969) by gas chromatography, functional group analysis, and spectroscopy.

$C_{10}H_{22}O_2$
Acetaldehyde diisobutyl acetal [*1,1-di(2-methylpropoxy)ethane*]
Found in rums by Liebich et al. (1970) using gas chromatography–mass spectrometry.

Acetaldehyde isopentyl propyl acetal [*1-(3-methylbutoxy)-1-propoxyethane*]
Found in Jamaica rum by Maarse and ten Noever de Brauw (1966) by gas chromatography–mass spectrometry.

Isobutyraldehyde ethyl isobutyl acetal [*1-ethoxy-1-(2-methylpropoxy)-2-methylpropane*]
Identified in rum by Liebich et al. (1970) by gas chromatography–mass spectrometry.

Propionaldehyde ethyl isopentyl acetal [*1-ethoxy-1-(3-methylbutoxy)-propane*]
Identified in rum by Liebich et al. (1970) by gas chromatography–mass spectrometry.

$C_{10}H_{22}O_3$
1-Ethoxy-1-(3-methylbutoxy)-3-hydroxypropane
Reported to be a likely component of bourbon whiskey by Kahn *et al.* (1969) on the basis of gas chromatography–mass spectrometry.

$C_{10}H_{22}S$
Di-(3-methylbutyl)sulfide
Identified in Russian brandy by Dellweg *et al.* (1969) using gas chromatography with a flame color detector.

$C_{11}H_{12}O_4$
3,5-Dimethoxy-4-hydroxycinnamaldehyde (sinapaldehyde)
Identified in spirits by Baldwin *et al.* (1967) in paper chromatography and ultraviolet spectroscopy.

$C_{11}H_{14}O_2$
2-Phenethyl propionate
Identified in corn fusel oil by Hirose *et al.* (1962) using distillation analysis, gas chromatography, and infrared.
Ethyl 2-phenylpropionate
Identified in rum by Liebich *et al.* (1970) using gas chromatography–mass spectrometry.

$C_{11}H_{14}O_3$
Ethyl 3-phenyllactate
Identified in Spanish and Californian flor sherries by Webb *et al.* (1967b), and by Webb *et al.* (1969) in Cabernet Sauvignon and Ruby Cabernet wines by gas chromatography and infrared techniques.

$C_{11}H_{22}O_2$
Ethyl nonanoate
Found by Kepner and Webb (1961) in muscat raisin fusel oil, and by Hirose *et al.* (1962) in corn fusel oil by distillation analysis and gas chromatographic techniques. Found in Sauvignon blanc wines by Chaudhary *et al.* (1968) by gas chromatography, and by Kahn *et al.* (1969) in bourbon whiskey, and in rum by Liebich *et al.* (1970), the latter two by gas chromatography–mass spectrometry.
Hexyl isovalerate
Found in Sauvignon blanc wines by Chaudhary *et al.* (1968) by gas chromatographic procedures.
Hexyl valerate
Identified in Sauvignon blanc wines by Chaudhary *et al.* (1968) by gas chromatography.

Isopentyl hexanoate
Found in fusel oils by Webb *et al.* (1952), and by Hirose *et al.* (1962) using distillation analysis and chemical techniques, and by Kepner and Webb (1961) using gas chromatographic procedures. Identified by Webb and Kepner (1962) in Australian flor sherry, by Webb *et al.* (1964a) in a Bordeaux Cabernet Sauvignon, by Webb *et al.* (1964b) in three types of Californian sherries, by Chaudhary *et al.* (1968) in Sauvignon blanc wines, and by Webb *et al.* (1969) in Cabernet Sauvignon and Ruby Cabernet wines, all using gas chromatographic and some, infrared, techniques.

2-Methyl-1-butyl hexanoate
Identified in fusel oils by Webb *et al.* (1952), and by Kepner and Webb (1961) by distillation analysis, gas chromatography, and preparation of derivatives.

Methyl decanoate
Identified in rum by Liebich *et al.* (1970) by gas chromatography–mass spectrometry.

1-Nonyl acetate
Found by Hirose *et al.* (1962) in corn fusel oil by distillation analysis and gas chromatographic techniques.

2-Nonyl acetate
Found in corn fusel oil by Hirose *et al.* (1962) by distillation analysis, gas chromatography, and infrared.

1-Propyl octanoate
Identified in muscat raisin fusel oil by Kepner and Webb (1961) using gas chromatography, and in wines by Hardy and Ramshaw (1970) by means of gas chromatography–mass spectrometry.

Undecanoic acid
Found by Chaudhary *et al.* (1968) in Sauvignon blanc wines by gas chromatography.

$C_{11}H_{24}O$

1-Undecanol
Found by Hirose *et al.* (1962) in corn fusel oil by distillation analysis, gas chromatography, and infrared, and in wines by Drawert and Rapp (1968) using gas chromatography on several columns.

$C_{11}H_{24}O_2$

Acetaldehyde butyl pentyl acetal (1-butoxy-1-pentoxyethane)
Found in Jamaica rum by Maarse and ten Noever de Brauw (1966) by gas chromatography–mass spectrometry.

Acetaldehyde butyl isopentyl acetal [1-butoxy-1-(3-methylbutoxy)-ethane]

Identified in rums by Liebich et al. (1970) by gas chromatography–mass spectrometry.

Acetaldehyde isobutyl isopentyl acetal [1-(2-methylpropoxy)-1-(3-methylbutoxy)ethane]

Identified in rum by Liebich et al. (1970) by gas chromatography–mass spectrometry.

Butyraldehyde ethyl isopentyl acetal [1-ethoxy-1-(3-methylbutoxy)butane]

Identified by Maarse and ten Noever de Brauw (1966) in Jamaica rum by gas chromatography–mass spectrometry.

Butryraldehyde ethyl pentyl acetal (1-ethoxy-1-pentoxybutane)

Identified in Jamaica rum by Maarse and ten Noever de Brauw (1966) by gas chromatography–mass spectrometry.

Isobutyraldehyde ethyl isopentyl acetal [1-ethoxy-1-(3-methylbutoxy)-2-methylpropane]

Identified by Liebich et al. (1970) in rum by gas chromatography–mass spectrometry.

Isobutyraldehyde isobutyl propyl acetal [1-(2-methylpropoxy)-1-propoxy-2-methylpropane]

Found in rum by Liebich et al. (1970) by gas chromatography–mass spectrometry.

Isovaleraldehyde dipropyl acetal (1,1-dipropoxy-3-methylbutane)

Identified in rum by Liebich et al. (1970) by gas chromatography–mass spectrometry.

Isovaleraldehyde ethyl isobutyl acetal [1-ethoxy-1-(2-methylpropoxy)-3-methylbutane]

Identified by Liebich et al. (1970) in rums by gas chromatography–mass spectrometry.

Propionaldehyde isopentyl propyl acetal (1-isopentoxy-1-propoxypropane)

Identified by Liebich et al. (1970) in rums by gas chromatography–mass spectrometry.

Propionaldehyde diisobutyl acetal [1,1-di-(2-methylpropoxy)propane]

Found in rum by Liebich et al. (1970) by gas chromatography–mass spectrometry.

Valeraldehyde ethyl isobutyl acetal [1-ethoxy-1-(2-methylpropoxy)pentane]

Identified by Liebich et al. (1970) in rum by gas chromatography–mass spectrometry.

$C_{12}H_{14}O_4$
Diethyl phthalate

Found by Van Wyk et al. (1967b) in White Riesling wine by gas chromatography and infrared techniques, and by Kahn *et al.* (1969) in bourbon whiskey by gas chromatography–mass spectrometry.

$C_{12}H_{16}O_2$
2-Phenethyl butyrate
Identified in wines by Hardy and Ramshaw (1970) by gas chromatography–mass spectrometry.

$C_{12}H_{18}O_2$
Acetaldehyde ethyl 2-phenethyl acetal [1-ethoxy-1-(2-phenethoxy)-ethane]
Identified by Galetto *et al.* (1966) in flor sherry by gas chromatography and preparation of derivatives.

$C_{12}H_{22}O_2$
Ethyl 9-decenoate
Identified by Van Wyk *et al.* (1967b) in White Riesling wine by gas chromatography and infrared techniques.
5-Hydroxydodecanoic acid δ-lactone (δ-dodecalactone)
Found in rum by Liebich *et al.* (1970) by gas chromatography–mass spectrometry.

$C_{12}H_{24}O_2$
1-Decyl acetate
Identified by Hirose *et al.* (1962) in corn fusel oil by distillation analysis, gas chromatography, and infrared.
Dodecanoic acid (lauric acid)
Found in bread pre-ferment by Hunter *et al.* (1961) by means of gas chromatography, and by Webb *et al.* (1964a) in a Bordeaux Cabernet Sauvignon by the same technique. Van Wyk *et al.* (1967a) found lauric acid in White Riesling wine by gas chromatography and infrared, while Chaudhary *et al.* (1968) found it in Sauvignon blanc wines by simple gas chromatography.
Ethyl decanoate
Found in fusel oils by Webb *et al.* (1952), by Kepner and Webb (1961), and by Hirose *et al.* (1962) by means of distillation analysis, and in the first case, preparation of derivatives, and the second two, gas chromatography. Using simple gas chromatography, Webb *et al.* (1964a) found ethyl decanoate in a Bordeaux Cabernet Sauvignon, Muller *et al.* (1964) found it in passion fruit wine, Webb *et al.* (1964b) found it in three types of Californian sherries, Chaudhary *et al.* (1968) found it in Sauvignon blanc wines, and Preobrazhenskii *et al.* (1969) found it in Russian sherries. Using gas chromatographic and

infrared techniques, Van Wyk et al. (1967b) found ethyl decanoate in White Riesling wine, and Webb et al. (1969) found it in Cabernet Sauvignon and Ruby Cabernet wines. It was found in bourbon whiskey by Kahn et al. (1969) using gas chromatography–mass spectrometry.

Hexyl hexanoate

Identified in trace quantities by Webb et al. (1964b) in three types of Californian sherries, by Muller et al. (1964) in passion fruit wine, and by Chaudhary et al. (1968) in Sauvignon blanc wines, all by gas chromatography.

Isobutyl octanoate

Identified by Kepner and Webb (1961) in fusel oil, by Webb et al. (1964b) in three types of Californian sherries, by Chaudhary et al. (1968) in Sauvignon blanc wines, and by Preobrazhenskii et al. (1969), all by gas chromatographic techniques.

$C_{12}H_{26}O$

1-Dodecanol

Identified in wines by Drawert and Rapp (1968) by gas chromatography on four different columns.

2-Dodecanol

Identified in wines by Drawert and Rapp (1968) and in beers by Drawert and Tressl (1969) by gas chromatography on several columns.

$C_{12}H_{26}O_2$

Acetaldehyde diisopentyl acetal (1,1-diisopentylethane)

Identified in flor sherry by Galetto et al. (1966) by gas chromatography, infrared, and preparation of derivatives. Infrared spectrum given. Found by Maarse and ten Noever de Brauw (1066) in Jamaica rum by gas chromatography–mass spectometry.

Acetaldehyde di-(2-methylbutyl) acetal [1,1-di-(2-methylbutoxy)-ethane]

Identified in flor sherry by Galetto et al. (1966) by gas chromatography, infrared, and preparation of derivatives. Infrared spectrum given.

Acetaldehyde isopentyl 2-methylbutyl acetal

Identified in flor sherry by Galetto et al. (1966) by gas chromatography, infrared, and preparation of derivatives. Infrared spectrum given. Also found in Jamaica rum by Maarse and ten Noever de Brauw (1966) by gas chromatography–mass spectrometry.

Acetaldehyde isopentyl pentyl acetal

Found in rum by Liebich et al. (1970) by gas chromatography–mass spectrometry.

Isobutyraldehyde diisobutyl acetal
Found by Liebich *et al.* (1970) in rum by gas chromatography–mass spectrometry.
Isobutyraldehyde isopentyl propyl acetal
Identified in rum by Liebich *et al.* (1970) using gas chromatography–mass spectrometry.
Isovaleraldehyde isobutyl propyl acetal
Identified in rum by Liebich *et al.* (1970) by gas chromatography–mass spectrometry.
Isovaleraldehyde ethyl isopentyl acetal
Identified by Liebich *et al.* (1970) in rum by gas chromatography–mass spectrometry.
Valeraldehyde ethyl pentyl acetal
Identified by Maarse and ten Noever de Brauw (1966) in rums from Jamaica by gas chromatography–mass spectrometry.

$C_{13}H_{16}$
3,8,8-Trimethyl-7,8-dihydronaphthalene
Identified by Liebich *et al.* (1970) in rums by gas chromatography–mass spectrometry.

$C_{13}H_{20}O$
4-(2,6,6-Trimethyl-2-cyclohexen-1-yl)-3-buten-2-one (α-ionone)
Identified in corn fusel oil by Hirose *et al.* (1962) by distillation analysis, gas chromatography, and infrared, and by LaRoe and Shipley (1970) in seven types of whiskies, brandy, and fusel oil by gas chromatography and mass spectrometry.
4-(2,6,6-Trimethyl-1-cyclohexen-1-yl)-3-buten-2-one (β-ionone)
Reported present in Cabernet and Riesling wines on the basis of thin-layer chromatography and color reactions by Pisarnitskii (1966), and by LaRoe and Shipley (1970) in seven types of whiskies, brandy, and fusel oil by gas chromatography and mass spectrometry.

$C_{13}H_{26}O_2$
Ethyl undecanoate
Found by Chaudhary *et al.* (1968) in Sauvignon blanc wines by gas chromatographic techniques, and in rum by Liebich *et al.* (1970) by gas chromatography–mass spectrometry.
Isobutyraldehyde isobutyl isopentyl acetal
Identified in rums by Liebich *et al.* (1970) by gas chromatography–mass spectrometry.

Isopentyl octanoate

Identified in fusel oil by Webb et al. (1952) by distillation analysis and preparation of derivatives. Found in fusel oil by Kepner and Webb (1961) and by Hirose et al. (1962) by distillation analysis and gas chromatographic techniques. Found by Webb and Kepner (1962) in Australian flor sherry, by Webb et al. (1964a) in Bordeaux Cabernet Sauvignon, by Webb et al. (1964b) in three types of Californian sherries, by Chaudhary et al. (1968) in Sauvignon blanc wines, and by Preobrazhenskii et al. (1969), all by gas chromatography. Identified by Webb et al. (1969) in Cabernet Sauvignon and Ruby Cabernet wines by gas chromatography and infrared, and in rums by Liebich et al. (1970) by gas chromatography–mass spectrometry.

Isovaleraldehyde diisobutyl acetal

Identified by Liebich et al. (1970) in rums by gas chromatography–mass spectrometry.

Isovaleraldehyde isopentyl propyl acetal

Found in rums by Liebich et al. (1970) by gas chromatography–mass spectrometry.

2-Methylbutyl octanoate

Identified in fusel oils by Webb et al. (1952), and by Kepner and Webb (1961) by distillation analysis, gas chromatography, and preparation of derivatives.

Methyl dodecanoate

Identified in rums by Liebich et al. (1970) by gas chromatography–mass spectrometry.

1-Undecyl acetate

Identified in corn fusel oil by Hirose et al. (1962) by means of distillation analysis, gas chromatography, and infrared.

Valeraldehyde diisobutyl acetal

Identified by Liebich et al. (1970) in rums by gas chromatography–mass spectrometry.

$C_{14}H_{20}O_2$

2-Phenethyl hexanoate

Identified by Webb and Kepner (1962) in Australian flor sherry, and by Webb et al. (1964a) in Bordeaux Cabernet Sauvignon by gas chromatographic and derivative preparation techniques.

$C_{14}H_{26}O_4$

Diisopentyl succinate

Identified as a constituent of wines by Hardy and Ramshaw (1970) by gas chromatography.

$C_{14}H_{28}O_2$
Ethyl dodecanoate (ethyl laurate)
Identified in fusel oils by Webb *et al.* (1952) by distillation analysis and preparation of derivatives. Found in fusel oils by Kepner and Webb (1961), and by Hirose *et al.* (1962) by distillation analysis, gas chromatography, and infrared. Identified by gas chromatography by Chaudhary *et al.* (1968) in Sauvignon blanc wines, and by Preobrazhenskii *et al.* (1969) in Russian sherries. Found in bourbon whiskey by Kahn *et al.* (1969) using gas chromatography–mass spectrometry.

Hexyl octanoate
Identified as a trace component in Bordeaux Cabernet Sauvignon by Webb *et al.* (1964a), and by Chaudhary *et al.* (1968) in Sauvignon blanc wines using gas chromatographic techniques.

Isobutyl decanoate
Identified in fusel oil by distillation analysis and preparation of derivatives by Webb *et al.* (1952). Found in muscat raisin fusel oil by Kepner and Webb (1961), and in Sauvignon blanc wines by Chaudhary *et al.* (1968) using gas chromatography. Found by Liebich *et al.* (1970) in rums by gas chromatography–mass spectrometry.

Tetradecanoic acid (myristic acid)
Found in bread pre-ferment by Hunter *et al.* (1961) using gas chromatography, and in Sauvignon blanc wines by Chaudhary *et al.* (1968) by the same techniques.

$C_{14}H_{30}O_2$
Isobutyraldehyde diisopentyl acetal
Identified in rums by Liebich *et al.* (1970) by gas chromatography–mass spectrometry.

Isovaleraldehyde isobutyl isopentyl acetal
Found in rums by Liebich *et al.* (1970) by gas chromatography–mass spectrometry.

$C_{15}H_{24}O_2$
Acetaldehyde isopentyl 2-phenethyl acetal
Identified in submerged-culture sherry by Galetto *et al.* (1966) by gas chromatography, infrared, and preparation of derivatives. Infrared spectrum given.

Acetaldehyde 2-methylbutyl 2-phenethyl acetal
Identified by Galetto *et al.* (1966) in submerged-culture flor sherry by gas chromatography, infrared, and preparation of derivatives.

$C_{15}H_{26}O$
3,7,11-Trimethyldodeca-2,6,10-trien-1-ol (farnesol)
Reported present in Cabernet and Riesling wines by Pisarnitskii (1966) on the basis of thin-layer chromatography and color reactions.

$C_{15}H_{30}O$
Methyl tridecyl ketone (2-pentadecanone)
Identified in corn fusel oil by Hirose *et al.* (1962) by distillation analysis, gas chromatography, and infrared.

$C_{15}H_{30}O_2$
Isopentyl decanoate
Identified in fusel oils by Webb *et al.* (1952), by Kepner and Webb (1961), and by Hirose *et al.* (1962), all by distillation analyses, the first by preparation of derivatives and the latter two by gas chromatography. Found in rums by Liebich *et al.* (1970) by gas chromatography–mass spectrometry.
2-Methylbutyl decanoate
Found in fusel oils by Webb *et al.* (1952) and by Kepner and Webb (1961) by distillation analyses and the preparation of derivatives.
Pentadecanoic acid
Identified in Sauvignon blanc wines by Chaudhary *et al.* (1968) by gas chromatography.

$C_{15}H_{32}O_2$
Isovaleraldehyde diisopentyl acetal
Identified in rums by Liebich *et al.* (1970) by gas chromatography–mass spectrometry.

$C_{16}H_{24}O_2$
2-Phenethyl octanoate
Identified in wines by Hardy and Ramshaw (1970) and in rums by Liebich *et al.* (1970) using gas chromatography–mass spectrometry.

$C_{16}H_{32}O_2$
Ethyl tetradecanoate (ethyl myristate)
Identified in fusel oils by Kepner and Webb (1961) and by Hirose *et al.* (1962) by means of distillation analysis, gas chromatography, and infrared. Found present in bourbon whiskey by Kahn *et al.* (1969) by gas chromatography–mass spectrometry, and in rums by Liebich *et al.* (1970) by the same techniques.
Hexadecanoic acid (palmitic acid)
Identified in bread pre-ferment by Hunter *et al.* (1961), and in

Sauvignon blanc wines by Chaudhary *et al.* (1968) using gas chromatography.

$C_{17}H_{34}O_2$
Ethyl pentadecanoate
Identified in muscat raisin fusel oil by Kepner and Webb (1961) by gas chromatography and chemical methods, and in bourbon whiskey by Kahn *et al.* (1969) by gas chromatography–mass spectrometry. Also found in rums by Liebich *et al.* (1970) by gas chromatography–mass spectrometry.
Heptadecanoic acid (margaric acid)
Found in Sauvignon blanc wines by Chaudhary *et al.* (1968) by gas chromatography, and in spirits by Nykänen *et al.* (1968) using gas chromatography.
Isopentyl dodecanoate (isoamyl laurate)
Found in fusel oils by Webb *et al.* (1952), and by Kepner and Webb (1961) using distillation analysis and gas chromatography in the latter case. Also found in rums by Liebich *et al.* (1970) using gas chromatography–mass spectrometry.
2-Methylbutyl dodecanoate (2-methylbutyl laurate)
Found in fusel oils by Webb *et al.* (1952), and by Kepner and Webb (1961) by distillation analysis and derivative preparation in the first case, and gas chromatography, in the second case.
Methyl hexadecanoate
Identified in rums by Liebich *et al.* (1970) by gas chromatography–mass spectrometry.

$C_{18}H_{28}O_2$
2-Phenethyl decanoate
Identified in rums by Liebich *et al.* (1970) by gas chromatography–mass spectrometry.

$C_{18}H_{34}O_2$
Ethyl 9-hexadecenoate
Found in bourbon whiskey by Kahn *et al.* (1969) using gas chromatography–mass spectrometry, and in rums by Liebich *et al.* (1970) by the same procedure.

$C_{18}H_{36}O_2$
Ethyl hexadecanoate (ethyl palmitate)
Found in fusel oils by Webb *et al.* (1952) using distillation analysis and preparation of derivatives, by Kepner and Webb (1961) using distillation analysis and gas chromatography, and by Kahn *et al.* (1969) in

bourbon whiskey using gas chromatography–mass spectrometry. Identified in rums by Liebich et al. (1970) by gas chromatography–mass spectrometry.

$C_{19}H_{38}O_2$
Ethyl heptadecanoate
Identified in rums by Liebich et al. (1970) by gas chromatography–mass spectrometry.
Isopentyl tetradecanoate (isoamyl myristate)
Reported a constituent of muscat raisin fusel oil by Kepner and Webb (1961) on the basis of gas chromatographic analysis, and found in rums by Liebich et al. (1970) by gas chromatography–mass spectrometry.
2-Methylbutyl tetradecanoate (act.-amyl myristate)
Reported present in muscat raisin fusel oils by Kepner and Webb (1961) on the basis of gas chromatographic analyses.
Propyl hexadecanoate
Identified in rums by Liebich et al. (1970) by gas chromatography–mass spectrometry.

$C_{20}H_{36}O_2$
Ethyl 9,12-octadecadienoate (ethyl linoleate)
Identified in bourbon whiskey by Kahn et al. (1969) using gas chromatography–mass spectrometry, and in rums by Liebich et al. (1970) by the same procedures.

$C_{20}H_{38}O_2$
Ethyl 9-octadecenoate (ethyl oleate)
Found in bourbon whiskey by Kahn et al. (1969) by gas chromatography–mass spectrometry, and in rums by Liebich et al. (1970) by the same technique.

$C_{20}H_{40}O_2$
Ethyl octadecanoate (ethyl stearate)
Identified by Liebich et al. (1970) in rums by gas chromatography–mass spectrometry.

$C_{21}H_{42}O_2$
Isopentyl hexadecanoate
Identified by Liebich et al. (1970) in rums by gas chromatography–mass spectrometry.

$C_{22}H_{44}O_2$
Docosanoic acid (behenic acid)

Identified in beer by Clarke *et al.* (1962) by gas chromatography of methyl ester on two different columns.

$C_{26}H_{52}O_2$
Hexacosanoic acid (cerotic acid)

Found in beer by Clarke *et al.* (1962) by gas chromatography of the methyl ester on two different columns.

O_2S
Sulfur dioxide

Found in seven different types of spirits by Dellweg *et al.* (1969) using gas chromatography with a flame color detector.

III. Biosynthetic Pathways and Interrelationships among Aroma Compounds

A. GENERAL CONSIDERATIONS

For purposes of classification, volatile aroma compounds in alcoholic beverages can be considered to arise from (a) components of the raw material, (b) the activity of yeast growth and fermentation of sugars, and (c) aroma substances produced during storage and processing. Special cases are the volatiles arising from bacterial activity and those of flor sherries resulting from the microaerophilic stage of yeast growing in a film on the wine's surface. In distilled beverages, volatiles also arise from heat-induced changes on those compounds already present and, in addition, from the cooperage by the solvent action of ethanol. Wood extractives are present in other beverages stored in barrels or wooden tanks, but in the case of spirits, because of the higher concentration of alcohol and the longer storage time the leached substances and their reaction products are more apparent. The spectrum of congeners in spirits is also directly affected by the degree of rectification as is readily apparent upon comparison of a heavy-bodied rum or cognac with a light spirit such as vodka.

At present it is generally assumed that most of the aroma constituents in wines arise through the action of yeast on sugars and on other constituents of the grapes, although it is recognized that, in the case of fine wines, the grapes do contribute importantly. In general, the aroma of wines is not due to an "impact" flavor compound (unless such compound is already present in the grapes and carries through the fermentation unchanged) but to a complex mixture of compounds. Ethyl esters predominate but also important are free acids, carbonyls, higher alcohols, etc., whose relative concentrations vary not only according to variety, yeast strain, vintage, handling of the raw material, and cellaring operations, but even more, according to fermentation conditions such as aeration, agitation, pH, and temperature. The volatile constituent makeup of a wine at any given time is only an indication of the substances which are present at the moment in greatest concentration; minor or trace constituents are used and formed continually as the various systems tend toward equilibrium. Some of these changes take years to result in differences detectable by sensory methods.

Whether the fermentation is carried out on the skins or not is obviously of importance as the skins can provide phenolic compounds (tannins and pigments) as well as other substances which are metabolized further by the yeast and which, by their chemical nature, take part in oxidation-reduction reactions both during and after fermentation. After the fermentation is finished and the yeast and enzymes are removed from the wine, purely chemical interactions occur among the aroma compounds. Transesterifications and the formation of acetals, for example, add to the variety and number of compounds present in wine and make the study of its flavor constituents an interesting endeavor.

Beers, in contrast to wines and spirits, seem to contain fewer and lower concentrations of volatile aroma materials. This may reflect the difficulties of study of a material with a higher water concentration rather than an actual paucity of aroma substances.

Odor thresholds of individual compounds vary over several orders of magnitude. While it might be assumed that the aroma of a beverage would be defined by the one or two compounds with low threshold present in sufficient concentration, in fact it is difficult to predict the contribution of an individual component to the total aroma picture because of the complicated odor interactions found. The total problem is further complicated by the fact that some compounds of low threshold value are difficult to determine quantitatively in complicated aroma mixtures, thus making an accurate assessment of their contribution to the total picture impossible.

Although a great amount of fine research work has been done in the field of identification of aroma compounds in alcoholic beverages, there is little doubt that many substances present in trace concentrations have eluded characterization. Some of these constituents are short lived and difficult if not impossible to isolate in the pure state due to their inherent instability or reactivity. Others may be involved in complex equilibria so that removal of their "parent" compounds during isolation operations precludes their recovery. These compounds may be intermediates in the several chemical and biochemical pathways by which flavor compounds are produced.

The basic biochemical pathways and systems are common to all plants, and although there are variations in the minor constituent pathways and in the number of systems involved among different plants, it is possible to recognize analogies in the biosynthesis of aroma materials. Yeasts have been studied extensively and, although a great deal is known concerning the volatile compounds produced by their metabolism, much remains to be elucidated. There is little doubt that research in this field will be exciting and rewarding in the near future.

For a recent review of flavor in fermented products (beer, bread pre-ferment, wine, and sake) the reader is referred to the excellent work by Margalith and Schwartz (1970). These authors cover the flavor role of alcohols, carbonyls, esters, and other metabolites, including the malo-lactic fermentation products. Suomalainen (1971) has reviewed the effect of yeast on the flavor of alcoholic beverages, also. As the biogenesis of various compounds is discussed in these reviews, only a few recently identified or peculiar substances will be considered here. Suggestions as to their origin will be presented in the light of currently available evidence.

For the flavor chemist it is of extreme importance that all compounds identified in his work or reported in the literature by others, fit into some known chemical or biochemical scheme. Compounds are interesting enough when they fit into a known pathway; they become fascinating when they do not. One always has visions of having found a compound which might give a clue to a yet unknown metabolic pathway; yet, more often than not, these turn out to be artifacts of one class or another or, as unfortunately is common, the compound was hastily listed from a table by another author which included not only the compounds actually isolated but standards used, for example, in calculating Kovat's indices. Such is the case of 3- and 4-heptanone which were used by Kepner *et al.* (1963) as standards, yet are reported in a recent review as constituents of beer. Often, a compound is reported present in a definite source while actually the source had been subjected to harsh treatment. One is likely to find just about anything in tars.

B. Higher Alcohols

Basically, metabolites in fermented products arise mainly from carbohydrates, amino acids, and lipids. Thus primary alcohols, higher than ethanol, generally arise via the Ehrlich-Neubauer-Fromherz mechanism (Ehrlich, 1906, 1907; Thoukis, 1958; SentheShanmuganthan, 1960; Yamada et al., 1962; Webb and Ingraham, 1963; Guymon, 1966; Äyräpää, 1967a, 1967b). Drawert and Tressl (1969) have shown that other primary alcohols, particularly straight-chain alcohols for which there is no amino acid parent, might be formed by the reducing action of yeast on aldehydes; reduction of keto compounds affording secondary alcohols. Both saturated and unsaturated primary alcohols have been found in grapes (Stevens et al., 1965, 1967; Van Wyk et al., 1967a; Webb and Kepner, 1957), in whiskey (Kahn et al., 1969), and in rum (Liebich et al., 1970). Reduction of aldehydes formed by oxidative degradation of lipids might account for these. Methyl carbinols, such as those found in corn fusel oil (Hirose et al., 1962), in wines (Webb et al., 1969), in Grenache grapes (Stevens et al., 1967), and reported in Vitis labrusca by Neudoerffer et al. (1965), probably arise by reduction of methyl ketones which in turn are formed from β-keto esters (Stokoe, 1928; Wong et al., 1958; Gehrig and Knight, 1963; Parks et al., 1964). Even- and odd-numbered ketones have been reported in fermented products (Hirose et al., 1962; Muller et al., 1964; Kahn et al., 1969).

C. Fatty Acids

Lipids constitute an important source of flavor compounds, and their role in various foods has been reviewed by Forss (1969). The thermal degradation of lipids by various chemical pathways is discussed by Nawar (1969). Methyl ketones, aldehydes, acids, and both Δ- and γ-lactones and other compounds have been reported (Nawar et al., 1962; Muck et al., 1963). The various aldehydes and acetals found in rum (Maarse and ten Noever de Brauw, 1966; Liebich et al., 1970) and in whiskey (Kahn et al., 1968, 1969; Kahn, 1969) might originate in this manner.

Since the nature of the lipids synthesized in a system and their mode of breakdown contribute to the flavor compounds produced, the study of the fatty acids formed by yeast under different growing conditions and whether β-oxidation pathways are indeed available for fatty acid degradation in yeast, merits some consideration.

α-Unsaturated and aliphatic monobasic acids can be formed by anabolic and catabolic pathways (Mudd, 1967). A profusion of aliphatic, mono- and polyunsaturated, as well as β-hydroxy acids arising through the β-oxidation scheme (Stumpf, 1965) have been isolated from plant tissues (Jennings et al., 1964; Creveling and Jennings, 1970). Acids and esters arising from β-oxidation often contribute "impact" flavor compounds of the type found in pears (Jennings, 1967). In fermented products however, compounds arising via the β-oxidation scheme are conspicuous by their paucity. Clarke *et al.* (1962) isolated 3-hexenoic and 3-heptenoic acids from beer. The acids seem to originate in the barley either by β-oxidation or by means of a chain elongation and desaturation mechanism as described by Hawke and Stumpf (1965) for *Hordeum vulgare* L. and other Gramineae rather than from yeast metabolism. According to Scheuerbrandt and Bloch (1962), *Saccharomyces cerevisiae* (incubated at 30°C) produces exclusively Δ-9 unsaturated fatty acids by a desaturation mechanism which requires oxygen. This view is shared by Suomalainen and Keränen (1968) in that unsaturated acids (e.g., oleic) are formed by yeast through desaturation of the corresponding saturated acids. Bloomfield and Bloch (1960) state that "To our knowledge, β-oxidation of higher fatty acids has never been demonstrated in yeast." Scheuerbrandt and Bloch (1962) further show that bacteria form β-unsaturated acids but yeasts do not.

Hunter and Rose (1971), in a review, state "long-chain fatty acids can be oxidized by many and probably all yeasts." They further assume that the oxidation proceeds by the β-oxidation pathway. This view is shared by Sols *et al.* (1971), who point out that although the β-oxidation scheme as they present it, is the one worked out in animal tissues (Lynen, 1967), the system in yeast should be similar. They further state that the situation is still unclear and that further study is needed.

At first sight, the paucity of compounds derived through the β-oxidation pathway (only 3-hydroxyoctanoic acid has been found in wine; Kepner *et al.*, 1968), seems to strengthen the argument that yeasts do not degrade fatty acids through a β-oxidation pathway. However, it should be pointed out that under the anaerobic (or microaerophilic) conditions present during fermentation these compounds might be thoroughly reduced to saturated acids. Therefore it seems obvious that in order to study and elucidate on the β-oxidation in yeast, conditions should be chosen carefully, particularly with respect to temperature and partial pressure of oxygen.

In this respect, several investigators (Kates, 1964, 1966; Kates and Baxter, 1962; Farrell and Rose (1967); Brown and Rose, 1969) have reported that the fatty acid composition of yeast (*Candida utilis*) changes from saturated to unsaturated as the incubation temperature is lowered. In addition, Brown and Rose (1969) found that there is a similarity between the effects of decreased oxygen partial pressure and decreased temperature on fatty acid syntheses. They explain this effect as either due to an increase in available oxygen at lower temperatures because of greater solubility, or "that the synthesis or activity, or both, of the desaturating enzyme(s) and of enzymes that catalyze reactions involved in lengthening the fatty-acid chain are very sensitive to temperature." Their results also show that synthesis of unsaturated fatty acids parallels synthesis of short-chain acids. Jollow et al. (1968) report that anaerobically grown *Saccharomyces cerevisiae* synthesizes shorter chain (C_{10}–C_{14}) saturated acids. Hunter and Rose (1971) state that when yeast cultures are grown at low temperatures, changes in fatty-acid composition are probably due to a decrease in growth rate and also to increase in dissolved oxygen partial pressure. Stokes (1971) states that it has been suggested that the fatty acid composition of the cell, in particular the cell membrane, might be responsible for setting the minimum temperature for yeast and bacterial growth. He also points out that there is little evidence to support this hypothesis. Evidence against it is reported by Marr and Ingraham (1962), whose results (on *Escherichia coli*) do not indicate a direct relationship between fatty acid composition and minimal growth temperature. Marr and Ingraham further state that physiological effects of fatty-acid composition appear to be trivial.

On the basis of evidence presented by Lyons and Asmundson (1965), and Lyons and Raison (1970), in which the authors effectively show that the nature of the mitochondrial membrane lipids change from saturated to unsaturated in plants capable of withstanding chilling, whereas in plants unable to withstand chilling, membrane fatty acids remain saturated, one is inclined to postulate here that a similar mechanism must exist in yeast and in psychrophilic organisms. The function of the unsaturated fatty acids in the membrane lipids is that of lowering the melting point of the lipid portion of the lipoprotein complex thus retaining the permeability of the membrane as suggested by Lyons and co-workers for plant tissue.

White wines and lager beers are customarily fermented under microaerophilic conditions and at low temperatures (about 10°C). Currently some Californian wineries are making some types of wines by even lower temperature fermentations. These wines have a pronounced "fruity" character and seem to be rich in esters (and possibly acetals). Although the exact nature of the volatiles produced under these conditions has not been studied in detail, one might be tempted to speculate at this time that a β-oxidation scheme does exist in yeast but that it is not readily apparent.

The apparent lack of β-oxidation metabolites in fermented products may be explained by either one or more of the following: (a) β-oxidation metabolites are formed initially during fermentation when oxygen partial pressure is relatively high but are reduced during the log phase of yeast growth when the oxygen partial pressure is lowered by the rapid cell multiplication and the higher temperature of the medium. (b) Yeast β-oxidation enzymes may be active only over a narrow temperature range or may require (in addition to molecular oxygen or hydrogen peroxide) the presence of unsaturated fatty acids to be active, or, as pointed out by Stumpf (1969), two enzyme systems may be involved in activating long-chain fatty acids, one for saturated and another for unsaturated acids. If *Saccharomyces cerevisiae* lacks the enzymes to activate saturated acids, β-oxidation products would not be produced at all at higher temperatures (about 30°C) when the lipids present are saturated. (c) Degradation of lipids through a β-oxidation pathway provides building blocks for further synthetic purposes, but under fermentation conditions, where relatively high levels of readily assimilable hexoses are available, the sugars are used preferentially.

An interesting compound, whose origin merits further study, 9-decenoic acid, has been isolated as its ethyl ester from beer by Strating and Venema (1961) and as the ester and the free acid by Van Wyk *et al.* (1967a,b) from Riesling wine. At first sight it is tempting to ascribe its origin to activity of Δ-9 desaturase on decanoic acid. However, as pointed out earlier, this mechanism requires molecular oxygen. An interesting alternative pathway which requires no molecular oxygen is that described by Bloomfield and Bloch (1960) for the biosynthesis of higher molecular weight unsaturated acids. A common yeast metabolite, 4-hydroxybutyrate, which has been isolated both as its ethyl ester and as its internal ester, γ-butyrolactone, by Van Wyk *et al.* (1967a) from White Riesling wine and by Webb *et al.* (1967b, 1969) could serve as the starting material. However, this scheme requires the dehydration of 4-hydroxybutyrate to 3-butenoate, while β-γ-enoyl-dehydrase, the enzyme responsible for the formation of *cis*-β, γ-decenoyl-ACP (acyl carrier protein) in Bloch's scheme acts on β-hydroxydecanoyl-ACP. The dehydrase responsible for dehydration of β-hydroxybutyrl-ACP produces *trans*-crotonyl-ACP instead of 3-butenoyl-ACP (Conn and Stumpf, 1967). In *Escherichia coli* there seem to be at least three different fatty acid dehydrase systems, each with a specific function (Stumpf, 1969). One acts on C_4–C_{10} substrates, another is involved with C_{10}–C_{18} saturated fatty acids, and a third is specific for C_{10} interconversions. A variety of dehydrases might be present in *Saccharomyces* as well.

It is interesting to note that 9-decenoic acid has been found by many workers in milk fat (L. M. Smith *et al.*, 1954) and in human depot fat (Jacob and Grimmer, 1968). The latter authors deduce from a plot of the ratio of saturated to monounsaturated fatty acids, that the biogenesis proceeds by the same mechanism, i.e., the 9-dehydrogenation of saturated fatty acids.

A third possibility for the production of 9-decenoic acid is by a mechanisms similar to that postulated by Choteau et al. (1962) in which a hydrocarbon (n-heptane) is oxidized by NAD^+ (nicotinamide-adenine dinucleotide) to the terminal olefin. Such a mechanism has not been demonstrated for *Saccharomyces*. ω-Oxidation, another potential mechanism involving the terminal carbon, requires oxygen (McKenna and Kallio, 1965; McKenna and Coon, 1970). This mechanism seems to proceed via an epoxide rather than an olefin, however. Primary or secondary alcohols can be produced depending on the mode of cleavage of the oxirane ring, and by further oxidation, even numbered carbon aldehydes, acids, and ketones. ω-Oxidations almost invariably produce oxygen-containing compounds. However, a sequential pathway can be visualized by which the aliphatic end of the molecule is desaturated first to the terminal olefin which then is oxidized to the 1,2-epoxide.

D. Ethers, Aldehydes, and Acetals

In addition to methoxy phenols (discussed by Singleton and Esau, 1969), various other ethers have been recently isolated from distilled products. Kahn et al. (1969) found ethyl 3-ethoxypropionate and 3-ethoxypropionaldehyde in whiskey. Liebich et al. (1970) isolated 4-ethoxy-2-butanone, 4-ethoxy-2-pentanone, and 2-methyl-3-tetrahydrofuranone. Ethers have been isolated from various plant and animal sources. Lennarz (1970) has suggested a mechanism for the *in vivo* synthesis of 1-alkoxyglycidyl-3-phosphate from glyceraldehyde-3-phosphate. An initially formed hemiacetal is subsequently dehydrated and the resulting 2-keto compound is reduced by NADPH (nicotinamide adenine dinucleotide phosphate). Similar mechanisms can be envisioned for the production of the above ethers. The respective precursors, or their oxidized products, for each ether (e.g., 3-hydroxypropionate, 2,3-pentanedione, 3-ketobutyrate) have been found in various other fermented products. The concurrent presence of 3-penten-2-one and 4-ethoxy-2-pentanone in rum seems to substantiate this possibility. It is interesting to notice in this respect the large number of acetals (and of aldehydes) that have been found recently in rum (Liebich et al., 1970), in Japanese and Scotch whisky (Nishimura and Masuda, 1971), in bourbon whiskey (Kahn et al., 1969), and in cognac (Schaefer and Timms, 1970). The study of the chemical changes among these aldehydes, acetals, and ethers during wood aging of the various spirits merits further detailed attention.

E. Phenolics

Phenolic substances in grapes and wines have been reviewed recently in a thorough fashion by Singleton and Esau (1969). They consider in detail how wood extractives, the oxidative reactions occurring during crushing, early fermentation, and storage can lead to aroma and bouquet materials in the aged product. LaRoe and Shipley (1970) have recently demonstrated that α- and β-ionone (which have pleasant odors and extremely low thresholds for sensory detection) are likely produced by the thermal degradation of β-carotene in the raw material, corn in the case of bourbon whiskey.

F. Nitrogen-Containing Compounds

The division of nitrogenous compounds of fermented beverages into groups on the basis of classical analytical methods has been reviewed by Amerine (1954), and the amino acids of wines have been considered by Ough and Bustos (1969). Simple amines have been reported in wines by Drawert (1965) and by Puputti and Suomalainen (1969). Most of the amines reported present can arise from decarboxylation of the corresponding amino acids, although, in some cases such as for hexyl amine, a transaminase acting on hexanal seems more likely.

Two interesting substituted amides found in sherries by Webb *et al.* (1966) were identified as N-isoamyl- and N-(2-phenethyl)acetamide. They proposed that the amides were formed during fermentation by reaction of acetyl-coenzyme A with the appropriate amine.

Liebich *et al.* (1970) have found a number of pyrazines in rum, and Nishimura and Masuda (1971), in addition to pyrazines, found a range of substituted pyridines and quinolines. Koehler *et al.* (1969) suggest that the pyrazines are formed by reaction of hexoses with amino acids, while Nishimura and Masuda demonstrate that the substituted pyridines and quinolines of whiskey come from the peat-fire smoke used in kilning malt.

It must be remembered that most of the nitrogen-containing compounds found so far in alcoholic beverages are weak to strong bases, and show their intense odors when in the free base form. At the pH levels of distilled beverages, and certainly for wines, the predominant form of these compounds would be the protonated cation, which as a salt in solution would have little odor.

G. Lactones

Both γ- and δ-lactones of up to about sixteen carbon atoms are important compounds in flavor. Some, such as C_8–C_{10} lactones, have pleasant coconutlike odors which can be desirable or undesirable depending on the product. Many are used in the flavor industry to provide "impact" flavors, a typical example being the use of γ-undecalactone for peach flavored foods and beverages.

Most small lactones (up to C_{16}) have been associated with unsaturated, hydroxy, and keto acids and particularly in milk and milk products (Boldingh and Taylor, 1962; Van der Ven, 1964; Jurriens and Oele, 1965; Forss et al., 1966; Kinsella et al., 1967). Various aliphatic and some unsaturated lactones (Δ-dodec-9-ene lactone, γ-dodec-6-ene lactone), including odd-numbered-carbon lactones have been isolated. Lactones have also been isolated from heated milk fat (Parliament et al., 1966), evaporated milk (Muck et al., 1963). Lactones have also been isolated from autooxidized vegetable oils (Fioriti et al., 1967) and, among various other plant sources, coconut meat (Lin and Wilkens, 1970). The general consensus is that these lactones arise from unsaturated, hydroxy, and keto acids. 4-Hydroxy acids lactonize spontaneously at room temperature. Unsaturated acids with double bonds on either positions 3, 4, or 5 lactonize easily, and preferentially to the γ-lactones, in the presence of acids. Even 2-enoic acids will form γ-lactones if in the presence of dilute acids (about pH 3.0–3.5) for some period of time (Kepner et al., 1972). The isomerization of the double bond to positions three or four is a relatively slow reaction whereas lactonization to the γ-position is very fast. Thermodynamic control thus produces the γ-lactones preferentially.

Not all the lactones encountered in natural products arise from unsaturated, hydroxy, or keto acids, however. Recent work by Sulser et al. (1967), on flavoring of protein hydrolyzates, indicates that lactones (α-keto-β-methyl-γ-caprolactone, in this instance) can be formed from α-keto acids, which are in turn the products of transamination of naturally occurring amino acids, via an acid-catalyzed aldol condensation. Aldol condensations are known to occur in nature not only by purely chemical means but also by enzyme catalysis.

Tatum et al. (1969), isolated γ-butyrolactone and γ-crotonolactone among many degradation products from heating an aqueous ascorbic acid solution in a model system. Pyrolysis of sucrose yielded γ-valerolactone and 5-methyl-2(5H)furanone (Johnson et al., 1969). Their results show that similar products are formed by pyrolysis of hexoses and by degradations in strongly reducing aqueous acid systems. These results are of importance in distilled products, particularly in heavy-bodied rums.

Recently, branched-chain γ-lactones, the two isomeric forms of 4-methyl-5-butyldihydro-2(3H)furanone, have been isolated from Scotch whisky by Suomalainen and Nykänen (1970), who named them "whisky lactones," and by Nishimura and Masuda (1971). The *trans* isomer has also been isolated by us from wood-aged Cabernet Sauvignon and Ruby Cabernet wines (Muller *et al.*, 1972). The same lactones were isolated from white oak and *Quercus mongolica*, among other *Quercus* species, by Masuda and Nishimura (1971) who named them "quercus lactones." The authors point out that the lactones have the "raw, woody aroma of whisky." These lactones, as isolated from oak are optically active and thus of biosynthetic origin. However, no rotations are reported for the lactones isolated from whiskey. It is interesting to note that the racemic mixture of one of the lactones synthesized by one of us (C. J. M.) has an odor reminiscent of coconut rather than whiskey.

It is tempting to suggest that the whiskey lactones arise as a result of an aldol condensation of 2-heptanone with either acetyl-coenzyme A or pyruvyl-coenzyme A as a detoxification mechanism in the plant.

Several seemingly biosynthetically related lactones and other compounds have been isolated from various wines in recent years. Sherry has yielded γ-butyrolactone, ethyl 4-hydroxybutyrate, γ-carboethoxybutyrolactone, ethyl pyroglutamate, diethyl succinate, diethyl malate, and diethyl tartrate (Webb *et al.*, 1967b). Also, two optically active diastereomers of the γ-lactone of 4,5-dihydroxyhexanoic acid (Muller *et al.*, 1969) and the γ-lactone of 4-hydroxy-5-ketohexanoic acid, "solerone" (Augustyn *et al.*, 1971) are sherry components. The latter lactone was also found in Cabernet Sauvignon wine (Webb *et al.*, 1969). Hirabayashi and Harada (1969) isolated 5-hydroxy-4-ketohexanoic acid from the fermentation of a synthetic medium by *Hansenula miso*. Lately ethyl 4-oxobutyrate and 5-ethoxy-γ-butyrolactone have been isolated from Ruby Cabernet wine (Muller *et al.*, 1972a). In addition, we have obtained evidence of the presence of the Δ-lactone of 4-keto-5-hydroxyhexanoic acid in the same wine (Muller *et al.* 1972b).

All of these compounds can be fitted neatly into a system of related biosynthetic pathways. See Muller *et al.* (1972a) for a diagram of the interrelationships and a detailed discussion. The recently isolated compound, solerone (4-hydroxy-5-oxohexanoic acid γ-lactone) has a pronounced winelike odor when pure, but it is not very stable.

H. Sulfur-Containing Compounds

It is well known that fermenting yeasts are capable of reducing both inorganic sulfates and sulfites as well as elemental sulfur to sulfide (Lawrence, 1964; Lewis and Wildenradt, 1969). In acid media such as wines and beers, the sulfide appears as hydrogen sulfide, which has a very low threshold value and a particularly objectionable aroma.

Dellweg et al. (1969) have given a list of organic sulfur-containing substances that they have identified in a number of different spirits. They show that these compounds arise from the sulfur-containing amino acids of the yeast protein through autolysis. Muller et al. (1971) have demonstrated that the Ehrlich mechanism for the production of fusel oils operates on methionine to produce methionol, a compound which has the odor of raw potato skins, and which was identified in Cabernet Sauvignon and Ruby Cabernet wines.

Levy and Stahl (1961) discuss the mass spectra of a number of aliphatic thiols and sulfides of importance in food aromas.

IV. Summary

The development of gas chromatography, and more recently the more powerful combination of gas chromatography–mass spectrometry for use in research on the trace aroma and flavor compounds of fermented and distilled beverages has provided us with extensive lists of components in these beverages. While much remains to be learned from this type of accumulation of lists of components, the next big steps forward in aroma and flavor research will likely come from elucidation of the biosynthetic pathways involved in the production of these trace components, and in studies of the processes by which these compounds interact with the human mouth and nose to create the impressions we get of fine wines, beers, and spirits.

Acknowledgments

We thank Mrs. Priscilla Douglas for her assistance in the literature search and for compiling the list of references. We also thank Richard E. Kepner for helpful discussions of the chemical aspects of this work.

References

Ahrenst-Larsen, B., and Hansen, H. L. (1964). *Wallerstein Lab. Commun.* **27**, 41.
Almashi, K. K. (1965). *Vinodel. Vinograd. SSSR* **25**, 7.
Amerine, M. A. (1954). *Advan. Food Res.* **5**, 353.
Augustyn, O. P. H., Van Wyk, C. J., Muller, C. J., Kepner, R. E., and Webb, A. D. (1971). *J. Agr. Food Chem.* **19**, 1128.
Äyräpää, T. (1965). *J. Inst. Brewing* **71**, 341.
Äyräpää, T. (1967a). *J. Inst. Brewing* **73**, 17.
Äyräpää, T. (1967b). *J. Inst. Brewing* **73**, 30.
Baldwin, S., Black, R. A., Andreasen, A. A., and Adams, S. L. (1967). *J. Agr. Food Chem.* **15**, 381.
Bloomfield, D. K., and Bloch, K. (1960). *J. Biol. Chem.* **235**, 337.
Boldingh, J., and Taylor, R. J. (1962). *Nature (London)* **194**, 909.
Brown, C. M., and Rose, A. H. (1969). *J. Bacteriol.* **99**, 371.
Chaudhary, S. S., Webb, A. D., and Kepner, R. E. (1968). *Amer. J. Enol. Viticult.* **19**, 6.

Choteau, J., Azoulay, E., and Senez, J. (1962). *Nature (London)* **194**, 576.
Clarke, B. J., Harold, F. V., Hildebrand, R. P., and Morieson, A. B. (1962). *Amer. Soc. Brewing Chem., Proc.* p. 179.
Cole, E. W., Helmke, V., and Pence, J. W. (1966). *Cereal Chem.* **43**, 357.
Conn, E. E., and Stumpf, P. K. (1967). "Outlines of Biochemistry," 2nd ed. Wiley, New York, pp. 287–289.
Creveling, R. K., and Jennings, W. G. (1970). *J. Agr. Food Chem.* **18**, 19.
Dellweg, H., Miglio, G., and Niefind, H.-J. (1969). *Branntweinwirtschaft* **109**, 445.
Drawert, F. (1965). *Vitis* **5**, 127.
Drawert, F., and Rapp, A. (1968). *Chromatographia* **1**, 446.
Drawert, F., and Tressl, R. (1969). *Brauwissenschaft* **22**, 169.
Ehrlich, F. (1906). *Biochem. Z.* **1**, 25.
Ehrlich, F. (1907). *Ber.* **40**, 1027.
Ehrlich, F. (1917). *Biochem. Z.* **79**, 232.
Farrell, J., and Rose, A. (1967). *Annu. Rev. Microbiol.* **21**, 101.
Fioriti, J. A., Krampl, V., and Sims, R. J. (1967). *J. Amer. Oil Chem. Soc.* **44**, 534.
Forss, D. A. (1969). *J. Agr. Food Chem.* **17**, 681.
Forss, D. A., Urbach, G., and Stark, W. (1966). *Proc. 17th Int. Dairy Congr.* p. 211.
Galetto, W. G., Webb, A. D., and Kepner, R. E. (1966). *Amer. J. Enol. Viticult.* **17**, 11.
Gehrig, R. F., and Knight, S. G. (1963). *Appl. Microbiol.* **11**, 166.
Guymon, J. F. (1966). In "Developments in Industrial Microbiology, Vol. 7," (C. F. Koda, ed.), American Institute of Biological Sciences, Washington, D.C., pp. 88–96.
Hardy, P. J., and Ramshaw, E. H. (1970). *J. Sci. Food Agr.* **21**, 39.
Harrison, G. A. F., and Collins, E. (1968). *Amer. Soc. Brewing Chem. Proc.* p. 101.
Hawke, J. C., and Stumpf, P. K. (1965). *Plant Physiol.* **40**, 1023.
Hennig, K., and Villforth, F. (1942). *Vorratspflege Lebensmittelforsch.* **5**, 181.
Hirabayashi, T., and Harada, T. (1969). *Agr. Biol. Chem.* **33**, 276.
Hirose, Y., Ogawa, M., and Kusuda, Y. (1962). *Agr. Biol. Chem.* **26**, 526.
Hrdlicka, J., Dyr, J., and Jely, E. (1968). *Brauwissenschaft* **21**, 333.
Hunter, I. R., Hawkins, N. G., and Pence, J. W. (1961). *J. Food Sci.* **26**, 578.
Hunter, K., and Rose, A. H. (1971). In "The Yeast" (A. H. Rose and J. S. Harrison, eds.), Vol. 2, Chapter 6. Academic Press, New York, p. 254.
Ikeda, R. M., Webb, A. D., and Kepner, R. E. (1956). *J. Agr. Food Chem.* **4**, 355.
Jacob, J., and Grimmer, G. (1968). *J. Lipid Res.* **9**, 730.
Jennings, W. G. (1967). In "Chemistry and Physiology of Flavors" (H. W. Schultz, E. A. Day, and L. M. Libbey, eds.), AVI, Westport, Connecticut, pp. 419–430.
Jennings, W. G., Creveling, R. K., and Heinz, D. E. (1964). *J. Food Sci.* **29**, 730.
Johnson, R. R., Alford, E. D., and Kinzer, G. W. (1969). *J. Agr. Food Chem.* **17**, 22.
Jollow, D., Kellerman, G. M., and Linnane, A. W. (1968). *J. Cell Biol.* **37**, 221.
Jurriens, G., and Oele, J. M. (1965). *J. Amer. Oil Chem. Soc.* **42**, 857.
Kahn, J. H. (1969). *J. Ass. Offic. Anal. Chem.* **52**, 1166.
Kahn, J. H., LaRoe, E. G., and Conner, H. A. (1968). *J. Food Sci.* **33**, 395.
Kahn, J. H., Shipley, P. A., LaRoe, E. G., and Conner, H. A. (1969). *J. Food Sci.* **34**, 587.
Kates, M. (1964). *Advan. Lipid Res.* **2**, 17.
Kates, M. (1966). *Annu. Rev. Microbiol.* **20**, 13.
Kates, M., and Baxter, R. M. (1962). *Can. J. Biochem. Physiol.* **40**, 1213.
Kayahara, K., Mori, S., Taguchi, T., and Miyachi, N. (1964). *J. Ferment. Technol.* **42**, 623.
Kepner, R. E., and Webb, A. D. (1956). *Amer. J. Enol.* **7**, 8.
Kepner, R. E., and Webb, A. D. (1961). *Amer. J. Enol. Viticult.* **12**, 159.
Kepner, R. E., Strating, J., and Weurman, C. (1963). *J. Inst. Brewing* **69**, 399.

Kepner, R. E., Webb, A. D., and Maggiora, L. (1968). *Amer. J. Enol. Viticult.* **19**, 116.
Kepner, R. E., Webb, A. D., and Maggiora, L. (1969). *Amer. J. Enol. Viticult.* **20**, 25.
Kepner, R. E., Hunter, R. J., Muller, C. J., and Webb, A. D. (1972). Unpublished data.
Kinsella, J. E., Patton, S., and Dimmick, P. S. (1967). *J. Amer. Oil. Chem. Soc.* **44**, 202.
Koehler, P. E., Mason, M. E., and Newell, J. A. (1969). *J. Agr. Food Chem.* **17**, 393.
LaRoe, E. G., and Shipley, P. A. (1970). *J. Agr. Food Chem.* **18**, 174.
Lawrence, W. C. (1964). *Wallerstein Lab. Commun.* **27**, 123.
Lennarz, W. J. (1970). *Annu. Rev. Biochem.* **39**, 364.
Levy, E. J., and Stahl, W. A. (1961). *Anal. Chem.* **33**, 707.
Lewis, M. J., and Wildenradt, H. L. (1969). *Brewers Dig.* **44**, 9.
Liebich, H. M., Koenig, W. A., and Bayer, E. (1970). *J. Chromatog. Sci.* **8**, 527.
Lin, F. M., and Wilkens, W. F. (1970). *J. Food Sci.* **35**, 538.
Lynen, F. (1967). *Biochem. J.* **102**, 381.
Lyons, J. M., and Asmundson, C. M. (1965). *J. Amer. Oil Chem. Soc.* **42**, 1056.
Lyons, J. M., and Raison, J. K. (1970). *Plant Physiol.* **45**, 386.
Maarse, H., and ten Noever de Brauw, M. C. (1966). *J. Food Sci.* **31**, 951.
McKenna, E. J., and Kallio, R. E. (1965). *Annu. Rev. Biochem.* **19**, 183.
McKenna, E., and Coon, M. J. (1970). *J. Biol. Chem.* **245**, 3882.
Margalith, P., and Schwartz, Y. (1970). *Advan. Appl. Microbiol.* **12**, 35.
Marinelli, L., Feil, M. F., and Schait, A. (1968). *Amer. Soc. Brewing Chem., Proc.* p. 113.
Marr, A. G., and Ingraham, J. L. (1962). *J. Bacteriol.* **84**, 1260.
Masuda, M., and Nishimura, K. (1971). *Phytochemistry* **10**, 1401.
Moshonas, M. G., and Lund, E. D. (1969). *J. Agr. Food Chem.* **17**, 802.
Muck, G. A., Tobias, J., and Whitney, R. McL. (1963). *J. Dairy Sci.* **46**, 774.
Mudd, J. B. (1967). *Annu. Rev. Plant Physiol.* **18**, 229.
Muller, C. J., Kepner, R. E., and Webb, A. D. (1964). *J. Food Sci.* **29**, 569.
Muller, C. J., Maggiora, L., Kepner, R. E., and Webb, A. D. (1969). *J. Agr. Food Chem.* **17**, 1373.
Muller, C. J., Kepner, R. E., and Webb, A. D. (1971). *Amer. J. Enol. Viticult.* **22**, 156.
Muller, C. J., Kepner, R. E., and Webb, A. D. (1972a). *J. Agr. Food Chem.* **20**, 193.
Muller, C. J., Kepner, R. E., Webb, A. D., and Douglas, P. S. (1972b). Unpublished data.
Nawar, W. W. (1969). *J. Agr. Food Chem.* **17**, 18.
Nawar, W. W., Cancel, L. E., and Fagerson, I. S. (1962). *J. Dairy Sci.* **45**, 1172.
Neudoerffer, T. S., Sandler, S., Zubeckis, E., and Smith, N. D. (1965). *J. Agr. Food Chem.* **13**, 584.
Nishimura, K., and Masuda, M. (1971). *J. Food Sci.* **36**, 819.
Nykänen, L., and Suomalainen, H. (1963). *Tek. Kem. Aikak.* **20**, 789.
Nykänen, L., Puputti, E., and Suomalainen, H. (1966). *J. Inst. Brewing* **72**, 24.
Nykänen, L., Puputti, E., and Suomalainen, H. (1968). *J. Food Sci.* **33**, 88.
Obata, Y., and Tanaka, H. (1965). *Agr. Biol. Chem.* **29**, 104.
Otsuka, K., and Imai, S. (1964). *Agr. Biol. Chem.* **28**, 356.
Ough, C. S. (1971). *J. Agr. Food Chem.* **19**, 241.
Ough, C. S., and Bustos, O. (1969). *Wines Vines* **50**, 50.
Palamand, S. R., Nelson, G. D., and Hardwick, W. A. (1970a). *Amer. Soc. Brewing Chem., Proc.* p. 186.
Palamand, S. R., Nelson, G. D., and Hardwick, W. A. (1970b). *Tech. Quart. Master Brewers Ass. Amer.* **7**, 111.
Parks, O. W., Keeney, M., Katz, I., and Schwartz, D. P. (1964). *J. Lipid Res.* **5**, 232.
Parliament, T. H., Nawar, W. W., and Fagerson, I. S. (1966). *J. Dairy Sci.* **49**, 1109.
Pisarnitskii, A. F. (1966). *Prikl. Biokhim. Mikrobiol.* **2**, 215.

Power, F. B., and Chesnut, V. K. (1921). *J. Amer. Chem. Soc.* **43**, 1741.
Preobrazhenskii, A. A., and Kornelli, B. M. (1969). *Sadovod. Vinograd. Vinodel. Mold.* **24**, 32.
Puputti, E., and Suomalainen, H. (1969). *Mitt. Hoeheren Bundeslehr Versuchsanst Wein Ostbau Klosterneuberg.* **19**, 184.
Ronkainen, P., Brummer, S., and Suomalainen, H. (1970). *Amer. J. Enol. Viticult.* **21**, 136.
Schaefer, J., and Timms, R. (1970). *J. Food Sci.* **35**, 10.
Scheuerbrandt, G., and Bloch, K. (1962). *J. Biol. Chem.* **237**, 2064.
SentheShanmuganathan, S. (1960). *Biochem. J.* **74**, 568.
Sihto, E., Nykänen, L., and Suomalainen, H. (1962). *Tek. Kemian Aikak.* **19**, 753.
Singleton, V. L., and Esau, P. (1969). *Advan. Food Res. Suppl.* **1**.
Slaughter, J. C. (1970). *J. Inst. Brewing* **76**, 22.
Smith, D. E., and Coffman, J. R. (1960). *Anal. Chem.* **32**, 1733.
Smith, L. M., Freeman, N. K., and Jack, E. L. (1954). *J. Dairy Sci.* **37**, 399.
Sols, A., Gancedo, C., and De La Fuente, G. (1971). In "The Yeasts" (A. H. Rose and J. S. Harrison, eds.), Vol. 2, pp. 271–307. Academic Press, New York.
Stevens, K. L., Lee, A., McFadden, W. H., and Teranishi, R. (1965). *J. Food Sci.* **30**, 1006.
Stevens, K. L., Bomben, J. L., and McFadden, W. H. (1967). *J. Agr. Food Chem.* **15**, 378.
Stevens, K. L., Flath, R. A., and Alson, L. (1969). *J. Agr. Food Chem.* **17**, 1102.
Stokes, J. L. (1971). In "The Yeasts" (A. H. Rose and J. S. Harrison, eds.), Vol. 2, pp. 119–134. Academic Press, New York.
Stokoe, W. N. (1928). *Biochem. Z.* **22**, 80.
Strating, J., and Venema, A. (1961). *J. Inst. Brewing* **67**, 525.
Stumpf, P. K. (1965). In "Plant Biochemistry" (J. Bonner and J. E. Varner, eds.), pp. 322–345. Academic Press, New York.
Stumpf, P. K. (1969). *Annu. Rev. Biochem.* **38**, 159.
Sulser, H., DePizzol, J., and Büchi, W. (1967). *J. Food Sci.* **32**, 611.
Suomalainen, H. (1971). *J. Inst. Brewing* **77**, 164.
Suomalainen, H., and Keränen, A. J. A. (1968). *Chem. Phys. Lipids* **2**, 296.
Suomalainen, H., and Nykänen, L. (1966). *Suom. Kemistilehti B* **39**, 252.
Suomalainen, H., and Nykänen, L. (1970). *Naeringsmiddelindustrien* (1/2), 1.
Taira, T. (1963). *Nippon Nogei Kagaku Kaishi* **37**, 630.
Takayama, T., and Mizuuchi, T. (1966). *Hakko Kyokaishi* **24**, 267.
Tatum, J. H., Shaw, P. E., and Berry, R. E. (1969). *J. Agr. Food Chem.* **17**, 38.
Terry, T. D., Kepner, R. E., and Webb, A. D. (1960). *J. Chem. Eng. Data* **5**, 403.
Thoukis, G. (1958). *Amer. J. Enol. Viticult.* **9**, 161.
Van der Ven, B. (1964). *Rec. Trav. Chim. Pays-Bas* **83**, 976.
Van Wyk, C. J., Webb, A. D., and Kepner, R. E. (1967a). *J. Food Sci.* **32**, 660.
Van Wyk, C. J., Kepner, R. E., and Webb, A. D. (1967b). *J. Food Sci.* **32**, 664.
Van Wyk, C. J., Kepner, R. E., and Webb, A. D. (1967c). *J. Food Sci.* **32**, 669.
Visser, M. K., and Lindsay, R. C. (1971). *W. A. Tech. Quart. Master Brewers Ass. Amer.* **8**, 123.
Wagener, W. W. D., and Wagener, G. W. W. (1968). *S. Afr. J. Agr. Sci.* **11**, 605.
Webb, A. D. (1967). In "Chemistry and Physiology of Flavors" (H. W. Schultz, E. A. Day, and L. M. Libbey, eds.), AVI, Westport, Connecticut, pp. 203–227.
Webb, A. D., and Ingraham, J. L. (1963). *Adv. Appl. Microbiol.* **5**, 317.
Webb, A. D., and Kepner, R. E. (1957). *Food Res.* **22**, 384.

Webb, A. D., and Kepner, R. E. (1962). *Amer. J. Enol. Viticult.* **13**, 1.
Webb, A. D., Kepner, R. E., and Ikeda, R. M. (1952). *Anal. Chem.* **24**, 1944.
Webb, A. D., Ribéreau-Gayon, P., and Boidron, J. N. (1964a). *Bull. Soc. Chim. F.* p. 1415.
Webb, A. D., Kepner, R. E., and Galetto, W. G. (1964b). *Amer. J. Enol. Viticult.* **15**, 1.
Webb, A. D., Kepner, R. E., and Galetto, W. G. (1966). *Amer. J. Enol. Viticult.* **17**, 1.
Webb, A. D., Kepner, R. E., and Maggiora, L. (1967a). *J. Agr. Food Chem.* **15**, 334.
Webb, A. D., Kepner, R. E., and Maggiora, L. (1967b). *Amer. J. Enol. Viticult.* **18**, 190.
Webb, A. D., Kepner, R. E., and Maggiora, L. (1969). *Amer. J. Enol. Viticult.* **20**, 16.
Wenzel, K. W. O., and DeVries, M. J. (1968). *S. Afr. J. Agr. Sci.* **11**, 273.
Weurman, C., and Van Straten, S. (1969). "Lists of Volatile Compounds in Food," 2nd ed., mimeo. Central Inst. Nutr. Food Res., Zeist, The Netherlands.
Williams, A. A., and Tucknott, O. G. (1971). *J. Sci. Food Agr.* **22**, 264.
Wong, N. P., Patton, S., and Forss, D. A. (1958). *J. Dairy Sci.* **41**, 1699.
Yamada, M., Shitara, J., Yogawa, H., Komoda, H., Kosaki, M., and Yoshizawa, K. (1962). *J. Agr. Sci.* **7**, 97.
Yamada, M., Komoda, H., Yoneyama, H., and Yoshizawa, K. (1963). *J. Agr. Sci.* **9**, 142.
Yamaguchi, M., Kamibayashi, A., and Ono, H. (1966). *Kogyo Gijutsuin Hakko Kenkyusho Kenkyu Hokoku* **29**, 45.

Correlative Microbiological Assays

LADISLAV J. HAŇKA

*Research Laboratories, The Upjohn Company,
Kalamazoo, Michigan*

I.	Introduction	147
II.	Materials and Methods	148
III.	Results	149
IV.	Discussion	154
V.	Summary	156
	References	156

I. Introduction

A highly sensitive and specific assay is a very important tool in pharmacological and clinical evaluation of any new drug. The investigator engaged in the isolation of new drugs produced by a fermentation process needs such an assay to monitor the titers of the drugs during the development studies. Different types of chemical and microbiological assays are most frequently used. The principal advantages of a microbiological assay are often its extreme sensitivity and a high degree of specificity. It is not unusual to find microorganisms that can detect accurately concentrations of 0.1 µg/ml or less of a drug in biological materials. Microbiological assays offer several additional advantages. As a rule, the amount of sample required is very small. Such assays are rather fast and easy to perform. Another advantage is that they help to recognize early any known drugs that might be present in fermentation liquors. This identification of drugs is done by paper chromatography of such samples in several solvent systems, followed by bioautography against a sensitive microorganism. The resulting R_f values are compared with the values of known antibiotics. Since this last step is usually done by means of a computer, it is very fast.

Through experience, it was learned that a large percentage of biologically active compounds, e.g., antitumor, antiviral, or antifertility agents, tranquilizers, and cholesterol-lowering compounds, had antimicrobial properties as well (Aaronson et al., 1962; Haňka and Smith, 1962; Hutner et al., 1958; Sokolski et al., 1959; West et al., 1962). For compounds that have shown such an antimicrobial activity and for which no rapid and sensitive assay is available, a correlative microbiological assay can be developed. When working with pure materials, any antimicrobial activity can be used as the basis of a future correlative assay. However, when working with samples containing other materials (e.g., fermentation liquors), it is necessary to demonstrate first that it is the same component that is responsible for the primary biological activity and also inhibits certain microorganisms. This point will be demonstrated using a specific example later in the text.

II. Materials and Methods

The first step in the development of a correlative assay involves the testing of the compounds of interest against a broad spectrum of different microorganisms. Since any antimicrobial activity to be found is really incidental, it is not necessary to employ microorganisms of any clinical significance. Thus, the following criteria should be considered:

A. The organisms should be taxonomically different; they should have different nutritional requirements and utilize different metabolic pathways.

B. They should be fairly easy to cultivate and the time required for an assay should be short (1 or 2 days).

C. Preferably, they should not be pathogens.

The inoculum of each microorganism is grown in the proper liquid medium and used to inoculate the agar. Plastic trays (20 × 50 cm) are poured with 125 ml of the molten, inoculated agar and the agar is allowed to solidify. The solutions of the compounds to be tested are prepared, preferably in water, at 1 mg/ml. The solutions (0.08 ml) are applied to 12.7-mm paper disks (Schleicher and Schuell Co., Keene, N.H.). The disks are placed on the surface of the agar and the trays are then incubated. After the incubation, the diameters of the zones of inhibition are measured and recorded. When working with materials poorly soluble in water, it is often convenient to dissolve them in some organic solvent and apply the 0.08-ml volumes to the paper disks. The disks are then completely dried, to avoid any possible interference from the solvent, and applied to the inoculated agar.

In some cases, a tube dilution microbiological assay is used instead of the disk-plate assay. Then the optical density is used to evaluate the degree of inhibition. While this type of assay is more sensitive, it is primarily suitable for work with pure compounds.

A list of the microorganisms used in the correlative testing program is presented in Table I. A summary of the specific cultivation conditions for each of the microorganisms used in the program is presented in Tables II and III.

TABLE I
MICROORGANISMS USED IN THE CORRELATIVE ASSAYS

Microorganism	Taxonomic designation	Notes
Bacillus subtilis	G + bacterium	—
Lactobacillus casei	G + bacterium	Anaerobic growth
Sarcina lutea	G + bacterium	—
Staphylococcus aureus	G + bacterium	—
Streptococcus faecalis	G + bacterium	—
Mycobacterium phlei	G + bacterium	—
Propionibacterium thoenii	G + bacterium	—
Klebsiella pneumoniae	G − bacterium	—
Salmonella gallinarum	G − bacterium	—
Azotobacter vinelandii	G − bacterium	Nitrogen-fixing microorganism
Rhodopseudomonas spheroides	G − bacterium	Photosynthetic microorganism
Chromobacterium violaceum	G − bacterium	—
Saccharomyces cerevisiae	Yeast	—
Trigonopsis variabilis	Yeast	—
Torulopsis albida	Yeast	—
Glomerella cingulata	Filamentous fungus	—
Penicillium oxalicum	Filamentous fungus	—
Ochromonas danica	Protozoan	Photosynthetic microorganism
Crithidia fasciculata	Protozoan	—
Chlorella vulgaris	Alga	Photosynthetic microorganism
Prototheca zopfii	Alga	Does not have chlorophyll; heterotropic growth

III. Results

A specific example will be used to demonstrate the method of establishing the correlation between the primary biological activity and the correlative antimicrobial activity.

The fermentation liquors of a certain *Actinomyces* inhibited *Vaccinia* virus when tested in a modified Dulbecco assay (Siminoff, 1961). This liquor was subsequently tested against the broad spectrum of microorganisms in the correlative testing program and it rather strongly inhibited a yeast, *Saccharomyces pastorianus*. A correlative assay with *S. pastorianus* was very desirable because of its relative simplicity and speed compared to the regular antiviral assay. However, before such a correlative microbiological assay could be considered, it was necessary to prove that the same component of the fermentation liquor inhibited the *Vaccinia* virus in the primary assay system and *S. pastorianus* in the correlative assay. This was done by

TABLE II
Cultivation Conditions of the Microorganisms from Table I

| Microorganism | Preparation of inoculum ||| Assay conditions |||
| | Cultivation medium | Cultivation conditions | Rate of inoculation (%) | Assay medium | Incubation ||
					Time (hr)	Temperature (°C)
Bacillus subtilis	Spore suspension used conc. 1.5×10^{10} spores/ml		0.05	Nutrient agar	18–24	32
Lactobacillus casei	Fluid thioglycolate medium (BBL)	37°C, 24 hr, shaker	0.5	Synthetic agar (No. 1)	18–24	37
				Thioglycolate agar	18–24	37
Sarcina lutea		See footnote[a]	0.5	Penassay base agar (Difco)	18–24	32
Staphylococcus aureus	Penassay seed broth (Difco)	37°C, 24 hr, shaker	0.2	Nutrient agar	18–24	37
Streptococcus faecalis	Brain heart infusion broth	37°C, 24 hr, shaker	0.4	Brain heart infusion agar	18–24	37
Mycobacterium phlei	Broth No. 2	37°C, 24 hr, shaker	3.0	Brain heart infusion agar	18–24	37
Klebsiella pneumoniae	Penassay seed broth (Difco)	37°C, 24 hr, shaker	0.05	Streptomycin assay agar (Difco)	18–24	37
Salmonella gallinarum	Penassay seed broth (Difco)	37°C, 24 hr, shaker	0.2	Nutrient agar	18–24	37

Organism	Broth	Growth conditions	Inoculum	Assay medium	Hr	°C
Azotobacter vinelandii	Broth No. 3	See footnote[b]	1	Agar No. 3	18–24	28
Propionibacterium thoenii	Broth No. 4	28°, 48 hr, stationary	0.5	Agar No. 4	18–24	28
Rhodopseudomonas spheroides	1% Yeast extract broth	See footnote[c]	0.5	1% Yeast extract agar	18–24	28
Chromobacterium violaceum	Broth No. 5	28°, 24–36 hr, stationary	1	Agar No. 5	18–24	28
Saccharomyces cerevisiae	Broth No. 6	28°, 24 hr, shaker	0.6	Agar No. 6	18–24	28
Trigonopsis variabilis	Broth No. 7	28°, 24 hr, shaker	0.6	Agar No. 7	18–24	28
Torulopsis albida	Broth No. 6	28°, 24 hr, shaker	0.6	Agar No. 6	18–24	28
Glomerella cingulata	Broth No. 8	28°, 72 hr, shaker	1	Agar No. 8	48	28
Penicillium oxalicum	Broth No. 8	28°, 72 hr, shaker	1	Agar No. 8	18–24	28
Ochromonas danica	Broth No. 9	See footnote[d]	8	Agar No. 9	24–36	28
Crithidia fasciculata	Broth No. 10	28°, 24 hr, shaker	8	Agar No. 10	24–36	28
Chlorella vulgaris	Broth No. 11	28°, 24 hr, shaker	1	Agar No. 11	24	28
Prototheca zopfii	Broth No. 7	28°, 48 hr, shaker[e]	0.6	Agar No. 7	18–24	28

[a] *Sarcina lutea* inoculum is grown on the surface of Penassay seed agar (Difco) in Roux bottles. It is prepared at a concentration of 1.6×10^{10} cells/ml and stored frozen in liquid nitrogen. The technique was described by Stapert et al. (1964).
[b] *Azotobacter vinelandii* inoculum is incubated in a 125-ml Erlenmeyer flask using a 50-ml volume of the Broth No. 3. It is cultivated at room temperature on a magnetic stirrer for 2 to 3 days.
[c] *Rhodopseudomonas spheroides* is cultivated in 1% yeast extract at room temperature under fluorescent light. It is incubated until a deep red color develops (2 to 3 days). The assay trays are incubated under fluorescent light also.
[d] *Ochromonas danica* is also capable of photosynthesis. The inoculum is cultivated stationary under fluorescent light at 25–28° until a dark green color is present (2 to 3 days). The assay trays are incubated under fluorescent light also.
[e] Reciprocating shaker, 100 strokes per minute.

TABLE III
Composition of Cultivation Media from Table II

Ingredients	\multicolumn{11}{c}{Grams per liter}

Ingredients	No. 1	No. 2	No. 3	No. 4	No. 5	No. 6	No. 7	No. 8	No. 9	No. 10	No. 11
Dextrose	2.0	10.0	20.0	—	4.0	30.0	10.0	20.0	—	10.0	10.0
Sucrose	—	—	—	—	—	—	—	—	—	—	—
Malt extract	—	—	—	—	—	—	—	20.0	—	—	—
Yeast extract	—	1.0	—	10.0	1.5	7.0	2.5	—	5.0	—	3.0
Yeastolate (Difco)	—	—	—	—	—	—	—	—	—	2.0	—
Casitone (Difco)	—	—	—	—	—	—	—	—	5.0	—	5.0
Peptone (Difco)	—	4.0	—	—	—	—	—	1.0	5.0	—	5.0
Proteose-peptone (Difco)	—	—	—	—	—	—	—	—	—	10.0	—
Beef extract	—	4.0	—	—	—	—	—	—	—	—	—
Dried blood— soluble (NBC)	—	—	—	—	—	—	—	—	—	0.2	—
Na-lactate	—	—	—	20.0	—	—	—	—	—	—	—
Tween 80	—	10 ml	—	—	—	—	—	—	—	—	—
NaCl	—	2.5	—	—	—	—	—	—	—	—	—
$Na_2HPO_4 \cdot 7H_2O$	1.7	—	—	—	—	—	—	—	—	1.2	—
KH_2PO_4	2.0	—	0.2	—	2.0	5.0	1.0	—	—	0.8	—
K_2HPO_4	—	—	0.8	—	—	—	—	—	—	—	—
$(NH_4)_2SO_4$	1.0	—	—	—	—	—	—	—	—	—	—
$MgSO_4$	0.1	—	—	—	0.2	—	—	—	—	—	—
$MgCl_2 \cdot 6H_2O$	—	—	0.2	—	—	—	—	—	—	—	—
$CaSO_4 \cdot 2H_2O$	—	—	0.1	—	—	—	—	—	—	—	—
Fe_1Mo solution[a]	—	—	1 ml	—	—	—	—	—	—	—	—
Metallic ions solution[b]	1 ml	—	—	—	—	—	—	—	—	—	—
Neomycin	—	—	—	—	—	—	—	—	50 µg/ml	—	—
Agar	—	—	—	—	—	—	—	—	—	—	—
Final pH adjust to:	6.2	7.0	7.85	—	—	—	—	—	—	—	—

[a] Fe_1Mo Stock Solution:
  1. $FeCl_3 \cdot 6H_2O$ — 8.1 gm, q.s. to 50 ml
  2. $Na_2MoO_4 \cdot 2H_2O$ — 0.5 gm, q.s. to 100 ml
  Combine 3 ml solution 1 and 2.5 ml solution 2 and q.s. to 50 ml.

[b] Metallic ion stock solution:

Compound	Concentration	Compound	Concentration
$NaMoO_4 \cdot 2H_2O$	200 µg/ml	$CaCl_2$	25 mg/ml
$CoCl_2$	100 µg/ml	$FeCl_2 \cdot 4H_2O$	5 mg/ml
$CuSo_4$	100 µg/ml	$ZnCl_2$	5 mg/ml
$MnSo_4$	2 mg/ml		

(Note: $ZnCl_2$ has to be dissolved separately using a drop of 0.1 N HCl for 10 ml of water.)

The stock solution is heated to bring all the compounds in solution, kept standing for 24 hr, and sterile-filtered.

proper use of paper chromatography. The fermentation liquor was applied to paper strips that were developed in five suitable solvent systems. The developed paper strips were cut in half lengthwise and plated both on agar trays inoculated with *S. pastorianus* and on *Vaccinia*-infected chick kidney monolayers. The results of this study are shown in Fig. 1. Since the R_f values of the principal component were comparable in both assay systems, it was possible to substitute the more desirable microbiological assay with *S. pastorianus* for the isolation work of this antiviral drug.

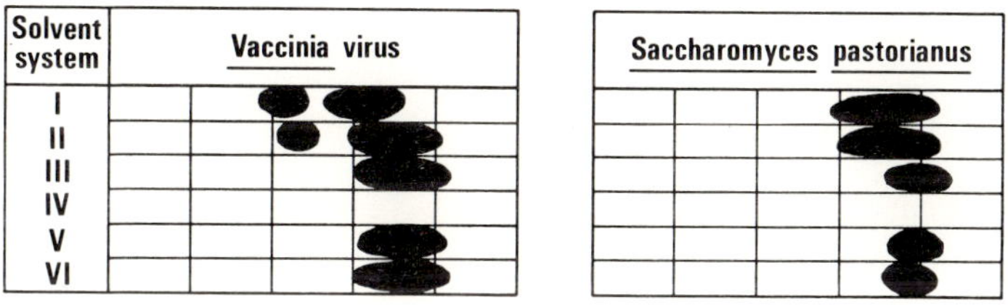

Solvent system	Vaccinia virus	Saccharomyces pastorianus
I		
II		
III		
IV		
V		
VI		

FIG. 1. Paper chromatograms of an antiviral activity on two assay systems (From Haňka and Smith, 1962).

Inhibition by different pharmacologically active compounds of several microorganisms is presented in Figs. 2, 3, and 4.

Since the sensitivity of the disc-plate assay for diethylstilbestrol with *Staphylococcus aureus* was only 100 µg/ml, a tube dilution assay

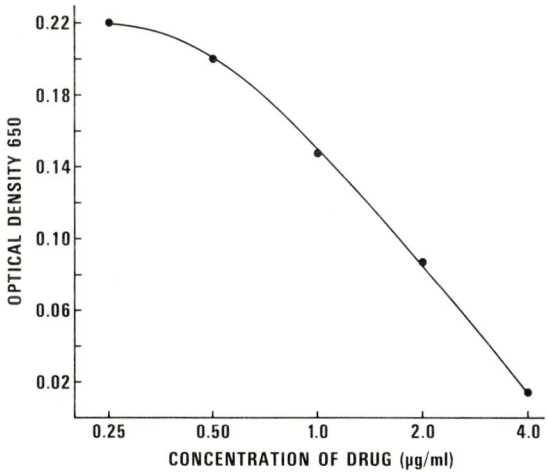

FIG. 2. Inhibition of *Propionibacterium thonii* in liquid medium by U-11, 555A, an antifertility agent (from Haňka and Smith, 1962).

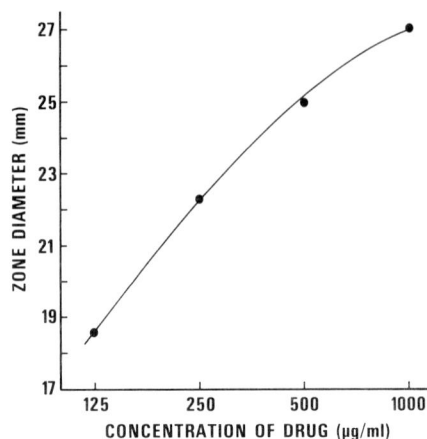

FIG. 3. Inhibition of a yeast, *Kloeckera apiculata*, on plates by a cholesterol-lowering agent (from Haňka and Smith, 1962).

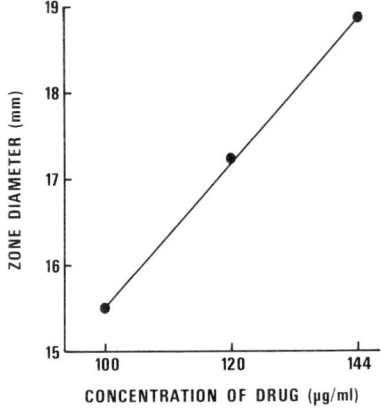

FIG. 4. Inhibition of *Staphylococcus aureus* ATCC 1359 on plates by diethylstilbestrol, a synthetic estrogen (from Haňka, 1958).

with the same microorganism was deemed desirable. As can be seen in Fig. 5, the sensitivity of *S. aureus* to the drug was substantially increased when it was cultivated in liquid medium (Haňka, 1958).

IV. Discussion

The potential usefulness of the correlative assays is best demonstrated in Figs. 2 through 4. The three compounds—an antifertility agent, a cholesterol depressant, and a synthetic estrogen—happen to

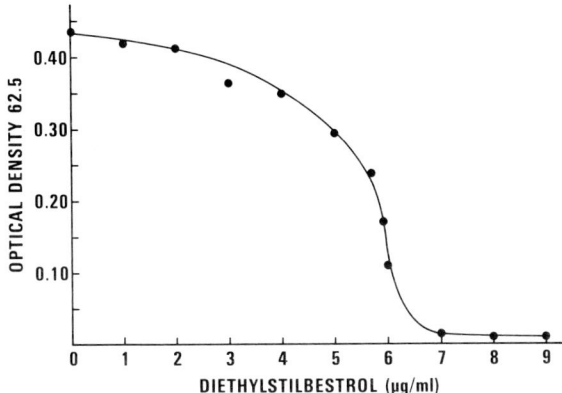

FIG. 5. Inhibition of broth cultures of *Staphylococcus aureus* by diethylstilbestrol (from Haňka, 1958).

demonstrate some antimicrobial properties. As a result, it was possible to develop convenient microbiological assays for such compounds.

The most sensitive among the microorganisms used in this program appears to be *Lactobacillus casei*. It is inhibited by over 50% of the samples tested. Furthermore, it tends to detect compounds that do not inhibit other microorganisms used in the program. A partial explanation might be that it is cultivated under anaerobic conditions. The next most sensitive microorganism was *Bacillus subtilis*, in particular when cultivated in a completely synthetic growth medium. The original reason for introducing *B. subtilis* and *Escherichia coli* cultivated in synthetic media into this program was their capacity to detect classes of compounds like antimetabolites, which do not inhibit the same test organism cultivated in complex nutrient media (Haňka, 1967). The least susceptible microorganisms to inhibition are the fungi and the protozoa.

The technique of correlative assays was first described in 1962 (Haňka and Smith, 1962) and has been used in our laboratory ever since. The spectrum of the microorganisms used was changed and improved. The correlative assays are used quite frequently to monitor the titers of fermentation liquors during culture development and during the isolation process. In particular, drugs with primary antitumor or antiviral activities, and produced by fermentation, were developed to pure materials, relying on correlative assays. Another area where correlative assays are used extensively involves the pharmacologic studies of pure drugs. A good recent example is the use of a microbiological assay with *Streptococcus faecalis* (Pittillo and

Hunt, 1967; Haňka et al., 1970) to monitor the blood and tissue levels of an antitumor drug, cytarabine.

V. Summary

Many pharmacologically active compounds (e.g., antiviral drugs, cholesterol depressants, tranquilizers, etc.) have often demonstrated antimicrobial activities as well. To detect a "correlative" antimicrobial activity, such compounds are tested against a broad spectrum of selected microorganisms. If a strong enough activity is found, a correlative microbiological assay can be developed to replace the original assay. Such microbiological assays offer several important advantages:

A. They are usually much faster than, for example, whole animal assays; they are easy to perform and to interpret; and they are, as a rule, very sensitive.

B. The amount of sample required for testing is very small.

C. The recognition of known compounds is greatly facilitated through the use of bioautography.

Acknowledgments

The author thanks the American Society for Microbiology for their permission to use Figs. 1, 2, and 3 which appeared in the 1962 issue of *Antimicrobial Agents and Chemotherapy* (pages 677–681).

References

Aaronson, S., Bensky, B., Shifrine, M., and Baker, H. (1962). *Proc. Soc. Expt. Biol. Med.* **109**, 130.
Haňka, L. J. (1958). Ph.D. Dissertation, Iowa State College.
Haňka, L. J. (1967). *Proc. Int. Congr. Chemother., 5th*, B9/2, p. 351
Haňka, L. J., and Smith, C. G. (1962). *Antimicrob. Ag. Chemother.* p. 677.
Haňka, L. J., Kuentzel, S. L., and Neil, G. L. (1970). *Cancer Chemother. Rept.* **54**, 353.
Hutner, S. H., Nathan, H. A., Aaronson, H., Baker, H., and Sher, S. (1958). *Ann. N.Y. Acad. Sci.* **76**, 457.
Pittillo, R. F., and Hunt, D. E. (1967). *Proc. Soc. Exp. Biol. Med.* **124**, 636.
Siminoff, P. (1961). *Appl. Microbiol.* **9**, 66.
Sokolski, W. T., Eillers, N. J., and Savage, G. M. (1959). *Antibiot. Annu.* p. 551.
Stapert, E. M., Sokolski, W. T., Kaneshiro, W. M., and Cole, R. J. (1964). *J. Bacteriol.* **88**, 532.
West, R. A., Jr., Barbera, P. W., Kolar, J. R., and Murrell, C. B. (1962). *J. Protozool.* **9**, 65.

Insect Tissue Culture

W. F. Hink

Department of Entomology,
The Ohio State University,
Columbus, Ohio

I.	Introduction	157
II.	Equipment and Sources of Supplies	158
III.	Media Formulation	160
IV.	Nutrition and Metabolism	166
V.	Primary Cultures	169
	A. Preparation of Insects	169
	B. Cell Dissociation	169
	C. Explant Technique	170
	D. Organs and Tissues for Primary Cultures	172
	E. Types of Cells and Outgrowths	174
	F. Differentiation and Maintenance of *in Vivo* Functions	177
VI.	Cell Lines	180
	A. Establishment of Cell Lines	180
	B. Characteristics of Cell Lines	185
VII.	Virus Studies	190
	A. Insect Viruses	190
	B. Arboviruses	197
	C. Other Animal Viruses	200
	D. Plant Viruses	201
VIII.	Protozoa Studies	202
IX.	Rickettsia Studies	202
X.	Interactions of Biologically Active Materials with Insect Cells and Tissues	203
	A. Hormones	203
	B. Insecticides	206
	C. Other Materials	207
XI.	Conclusion	207
	References	209

I. Introduction

During the 1960's and the first two years of the 1970's, the field of insect tissue culture has emerged as an important discipline. Evidence for recent progress in this area is that since the first established insect cell line was described in 1962 (Grace, 1962a), at least 81 cell lines from 34 different insect species have become available (Hink, 1972). These cell lines, along with primary explants, serve as excellent systems for investigating many biological phenomena. Insect tissue culture is no longer "an art in itself" but is a "tool" to be used in cellular biology, developmental biology, molecular biology, entomology, parasitology, microbiology, virology, toxicology, biochemistry, and genetics.

In this paper the terminology as defined by the Committee on Terminology of the Tissue Culture Association will be used (Fedoroff, 1967). Before beginning this discussion, a few definitions may be helpful. Tissue culture involves the study of organs, tissues, and cells removed from the body of the animal and grown or maintained *in vitro* for more than 24 hr. A primary explant or primary culture refers to cells, tissues, or organs taken directly from the animal and placed into culture media. A cell line develops when cells of the primary culture multiply and are transferred to another vessel. This may become an established cell line that is capable of multiplying indefinitely *in vitro*.

II. Equipment and Sources of Supplies

The maintenance and operation of a laboratory devoted to insect tissue culture requires the following equipment and supplies. Only the more important items that are in general use will be discussed. Mention of a specific product, manufacturer, and/or supplier does not indicate endorsement by the author.

Since tissue culture media consist of a rich mixture of nutrients, contamination of cultures with bacteria, fungi, and yeasts can be a problem. The use of antibiotics in the media is recommended only when absolutely necessary because their presence often masks microbial contamination. To minimize microbial contamination, all procedures must be done under aseptic conditions. A small room equipped with filters, for removing particulate material from the air, and germicidal ceiling lights should be used. As a further precaution, all manipulations that expose media bottles, cultures, or primary explants to the air should be done under a sterile transfer hood or, if done on the bench top, a mask should be worn.

Insect cells and tissues are usually cultured at 26–30°C. For maintenance of a specific temperature within this range, a water-jacketed incubator is preferred since this type provides a uniform temperature throughout the chamber.

The tissue culture media are sterilized by filtration. Pressure filtration is preferred and filters with a pore size of 0.22 μ or less should be employed. In many instances, relatively small amounts of media are required so a stainless steel unit with a volume of 100 ml is useful.[1] For smaller volumes of media, the Swinny filter holder from the above company is recommended.

---

[1] This is available from Millipore Corporation, Bedford, Massachusetts 01730.

Many insect tissue culture media must be made up by the individual investigator. However, Grace's (1962a) medium, Grace's medium as modified by Yunker et al. (1967), Schneider's (1964) revised Drosophila medium, Mitsuhashi and Maramorosch's (1964) medium, and Mitsuhashi's (1965) leafhopper medium are commercially available.[2]

Excised insect organs and tissues are usually quite small and therefore special culture vessels are often used. The standing drop technique is popular. A drop of medium is put on a depression plate or slide, the explant is introduced, a cover slip is placed over the depression, and it is sealed with sterile paraffin, wax, or petroleum jelly. A tissue culture chamber composed of a glass ring sandwiched between two cover slips is of extreme value if detailed observations of living cells at high magnification are desired. A sterile micro slide ring, 25 mm in diameter and 10 mm in height, is attached by means of melted paraffin to a sterile 30- × 35-mm No. 2 glass cover. This forms the chamber and a standing drop of medium is placed in the center. To minimize loss of H_2O from the drop during equilibration of the atmosphere in the chamber, several other drops should be placed in the vessel. The tissue or cells are placed in the drop of medium and the chamber is closed by adding the top cover glass. This glass is sealed onto the ring by petroleum jelly.[3] Two other commercially available chambers designed to permit study of cells at high magnifications are the Sykes-Moore (1959) tissue culture chamber and the Rose (1954) multipurpose tissue culture chamber. If high resolution is not necessary, glass or plastic petri dishes can be used as vessels for the standing drop technique. Of course, they must be sealed or incubated under a high humidity. Primary tissues have also been cultured on cover slips in Leighton tubes, T-flasks, glass prescription bottles, hanging drops, carrel flasks, and sealable glass culture dishes that resemble small petri dishes.[4] Most established insect cell lines are grown in glass T-flasks or polystyrene disposable tissue culture flasks.[5]

---

[2] Grand Island Biological Company, 3175 Staley Road, Grand Island, New York 14072. The vertebrate sera and extracts used to supplement the media may be purchased from the above company or: Microbiological Associates, Inc., 4733 Bethesda Avenue, Bethesda, Maryland 20014; Flow Laboratories, Inc., 12601 Twinbrook Parkway, Rockville, Maryland 20852; Difco Laboratories, Detroit, Michigan 48201; or other tissue culture supply houses.

[3] These aforementioned items are available from Arthur H. Thomas Company, Vine Street at Third, Philadelphia, Pennsylvania 19105.

[4] Most of these items are distributed by Bellco Glass, Inc., Vineland, New Jersey 08360.

[5] Falcon Plastics, 5500 West 83rd Street, Los Angeles, California 90045.

Since these cells are very sensitive to foreign organic and inorganic materials, a detergent designed for tissue culture should be used. Micro-Solv is one of these products and is distributed by Grand Island Biological Company. The water used for media formulation and rinsing glassware must contain as few impurities as possible. Pyrogen-free glass distilled water which has no organic material is recommended. Sterilize as much lab ware as possible with dry heat and autoclave only those items that would be damaged by dry heat. The reason for this is that steam of most autoclaves contains potentially toxic metal ions that are deposited on the lab ware.

In addition to the items already mentioned, a complete tissue culture laboratory should have an inverted phase-contrast microscope for observing cells that adhere to the bottom of tissue culture flasks and a standard phase-contrast microscope for routine examination. A centrifuge is needed for sedimenting cells when cultures are refed, for removing trypsin, and for many other techniques. A refrigerator and balance are also necessary.

III. Media Formulation

Insect tissue culture originated with the work of Goldschmidt (1915) when he observed spermatogenesis in testes of the moth, *Samia cecropia, in vitro*. The growth medium was a hanging drop of hemolymph in which the cells remained viable for 3 weeks. Historically, insect cells, tissues, or organs have been cultured in media consisting of: (1) hemolymph; (2) saline solutions plus hemolymph; (3) salts, sugars, and hemolymph; (4) minimal media containing amino acids, sugars, salts, organic acids, and vitamins at concentrations based on biochemical analyses of the hemolymph with supplements of insect tissue extracts, growth hormones, or hemolymph; (5) minimal media as above supplemented with vertebrate sera and extracts, protein fractions and hydrolyzates, yeast extracts, or tissue culture media designed for vertebrates; and (6) chemically defined media. Currently, many insect cells and tissues are cultured in media described in (5) above.

Insects possess an open circulatory system and organs are directly bathed in the hemolymph. Therefore, formulation of complex mixtures of organic and inorganic compounds in efforts to duplicate hemolymph seems to be a reasonably scientific approach to arriving at a tissue culture medium capable of supporting cell proliferation. However, this approach of attempting to biochemically duplicate hemolymph may be unsound because cells from one species will often multiply in media based upon the composition of hemolymph from another species. Insect hemolymph differs significantly from mammalian blood and therefore the tissue culture media differ. The concentrations of hemolymph amino acids are comparatively high. They may contribute up to 40% of the total osmotic pressure, while in man they contribute only 1%. Even though they are present in large quantities, many may not be required for *in vitro* cell growth (Hink *et al.*, 1972). They apparently function as buffers and reserves to be drawn upon when needed. Organic acids are present at high levels and these also have an important buffering capacity. Uric acid and organic phosphate values are higher than in mammalian blood. In some insects the Na:K ratio is inverted, i.e., there is more K^+ than Na^+. An important consideration is that the levels of some biochemical parameters fluctuate during development of the insect. The following examples will demonstrate the tremendous magnitude of these variations.

In the silkworm, *Bombyx mori*, glutamic acid is present at 230 mg/100 ml in last instar larval hemolymph, it falls to about 30 mg/100 ml when larvae are spinning their cocoons, and it rises again to 270 mg/100 ml in the pupal stage (Jeuniaux *et al.*, 1961). Over this same developmental period methionine levels rise from about 20 mg/100 ml to 110 mg/100 ml. In the wax moth, *Galleria mellonella*, hemolymph K^+ in sixth instar larvae is 105.2 mmoles, it decreases to 67.4 mmoles in last (seventh) instar larvae, reaches the lowest level of 45.6 mmoles in the prepuppae, and rises again to a value of 93.5 mmoles during the pupal stage (Hink, 1965).

Grace's (1962a) medium is an example of a medium that resembles hemolymph (Table I). It is based upon the composition of *B. mori* hemolymph as described by G. R. Wyatt *et al.* (1956). Before use, this medium was supplemented with hemolymph from heterologous or homologous insects. Currently, this medium is in common use and the hemolymph supplement is usually replaced with vertebrate sera, tissue extracts, or hydrolyzates. An example of this type of medium is shown in Table II.

The replacement of the hemolymph supplement with vertebrate sera and extracts was an achievement of practical importance because it eliminated the tedious job of bleeding insects and it also removed a potential source of viral contamination. Sen Gupta (1961) was one of the first investigators to suggest that hemolymph may not be required in culture media. The survival and growth of *Galleria mellonella* intestinal, ovarial, and testicular tissues were compared in hemolymph-supplemented media and similar medium without hemolymph

TABLE I
Grace (1962a) Insect Tissue Culture Medium (G.M.A.) for Moth, *Antheraea eucalypti*, Ovarian Cells

Ingredient	Amount (mg/100 ml)	Ingredient	Amount (mg/100 ml)
Salts		Amino acids (*Continued*)	
$NaH_2PO_4 \cdot 2H_2O$	114	L-Glutamine	60
$NaHCO_3$	35	L-Threonine	17.5
KCl	224	L.Valine	10
$CaCl_2$	100	Sugars	
$MgCl_2 \cdot 6H_2O$	228	Sucrose (in gm)	2.668
$MgSO_4 \cdot 7H_2O$	278	Fructose	40
Amino acids		Glucose	70
L-Arginine HCl	70	Organic acids	
L-Aspartic acid	35	Malic	67
L-Asparagine	35	α-Ketoglutaric	37
L-Alanine	22.5	Succinic	6
β-Alanine	20	Fumaric	5.5
L-Cystine HCl	2.5	Vitamins	
L-Glutamic acid	60	Thiamine HCl	0.002
L-Glycine	65	Riboflavin	0.002
L-Histidine	250	Ca pantothenate	0.002
L-Isoleucine	5	Pyridoxine HCl	0.002
L-Leucine	7.5	*p*-Aminobenzoic acid	0.002
L-Lysine HCl	62.5	Folic acid	0.002
L-Methionine	5	Niacin	0.002
L-Proline	35	Isoinositol	0.002
L-Phenylalanine	15	Biotin	0.001
DL-Serine	110	Choline chloride	0.02
L-Tyrosine	5	Antibiotics	
L-Tryptophan	10	Penicillin G, Na salt	3
		Streptomycin sulfate	10

but supplemented with calf serum, yeast extract, trehalose, organic acids, and glutathione. The growth in media supplemented with 5–20% hemolymph was more pronounced but the calf-serum-supplemented medium also supported growth for 1–3 weeks and cells survived for up to 5 weeks. The conclusion that hemolymph may not be an essential factor was arrived at because cells grew in the absence of hemolymph and variation in the hemolymph supplement between 5 and 20% did not greatly affect growth.

Vago and Chastang (1962a) also demonstrated that calf serum could be used as a replacement for hemolymph. *Bombyx mori* ovarian tissues were placed in different media supplemented with: (1) 15% hemolymph, (2) 10% hemolymph and 5% calf serum, (3) 5% hemo-

TABLE II
HINK (1970) INSECT TISSUE CULTURE MEDIUM
(TNM-FH) FOR CABBAGE LOOPER,
Trichoplusia ni, OVARIAN CELLS

Ingredient	Amount
Grace's insect tissue culture medium	90.0 ml
Fetal bovine serum (FBS)	8.0 ml
Chicken egg ultrafiltrate	8.0 ml
TC yeastolate	0.3 gm
Lactalbumin hydrolyzate	0.3 gm
Bovine plasma albumin, crystallized	0.5 gm

lymph and 10% serum, (4) 2% hemolymph and 13% serum, and (5) 15% serum. In the first four media, cell outgrowth began within 8 hr, cells were present in large numbers on the third day, and they continued to grow for 16–34 days. In the hemolymph-free medium, cell multiplication was slower and growth ceased after 11–12 days. Calf serum is also beneficial in the culture medium for ovarian explants of the mosquito, *Culex pipiens* (Kitamura, 1965). In minimal medium, ovaries survived for more than 3 months but no cell division occurred. Addition of calf serum to the minimal medium was effective in stimulating cellular migration and maintaining healthy cell sheets. Calf serum gave better results than chicken plasma or ovine serum.

Other substances, such as chick embryo extract, may be beneficial to cultured cells. Heart, gonadal, and imaginal tissues of the blowfly, *Calliphora erythrocephala*, and salivary glands of *Drosophila melanogaster* were cultured in a simple medium of salts, trehalose, oestradiol, and chick embryo extract (Demal and Leloup, 1963). The chick embryo extract was thought to be the growth-stimulating constituent. Vago and Chastang (1962b) also concluded that it was possible to grow primary cultures in media that lacked hemolymph or sera but contained lyophilized chick embryo extract. However, the cultures in medium with extract only did not live as long or multiply as rapidly. Lactalbumin hydrolyzate and yeast extract are other compounds that may efficiently replace hemolymph (Aizawa and Sato, 1963).

Apparently the first long-term culture of insect cells in hemolymph-free medium was achieved by Horikawa *et al.* (1966). They succeeded in growing cells from embryos of *D. melanogaster* for more than a year through 43 subcultures. Since then most insect cell lines have been adapted to, or were established directly in, hemolymph-free media. Most of the hemolymph-free media contain fetal bovine serum (FBS) as a hemolymph substitute.

In recent years, research in formulating new media has proceeded along two different routes. In some instances, the individual amino acids are replaced with hydrolyzates, and vitamins are replaced with yeastolate, a soluble portion of autolyzed yeast (Table III). These

TABLE III
MITSUHASHI (1967a) INSECT TISSUE CULTURE MEDIUM (CSM-2F) FOR RICE STEM BORER, *Chilo suppressalis*, HEMOCYTES

Ingredient	Amount (mg/100 ml)	Ingredient	Amount (mg/100 ml)
$NaH_2PO_4 \cdot H_2O$	50	Lactalbumin hydrolyzate	520
$MgCl_2 \cdot 6H_2O$	120	Bacto-peptone	520
$MgSO_4 \cdot 7H_2O$	160	TC-yeastolate	200
KCl	120	Choline chloride	40
$CaCl_2 \cdot 2H_2O$	40	TC-199	20 ml
Glucose	80	Fetal bovine serum	20 ml
Fructose	80	Dihydrostreptomycin sulfate	10

HSU *et al.* (1970) INSECT TISSUE CULTURE MEDIUM (721) FOR MOSQUITO, *Culex quinquefasciatus*, OVARIAN CELLS

Ingredient	Amount (mg/100 ml)		Amount (mg/100 ml)
KCl	80	Lactalbumin hydrolyzate	2000
NaCl	450	Bacto-peptone	500
$CaCl_2$	35	TC-yeastolate	200
$MgSO_4 \cdot 7H_2O$	40	l-Malic acid	60
KH_2PO_4	25	α-Ketoglutaric acid	40
$NaHCO_3$	100	Succinic acid	6
D-glucose	160	Fumaric acid	6
Sucrose	600	Medium No. TC-199 (IX)	20%
		FBS	10%

media still contain comparatively small amounts of individual amino acids and vitamins from the TC-199. This makes a more simple medium in terms of individual compounds but since these lysates are complex mixtures of amino acids, vitamins, etc., it introduces more chemically undefined substances. The other approach is to develop chemically defined media. Landureau and Jolles (1969) succeeded in growing the cockroach *Periplaneta americana* (EPa) cell line in a serum-free medium in which serum was replaced by protein fractions

(Table IV). As a further step toward a chemically defined medium, he replaced the protein fractions with cyanocobalamine at a level of 10^{-8} moles and 5 gm Ficoll per liter of medium (Landureau, 1970).

TABLE IV
LANDUREAU AND JOLLES (1969) INSECT TISSUE CULTURE MEDIUM (S 19)
FOR COCKROACH, *Periplaneta americana*, EMBRYONIC CELLS

Ingredient	Amount (mmoles/liter)	Ingredient	Amount (mmoles/liter)
Salts		Amino acids (*Continued*)	
$CaCl_2$	4.4	L-Threonine	1.68
KCl	14.0	L-Tryptophan	0.99
$MgSO_4 \cdot 7H_2O$	5.1	L-Tyrosine	1.0
$MnSO_4 \cdot H_2O$	0.38	L-Valine	1.28
NaCl	145.4	Sugars	
$NaHCO_3$	4.3	Glucose	16.7
PO_3H_3	11.0	Vitamins	
Amino acids		d-Biotin	0.00004
α-Alanine	1.35	Ca pantothenate	0.0002
L-Arginine HCl	3.8	Choline HCl	0.00254
L-Aspartic acid	1.88	Inositol	0.0003
L-Cysteine HCl	1.65	Nicotinamide	0.000025
L-Glutamine	2.06	Pyridoxine HCl	0.00015
L-Glutamic acid	10.2	Riboflavin	0.00013
L-Glycine	10.0	Thiamine HCl	0.00003
L-Histidine	1.94	Protein fractions	
L-Isoleucine	0.92	Human serum albumin,	
L-Leucine	1.91	fraction V	4.0 g/l
L-Lysine HCl	0.88		
L-Methionine	3.35	Human α-2-	
L-Phenylalanine	1.21	macroglobulin	50 mg/l
L-Proline	6.53		
L-Serine	0.76		

Osmotic pressure, pH, and oxygen tension are important parameters to consider in media formulation. The osmotic pressure should approximate that of the hemolymph. This varies intraspecifically and within the same species depending upon stage and physiological condition. The pH should also be the same as the hemolymph, which for most insects is slightly acid. Insect blood does not function in oxygen transport and oxygen tension of media should be low to reflect this characteristic.

In the preparation of primary explants and the handling of cell lines, saline solutions are often employed (Table V). These salines should be buffered and should contain ions at the levels present in hemolymph. Their osmotic pressures should also approach that of the blood.

TABLE V
INSECT SALINES

Compound (mg/liter)	PBS[a]	Drosophila PBS[b]	P. saucia saline[c]	Rinaldini's saline[d]
NaCl	8000.0	3038.88	–	8000.0
KCl	200.0	2984.0	6262.0	200.0
$CaCl_2$	100.0	110.99	2553.0	–
$MgCl_2 \cdot 6H_2O$	100.0	243.97	3660.0	–
$H_2(HPO_3)$	–	–	1149.0	–
$MgSO_4 \cdot 7H_2O$	–	295.78	–	–
KH_2PO_4	200.0	503.55	–	–
Na_2HPO_4	1150.0	716.28	–	–
NaH_2PO_4	–	–	–	40.0
Sodium citrate	–	–	–	676.0
$NaHCO_3$	–	–	–	1000.0
NaOH	–	–	1068.0	–
L-Ascorbic acid	–	–	1000.0	–
Glucose	–	1802.6	–	1000.0
Sucrose	–	34230.0	–	–
H_2O	to 1000 ml	to 1000 ml	to 1000 ml	1000 ml

[a] Dulbecco and Vogt (1954).
[b] Robb (1969).
[c] Martignoni and Scallion (1961a).
[d] Rinaldini (1953).

IV. Nutrition and Metabolism

In vitro cellular metabolism and nutrient utilization must be investigated and understood so that the essential factors in media may be delineated and chemically defined or more adequate media developed. Understanding metabolic conversions would lead to distinguishing the compounds that are synthesized by cells and therefore unnecessary media constituents. The noncatabolized substances could also be eliminated. Information of this type will also provide data on basic biochemical processes and point out metabolic differences between cells from different tissues, different insect species, and invertebrate and vertebrate cells.

Insects, unlike vertebrates, are incapable of synthesizing sterols from precursor compounds. Sterols in most insect cell culture media are in the FBS and bound to bovine serum albumin. The only fact available concerning sterol metabolism *in vitro* is that cholesterol is taken in by the *Antheraea eucalypti* cells (Vaughn *et al.*, 1971).

Most culture media contain three sugars: glucose, fructose, and sucrose. A few formulas include trehalose, a sugar present in insect hemolymph that may constitute up to 90% of the total hemolymph sugars. The *A. eucalypti* cells metabolize glucose at approximately the same rate as trehalose. Glucose is used more rapidly than fructose and has disappeared by the fifth day of culture. Sucrose utilization begins, in growing cultures, after the glucose level has fallen to about one-half of its original level (Clements and Grace, 1967). In contrast to the *A. eucalypti* cell line, two lines of *Carpocapsa pomonella* cells, cultured in medium with similar sugar levels, do not use sucrose (Hink *et al.*, 1972). With these two lines, fructose and glucose were depleted after 8 days in culture. These two sugars are growth limiting as omission of them caused a drastic reduction in cell multiplication with vacuolation, clumping, and lysing of cells. A twofold increase in the concentration of medium fructose and glucose resulted in higher cell populations with an extension in the log phase of growth.

Amino acid requirements and utilization have been studied in several cell lines. The EPa line of *Periplaneta americana* cells requires 16 amino acids to ensure continued survival (Landureau and Jolles, 1969). All but one of the amino acids in the medium decreased as the cultures grew. At the end of a 7-day growth period, the α-alanine value had increased about 5 times over the original value in fresh medium. Extensive transamination occurred between α-alanine, aspartic acid, and glutamic acid. Grace and Brzostowski (1966) reported that the levels of 14 of 21 amino acids decreased as the *A. eucalypti* cell population increased. L-Alanine increased in the medium and the other 6 amino acids remained at their original levels. The *Carpocapsa pomonella* cells probably require fewer amino acids than the two lines previously mentioned since fewer amino acids decrease during growth of these cells (Hink *et al.*, 1972). Only aspartic acid, glutamic acid, 1/2 cystine, methionine, and tyrosine disappear from media supporting growth of both *C. pomonella* lines. The patterns of methionine decrease from the media are different in the two *C. pomonella* lines and proline is utilized by one line and not by the other. While these studies are rather preliminary and different media were used for lines from different insects, these results suggest differences in amino acid metabolism and therefore biochemical differences among established insect cell lines.

Most insects require the B vitamins in their diet and this suggests they might be essential in tissue culture medium. The *P. americana*

(EPa) line requires Ca pantothenate, cyanocobalamine (vitamin B_{12}), riboflavin (vitamin B_2), inositol, and thiamine for cell survival. These vitamins plus choline, pyridoxine (vitamin B_6), folic acid, nicotinic acid, and biotin must be incorporated into the medium to support optimum cell multiplication (Landureau, 1969). Cyanocobalamine is one of the more important vitamins and EPa cells growing in serum-free medium were stimulated by addition of this compound (Landureau and Steinbuck, 1969). In very slow growing cultures, that have been deprived of serum and protein fraction V, the cells begin more active growth upon addition of cyanocobalamine. Choline is another extremely beneficial vitamin. At the concentration of 50 mg/liter, the maximum population of Grace's *Aedes aegypti* cells was increased by 57% over that obtained in medium with 1 mg/liter (Nagle, 1969). Folate also has pronounced effects on insect cells. This compound and an analogue, N-10-formylate, promoted cell migration and outgrowth from primary cultures of wax moth and silkworm ovarian tissue (Zielinska and Saska, 1972).

Certain proteins are essential for supporting the growth of cultured insect cells. The protein source is often supplied by FBS. Using the *Aedes aegypti* cells, Kuno et al. (1971) demonstrated that the non-dialyzable fraction of FBS retained approximately 50% of the growth-promoting activity present in complete FBS. Gel filtration techniques revealed that fractions with high concentrations of proteins were most active in stimulating cell multiplication. Treatment of protein fractions with 6 M urea and pronase caused a loss of stimulatory activity. Cohn's cold ethanol fractionation of FBS proteins yielded only one active fraction, precipitate V. Precipitates I–IV, α-globulin, and transferrin did not promote cell multiplication. Disc electrophoresis of fresh medium and medium from 7-day-old cultures showed that 13% of proteins in the albumin region and 67% of proteins in the fetuin region had disappeared. These results demonstrated that specific proteins or moieties bound to the proteins are growth promoting.

The apparent need for protein is also exhibited by EPa cells (Landureau and Steinbuck, 1969). In serum-free medium, these cells required 4 gm human serum fraction V per liter to sustain cell growth. However, this protein fraction contained impurities. Crystallized pure albumin would not support cell growth. This, along with other evidence, suggests that the activity of this fraction may reside in substances such as vitamins that are bound to the proteins. The serum proteins may also serve as antiproteases and protect the cells by virtue of this function.

V. Primary Cultures

A. PREPARATION OF INSECTS

The tissues employed for primary cultures must be sterile or have a very low microbial population. The inclusion of antibiotics, usually penicillin and streptomycin, in culture media will reduce contamination problems but antibiotics should be used only when absolutely necessary. Yeasts and fungi are the most difficult contaminants to control.

Aseptically reared insects are ideal sources of tissues. However, if they are unavailable, the surface of the donor must be sterilized prior to making the incision for removing the tissue. If, after surface sterilization, the insect is placed in saline during the dissection procedure it may be desirable to seal the mouth and anus by ligation or melted wax so that regurgitation or defecation will not contaminate the tissues.

Many disinfectants are used for surface sterilization. The treatments range from immersing insects in 70 to 95% EtOH to involved procedures in which housefly, *Musca domestica,* eggs are disinfected and dechorinated by immersion in a 1/750 solution of benzalkonium chloride, transferred to 5% formalin for 5 minutes, rinsed in sterile H_2O, put in 0.005% NaOCl for 2 minutes, and lastly treated with White's solution (0.25 gm $HgCl_2$, 6.5 mg NaCl, 1.25 ml HCl, 250 ml EtOH, and 750 ml H_2O) for 1 hr (Wallis and Lite, 1970). Sodium hypochlorite (NaOCl) is a widely used compound and addition of 1–2 drops of a wetting agent, such as Triton X-100 (Emulsion Engineering, Inc.), to 100 ml disinfectant greatly increases its effectiveness. At concentrations of 1.0–2.5% it sterilizes and removes the egg chorion all in one step (Greenberg and Archetti, 1969; Echalier and Ohanessian, 1970). Other materials and insects treated include: adult mosquitoes immersed in 0.25% $HgCl_2$ in 70% EtOH for 5 minutes (Kitamura, 1965), leafhopper eggs treated with 70% EtOH for 1 minute followed by 0.1% Hyamine 2389 (Rohm and Haas Co.) for 10 minutes (Hirumi and Maramorosch, 1964), moth larvae immersed in 0.1% Quadramine X-100 (Soden Chemicals) for 1 minute (Sohi, 1968), mosquito eggs in 10% benzalkonium chloride for 1 hr and into 70% EtOH for 20 minutes (Schneider, 1971a), and adult moths exposed to 0.5% NaOCl for 5 minutes (Hink, 1970; Hink and Ignoffo, 1970).

B. CELL DISSOCIATION

Depending upon the objectives of the study, the tissues may or may not be treated with enzymes to disperse the cells. The larval integument of the variegated cutworm, *Peridroma saucia,* was treated with trypsin and Versene (ethylenediaminetetraacetate) but the liberated cells were damaged (Martignoni *et al.,* 1958). Undamaged cells were obtained by using an extract from the hepatopancreas of the snail, *Helix aspersa.* In another study, agitation, trypsin, Versene, hyaluronidase, and hepatopancreas extract of *H. aspersa* were all evaluated for separating cells from epidermal, intestinal, and gonadal tissues of 4 species of Lepidoptera (Aizawa and Vago, 1959a). Trypsin, hyaluronidase, and Versene were destructive at concentrations used in vertebrate tissue culture. Weaker concentrations were not detrimental and dispersed cells formed monolayers. Cells dispersed by agitation also attached to the culture vessel and those treated with snail extract produced irregular results. *Drosophila melanogaster* embryonic tissues were dissociated by placing them in Ca- and Mg-free saline containing 0.12% Versene and 0.15% crude trypsin for 15 minutes (Lesseps, 1965). The age of the embryo was important. Embryos up to 11 hr old yielded uniform suspensions of single cells and older embryos produced large undissociated groups of cells. Treatment of leafhopper, *Agallia constricta,* embryos with 0.25% trypsin did not produce much cell dissociation but improved initial outgrowth of cells from primary explants (Chiu *et al.,* 1966).

Many investigators use various concentrations of trypsin in Rinaldini (1953) salt solution (RSS). For composition of RSS see Table V. Leafhopper embryos were exposed to 0.25% trypsin in RSS for 15 minutes (Mitsuhashi and Maramorosch, 1964), newly hatched mosquito larvae were minced in 0.2% trypsin in RSS and incubated for 15–30 minutes in fresh trypsin-RSS solution (Schneider, 1971a), and mosquito eggs were minced in 0.1% trypsin in RSS (Singh and Bhat, 1969). Of course, care must be taken to ensure that the enzyme is removed or inactivated by addition of excess FBS before placing the cells in the culture vessel. For some purposes, disruption of tissues by mincing, trituration, or pipetting is as effective as enzyme treatment. Long-term cell cultures from several insect species have been derived from such mechanically dissociated tissues.

C. Explant Technique

The following is a stepwise detailed procedure for setting up a primary culture in 5.0 ml of culture medium. The example chosen is ovarian tissue from an adult moth, but the technique is applicable, with some modifications, to other tissues. All procedures are done under a sterile hood and dissecting tools are dipped in 95% EtOH and flamed after each manipulation.

I. Preparation of insect
   A. Cut off wings and legs of newly emerged adult moth
   B. Submerge the abdomen and thorax in 0.5% NaOCl + wetting agent for 5 minutes.
   C. Pin insect on a sterile paraffin-filled petri dish with the ventral side up.
II. Dissection
   A. Make incision along ventral midline of the abdomen. Start at the tip of the abdomen and cut up between the leg stumps.
   B. Pull epidermis back so that internal organs are exposed. Pin epidermis to paraffin.
   C. Sever the lateral oviducts and remove both ovaries.
III. Treatment of ovaries
   A. Place both ovaries in a depression plate filled with a balanced salt solution (BSS) or tissue culture medium.
   B. Remove adhering tissue debris from ovaries.
   C. Transfer ovaries to new depression that contains 0.25% trypsin in BSS or RSS.
   D. Mince ovaries with scissors.
   E. Pipette minced ovaries up and down in a Pasteur pipette for 10 minutes. This facilitates dissociation by flushing the tissues through the tip of the pipette.
   F. Transfer suspended tissues and cells to a 12- × 75-mm test tube.
IV. Removal of trypsin
   A. Centrifuge suspension at 650 R.C.F. for 7 minutes.
   B. Remove supernatant and resuspend tissues and cells in 1.5 ml tissue culture medium.
   C. Repeat washing by centrifugation 2 more times.
   D. Resuspend the tissues in 1.5 ml medium and transfer to a 30-ml disposable tissue culture flask (Falcon Plastics). Add 3.5 ml more medium to the culture flask which brings the volume to 5.0 ml.
V. Refeeding the primary cultures
   A. Four-fifths of the old medium is replaced with fresh medium at 10- to 14-day intervals.
   B. Pipette 4.0 ml medium from the tissue culture flask and transfer it to a 16- × 100-mm test tube.
   C. Add 2.0 ml fresh medium to the culture flask to ensure that attached cells remain covered with medium.
   D. Centrifuge the old medium, which contains some cells and tissues that haven't attached, at 650 R.C.F. for 10 minutes.

E. Remove the supernatant and resuspend tissues in 2.0 ml medium.

F. Transfer suspension back to the tissue culture flask.

These techniques have been used by this author to establish long-term cultures from 3 species of insects.

D. ORGANS AND TISSUES FOR PRIMARY CULTURES

1. Embryonic Tissue

The culture of intact whole embryos was reviewed by Counce (1966). This discussion will not center around development and survival of intact embryos *in vitro*, but will be restricted to growth of the embryonic cells originating from dissociated embryonic tissues. These tissues are prime candidates for *in vitro* growth because they are in a state of active multiplication.

Six-hour old eggs from the housefly, *Musca domestica,* were crushed and put through a procedure which furnished a suspension of single cells (Eide and Chang, 1969). Cells attached to the culture vessel within 4–6 hr after initiation of the culture and three morphologically distinct cell types were present. The cells aggregated into clumps of muscle-like and nerve-like cells. They were subcultured and maintained for a maximum of 1 year. Four-hour-old eggs produced outgrowths with a greater variety of cell types than the larvae, pupae, or adults (Greenberg and Archetti, 1969). Wallis and Lite (1970) also cultured housefly embryonic cells but noted that pupal tissues grew better.

Cells from *D. melanogaster* eggs continued to multiply and were subcultured (Horikawa and Fox, 1964; Horikawa *et al.*, 1966). Growth potential of cells is related to the age of the embryo as 2-hr embryos produced no growth and 8-hr embryonic cells entered logarithmic growth. These cells apparently dedifferentiate *in vitro* while Seecof and Unanue (1968) observed that dispersed morphologically simple cells from 3- to 7-hr embryos differentiate into nerve-like and muscle-like cells. Echalier and Ohanessian (1969) isolated eight cell lines from 12-hr mechanically dissociated embryos.

Eggs with fully formed embryos of the mosquito, *Anopheles stephensi,* provide an excellent source of cells (Singh and Bhat, 1969). Within several hours the cells had attached to the vessel, after 24 hr many had produced thin protoplasmic extensions, by day three or four different morphological types were present, and after 10 days most of the glass surface was covered with cells.

Leafhopper embryonic tissues have been successfully cultivated as primary cultures (Hirumi and Maramorosch, 1964) and developed into cell lines (Chiu and Black, 1967; Richardson and Jensen, 1971; Peters and Spaasen, 1972). Primary explants from trypsinized cockroach embryos, *Periplaneta americana, Blabera fusca,* and *Blattella germanica,* were subcultured and established as cell lines (Landureau, 1966, 1968). Embryonic tissues from such diverse species as the yellow-fever mosquito, honeybee, spruce budworm, and conenose bug have also been cultured (Peleg, 1966; Giauffret *et al.,* 1967; Sohi, 1968; Bhat and Singh, 1970).

In most primary cultures, the cells begin to attach to the culture vessels within several hours and cell migration and multiplication is followed by observing these cells. The embryonic cells from the codling moth, *Carpocapsa pomonella,* did not follow this pattern (Hink and Ellis, 1971). They did not attach to the culture flasks but remained suspended. After 1 month, many layers of cells had formed on the exterior of the suspended tissue fragments and they then descended to the bottom of the flask and became attached. These cells multiplied and produced monolayers and cell lines.

2. Gonadal Tissue

With Lepidopterous insects, of all tissues tested, the ovarian tissues have been most successfully cultured. The tissues may come from larvae, pupae, or adults. The monumental and pioneering work of S. S. Wyatt (1956) demonstrated that *Bombyx mori* ovarian tissue cells would exhibit rapid mitosis in medium formulated to resemble the composition of hemolymph. This medium formulation, as mentioned in Section III, has been used as a guideline for many subsequent insect tissue culture media. Following Wyatt's findings, long-term cultures and cell lines were established from larval ovaries of *B. mori* (Gaw *et al.,* 1959; Grace, 1967; Vago and Chastang, 1958). Hink (1970) and Hink and Ignoffo (1970) succeeded in establishing cell lines from adult ovaries of the cabbage looper, *Trichoplusia ni,* and the corn earworm, *Heliothis zea.* The cells proliferating from ovarian explants appear to originate from the intermediate layer and follicular epithelium (Stanley and Vaughn, 1968).

Ovaries from 3 species of adult mosquitoes produce proliferating hemocyte-like cells that probably adhered to the surface of the explants when they were removed from the donor (Kitamura, 1966). Undissociated ovaries of the mosquito, *Culex molestus,* began cellular growth 20 hr after the culture was started (Kitamura, 1970). About 30 days after initiation of the culture, cell growth stopped and cultures remained inactive for about 5 months. They were refed weekly during this period. Six months after setup of the culture, cell multiplication began again and by the eighth month the first subculture was made. Ovaries from other moths, mosquitoes, plant bugs, and beetles have also been cultured (Grace, 1959; Hsu et al., 1970; Ittycheriah and Stephanos, 1969; Lender and Laverdure, 1967).

3. Hemocytes

In most insects, the blood cells multiply throughout the life span of the immature stages and therefore they are promising sources of proliferating cells. Hemocytes from *Peridroma saucia* formed monolayers that survived for 10–15 days (Martignoni and Scallion, 1961a). The prohemocytes, which are probably the stem cells, from 3 species of lepidopterous larvae were the only type of hemocyte that multiplied *in vitro* (Kurtti and Brooks, 1970a). The other three morphologically distinguishable kinds of blood cells never divided but the mixed populations were maintained for 100 days. Hemocytes from the gypsy moth, *Lymantria dispar,* and the earwig, *Forficula auricularia,* have also been maintained *in vitro* (Mazzone, 1971; Arvy and Gabe, 1946). Several of these investigators mentioned the problem of melanization which is a product of tyrosinase activity and results in the production of potentially toxic quinones. Melanization of cultures should generally be avoided and the use of 5×10^{-3} M cysteine (Mazzone, 1968) or 6 mM l-ascorbic acid (Martignoni and Scallion, 1961a) is recommended. Cell lines have been established from hemocytes of the rice stem borer, *Chilo suppressalis* (Mitsuhashi, 1967a), the forest tent caterpillar, *Malacosoma disstria* (Sohi, 1972b), and the cynthia moth, *Samia cynthia* (Chao and Ball, 1971).

4. Macerated Insects

If the objective of the individual study is to obtain cells capable of long-term growth, culture of macerated insects may be the most rewarding approach. Reports within the last 6 years describe cell lines from 10 species of mosquitoes that have originated from minced larvae, pupae, or adults (Table V and Section VI, A). One of the major problems with this approach is that the origin of the cells is unknown.

Primary cultures of imaginal discs (see also Section X, A), cuticle, trachea, nerve ganglia, differentiating single cells, etc., will be discussed in Section V, F under Differentiation and Maintenance of *in vivo* Functions.

E. Types of Cells and Outgrowths

Different types of cells are observed in primary cultures and they are often classified according to morphological characteristics. However, one must remember that the shape and general appearance of these cells are constantly changing so grouping of cells according to morphology has some obvious limitations. Examples of these dynamic processes were recorded by Suitor *et al.* (1966) and Eide and Chang (1969) with *Aedes aegypti* and *Musca domestica* cells respectively. Many cells are morphologically similar to those seen in cultures of vertebrate cells and tissues but in most cases their homologies are uncertain.

Fibroblast cells are mesenchymal, form a loose network, are polarized, and partly contiguous in vertebrate tissue culture (Willmer, 1965). Insect cultures contain attached fibroblast-like cells that are often the predominant type. These cells are usually spindle-shaped and they may or may not form an oriented pattern during migration from the explant (Fig. 1a). They may produce cytoplasmic cross-bridges and form an open cell network. These cells were usually the first type to migrate from embryonic tissue of leafhoppers and were followed by epithelial-like cells (Mitsuhashi and Maramorosch, 1964). The epithelial-like cells usually outnumbered the fibroblast-like cells and the former produced contiguous cell sheets while the latter formed networks. These fibroblast-like cells were often in close association with muscle cells in primary cultures of *Anopheles stephensi* embryos (Singh and Bhat, 1969).

Epithelial-like cells are common. They are flattened and their cell membranes are in contact with one another thereby forming a sheet of cells (Fig. 1b). This epithelial morphology is not a valid criterion for assigning an epithelial origin to these cells.

Nerve-like cells are rather spectacular in appearance (Fig. 1c). They possess extremely long thin protoplasmic extensions that may extend up to 1000 μ. The main fibers branch repeatedly and form many cross connections. The developmental pattern of similar nerve fibers was described by Seecof and Unanue (1968), Chen and Levi-Montalcini (1969), and Shields and Sang (1970).

Muscle-like cells were the principal cell type in primary cultures of *Anopheles stephensi* embryos (Singh and Bhat, 1969) and they are also present in cultures of other insect tissue (Kurtti and Brooks, 1970b). Those from *A. stephensi* have thin protoplasmic processes and one to many nuclei. They grow over one another and form extensive networks. Dark and light cross bands were observed, and the cells exhibit rhythmic or spontaneous contractions. In 10- to 15-day-old cultures, these cells had spread out over the surface of the culture flask and attained lengths of 80–300 μ.

Wandering cells that resemble hemocytes migrate from leafhopper embryos (Mitsuhashi and Maramorosch, 1964). Many cultures also

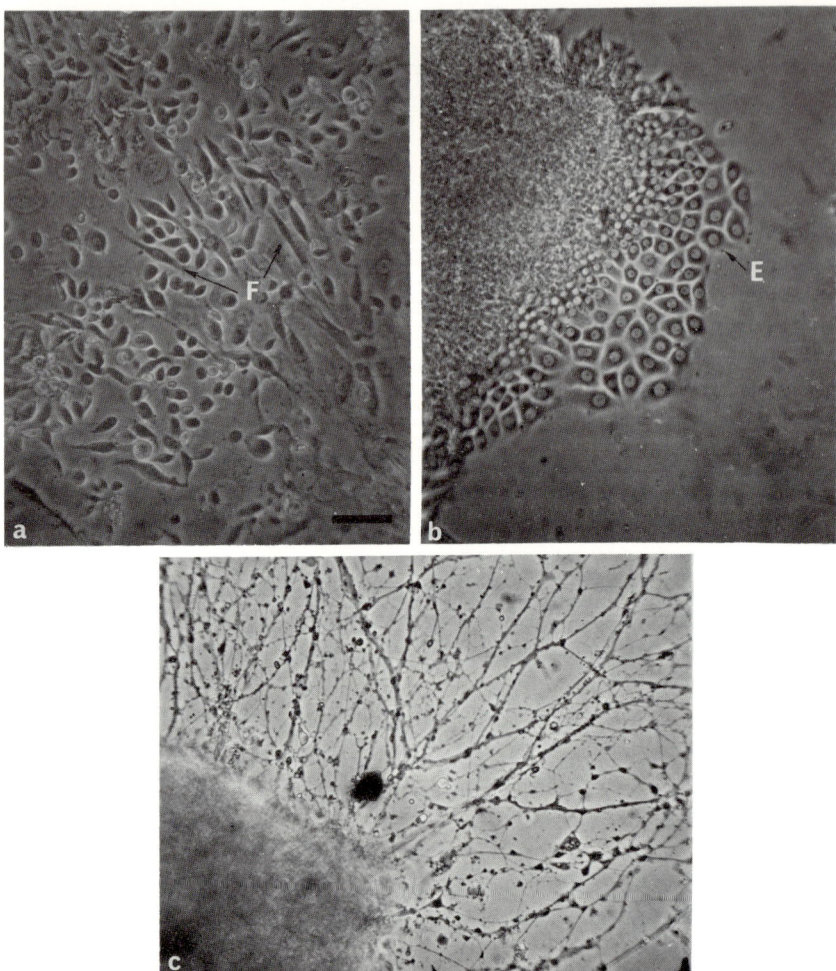

FIG. 1. Phase contrast appearance of different cell types. All photographs are the same magnification, bar equals 50 μ. a, Fibroblast-like cells (F) in 12-day-old culture from ovaries of the adult cabbage looper, *Trichoplusia ni*. b, Epithelial-like cells (E) migrating from a 1-day-old primary culture of honeybee, *Apis mellifera*, embryo. c, Nerve-like cells in a 27-day-old primary culture of minced mosquito, *Anopheles quadrimaculatus*, pupa.

contain spherical, subspherical, polygonal, and spindle-shaped free floating cells. They may be mono- or binucleate, possess comparatively large or small amounts of cytoplasm, and multiply while in suspension.

Explants may produce the following general patterns of cell migration and/or growth: (1) cell sheets or monolayers where cells are in close contact with each other, (2) cell networks in which cells are less intimately associated and may be connected by thin strands of protoplasm, (3) single attached cells, (4) free cell clumps, (5) free single cells, (6) vesicles, or (7) combinations of the above. The vesicles are hollow, multicellular, fluid-filled structures observed in primary cultures of leafhopper, cockroach, moth, mosquito, planthopper, and housefly tissues (Mitsuhashi and Maramorosch, 1964; Larsen, 1967; Sohi, 1968; Bhat and Singh, 1969, 1970; Mitsuhashi, 1969; Wallis and Lite, 1970). The vesicles in larval *Aedes aegypti* and cockroach heart tissue cultures are composed of a single layer of epithelial-like cells that have a cuticular covering on the outside. They originate from cut ends of the tissue, enlarge by increase in the number of cells, and reach a diameter of 2–3 mm. Vesicles from *Aedes aegypti* and *A. vittatus* multiply and have been subcultured up to 23 times. Vesicles in housefly cultures reproduce by budding and daughter vesicles detach from each other and float free. The hollow vesicles in leafhopper cultures are formed by part of the explant tissue becoming swollen and this area developing into the multicellular vesicle. When first formed, the wall is a single layer thick but more cell layers build up as the structure enlarges and becomes older. This type of vesicle may differ from the others in respect to formation and structure.

F. DIFFERENTIATION AND MAINTENANCE OF *in Vivo* FUNCTIONS

Grobstein (1965) pointed out that a standard of reference is necessary when deciding what changes in cultured tissues or cells are to be regarded as differentiative. According to his definition "differentiation refers not to any alteration and diversification of properties which may appear or exist in cultured cells and tissues, but only to those which probably repeat processes occurring in development *in vivo*." The following discussion of differentiation will center around this definition.

Imaginal discs are aggregations of undifferentiated cells in immature stages of insects (larvae or pupae) that differentiate during pupal development to become adult structures such as eyes, legs, wings, antennae, or mouthparts. Each adult structure develops from a specific disc referred to as antennal disc, wing disc, etc. Information on the effects of hormones on imaginal discs that has contributed to our understanding of differentiation in these tissues is discussed in Section X,A. In this section, we will briefly examine morphological development.

Larval *Drosophila melanogaster* eye-antennal discs, with attached brain and associated structures, were cultured *in vitro* and the daily progress of differentiation followed by observing morphological changes and pigment deposition (Schneider, 1964). In brief, the *in vitro* differentiative processes in eye discs consisted of eyes being properly positioned and shaped, pigments being deposited, and the formation of cornea, ommatidia, and lenses. The *in vitro* differentiation of lenses and ommatidia was also recorded by Sengel and Mandron (1969) and Kuroda (1970). The antennal discs differentiated into 3-segmented antenna-like structures that resemble adult antennae. Sengel and Mandron (1969) reported that the first and second antennal segments became distinct individual structures. Leg discs underwent development into femur, tibia, tarsus, and claws (the distal segments) but differentiation of the two proximal leg segments was incomplete. The wing discs formed structures that resembled the general appearance of wings. As might be expected, discs differentiate at a slower rate *in vitro* than *in vivo*.

Explanted wing discs from 4 species of moth larvae continued mitotic activity for 3–4 weeks, survived for 80 days, and probably produced cuticle (Kurtti and Brooks, 1970a). In contrast to this, cells from differentiated organized larval tissues such as fat body, salivary gland, Malpighian tubule, nerve cord, and prothoracic gland exhibited no mitosis and remained viable for only 3 weeks. The authors offered this generalization: "with the acquisition of differentiation *in vivo* the potential for growth *in vitro* is reduced." Their study suggests this conclusion, but cells from adult tissues, which are more highly differentiated than those from larvae, will proliferate and have become established as cell lines (Hink, 1970; Hink and Ignoffo, 1970; Hsu *et al.*, 1970; Kitamura, 1970; Sweet and McHale, 1970). The process of differentiation apparently reduces the capacity for *in vitro* growth of some tissues and cells but does not have this effect on all tissues.

The *in vitro* growth and migration of axons from embryonic cockroach brains is remarkable (Chen and Levi-Montalcini, 1969). When five or six brains and ganglia from 16- to 18-day-old embryos of *Periplaneta americana* were pooled and cultured, axons began to emerge from the explants within 48 hr. Some outgrowths consisted of tiny filaments that joined together to form larger fibers while other outgrowths were thick fibers that branched into thin filaments. The emerging axons branched extensively and became connected with other explants and axons. The nervous tissue outgrowths also formed glial cells, tracheoblasts, and neurons. It was postulated that the migrating glial cells pulled the neurons into the surrounding medium.

Mechanical dissociation of brains and ganglia provided single cells that were cultured with foregut explants (Chen and Levi-Montalcini, 1970). The nerve cells produced very dense fibrillar networks around the explants. Fibers emerged from these networks and developed synaptic connections with the foregut. Neurons cultured in the presence of foregut survive much longer than when cultured alone. Also, in the absence of explants, neurons do not develop the dense fibrillar network. These observations demonstrate that tissue explants enhance growth and differentiation of embryological nerve cells. However, the mechanism for such activity remains unknown. The morphological development of neurons within the central nervous system of embryonic cockroaches was also observed by tissue culture techniques (Reinecke, 1970). The embryonic neuroblasts divided to produce ganglion cells which subsequently underwent further divisions. Single neuroblasts produced groups of neurons within the ganglia. Axons and dendrites emerged from developing neurons. In the peripheral nervous system, sensory structures developed between the epithelium and central nervous system. These phenomena were probably representative of *in vivo* processes.

Embryonic cells of *Drosophila melanogaster* also differentiated into nervelike and musclelike cells *in vitro*. Immediately after dissociation of whole 3- to 7-hr-old embryos, the cultured cells are ovoid, spherical, or teardrop shaped. During the first few days of culture some of these morphologically simple cells develop protoplasmic extensions, some of which are branched and possess beadlike swellings. Electron microscopic examination of these cultures confirmed the presence of axons (Seecof and Teplitz, 1971). Other cells differentiated into spindle-shaped, contractile, multinucleated muscle cells. As the cultures age, these muscle cells become attached to the nervelike cells. It was not determined whether the cells became committed to differentiation while in the embryos or while in culture.

Intact brains and ganglia retain their structural integrity and continue some of their functions when placed in culture media. The neurosecretory cells of the cultured cockroach brain synthesize neurosecretory material, and fairly convincing evidence suggests that the material moves along the median fiber tracts to the corpora cardiaca (Marks, 1971). The material apparently enters the tissue culture medium since normal target glands such as prothoracic glands become activated when cultured with corpora cardiaca (Marks, 1968). Neurosecretory and neuroglandular cells from explanted brains of other cockroach species, tobacco hornworm, and a dragonfly continue functioning and store or release their products (Seshan and

Levi-Montalcini, 1971; Leloup and Marks, 1972; Schaller and Meunier, 1967).

The differentiation of other tissues and cells has received less attention. Martin and Nishiitsutsuji-Uwo (1967) and Pihan (1969) observed trachael morphogenesis and development of tracheoblasts *in vitro*. Cuticle deposition, which has been described in leg regenerates, leg imaginal discs, and vesicles, is discussed in Sections V, E and X, A. Fragments of larval or nymphal integument formed new cuticle *in vitro* (Miciarelli *et al.*, 1967; Agui *et al.*, 1969b). Myoblasts, in culture with nonmyoblastic cells, from pupae of 3 species of moths exhibit remarkable differentiation *in vitro*. The freshly isolated, spindle-shaped myoblasts rapidly attached to the flasks and elongated to lengths of 70–100 μ. Cell fusion to form elongated muscle rudiments (myotubes) was the first suggestion of muscle morphogenesis. This was a continuing process in which cells divided, aggregated, and fused over a period of several days. The myotubes became multinucleated, cross bands formed, and they began to contract (Kurtti and Brooks, 1970b).

VI. Cell Lines

A. ESTABLISHMENT OF CELL LINES

During the last several years many new insect cell lines have been described (Table VI). Published and unpublished communications have described lines from 13 species of mosquitoes, 10 species of moths, 7 species of leafhoppers, 4 species of cockroaches, and the fruit fly. To my knowledge, the first reported line was from ovaries of the silkworm, *Bombyx mori*, and it was subcultured 69 times before it died out (Vago and Chastang, 1958). Grace (1958a) also obtained long-term culture (survival for 186 days) of insect ovarian tissues. The line that has been cultured for the longest period of time was established 10 years ago by Grace (1962a) and is still multiplying. The lines originated from various primary explants: minced and trypsinized embryos; pupal and larval hemocytes; ovaries from larvae, pupae, or adults; minced larvae, pupae, or adults; imaginal discs; and dorsal vessel (heart). Most mosquito lines originated from minced first instar larvae, most moth lines were established from ovarian tissues of larvae, pupae, or adults, and the majority of the cockroach and leafhopper lines developed from embryonic tissues. In most cases we do not know the tissue origin of the cells. This is important if one wants to compare *in vitro* and *in vivo* parameters such as multiplication of intracellular parasites, cell physiology, cell nutrition, etc. The lines derived from ovarian tissue may come from cells of the intermediate layer of the ovariole or hemocytes that have adhered to the surface of the ovary (Kitamura, 1966; Stanley and Vaughn, 1968; White, 1971). Electron microscopic evidence suggests that the embryonic cockroach line (EPa) may have a hemocyte origin (Deutsch and Landureau, 1970). Karyotypic analyses of the *D. melanogaster* cell line originated from embryonic tissues by Echalier and Ohanessian (1970) strongly support the hypothesis that the line developed from a single cell (Dolfini, 1972).

During the process of establishing cell lines, a common generalized growth pattern is often observed. Cell multiplication or migration from the explant begins one to several days after initiation of the culture and continues for up to 4–6 weeks. Proliferation ceases but cells remain viable for weeks, months, or at most, a year. During this period of adaptation or selection, the cultures are refed at 1- to 3-week intervals. Finally, the cells exhibit renewed growth. An example of this growth pattern is illustrated in Fig. 2. Most cells are attached to the culture flask by the fifth day and multiplication is occurring (Fig. 2a). By the twelfth day, large homogeneous colonies of cells have formed (Fig. 2b) or morphologically diverse cells near the ovariole have multiplied (Fig. 2c). Following this initial 2-week period of growth, there was an inactive period of no cell multiplication that lasted for 6½ months. Proliferation resumed after the lag period and the cells are currently subcultured at 2- to 3-day intervals and have a population doubling time of 16 hr during their logarithmic phase of growth (Fig. 2d). There are some exceptions to this general pattern in which cells either multiply very slowly or not at all for the initial several weeks and then exhibit increased proliferation. Also, there may be no pronounced lag period of growth and subculturing is performed within 1–2 weeks after primary culture initiation (Schneider, 1972).

Some of the earlier cell lines were established in hemolymph-supplemented (HS) media and were later adapted to hemolymph-free (HF) media (see Section III for media formulas). The *Aedes aegypti* cell line developed by Grace (1966) in HS medium has been adapted to grow in media in which FBS has replaced the hemolymph (Sohi, 1969; Mitsuhashi and Grace, 1969; Hsu *et al.*, 1969). Likewise, the *Antheraea eucalypti* line was established in HS medium but is now cultured in HF medium (Yunker *et al.*, 1967; Nagle *et al.*, 1967; Mitsuhashi and Grace, 1969). If FBS is omitted from media that support growth of either cell line, cell viability is reduced and there is no cell multiplication (Sohi and Smith, 1970). Cell growth of *Aedes*

TABLE VI
Insect Cell Lines

Insect species[a]	Common name	Primary explant	Reference
Antheraea eucalypti	Australian emperor gum moth	Ovaries	Grace (1962a)
Bombyx mori	Silkworm (moth)	Ovaries	Vago and Chastang (1958)
Bombyx mori	Silkworm (moth)	Gonad	Gaw et al. (1959)
Bombyx mori	Silkworm (moth)	Ovaries	Grace (1967)
Bombyx mori	Silkworm (moth)	Ovary	Grace (1970)
Carpocapsa pomonella	Codling moth	Embryos	Hink and Ellis (1971)
Chilo suppressalis	Rice stem borer (moth)	Hemocytes	Mitsuhashi (1967a)
Choristoneura fumiferana	Spruce budworm (moth)	Larvae	Sohi (1972a)
Heliothis zea	Corn earworm (moth)	Ovary	Hink and Ignoffo (1970)
Malacosoma disstria	Forest tent caterpillar (moth)	Hemocytes	Sohi (1972b)
Samia cynthia	Cynthia moth	Hemocytes	Chao and Ball (1971)
Spodoptera frugiperda	Fall armyworm (moth)	Ovaries	Vaughn (1970)
Trichoplusia ni	Cabbage looper (moth)	Ovaries	Hink (1970)
Aedes aegypti	Yellow fever mosquito	Larvae	Grace (1966)
Aedes aegypti	Yellow fever mosquito	Larvae	Singh (1967)
Aedes aegypti	Yellow fever mosquito	Embryos	Peleg (1969b)
Aedes aegypti	Yellow fever mosquito	Larvae	Varma and Pudney (1969)
Aedes albopictus	Mosquito	Larvae	Singh (1967)
Aedes novo-albopictus	Mosquito	Larvae	Bhat (1971)
Aedes vexans	Mosquito	Pupae	Sweet and McHale (1970)
Aedes vittatus	Mosquito	Larvae	Bhat and Singh (1970)
Aedes w-albus	Mosquito	—	Singh and Bhat (1971)
Anopheles gambiae	Mosquito	Larvae	Varma and Pudney (1971)
Anopheles stephensi	Mosquito	Larvae	Schneider (1969)

Species	Common name	Tissue	Reference
Anopheles stephensi	Mosquito	Larvae	Pudney and Varma (1971)
Anopheles stephensi	Mosquito	Larvae	Varma (1971)
Culex molestus	Mosquito	Ovaries	Kitamura (1970)
Culex quinquefasciatus	Mosquito	Ovaries	Hsu et al. (1970)
Culex salinarius	Mosquito	Larvae	Schneider (1972)
Culex tritaeniorhynchus	Mosquito	Ovary	Hsu (1971)
Culex tritaeniorhynchus	Mosquito	Larvae	Schneider (1972)
Culiseta inornata	Mosquito	Adult	Sweet and McHale (1970)
Drosophila melanogaster	Fruit fly	Embryos	Horikawa et al. (1966)
D. melanogaster	Fruit fly	Embryos	Kakpakov et al. (1969)
D. melanogaster	Fruit fly	Embryos	Echalier and Ohanessian (1970)
D. melanogaster	Fruit fly	—	Richard-Molard (1971)
D. melanogaster	Fruit fly	Embryo	Schneider (1971b)
D. melanogaster	Fruit fly	Imaginal discs	Schneider (1971b)
D. melanogaster	Fruit fly	Embryos	Dolfini and Mosna (1971)
Blabera fusca	Cockroach	Embryos	Landureau (1968)
Blattella germanica	German cockroach	Embryos	Landureau (1966)
Leucophaea maderae	Cockroach	Dorsal vessel	Vago and Quiot (1969)
L. maderae	Cockroach	Ovaries	Quiot and Vago (1971)
Periplaneta americana	American cockroach	Embryos	Landureau (1968)
Aceratagallia sanquinolenta	Clover leafhopper	Embryos	Chiu and Black (1969)
Agallia constricta	Leafhopper	Embryos	Chiu and Black (1967)
Agallia quadripunctata	Leafhopper	Embryos	Chiu and Black (1967)
Colladonus montanus	Leafhopper	Embryos	Richardson and Jensen (1971)
Macrosteles sexnotatus	Leafhopper	Embryos	Peters and Spaansen (1972)
Nephotettix apicalis	Leafhopper	Embryos	Mitsuhashi (1971)
N. cincticeps	Leafhopper	Embryos	Mitsuhashi (1971)

[a]In some papers more than one line was established from a given species.

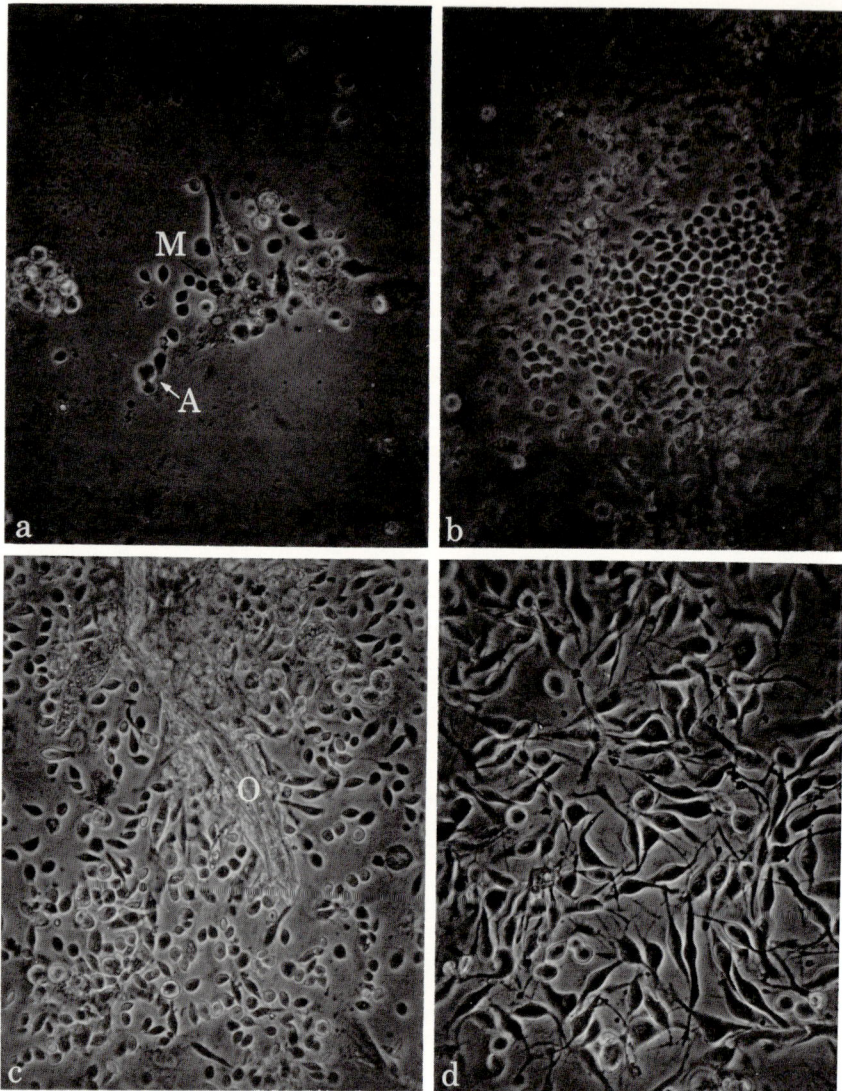

FIG. 2. Phase contrast appearance of living cabbage looper, *Trichoplusia ni*, cells. All photographs are the same magnification, bar equals 50 μ. a, A representative isolated colony of cells that originated from finely minced *T. ni* adult ovaries after 7 days *in vitro*. Note cells in metaphase (M) and anaphase (A). b, isolated colony of morphologically homogeneous cells that probably originated from a single cell after 12 days in culture. c, Cells near the remains of an ovariole (O) after 12 days *in vitro*. Notice the increase in cell numbers between 7- and 12-day-old cultures. d, *Trichoplusia ni* (TN-368) cell line 3 years and 8 months after initiation of primary culture and in its 432nd subculture.

aegypti cells was maximum with a 10% FBS supplement and *Antheraea eucalypti* grew best with 5% FBS. Nearly all cell lines are currently cultured in HF media.

B. Characteristics of Cell Lines

1. Morphology

The morphological description of cell populations is of questionable value in characterizing cell lines because cellular appearances change with age of the culture, type of growth medium, temperature of incubation, and cell density. Since these and other environmental factors affect cell morphology, when describing the appearance of a cell line the cultural conditions must be completely enumerated. While many lines have similar cells, the trained eye can detect differences in cell populations (Fig. 3). The *Antheraea eucalypti* (Fig. 3a), *Aedes aegypti* (Fig. 3b), and *Heliothis zea* (Fig. 3e) are suspension cultures and cells multiply while suspended in the medium. A few cells may be very loosely attached. The two *Carpocapsa pomonella* lines (Fig. 3c,d) and the *Trichoplusia ni* line (Fig. 3f) are attached to the culture vessels while the populations are relatively sparse, but in dense populations they also multiply in suspension. There are obvious size differences among the spherical and spindle-shaped cells of each line. Spherical cells are usually 15–25 μ in diameter. In some insect lines (not shown) the round cells attain 60 μ diameters. All lines in Fig. 3 have spindle-shaped cells but of all lines, the CP-169 line has the smallest percentage of these cells. These spindle-shaped cells are 28–68 μ long in five of the lines. The *T. ni* cells average 103 μ in length and some have extremely long protoplasmic extensions. As seen in Fig. 3f, spindle-shaped cells make up the majority of the population, but to give an idea of changes in cell morphology the following fluctuations in percentages have been observed. During a culture period of 5 days, the spindle-shaped cells varied from 12 to 64% and spherical plus subspherical cells made up 14–60% of the populations. Other cell lines, not in Fig. 3, are composed of epithelial-like cells, vesicles, mixtures of epithelial-like, spindle-shaped, and round cells, or heterogeneous populations of cells with diverse outlines, occasional binucleate cells, and no particular type predominating.

2. Growth Patterns

Some lines multiply only while attached and others while suspended in media. This latter characteristic makes them prime candidates for mass propagation in submerged culture. To subculture attached cells, they must first be loosened by agitation, scraping, or treatment with trypsin. An aliquot of the resulting cell suspension is

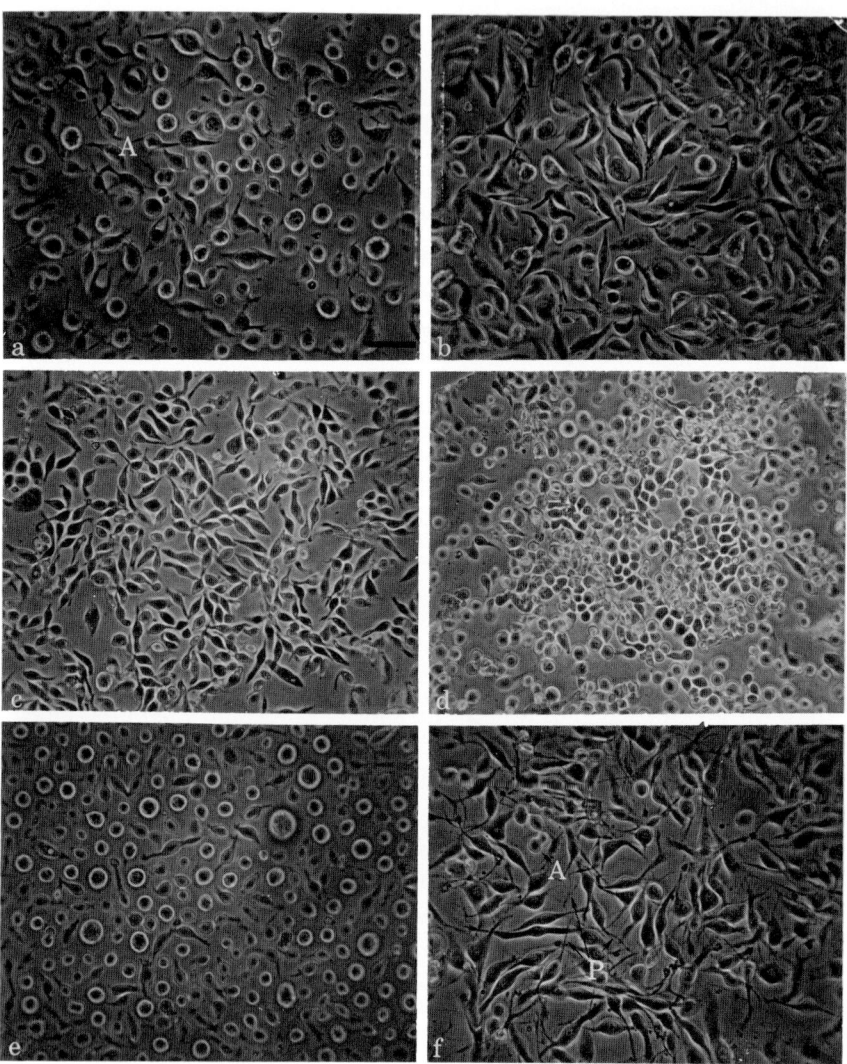

FIG. 3. Phase contrast appearance of living cell lines maintained in the author's laboratory. All lines photographed at the same magnification, bar equals 50 μ. a, A 4-day-old culture of Grace's *Antheraea eucalypti* line. b, A 6-day-old culture of Grace's *Aedes aegypti* line. c, A 2-day-old culture of the *Carpocapsa pomonella* (CP-1268) line. d, A 5-day-old culture of the *C. pomonella* (CP-169) line. e, A 4-day-old culture of the *Heliothis zea* (IMC-HZ-1) line. f, A 1-day-old culture of the *Trichoplusia ni* (TN-368) line; A, cell in anaphase; P, protoplasmic extensions.

transferred to fresh medium in a new flask. The ratio of the volume transferred from the parent culture to the volume of the new culture is termed the split ratio. A 0.2-ml aliquot of suspended cells from a 5.0-ml parent culture that is transferred to 4.8 ml of fresh medium is a 1:25 split ratio. Depending on the line, split ratios vary between 1:2 and 1:25 and subculture intervals range from 2 days to 2 weeks.

Figure 4 depicts growth curves of several lines. After initiation of new cultures there are lag periods ranging from several hours to 3–4 days. Logarithmic growth begins and may last 2–7 days after which time the number of cells levels off and may even drop. In Fig. 4, the population doubling times during logarithmic phases of growth range from 16 hr for *Trichoplusia ni* to 48 hr for *Antheraea eucalypti*. The maximum populations are 1×10^7 cells/ml for CP-1268 to 1×10^6 cells/ml for *A. eucalypti*. The growth patterns for other insect cell lines, in general, fall within these ranges.

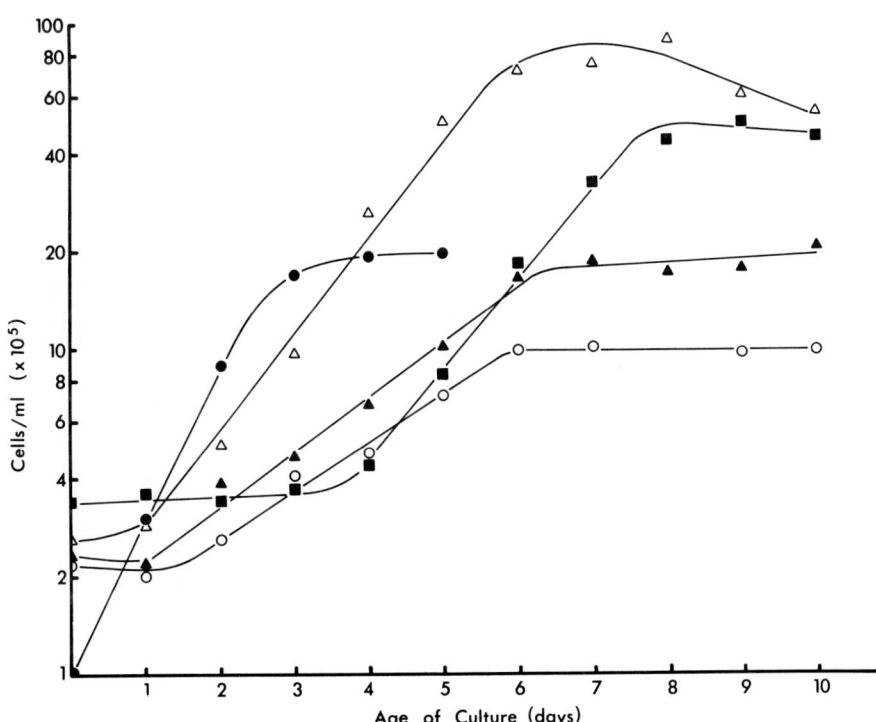

FIG. 4. Growth curves of insect cell lines.
Key: ● *Trichoplusia ni* (TN-368)   △ *Carpocapsa pomonella* (CP-1268)
○ *Antheraea eucalypti*
▲ *Aedes aegypti*   ■ *C. pomonella* (CP-169)

Long-term storage of cells is essential because it: (1) reduces incidence of microbial contamination that might occur during frequent subculturing, (2) eliminates mutations that may occur in continuous culture, (3) prevents loss of cell lines, and (4) is more economical because less media and fewer man-hours are required to maintain many cell lines (Greene and Charney, 1971). The *Antheraea eucalypti* cells may be frozen in liquid nitrogen in medium containing 5-20% glycerol for 9 months to 4 years and retain viability (Rahman *et al.*, 1966; Greene and Charney, 1971). This same line may also be stored at $-68°C$ in medium supplemented with 10% glycerol (Grace, 1962a). Other lines such as *Bombyx mori, Anopheles stephensi, Aedes aegypti, Aedes vittatus,* and *Aedes w-albus* were successfully stored at $-180°C$ to $-196°C$ in media with 10% glycerol or 10% dimethyl sulfoxide (Grace, 1967; Schneider, 1969; Varma and Pudney, 1969; Bhat and Singh, 1970; Singh and Bhat, 1971).

For shorter storage periods, 5°C may be used. The *Chilo suppressalis* line may be stored at this temperature for 120 days and recover with slightly reduced growth (Mitsuhashi, 1968). The maximum period for retaining viability at 5°C is different for each line (Hink, 1970). For the *Carpocapsa pomonella* lines it is 90 days, for Grace's *Aedes aegypti* it is 40 days, for *Antheraea eucalypti* it is 21 days, and for *Trichoplusia ni* the maximum storage period is 14 days.

3. Cytogenetics

Many mosquito cell lines are predominately diploid and in all species of mosquitoes studied the diploid number is 6. Singh's *Aedes aegypti* and *Aedes albopictus* lines have distinct stemline chromosome numbers of 6 (Nichols *et al.*, 1971). The *Anopheles stephensi* line is predominately diploid with approximately 5% of the cells having 7 or 8 chromosomes (Schneider, 1969). The *Aedes aegypti* lines of Peleg (1968) and Varma and Pudney (1969) are diploid while Grace's *A. aegypti* line is heteroploid. The majority of cells in this line are $16n$ and $32n$ (Grace, 1966). The *Culex molestus* line (Kitamura, 1970), *Culex quinquefasciatus* line (Hsu *et al.*, 1970). *Aedes vittatus* line (Bhat and Singh, 1970), and *Anopheles stephensi* line (Pudney and Varma, 1971) are all predominately diploid.

Most cell lines from moths are heteroploid. Many of the chromosomes are minute and lack discrete localized centromeres which makes karyotypic analyses impractical or probably impossible. The *Antheraea euclypti* line has no distinct stemline number of chromosomes but about 30% of the cells have 120-149 chromosomes (Nichols *et al.*, 1971). Thompson and Grace (1963) reported that chromosome counts for this line ranged from $2n$ to $128n$. In this species $n = 25$, so there would be 50 to 3,200 chromosomes. The lines from the moth *Carpocapsa pomonella* exhibit less pronounced heteroploidy. The chromosomes in the CP-1268 line have a bimodal distribution with 51% of cells having 51–57 chromosomes, which is near the diploid number of 54, and 35% with 102–110 chromosomes, which is in the tetraploid range. The other line, CP-169, has a higher degree of heteroploidy and does not have a bimodal chromosome distribution. About 72% of the cells in this population have more than 100 chromosomes (Hink and Ellis, 1971). The *Bombyx mori, Heliothis zea, Trichoplusia ni*, and *Spodoptera frugiperda* lines are also heteroploid (Grace, 1967; Hink and Ignoffo, 1970; Hink, 1970; Vaughn, 1971). This wide variation in chromosome numbers may reflect the fact that some somatic tissues from normal insects exhibit extensive polyploidy.

4. Biochemical and Immunological Characterization

Along with morphological information, growth characteristics, and chromosomal data, we should have biochemical and immunological knowledge for characterizing each cell line. Greene and Charney (1971) demonstrated that immunodiffusion techniques can be used to distinguish moth (*Antheraea eucalypti*) and mosquito (*Aedes albopictus*) cell lines from each other and from warm- and cold-blooded animal cell lines. Polyacrylamide gel electrophoresis revealed characteristic isoenzyme patterns. Moth, mosquito, mouse, and human cells were distinguished from each other by acid phosphatase, glucose 6-phosphate dehydrogenase, lactate dehydrogenase, and malate dehydrogenase patterns. These techniques also revealed that the original Grace's *Aedes aegypti* mosquito line had become contaminated with *Antheraea eucalypti* cells. In an extension of this study, Greene *et al.* (1972) pointed out that three methods may be employed to confirm the Order to which insect cells belong. These techniques are: (1) cytogenetics, (2) immunological reactions, and (3) isozyme analyses. By these methods they produced convincing evidence that mosquito cells designated as *Aedes aegypti, Culiseta inornata*, and *Aedes vexans* were moth cells. While the circumstances behind the incident of initial contamination could not be ascertained, this points out the absolute necessity for strict isolation techniques while handling cultures. These phenomena are not unique to insect tissue culture, as intraspecific contamination of vertebrate cell cultures has all too frequently occurred (McGarrity and Coriell, 1971).

VII. Virus Studies

A. INSECT VIRUSES

Insects are susceptible to different groups of viruses that are host specific or will only infect closely related insect species. The nuclear polyhedrosis viruses (NPV) are DNA viruses that multiply in the nuclei of the host cell; the virions are rod-shaped, and 10 to 100 virions are occluded in protein polyhedral inclusion bodies that range from 200 to 2000 mµ in size. There are approximately 140 described NPV diseases of insects. The cytoplasmic polyhedrosis viruses (CPV) possess spherical-shaped virions with RNA cores, the principal site of multiplication is the cytoplasm, and 100 to 1000 virions are present in 500- to 2500-mµ polyhedrae. About 110 CPV diseases have been described. The granulosis viruses (GV) are composed of one or two rod-shaped virions embedded within oval protein inclusion bodies of about 50×350 mµ. These are DNA viruses that replicate in cytoplasm and nucleus. The fourth major group of insect viruses is the noninclusion viruses (NIV). As the name implies, these virions are not occluded in protein bodies. About 30 insect viruses are placed in this group. The nucleic acid is DNA or RNA, the virions are icosahedral or rod shaped, and they multiply in the nuclei and cytoplasm. These four groups of viruses cause their respective typical symptoms and infection usually results in death of the host.

Cultured insect cells are used to investigate the infectivity of these viruses, the kinetics of virus multiplication, the assay of virus titers, and fundamental virus-cell interactions. In the future, these systems may be developed for propagating insect viruses for use as microbial insecticides and assay of viral residues.

The first successful infection of cultured insect cells with a pathogenic insect virus was by Trager (1935). At this point in the history of insect tissue culture, relatively little success had been obtained in culturing cells. Therefore, the task was not only to obtain infection but also to culture metabolically active cells capable of manufacturing virus. No work was reported in this area until 23 years later when Grace (1958b) observed NPV infection in cultured ovarian tissues of the tussock moth, *Hemerocampa leucostigma*. These cells apparently carried a latent infection, and polyhedra formation occurred after shocking the cells by changing the composition of the growth medium. *In vitro* NPV infection of *Bombyx mori* tissues was reported by three investigators some 24 years after Trager's (1935) initial success with the same virus and insect tissue (Aizawa and Vago, 1959b; Gaw *et al.*,

1959; Medvedeva, 1959). Since 1959, at least twenty studies have been made on virus infection and/or multiplication in insect cells *in vitro* (Table VII).

1. Infection of Cultures

The viral preparation used as inoculum must not contain bacterial or fungal contaminants as these will rapidly contaminate the cell cultures. Therefore, the inoculum must be filtered or obtained aseptically from diseased insects. For infecting primary explants, the virus is usually introduced when the culture is set up or within 2–3 days. The most often used preparation for infecting cells is hemolymph from virus-diseased insects. If taken during the proper stage of the disease, hemolymph contains infectious material in the form of nucleic acid, naked virions, and polyhedral inclusion bodies. Other viral inocula used for infecting cells are: DNA, purified virus particles, explants from diseased insects, infected cell cultures, and triturated whole diseased insects.

Infection of *Bombyx mori* ovarian tissue with *B. mori* NPV is the most completely investigated insect virus-cell culture system. Results on infectivity of various inocula are somewhat contradictory. Vaughn and Faulkner (1963) obtained infection by using hemolymph from diseased insects or purified virus particles combined with a solution of polyhedral protein. Purified virus particles, released from polyhedra by Na_2CO_3 treatment and containing no polyhedral protein, were not infective. Miloserdova (1965) also found diseased hemolymph infectious while purified virus failed to produce infection. It appeared that polyhedral protein played some, as yet unknown, role in the infective process. Diseased hemolymph, supernatant from crushed diseased adipose tissue, and fragments of infected tissue were effective in initiating viral infection of cultured *Lymantria dispar* ovaries (Vago and Bergoin, 1963). Free virus rods seldom produced infection. Mazzone (1971) also found that gypsy moth hemocytes were less susceptible to purified virus rods than to diseased hemolymph. Cells exposed to virus rods contained fewer and smaller polyhedra. Other studies have shown that purified virus particles will infect *B. mori* ovarian cultures (Krywienczyk and Sohi, 1967). However, the susceptibility of these tissues to viral rods free of polyhedral protein varies and high virus titers sometimes failed to infect while lower titers were infectious (Faulkner and Vaughn, 1965). These variations in infectivity may be the result of slight differences in viral preparation procedures, composition of culture media, or differences in susceptibility of tissues from different insects. If the specific type of

TABLE VII
INSECT VIRUS INFECTION AND MULTIPLICATION IN INSECT TISSUE OR CELL CULTURES

Insect and tissue or cell	Virus[a]	Viral inocula[b]	Reference
Bombyx mori: Silkworm			
Larval ovaries	*B. mori* NPV	H, T	Trager (1935)
Larval ovaries	*B. mori* NPV	H, T, E	Aizawa and Vago (1959b)
Larval ovaries	*B. mori* NPV	H	Vaughn and Faulkner (1963)
Larval ovaries	*B. mori* NPV	H, P	Faulkner and Vaughn (1965)
Larval ovaries	*B. mori* NPV	P	Krywienczyk and Sohi (1967)
Larval ovaries	*Nudaurelia cytherea capensis* NIV	P	Tripconey (1970)
Long-term cell cultures from larval ovaries	*B. mori* NPV	H	Vago and Croissant (1963)
Larval ovaries and testes	*B. mori* NPV	H	Gaw et al. (1959)
Larval ovaries and testes	*B. mori* NPV	H	Medvedeva (1959)
Larval hemocytes and gonads	*B. mori* NPV	H	Miloserdova (1965)
Larval ovaries, trachea, dorsal vessel, fat body, and silk glands	*B. mori* NPV	H	Vago et al. (1969)
Larval and pupal ovaries	*B. mori* NPV	H, P	Vago and Chastang (1960)
Pupal ovaries	Dersonucleous virus	P	Vago et al. (1966)
Pupal ovaries	*B. mori* NPV	H	Vaughn and Stanley (1970)
Intact developing embryo	*B. mori* NPV		Takami et al. (1966)
Galleria mellonella: Wax moth			
Larval ovaries	*G. mellonella* NPV	H	Heitor (1963)
Larval fat body and hemocytes	*G. mellonella* NPV	H	Sen Gupta (1963)
Larval ovaries	Densonucleous virus	H	Vago and Luciani (1965)
Larval ovaries and hemocytes	Densonucleous virus	P	Kurstak et al. (1971)
Larval ovaries, trachea, dorsal vessel, fat body, and silk glands	Densonucleous virus	H	Vago et al. (1969)

INSECT TISSUE CULTURE 193

Larval fat body, silk gland, dorsal vessel, and pupal ovaries	Densonucleosis virus	H, P	Quiot et al. (1970)
Antheraea eucalypti: Emperor gum moth			
Pupal ovaries	CPV	L	Grace (1962b)
Cell line	*B. mori* NPV	P	Grace (1969)
Cell line	SIV	P	Bellett and Mercer (1964)
Cell line	TIV and CIV	H	Hukuhara and Hashimoto (1967)
Lymantria dispar: Gypsy moth			
Larval and pupal ovaries	*L. dispar* NPV and CPV; *B. mori* NPV, *Antheraea pernyi* NPV, *Pieris brassicae* GV	H, P, E	Vago and Bergoin (1963)
Hemocytes	*L. dispar* NPV	H, P	Mazzone (1971)
Hemerocampa leucostigma: Tussock moth			
Larval ovaries	NPV	L, T	Grace (1958b)
Antheraea pernyi: Oak silkworm			
Pupal testes	*A. pernyi* NPV	H	Medvedeva (1959)
Chilo suppressalis: Rice stem borer			
Cell line	CIV	A	Mitsuhashi (1967a)
Spodoptera frugiperda: Fall armyworm			
Cell line	*S. frugiperda* NPV	H, T	Goodwin et al. (1970)
Heliothis zea: Corn earworm			
Cell line	*H. zea* NPV	M, T	Ignoffo et al. (1971)

(Continued)

TABLE VII (Continued)

Insect and tissue or cell	Virus[a]	Viral inocula[b]	Reference
Drosophila melanogaster: Fruit fly			
Embryo	sigma virus	M	Ohanessian and Echalier (1967)
Larvae	sigma virus	P	Seecof (1969)
Peridroma saucia: Variegated cutworm			
Larval hemocytes	*P. saucia* NPV	M	Martignoni and Scallion (1961b)
Nephotettix cincticeps: Green rice leafhopper, and *Laodelphax striatella*: Small brown planthopper			
Embryos	CIV	P	Mitsuhashi (1967b)

[a] NPV, nuclear polyhedrosis virus; CPV, cytoplasmic polyhedrosis virus; GV, granulosis virus; SIV, *Sericesthis* iridescent virus; TIV, *Tipula* iridescent virus; CIV, *Chilo* iridescent virus.

[b] H, hemolymph from virus diseased insects; E, explants from diseased insects; P, purified virus particles; M, triturated diseased insects; T, virus infected tissue culture; A, accidertly infected; L, latent infection.

inoculum is held constant, these cultures respond in a predictable manner and may be used to titrate hemolymph virus (Vaughn and Stanley, 1970).

There are definite differences in the properties of purified virus rods versus the virus from diseased hemolymph, as demonstrated by ion-exchange chromatography (Vaughn, 1965). Whether or not these differences are related to tissue culture infectivity has not been determined. Purification of virus with Na_2CO_3 may alter the electrical potential on the viral particle and interfere with absorption of virus to the cells.

Compared to primary explants, few results are available on susceptibility of cell lines to viral inocula. All lines are quite easily infected with noninclusion insect viruses. Grace's *Antheraea eucalypti* cells support growth of *Sericesthis* iridescent virus (SIV) (Bellett and Mercer, 1964), *Tipula* iridescent virus (TIV), and *Chilo* iridescent virus (CIV) (Hukuhara and Hashimoto, 1967). The inocula used for SIV infection was purified virus and for TIV and CIV it was hemolymph from infected wax moth, *Galleria mellonella*, larvae. Infection of cell lines with viruses embedded in protein inclusion bodies (NPV, CPV, and granulosis), many of which offer promise as microbial control agents, has met with little success. Grace (1969) infected his *A. eucalypti* cell line with an inoculum of purified virus particles from *B. mori* NPV. Approximately 20% of the cells are infected and polyhedra appear 2½ days after infection. Goodwin et al. (1970) employed NPV-diseased hemolymph and the supernatant from infected primary cultures of *Spodoptera frugiperda* (fall armyworm) pupal ovaries to infect the *S. frugiperda* cell line. Ignoffo et al. (1971) used triturated diseased insects to infect the *Heliothis zea* cell line with *H. zea* NPV.

DNA or RNA, isolated from viruses or diseased insect tissue, may provide an effective means for infecting cell cultures. Infectious DNA has been isolated from *B. mori* NPV (Onodera et al., 1965), tissues of NPV-infected *B. mori* larvae (Dobrovol's'ka et al., 1965), and NPV-diseased *H. zea* larvae (Ignoffo et al., 1971). Infectious RNA has also been isolated from *B. mori* CPV and midgut epithelium of infected larvae (Kawase and Miyajima, 1968). Viral DNA from *B. mori* NPV initiated inclusion body and virus formation in an established cell line (FL cells) derived from human amnionic tissue (Himeno et. al., 1967).

2. Incubation and Replication

Following inoculation of primary cell cultures with virus there is an incubation period of 1-14 days, depending on virus and cells, before signs of infection are evident. *Bombyx mori* ovarian tissues develop nuclear polyhedral inclusion bodies 1–4 days after infection. There are usually 1–4 inclusion bodies per nucleus with a maximum of 7 (Gaw *et al.*, 1959; Vago and Chastang, 1960; Vaughn and Faulkner, 1963). Of virus-cell systems studied, this one produces the fewest inclusion bodies per cell. In cultured fat body of *B. mori*, the polyhedra appear 6 days after infection and by the eighth day most infected cells have lysed with liberation of inclusion bodies (Vago *et al.*, 1969). The infectious process varies among neighboring cells as some have several inclusion bodies, others have one inclusion body, and some cells have no inclusion bodies but the nuclei are hypertrophied. Cultured hemocytes of *Peridroma saucia*, infected with *P. saucia* NPV, contain an average of 50 inclusion bodies per cell 5 days after infection (Martignoni and Scallion, 1961b). A latent infection of cultured tussock moth ovaries produced infected cells containing 10–20 inclusion bodies per cell 7 days postinfection (Grace, 1958b). *Antheraea eucalypti* ovarian tissue developed a CPV infection after the cultures were subjected to *B. mori* NPV. The cytoplasm of infected cells contained several hundred polyhedra (Grace, 1962b). The aforementioned numerical estimations of inclusion bodies per cell were obtained by counting inclusion bodies in the published photographs. After 20 days incubation, a maximum of 6 polyhedra per cell nucleus were observed in the *Spodoptera frugiperda* cell line (Goodwin *et al.*, 1970). The tissue culture medium was modified in an effort to make it nutritionally more adequate and this improved viral replication. Inclusion bodies were first seen 5 days postinoculation and up to 15 polyhedra were present in cell nuclei. Generally, cultured cells produce fewer inclusion bodies per infected cell than cells of living insects (Vago and Bergoin, 1963). Also, some cells in a given culture are infected and others are not. This inability of cells to produce quantities of viruses comparable to living insects may be a manifestation of reduced metabolic functions *in vitro*.

The development of NPV within primary cell cultures is similar to *in vivo* viral synthesis. There is an eclipse phase after absorption followed by viral synthesis and subsequent release of virus. Cytological changes, resulting from *in vitro* infection, follow sequences similar to those observed in tissues of living insects (Krywienczyk and Sohi, 1967; Martignoni and Scallion, 1961b).

Quantitative estimates of increases in virus titers are few. SIV multiplication in the *Antheraea eucalypti* cell line was investigated qualitatively and quantitatively by Bellett and Mercer (1964) and Bellett (1965a,b). Virus antigen appeared in the cytoplasm 2–3 days after infection with approximately 90% of the cells infected. Cells began to release virus into the medium 4½ days after infection and continued production until the eighth day. Maximum viral concentration was reached 7 days postinfection with a yield of 495 infectious units per cell. In three other quantitative estimates of virus multiplication, Ignoffo *et al.* (1971) found that infectious material from *Heliothis zea* NPV increased about 10^8 times after seven consecutive serial passages in the *H. zea* cell line, Grace (1962b) reported that CPV concentration increased 300 times in *A. eucalypti* ovarian tissues and Trager (1935) obtained a 10,000-fold multiplication of *Bombyx mori* NPV in ovarian cultures after an incubation period of 9 days. This 10,000-fold increase in virus titer is comparable to the titer increase in hemolymph of NPV-infected *B. mori* larvae (Aizawa, 1959). However, this *in vivo* increase in hemolymph virus occurred after an incubation period of 50 hr.

B. Arboviruses

A large proportion of papers on arbovirus-insect cell interactions are concerned with determining the susceptibility of various primary cultures and cell lines to different viruses and evaluating titer increases. The reader is referred to an excellent review by Yunker (1971) for a thorough discussion of infection and multiplication of many arboviruses in arthropod tissue culture. For recent developments in this field consult the proceedings of a symposium entitled "Arthropod Cell Cultures and Their Application to the Study of Viruses" (Weiss, 1971) and the proceedings of The Third International Colloquium on Invertebrate Tissue Culture to be published in 1972. Insect tissue culture is of value for the study of arbovirus epidemiology, determining virus-vector relationships at the cellular and subcellular level, and titration and identification of viruses.

1. Isolation and Titration

The *Aedes albopictus* cell line was used for primary isolation and identification of dengue virus serotypes from human sera and triturated mosquitoes (Singh and Paul, 1969). Dengue types 1 and 2 produced a characteristic CPE (cytopathic effect) of progressive syncytial formation that eventually led to complete lysis of the entire cell monolayer. Dengue types 3 and 4 caused a different CPE that was nonprogressive with small foci of syncytial formations that did not spread to all cells of the monolayer. Several other group B mosquito-borne arboviruses produce CPE similar to that caused by dengue 2 (Paul *et al.*, 1969). Arbovirus representatives from serogroup A or several other arbovirus groups do not cause CPE. The progression of the CPE consists of lysis of individual cells, cell fusion resulting in the formation of syncytial masses, giant multinucleate cells increasing in number, phagocytosis of dead cells, and the eventual recovery of the infected cell population. Since this CPE is easy to observe, end points may be conveniently determined (Suitor and Paul, 1969). This system appears more sensitive to dengue virus infection than PS(Y-15') cells or suckling mice.

Plaque formation is a valuable technique for quantitative assay of many viruses and has been reported a few times in mosquito cell lines. The reason for the general lack of plaque assays is that CPE is not produced in most arbovirus-infected cultured insect cells. Therefore, virus titers must often be determined in non-insect systems.

Suitor (1969) obtained plaque formation in *Aedes albopictus* cells infected with Japanese encephalitis virus. The plaque numbers were directly proportional to the virus concentration but the overall technique was less sensitive than other tissue culture systems. Vesicular stomatitis, dengue-1, and West Nile viruses also produce plaques in this cell line (Yunker, 1971). CPE and plaques were also produced by West Nile virus in a cloned strain of Varma and Pudney's (1969) M29 *Aedes aegypti* cell line (Porterfield and de Madrid, 1972). Some of the clones yielded specific West Nile hemagglutinin. No CPE was observed in the uncloned parent line. Thus, the cell cloning technique produced cell populations that differ in their sensitivity to this arbovirus.

2. Specificity of in Vitro Infections

This area of study has received attention because cultured vector cells may be used to differentiate between arbo and other viruses, or to carry it a step further, between various arboviruses. Specific mosquito cell lines may be of diagnostic value with only mosquito-borne viruses multiplying in these lines. We have evidence that some viral specificity exhibited in nature is reduced when cells are grown *in vitro*. Tick-borne viruses will multiply in mosquito, *Aedes albopictus*, cells (Buckley, 1969; Yunker and Cory, 1969), and viruses normally transmitted by mosquitoes will grow in the *Antheraea eucalypti* moth cell line (Suitor, 1966; Converse and Nagle, 1967; Yunker and Cory, 1968) and a *Drosophila melanogaster* cell line (Hannoun and Echalier, 1971). However, in many studies there is specificity of infection that reflects the natural vector relationship and a few selected recent examples will be discussed.

The susceptibility of Singh's *Aedes albopictus* and *A. aegypti* cell lines to Junin and Portillo, two arenoviruses, and to Saint Louis encephalitis, an arbovirus, was determined by Mettler and Buckley (1972). The two arenoviruses will not multiply in either line but the arbovirus grows in both lines.

Ťahyňa virus, an arbovirus transmitted in nature by *Aedes* mosquitoes, will grow in *Aedes aegypti* cells but two viruses transmitted by ticks and one arbovirus vectored by mosquitoes of genera other than *Aedes* will not multiply in these cells (Málková and Marhoul, 1972). Four viruses originally isolated from anopheline mosquitoes will grow in the *Anopheles stephensi* cell line but not in the *Aedes aegypti* cell line (Varma and Pudney, 1971). These cells may therefore be useful in differential diagnosis between these arboviruses.

Peleg's *A. aegypti* cell line supports proliferation of two group A and one group B arboviruses but is refractive to encephalomyocarditis (EMC) and polio viruses (Peleg, 1968). In an effort to circumvent this resistance, efforts were made to infect these cells with viral RNA. The RNA extracted from two arboviruses, Semliki Forest and West Nile, were infectious and treated cells yielded viruses identical to those from which the nucleic acid was extracted. Cells exposed to EMC and polio virus RNA were not infected (Peleg, 1969a). This author speculated that these cells are resistant because RNA doesn't reach the replication site within the cell or these cells do not possess metabolic pathways necessary for production of these viruses. Real evidence that different biochemical mechanisms are operative in the production of viruses by vertebrate and mosquito cells was presented by Stollar *et al.* (1972). The major finding was that different species of viral RNA are present in vertebrate (BHK-21 and chick embryo) and mosquito (*A. albopictus*) cells infected with Sindbis virus.

We must conclude, from the present state of knowledge, that in some instances the susceptibility of *in vitro* cultured cells to virus reflects their relationship in nature while in other virus-cell systems the arboviruses will multiply in cells from unnatural hosts. Each cell line and virus must be individually evaluated for diagnostic possibilities.

3. Persistent Infections of Cell Lines

As opposed to vertebrate cell cultures where infection causes death of most cells, arboviruses often become established in infected cell lines and the virus is maintained through prolonged serial subcultures. These persistent infections may represent a characteristic of the "mosquito phase" of virus replication, for in nature the mosquito remains infectious for long periods in which cells continue to produce virus.

Persistent infections have been observed in several lines. Grace's *Aedes aegypti* line supports growth of four arboviruses (Murray Valley encephalitis, Japanese encephalitis, West Nile, and Kunjin) without any obvious adverse effects on the cells, and persistent infections were initiated (Řeháček, 1968). The infection of the *A. albopictus* line with Chikungunya virus was maintained in serial subcultures for several years (Buckley, 1972). During these transfers the virus became attenuated as reflected by loss of pathogenicity to mice and reduction of CPE in BHK-21 cells. This same line was persistently infected with Kemerovo virus for 1½ years. There was no CPE, and interferon production could not be demonstrated (Libíková and Buckley, 1972).

The mechanisms by which virus and cells coexist were investigated by Peleg (1969b). His *A. aegypti* line (59) was infected with the mosquito-borne Semliki Forest virus. The cells began to release virus 8 hr after exposure to virus, the peak titer was reached 2 days postinfection, and by day 10 the titer decreased to a level below that of the initial dose. The virus concentration remained at this level for 150–210 days and no CPE was observed. It was postulated that the long-term association of virus and cells may be explained by: (1) the low temperature of incubation, (2) antiviral interferon-like substances being present in the cultures, or (3) the virus-yielding cells constitute a small percentage of the cell population and infected cells release a small number of viruses. Experiments revealed that the latter statement was the most likely explanation because only about 8% of the cells produced virus with a yield of 1 to 6 plaque-forming units (p.f.u.) per infected cell. This virus became attenuated in this line and the possible use of arthropod cells for production of viral vaccines was discussed by Peleg (1971).

C. Other Animal Viruses

The *Aedes albopictus* cell line was refractive to infection by three oncogenic viruses: Friend leukemia, murine sarcoma, and reticuloendotheliosis (Řeháček *et al.*, 1971). Myxoma virus also failed to replicate in this cell line (Hagen *et al.*, 1971). Another tumor-producing virus, simian virus 40, in the form of virions or infectious DNA produced no CPE or detectable viral antigens in *Trichoplusia ni* or *A. aegypti* cell lines (Hink and Shadduck, 1970). Three nonarboviruses

—pseudorabies, herpes simplex, and encephalomyocarditis—were neither sustained nor propagated in *Antheraea eucalypti* cells but 10 of 34 arboviruses were (Yunker and Cory, 1968). We may conclude from present information that insect cell lines are susceptible to representatives from these groups of viruses: arborviruses, insect viruses, and plant viruses. In nature, insects are intimately associated with these viruses and their relationships are carried over and reflected *in vitro*.

Several investigators have reported viruslike particles in what were thought to be normal cell lines. In the *Antheraea eucalypti* and *Aedes aegypti* lines established by Grace (1962a, 1966), spherical 420- to 460-Å particles are present after exposing the cells to arboviruses or *Bacillus thuringiensis* crystals. This may indicate activation of a latent infection (Filshie *et al.*, 1967). Rod-shaped particles that resemble tobacco mosaic virus were observed in the *Anopheles stephensi* (Mos 43) cell line (Pudney *et al.*, 1972) and particles morphologically similar to virions of NPV were seen in the *Heliothis zea* (IMC-HZ-1) cell line (Roy, 1971). The significance of these particles is most intriguing. They may be latent viruses, their pathogenicity is unknown, and their role in interactions with other viruses has not been determined. Their presence may explain why some insect cell lines, in comparison with vertebrate cells, are less susceptible to virus infection.

D. Plant Viruses

Insects transmit several virus diseases of economically important plants and cultured vector cells have been used to study cell-virus interactions. The rice dwarf virus, vectored by leafhoppers, will multiply in leafhopper tissue *in vitro* (Mitsuhashi and Nasu, 1967). Wound tumor virus, an agent that experimentally produces tumors in host plants and is transmitted by leafhoppers, will replicate in primary cultures of leafhopper embryonic tissue (Chiu *et al.*, 1966). The virus was serially passed *in vitro* and no CPE were observed in treated cultures. This virus also infects leafhopper cell lines and will replicate to higher titers in vector than in nonvector cells (Chiu and Black, 1967, 1969). Another leafhopper-borne virus, potato yellow drawf, infected vector cells *in vitro* and thus provided evidence that virus transmission in nature was not merely mechanical (Chiu *et al.*, 1970). In another effort to understand the nature of virus transmission, Peters and Black (1970) demonstrated that sowthistle yellow vein virus infects primary tissue cultures of its aphid vector and these cultures may be useful in assaying viral infectivity.

For detailed discussions of plant viruses in tissue culture refer to the reviews by Black (1969) and Hirumi (1971).

VIII. Protozoa Studies

The development of the causative organisms of animal malaria was followed in mosquito tissue culture. Since the avian malarial parasite, *Plasmodium relictum*, attaches to the stomach and development occurs here *in vivo*, it was postulated that this tissue would support such development *in vitro*. However, in culture medium without tissue present, the oocysts grew and attained stages similar to those oocysts cultured in association with stomachs. The immature oocysts do not require salivary glands, a tissue normally invaded *in vivo*, to complete their *in vitro* development into mature infective sporozoites (Ball and Chao, 1963). Schneider (1968) also obtained sporozoite formation from oocysts of *Plasmodium gallinaceum* in tissue culture medium.

The growth of trypanosomes that are causative agents of African sleeping sickness and Nagana was demonstrated in tsetse fly tissue culture (Trager, 1959; Nicoli and Vattier, 1964; Cunningham, 1972). Some species of parasites developed to infective forms and others did not. They appear to multiply better in the presence of vector tissues than in media alone.

The obligate intracellular protozoan, *Nosema bombycis*, a microsporidian parasite of the silkworm, *Bombyx mori*, has received considerable attention (Trager, 1937; Ishihara and Sohi, 1966; Ishihara, 1968, 1969). The parasite completes its life cycle and infective spores are produced by cultured cells. It will multiply in many different cells and tissues, i.e., *B. mori* ovarian and rat embryo primary cultures, and *Antheraea eucalypti*, *Aedes aegypti*, and human amnion cell lines. *Nosema mesnili* and an unnamed microsporidian of the tent caterpillar, *Malacosoma disstria*, will also grow in primary insect tissue cultures (Sen Gupta, 1964; Kurtti and Brooks, 1971).

IX. Rickettsia Studies

Rickettsiae pathogenic to vertebrates will multiply in insect cell lines. Two strains of the Q-fever agent, *Coxiella burneti*, grow well in *Antheraea eucalypti* and *Aedes albopictus* cell lines (Yunker *et al.*, 1970). In the *Antheraea eucalypti* cells, the rickettsia continued to multiply for 16–23 weeks. At the end of this period about 99% of the cells were infected but viable cells were also present. Most *Aedes albopictus* cells were infected by the tenth week postinoculation and by the fifteenth week few viable cells were seen. The authors suggested that insect cells offer advantages over other methods for production of antigens and vaccines in that the preparations would be free of many contaminating antigens. The etiological agent of scrub typhus, *Rickettsia tsutsugamushi*, probably does not grow in *Antheraea eucalypti* cells (Yunker and Cory, 1972). It multiplies in several mosquito cell lines to levels of 3.5 to 4.5 dex over that originally inoculated.

Řeháček and co-workers have propagated *Coxiella burneti, Rickettsia prowazeki, R. mooseri, R. akari, R. conori,* and rickettsiae of the Rocky Mountain spotted fever (RMSF) group in primary cultures of tick tissues. The RMSF rickettsiae multiply better in tick tissue culture than in vertebrate cells *in vitro* or *in vivo* (Řeháček and Brezina, 1964; Řeháček *et al.*, 1968, 1972).

The multiplication of insect rickettsiae occurs in cultured vertebrate and insect cells. *Rickettsiella melolonthae*, a pathogen of the beetle, *Melolontha melolontha*, will grow in KB, Hep^2, chicken embryo, and rabbit kidney cells. The rickettsial elements appear in the cytoplasm 3 days after infection, accumulate in cytoplasmic vacuoles, and do not form crystalloid inclusion bodies characteristic in infected living insects. Cultures infected for 10 days produce rickettsiosis when injected into *M. mololontha* larvae. The organism retained pathogenicity during three *in vitro* passages (Pourquier *et al.*, 1963). *Rickettsiella popilliae*, a pathogen of the Japanese beetle, *Popillia japonica*, multiplies in chick embryo, McCoy, and *Antheraea eucalypti* cell lines (Suitor, 1964; Suitor and Paul, 1968). The organisms appear in cells 1–2 weeks postinfection and morphologically resemble those growing *in vivo*.

X. Interactions of Biologically Active Materials with Insect Cells and Tissues

A. HORMONES

Tissue and cell cultures are being employed in efforts to understand the action of ecdysone on cells and tissues. Ecdysone is the term generally used to denote a group of steroid insect molting hormones that exist as different analogues. The analogues that have received the most attention are α-ecdysone (ecdysone), β-ecdysone (ecdysterone), 22-isoecdysone, and inokosterone. These molting hormones initiate metamorphosis, i.e., the transformation of larva to pupa or pupa to adult insect. They also function as stimuli for molts between successive nymphal or larval stages.

The premise that ecdysone was "the only extrinsic agent necessary for metamorphosis" was tested by Oberlander and Fulco (1967). Imaginal wing discs of wax moth larvae, *Galleria mellonella*, were cultured *in vitro* in the presence and absence of this hormone. In cultures with the hormone 26 of 35 discs initiated metamorphosis, i.e., they increased in length, the tracheae migrated and spread out over the discs, and the discs became more winglike. In controls without ecdysone, only 3 of 34 discs began to metamorphose. Thus, ecdysone is responsible for the initiation of metamorphosis *in vitro*.

Two ecdysone analogues, ecdysterone and inokosterone, will also initiate morphological changes indicative of metamorphosis *in vitro* (Oberlander, 1969a). However, the discs are more sensitive to ecdysone than the analogues. Discs cultured without ecdysone or in the presence of ecdysterone and inokosterone ceased or drastically reduced DNA synthesis after approximately 24 hr in culture. If ecdysone was in the medium, the discs continued DNA synthesis for 6 days. In the ecdysterone- and inokosterone-supplemented cultures, the impetus for DNA synthesis probably carried over from the *in vivo* environment but ecdysone is necessary for continued synthetic activity. Not only does ecdysone promote continued DNA synthesis, it also turns on DNA synthesis in metabolically inactive cells (Oberlander, 1969b). Evidence suggests that this DNA synthesis is not a necessary prerequisite for the morphological changes that indicate metamorphosis. These events appear to be independent responses to this hormone.

Employing an *in vitro* system of minced wing tissue from diapausing pupae of the cecropia moth, *Hyalophora cecropia*, S. S. Wyatt and Wyatt (1971) demonstrated that ecdysterone was more effective than ecdysone in stimulating protein and RNA synthesis. In the presence of ecdysterone, the rates of wing tissue protein and RNA syntheses were raised about 65% over controls. Ecdysone stimulated protein and RNA synthesis approximately 30% and in some experiments there was no stimulation. These *in vitro* studies complement *in vivo* findings that suggest the two analogues (ecdysone and ecdysterone) trigger different biochemical processes involved in metamorphosis.

The physiological condition of the donor insect influences the *in vitro* response of imaginal discs to ecdysterone (Agui *et al.*, 1969a). Wing discs from rice stem borer larvae, *Chilo suppressalis*, within 1 day after the final molt did not respond to ecdysterone. Wing discs from larvae in the 2- to 3-day stage after the final molt metamorphosed and cuticle formation occurred in the presence of 3 μg ecdysterone per milliliter medium. If ecdysterone was absent these discs did not develop. Wing discs from a later larval period, 1–2 days before pupation, differentiated in the absence of the hormone. These discs were probably stimulated *in vivo* at a sensitive period and therefore did not require the continued presence of the hormone.

Ecdysterone will stimulate morphological changes in other tissues. The hindgut is another organ that undergoes extensive reorganization during metamorphosis from pupa to adult. Judy (1969) employed tissue culture techniques in an effort to understand the mechanisms involved in the morphogenic changes stimulated by ecdysterone. Hindguts from diapausing pupae of the tobacco hornworm, *Manduca sexta*, were cultured in Grace's insect tissue culture medium (Grace, 1962a). In medium without the hormone, the cells migrated from the explant, flattened out, and extended long cytoplasmic processes. By time-lapse photomicrography the cells were observed to move by slow gliding processes and have little peripheral cytoplasmic activity. After these cultures were maintained *in vitro* for several weeks they were treated with ecdysterone or 22-isoecdysone. The cultures did not respond to 22-isoecdysone but those treated with ecdysterone exhibited increased movement of cytoplasmic peripheral membranes, cells moved more rapidly, and there was a tendency for cells to dissociate from each other. This hormone produced the same general effect in hemocyte cultures. This *in vitro* stimulation of cells duplicates some of their *in vivo* activities during the course of metamorphosis.

The molting hormones also influence development of gonadal tissue. In the presence of ecdysterone, testes from diapausing larvae of *Chilo suppressalis* increased in volume and cell migration was more rapid than in controls without this hormone (Yagi *et al.*, 1969). Spermatocytes apparently developed into spermatids in hormone-treated cultures but in cultures without ecdysterone there was no development within the spermatocysts. These results substantiate studies with living insects that molting hormones influence germ cell development.

The hormonal control over cuticle deposition in regenerative tissue has been studied in a series of excellent papers (Marks and Leopold, 1971; Marks, 1972). The legs of fifth instar nymphs of the cockroach, *Leucophaea maderae*, are severed at the trochanter-femoral joint. The wound closes and a regenerative blastema appears just proximal to the wound. This grows and forms a new limb at the next molt. The regenerating tissue is obtained by severing the leg, permitting the tissue to develop for 10–21 days *in vivo*, and then dissecting it out and placing it in Rose (1954) tissue chambers. This tissue is termed the "leg regenerate" and it will deposit cuticle in response to ecdysterone or ecdysone stimulation. The response varies depending on the concentration of the dose, the length of exposure, and the age of the leg regenerate. With large doses, cuticle deposit is thick and bears spins and with smaller doses a thin membrane is produced. By administering repeated doses of ecdysterone, up to five layers of cuticle are secreted by the regenerates. Ecdysterone is more active than ecdysone in stimulating the complete process of cuticle deposition. Ecdysone is more active in the initial activation of cells but frequently fails to stimulate the entire process of cuticle deposition. Ecdysone may be more effective in stimulating DNA synthesis, as previously mentioned with wing imaginal discs, but the latter stages of deposition are more dependent on stimulation by ecdysterone.

The effects of ecdysterone on established insect cell lines are contradictory. Mitsuhashi and Grace (1970) reported that at a concentration of 0.1 µg/ml, ecdysterone stimulated multiplication of cells of the *Antheraea eucalypti* cell line when they are grown in hemolymph-supplemented medium. Cells cultured in the same ecdysterone-containing medium but without the hemolymph supplement do not grow more rapidly than the hemolymph-supplemented controls. Judy (1969) also found these cells unresponsive to ecdysterone in hemolymph-free medium. This suggests that important interactions occur between hormones and insect blood. In contrast to these two studies, Reinecke and Robbins (1971) demonstrated that ecdysterone affects this same cell line cultured in hemolymph-free medium. This hormone induced an increase in cell volume and cell membrane movement. Also, the growth rate was reduced in the presence of 0.1–2.0 µg ecdysterone per milliliter medium. RNA synthesis was probably stimulated but DNA synthesis was not. These particular results concerning nucleic acid syntheses are in general agreement with the previously mentioned work with wing discs.

B. Insecticides

The effects of insecticides on growth rates of *Antheraea eucalypti* and *Aedes aegypti* cell lines were reported by Mitsuhashi *et al.* (1970a,b) and Grace and Mitsuhashi (1971). Both cell lines responded similarly to many classes of insecticides. Of 20 insecticides assayed, 13 were 100 to 10,000 times more toxic to last instar mosquito larvae than to the cell lines. These findings might be expected with many insecticides, because their modes of action do not interfere with vital processes of isolated somatic cells. However, for those that act on basic common biochemical pathways such as oxidative phosphorylation, the comparative resistance of these cultured cells is more difficult to explain. Rotenone was the most toxic insecticide and inhibited cell division at 0.001 µg/ml medium. The *A. aegypti* cells recovered from brief rotenone intoxication, cell clones exhibited different sensitivities, and a slightly rotenone-resistant strain of cells was obtained. Known "inducing agents," that function at sublethal doses to stimulate protein synthesis in insects and cultured mammalian cells, include several insecticides. DDT, dieldrin, and phenobarbital stimulate protein synthesis in Grace's *A. aegypti* cell line but RNA and DNA syntheses were not stimulated (Wang et al., 1970).

C. Other Materials

The effects of various X-ray exposures on the mitotic cycle of cultured grasshopper neuroblasts were reported by Carlson (1969). The dose of X-rays or gamma rays necessary to kill an adult insect is, in general, 100 times greater than that required to kill an adult mammal (Day and Oster, 1963). This radioresistance has been noted in isolated cultured hearts, Malpighian tubules, and guts of a cockroach (Larsen, 1963), and expanded use of tissue culture would be useful in gaining an understanding of mechanisms responsible for such resistance.

Studies on the activity of colchicine in cultured grasshopper neuroblasts and embryos demonstrated that, as in other living cells, this compound blocks mitosis by affecting the spindle (Gaulden and Carlson, 1951; Bergerard and Morio, 1963; Mueller et al., 1971). The action of actinomycin-D and Puromycin on grasshopper embryos *in vitro* has also been examined (Cole et al., 1967, 1968).

XI. Conclusion

It would be redundant to review the advances and contributions just mentioned. Therefore, I wish to conclude by suggesting areas that need increased study and emphasis. This may be done by looking at tissue culture from two different points of view: (1) the further development of insect tissue culture as a subject in itself, and (2) the advancement of studies involving applied problems or other disciplines that use insect tissue cultures as experimental systems. Of course, these two general headings are not mutually exclusive and they are often interrelated with one complementing the other.

We need to develop more rational approaches to media formulation. There is increasing evidence that media formulations based on biochemical duplication of hemolymph may not necessarily provide the best *in vitro* environment. To develop more adequate and chemically defined media we must become involved with more sophisticated metabolic studies. We should probably consider the microenvironment of the cells *in vivo*. We do not understand why cells from one species multiply *in vitro* while those from other species within the same genus fail to grow. When media and techniques are perfected so that explants will continue more *in vivo* functions, insect tissue culture will certainly contribute to our understanding of many biological phenomena.

Further work in the characterization of cell lines by biochemical, immunological, and karyological techniques is imperative so that investigators may be certain they are working with properly designated cellular material.

The culture of specific tissues should receive more attention. Cultured vector salivary glands, where some pathogens multiply *in vivo*, would be a more valuable substrate for study of vector-pathogen interactions than cultured vector cells from unknown origins. We could determine whether different tissues from different insect stages exhibit varying susceptibilities to intracellular parasites.

Utilization of cell cultures for basic genetic studies and the transfer of genetic material (viral genes) from insect vector to either plants or animals are exciting fields we have only touched upon.

We must look at the "insect phase" of vectored pathogens and determine differences between maturation or multiplication in the vector and the non-arthropod host. An important question to be answered is: what are the mechanisms that render insect cells susceptible to insect viruses and plant and animal viruses vectored by insects but not to many other animal viruses? Also, we should concentrate on efforts to gain an understanding of why cultured insect cells are generally more resistant or produce less virus than cells of the living host. Are interferon-like compounds produced? Are receptor sites lacking? Are biochemical pathways missing or do they become inoperative when cells are placed in culture?

Research in insecticide mode of action and development of resistance at the cellular and molecular level may be advanced by use of tissue culture. To investigate compounds that interfere with functions of organized tissues, we must become more experienced with *in vitro* maintenance of differentiated organs and tissues.

Insect cell lines will undoubtedly be employed for arbovirus diagnostic purposes and research will continue toward such goals. Vaccine production in insect cells *in vitro* should receive more attention because insects are so distantly related to mammals that these cells probably would not contaminate vaccines with carcinogenic materials, and many lines multiply in suspension and therefore could be mass produced.

REFERENCES

Aizawa, K. (1959). *J. Insect Pathol.* **1**, 67.
Aizawa, K., and Sato, F. (1963). *Ann. Epiphyt.* **14**, 125.
Aizawa, K., and Vago, C. (1959a). *C. R. Acad. Sci.* **249**, 928.
Aizawa, K., and Vago, C. (1959b). *Ann. Inst. Pasteur, Paris* **96**, 455.
Agui, N., Yagi, S., and Fukaya, M. (1969a). *Appl. Entomol. Zool.* **4**, 158.
Agui, N., Yagi, S., and Fukaya, M. (1969b). *Appl. Antomol. Zool.* **4**, 156.
Arvy, L., and Gabe, M. (1946). *C. R. Soc. Biol.* **140**, 787.
Ball, G. H., and Chao, J. (1963). *Ann. Epiphyt.* **14**, 205.
Bellett, A. J. D. (1965a). *Virology* **26**, 127.
Bellett, A. J. D. (1965b). *Virology* **26**, 132.
Bellett, A. J. D., and Mercer, E. H. (1964). *Virology* **24**, 645.
Bergerard, J., and Morio, H. (1963). *Ann. Epiphyt.* **14**, 55.
Bhat, U. K. M. (1971). Unpublished data.
Bhat, U. K. M., and Singh, K. R. P. (1969). *J. Med. Entomol.* **6**, 71.
Bhat, U. K. M., and Singh, K. R. P. (1970). *Curr. Sci.* **39**, 388.
Black, L. M. (1969). *Annu. Rev. Phytopathol.* **7**, 73.
Buckley, S. M. (1969). *Proc. Soc Exp. Biol. Med.* **131**, 625.
Buckley, S. M. (1972). *Proc. Int. Colloq. Invertebr. Tissue Cult., 3rd, 1971* (in press).
Carlson, J. G. (1969). *Radiat. Res.* **37**, 1.
Chao, J., and Ball, G. H. (1971). *Curr. Top. Microbiol. Immunol.* **55**, 28.
Chen, J. S., and Levi-Montalcini, R. (1969). *Science* **166**, 631.
Chen, J. S., and Levi-Montalcini, R. (1970). *Arch. Ital. Biol.* **108**, 503.
Chiu, R.-J., and Black, L. M. (1967). *Nature (London)* **215**, 1076.
Chiu, R.-J., and Black, L. M. (1969). *Virology* **37**, 667.
Chiu, R.-J., Reddy, D. V. R., and Black, L. M. (1966). *Virology* **30**, 562.
Chiu, R.-J., Liu, H.-Y., MacLeod, R., and Black, L. M. (1970). *Virology* **40**, 387.
Clements, A. N., and Grace, T. D. C. (1967). *J. Insect Physiol.* **13**, 1327.
Cole, M. B., Stephens, R. E., and Carlson, J. G. (1967). *A. S. B. Bull.* **14**, 25 (abstr.).
Cole, M. B., Cole, A. A., and Carlson, J. G. (1968). *A. S. B. Bull.* **15**, 35 (abstr.).
Converse, J. L., and Nagle, S. C. (1967). *J. Virol.* **1**, 1096.
Counce, S. J. (1966). *Ann. N. Y. Acad. Sci.* **139**, 65.
Cunningham, I. (1972). *Proc. Int. Colloq. Invertebr. Tissue Cult., 3rd, 1971* (in press).
Day, M. F., and Oster, I. I. (1963). *Insect Pathol.* **1**, 29–63.
Demal, J., and Leloup, A. M. (1963). *Ann. Epiphyt.* **14**, 91.
Deutsch, V., and Landureau, J. C. (1970). *C. R. Acad. Sci.* **270**, 1491.
Dobrovol's'ka, H. M., Kok, I. P., Smirnova, I. A., and Chistyakova, A. V. (1965). *Mikrobiol. Zh. (Kiev)* **27**, 73.
Dolfini, S. (1972). *Proc. Int. Colloq. Invertebr. Tissue Cult., 3rd, 1971* (in press).
Dolfini, S., and Mosna, G. (1971). Unpublished data.
Dulbecco, R., and Vogt, M. (1954). *J. Exp. Med.* **99**, 167.
Echalier, G., and Ohanessian, A. (1969). *C. R. Acad. Sci.* **268**, 1771.
Echalier, G., and Ohanessian, A. (1970). *In Vitro* **6**, 162.
Eide, P. E., and Chang, T.-H. (1969). *Exp. Cell Res.* **54**, 302.

Faulkner, P., and Vaughn, J. L. (1965). *Proc. Int. Congr. Entomol., 12th, 1964* p. 717.
Fedoroff, S. (1967). *J. Nat. Cancer Inst.* **38**, 607.
Filshie, B. K., Grace, T. D. C., Poulson, D. F., and Řeháček, J. (1967). *J. Invertebr. Pathol.* **9**, 271.
Gaulden, M. E., and Carlson, J. G. (1951). *Exp. Cell. Res.* **2**, 416.
Gaw, Z.-Y., Liu, N. T., and Zia, T. U. (1959). *Acta Virol. (Prague), Engl. Ed.* **3**, Suppl., 55.
Giauffret, A., Quiot, J.-M., Vago, C., and Poutier, F. (1967). *C. R. Acad. Sci.* **265**, 800.
Goldschmidt, R. (1915). *Proc. Nat. Acad. Sci. U.S.* **1**, 220.
Goodwin, R. H., Vaughn, J. L., Adams, J. R., and Louloudes, S. J. (1970). *J. Invertebr. Pathol.* **16**, 284.
Grace, T. D. C. (1958a). *J. Gen. Physiol.* **41**, 1027.
Grace, T. D. C. (1958b). *Science* **128**, 249.
Grace, T. D. C. (1959). *Ann. N.Y. Acad. Sci.* **77**, 275.
Grace, T. D. C. (1962a). *Nature (London)* **195**, 788.
Grace, T. D. C. (1962b). *Virology* **18**, 33.
Grace, T. D. C. (1966). *Nature (London)* **211**, 366.
Grace, T. D. C. (1967). *Nature (London)* **216**, 613.
Grace, T. D. C. (1969). Personal communication.
Grace, T. D. C. (1970). Unpublished data.
Grace, T. D. C., and Brzostowski, H. W. (1966). *J. Insect Physiol.* **12**, 625.
Grace, T. D. C., and Mitsuhashi, J. (1971). *Curr. Top. Microbiol. Immunol.* **55**, 108.
Greenberg, B., and Archetti, I. (1969). *Exp. Cell Res.* **54**, 284.
Greene, A. E., and Charney, J. (1971). *Curr. Top. Microbiol. Immunol.* **55**, 51.
Greene, A. E., Charney, J., Nichols, W. W., and Coriell, L. L. (1972). In press.
Grobstein, C. (1965). *In* "Cells and Tissues in Culture" (E. N. Willmer, ed.), Vol. 1, pp. 463–488. Academic Press, New York.
Hagen, K. W., Vavra, R. W., and Harwood, R. F. (1971). *Calif. Vector Views* **18**, 7.
Hannoun, C., and Echalier, G. (1971). *Curr. Top. Microbiol. Immunol.* **55**, 227.
Heitor, F. (1963). *Ann. Epiphyt.* **14**, 213.
Himeno, M., Sakai, F., Onodera, K., Nakai, H., Fukada, T., and Kawade, Y. (1967). *Virology* **33**, 507.
Hink, W. F. (1965). *Proc. N. Cent. Br. Entomol. Soc. Amer.* **20**, 71.
Hink, W. F. (1970). *Nature (London)* **226**, 466.
Hink, W. F. (1970). Unpublished data.
Hink, W. F. (1972). *In* "Invertebrate Tissue Culture" (C. Vago, ed.), Vol. 2, Chapter 11, Academic Press, New York.
Hink, W. F., and Ellis, B. J. (1971). *Curr. Top. Microbiol. Immunol.* **55**, 19.
Hink, W. F., and Ignoffo, C. M. (1970). *Exp. Cell Res.* **60**, 307.
Hink, W. F., and Shadduck, J. A. (1970). Unpublished data.
Hink, W. F., Richardson, B. L., Schenk, D. K., and Ellis, B. J. (1972). *Proc. Int. Colloq. Invertebr. Tissue Cult., 3rd, 1971* (in press).
Hirumi, H. (1971). *Curr. Top. Microbiol. Immunol.* **55**, 170.
Hirumi, H., and Maramorosch, K. (1964). *Exp. Cell Res.* **36**, 625.
Horikawa, M., and Fox, A. S. (1964). *Science* **145**, 1437.
Horikawa, M., Ling, L.-N., and Fox, A. S. (1966). *Nature (London)* **210**, 183.
Hsu, S. H. (1971). Unpublished data.
Hsu, S. H., Liu, H. H., and Suitor, E. C. (1969). *Mosquito News* **29**, 439.
Hsu, S. H., Mao, W. H., and Cross, J. H. (1970). *J. Med. Entomol.* **7**, 703.
Hukuhara, T., and Hashimoto, Y. (1967). *J. Invertebr. Pathol.* **9**, 278.

Ignoffo, C. M., Shapiro, M., and Hink, W. F. (1971). *J. Invertebr. Pathol.* **18,** 131.
Ishihara, R. (1968). *J. Invertebr. Pathol.* **11,** 328.
Ishihara, R. (1969). *J. Invertebr. Pathol.* **14,** 316.
Ishihara, R., and Sohi, S. S. (1966). *J. Invertebr. Pathol.* **8,** 538.
Ittycheriah, P. I., and Stephanos, S. (1969). *Indian J. Exp. Biol.* **7,** 17.
Jeuniaux, C., Duchâteau-Bosson, G., and Florkin, M. (1961). *Arch. Int. Physiol. Biochim.* **69,** 617.
Judy, K. J. (1969). *Science* **165,** 1374.
Kakpakov, V. T., Goosdev, V. A., Platova, T. P., and Polukarova, L. G. (1969). *Genetika* **5,** 67.
Kawase, S., and Miyajima, S. (1968). *J. Invertebr. Pathol.* **11,** 63.
Kitamura, S. (1965). *Kobe J. Med. Sci.* **11,** 23.
Kitamura, S. (1966). *Kobe J. Med. Sci.* **12,** 63.
Kitamura, S. (1970). *Kobe J. Med. Sci.* **16,** 41.
Krywienczyk, J., and Sohi, S. S. (1967). *J. Invertebr. Pathol.* **9,** 568.
Kuno, G., Hink, W. F., and Briggs, J. D. (1971). *J. Insect Physiol.* **17,** 1865.
Kuroda, Y. (1970). *Exp. Cell Res.* **59,** 429.
Kurstak, E., Belloncik, S., and Garzon, S. (1971). *Curr. Top. Microbiol. Immunol.* **55,** 200.
Kurtti, T. J., and Brooks, M. A. (1970a). *J. Invertebr. Pathol.* **15,** 341.
Kurtti, T. J., and Brooks, M. A. (1970b). *Exp. Cell Res.* **61,** 407.
Kurtti, T. J., and Brooks, M. A. (1971). *Curr. Top. Microbiol. Immunol.* **55,** 204.
Landureau, J.-C. (1966). *Exp. Cell Res.* **41,** 545.
Landureau, J.-C. (1968). *Exp. Cell Res.* **50,** 323.
Landureau, J.-C. (1969). *Exp. Cell Res.* **54,** 399.
Landureau, J.-C. (1970). *C. R. Acad. Sci.* **270,** 3288.
Landureau, J.-C., and Jolles, P. (1969). *Exp. Cell Res.* **54,** 391.
Landureau, J.-C., and Steinbuch, M. (1969). *Experientia* **25,** 1078.
Larsen, W. P. (1963). *Ann. Entomol. Soc. Amer.* **56,** 720.
Larsen, W. P. (1967). *J. Insect Physiol.* **13,** 613.
Leloup, A. M., and Marks, E. P. (1972). *Proc. Int. Colloq. Invertebr. Tissue Cult., 3rd, 1971* (in press).
Lender, T., and Laverdure, A.-M. (1967). *C. R. Acad. Sci.* **265.** 451.
Lesseps, R. J. (1965). *Science* **148,** 502.
Libíková, H., and Buckley, S. M. (1972). *Proc. Int. Colloq. Invertebr. Tissue Cult., 3rd, 1971* (in press).
McGarrity, G. J., and Coriell, L. L. (1971). *In Vitro.* **6,** 257.
Málková, D., and Marhoul, Z. (1972). *Proc. Int. Colloq. Invertebr. Tissue Cult., 3rd, 1971* (in press).
Marks, E. P. (1968). *Gen. Comp. Endocrinol.* **11,** 31.
Marks, E. P. (1971). *Curr. Top. Microbiol. Immunol.* **55,** 75.
Marks, E. P. (1972). *Proc. Int. Colloq. Invertebr. Tissue Cult., 3rd, 1971* (in press).
Marks, E. P., and Leopold, R. A. (1971). *Biol. Bull.* **140,** 73.
Martignoni, M. E., and Scallion, R. J. (1961a). *Biol. Bull.* **121,** 507.
Martignoni, M. E., and Scallion, R. J. (1961b). *Nature (London)* **190,** 1133.
Martignoni, M. E., Zitcer, E. M., and Wagner, R. P. (1958). *Science* **128,** 360.
Martin, B., and Nishiitsutsuji-Uwo, J. (1967). *Biochem. Z.* **346,** 491.
Mazzone, H. M. (1968). *Proc. Int. Colloq. Invertebr. Tissue Cult., 2nd, 1967* p. 14.
Mazzone, H. M. (1971). *Curr. Top. Microbiol. Immunol.* **55,** 196.
Medvedeva, N. B. (1959). *Probl. Virol. (USSR)* **4,** 64.

Mettler, N. E., and Buckley, S. M. (1972). *Proc. Int. Colloq. Invertebr. Tissue Cult., 3rd, 1971* (in press).
Miciarelli, A., Sbrenna, G., and Colombo, G. (1967). *Experientia* **23**, 64.
Miloserdova, V. D. (1965). *Vop. Virusol.* **10**, 417.
Mitsuhashi, J. (1965). *Jap. J. Appl. Entomol. Zool.* **9**, 107.
Mitsuhashi, J. (1967a). *Nature (London)* **215**, 863.
Mitsuhashi, J. (1967b). *J. Invertebr. Pathol.* **9**, 432.
Mitsuhashi, J. (1968). *Appl. Entomol. Zool.* **3**, 1.
Mitsuhashi, J. (1969). *Appl. Entomol. Zool.* **4**, 151.
Mitsuhashi, J. (1971). Unpublished data.
Mitsuhashi, J., and Grace, T. D. C. (1969). *Appl. Entomol. Zool.* **4**, 121.
Mitsuhashi, J., and Grace, T. D. C. (1970). *Appl. Entomol. Zool.* **5**, 182.
Mitsuhashi, J., and Maramorosch, K. (1964). *Contrib. Boyce Thompson Inst.* **22**, 435.
Mitsuhashi, J., and Nasu, S. (1967). *Appl. Entomol. Zool.* **2**, 113.
Mitsuhashi, J., Grace, T. D. C., and Waterhouse, D. F. (1970a). *Entomol. Exp. Appl.* **13**, 327.
Mitsuhashi, J., Grace, T. D. C., and Waterhouse, D. F. (1970b). *Entomol. Exp. Appl.* **13**, 467.
Mueller, G. A., Gaulden, M. E., and Drane, W. (1971). *J. Cell Biol.* **48**, 253.
Nagle S. C. (1969). *Appl. Microbiol.* **17**, 318.
Nagle, S. C., Crothers, W. C., and Hall, N. L. (1967). *Appl. Microbiol.* **15**, 1497.
Nichols, W. W., Bradt, C., and Browne, W. (1971). *Curr. Top. Microbiol. Immunol.* **55**, 61.
Nicoli, J., and Vattier, G. (1964). *Bull. Soc. Pathol. Exot.* **57**, 213.
Oberlander, H. (1969a). *J. Insect Physiol.* **15**, 297.
Oberlander, H. (1969b). *J. Insect Physiol.* **15**, 1803.
Oberlander, H., and Fulco, L. (1967). *Nature (London)* **216**, 1141.
Ohanessian, A., and Echalier, G. (1967). *Nature (London)* **213**, 1049.
Onodera, K., Komano, T., Himeno, M., and Sakai, F. (1965). *J. Mol. Biol.* **13**, 532.
Paul, S. D., Singh, K. R. P., and Bhat, U. K. M. (1969). *Indian J. Med. Res.* **57**, 339.
Peleg, J. (1966). *Experientia* **22**, 555.
Peleg, J. (1968). *Virology* **35**, 617.
Peleg, J. (1969a). *Nature (London)* **221**, 193.
Peleg, J. (1969b). *J. Gen. Virol.* **5**, 463.
Peleg, J. (1971). *Curr. Top. Microbiol. Immunol.* **55**, 155.
Peters, D., and Black, L. M. (1970). *Virology* **40**, 847.
Peters, D., and Spaansen, C. H. (1972). *Proc. Int. Colloq. Invertebr. Tissue Cult., 3rd, 1971* (in press).
Pihan, J.-C. (1969). *C. R. Acad. Sci.* **268**, 2074.
Porterfield, J. D., and de Madrid, A. T. (1972). *Proc. Int. Colloq. Invertebr. Tissue Cult., 3rd, 1971* (in press).
Pourquier, M., Mandin, J., and Vago, C. (1963). *Ann. Epiphyt.* **14**, 193.
Pudney, M., and Varma, M. G. R. (1971). *Exp. Parasitol.* **29**, 7.
Pudney, M., McCarthy, D., and Shortridge, K. F. (1972). *Proc. Int Colloq. Invertebr. Tissue Cult., 3rd, 1971* (in press).
Quiot, J.-M., and Vago, C. (1971). Unpublished data.
Quiot, J.-M., Vago, C., Luciani, J., and Amargier, A. (1970). *Bull. Soc. Zool. Fr.* **95**, 341.
Rahman, S. B., Perlman, D., and Ristich, S. S. (1966). *Proc. Soc. Exp. Biol. Med.* **123**, 711.
Řeháček, J. (1968). *Acta Virol. (Prague), Engl. Ed.* **12**, 241.

Řeháček, J., and Brezina, R. (1964). *Acta Virol. (Prague), Engl. Ed.* **8**, 380.
Řeháček, J., Brezina, R., and Majerska, M. (1968). *Acta Virol. (Prague), Engl. Ed.* **12**, 41.
Řeháček, J., Dolan, T., Thompson, K., Fischer, R. G., Řeháček, Z., and Johnson, H. (1971). *Curr. Top. Microbiol. Immunol.* **55**, 161.
Řeháček, J., Župančičová, M., Brezina, R., and Urvölgyi, J. (1972). *Proc. Int. Colloq. Invertebr. Tissue Cult., 3rd, 1971* (in press).
Reinecke, J. P. (1970). *Ann. Entomol. Soc. Amer.* **63**, 1667.
Reinecke, J. P., and Robbins, J. D. (1971). *Exp. Cell Res.* **64**, 335.
Richard-Molard, C. (1971). Unpublished data.
Richardson, J., and Jensen, D. D. (1971). *Ann. Entomol. Soc. Amer.* **64**, 722.
Rinaldini, L. M. (1953). *J. Physiol. (London)* **123**, 20P.
Robb, J. A. (1969). *J. Cell Biol.* **41**, 876.
Rose, G. (1954). *Tex. Rep. Biol. Med.* **12**, 1074.
Roy, K. L. (1971). Personal communication.
Schaller, F., and Meunier, J. (1967). *C. R. Acad. Sci.* **264**, 1441.
Schneider, I. (1964). *J. Exp. Zool.* **156**, 91.
Schneider, I. (1968). *Exp. Parasitol.* **22**, 178.
Schneider, I. (1969). *J. Cell Biol.* **42**, 603.
Schneider, I. (1971a). *Curr. Top. Microbiol. Immunol.* **55**, 1.
Schneider, I. (1971b). Unpublished data.
Schneider, I. (1972). *Proc. Int. Colloq. Invertebr. Tissue Cult., 3rd, 1971* (in press).
Seecof, R. L. (1969). *Virology* **38**, 134.
Seecof, R. L., and Teplitz, R. L. (1971). *Curr. Top. Microbiol. Immunol.* **55**, 71.
Seecof, R. L., and Unanue, R. L. (1968). *Exp. Cell Res.* **50**, 654.
Sengel, P., and Mandron, P. (1969). *C. R. Acad. Sci.* **268**, 405.
Sen Gupta, K. (1961). *Folia Biol. (Prague)* **7**, 400.
Sen Gupta, K. (1963). *Indian J. Exp. Biol.* **1**, 222.
Sen Gupta, K. (1964). *Curr. Sci.* **33**, 407.
Seshan, K. R., and Levi-Montalcini, R. (1971). *Arch. Ital. Biol.* **109**, 81.
Shields, G., and Sang, J. H. (1970). *J. Embryol. Exp. Morphol.* **23**, 53.
Singh, K. R. P. (1967). *Curr. Sci.* **36**, 506.
Singh, K. R. P., and Bhat, U. K. M. (1969). *Indian J. Med. Res.* **57**, 52.
Singh, K. R. P., and Bhat, U. K. M. (1971). *Experientia* **27**, 142.
Singh, K. R. P., and Paul, S. D. (1969). *Bull. WHO* **40**, 982.
Sohi, S. S. (1968). *Can. J. Zool.* **46**, 11.
Sohi, S. S. (1969). *Can. J. Microbiol.* **15**, 1197.
Sohi, S. S. (1972a). *Proc. Int. Colloq. Invertebr. Tissue Cult., 3rd, 1971* (in press).
Sohi, S. S. (1972b). *Proc. Int. Colloq. Invertebr. Tissue Cult., 3rd, 1971* (in press).
Sohi, S. S., and Smith, C. (1970). *Can. J. Zool.* **48**, 427.
Stanley, M. S. M., and Vaughn, J. L. (1968). *Ann. Entomol. Soc. Amer.* **61**, 1067.
Stollar, V., Schenk, T. E., Stollar, B. D., Stevens, T. M., and Schlesinger, R. W. (1972). *Proc. Int. Colloq. Invertebr. Tissue Cult., 3rd, 1971* (in press).
Suitor, E. C. (1964). *J. Insect Pathol.* **6**, 31.
Suitor, E. C. (1966). *Virology* **30**, 143.
Suitor, E. C. (1969). *J. Gen. Virol.* **5**, 545.
Suitor, E. C., and Paul, F. (1968). *Proc. Soc. Invertebr. Pathol., 1968* Abstr., p. 2.
Suitor, E. C., and Paul, F. J. (1969). *Virology* **38**, 482.
Suitor, E. C., Chang, L. L., and Liu, H. H. (1966). *Exp. Cell Res.* **44**, 572.
Sweet, B. H., and McHale, J. S. (1970). *Exp. Cell Res.* **61**, 51.
Sykes, J. A., and Moore, E. B. (1959). *Proc. Soc. Exp. Biol. Med.* **100**, 125.

Takami, H. S., Kitazawa, T., and Kanda, T. (1966). *Jap. J. Appl. Entomol. Zool.* **10**, 197.
Thompson, J. A., and Grace, T. D. C. (1963). *Aust. J. Biol. Sci.* **16**, 869.
Trager, W. (1935). *J. Exp. Med.* **61**, 501.
Trager, W. (1937). *J. Parasitol.* **23**, 226.
Trager, W. (1959). *Ann. Trop. Med. Parasitol.* **53**, 473.
Tripconey, D. (1970). *J. Invertebr. Pathol.* **15**, 268.
Vago, C., and Bergoin, M. (1963). *Entomophaga* **8**, 253.
Vago, C., and Chastang, S. (1958). *Experientia* **14**, 110.
Vago, C., and Chastang, S. (1960). *C. R. Acad. Sci.* **251**, 903.
Vago, C., and Chastang, S. (1962a). *Entomophaga* **7**, 175.
Vago, C., and Chastang, S. (1962b). *C. R. Acad. Sci.* **255**, 3226.
Vago, C., and Croissant, O. (1963). *Ann. Epiphyt.* **14**, 43.
Vago, C., and Luciani, J. (1965). *Experientia* **21**, 393.
Vago, C., and Quiot, J.-M. (1969). *Ann. Zool. Ecol. Anim.* **1**, 231.
Vago, C., Quiot, J.-M., and Luciani, J. (1966). *C. R. Acad. Sci.* **263**, 799.
Vago, C., Quiot, J.-M., and Amargier, A. (1969). *C. R. Acad. Sci.* **269**, 978.
Varma, M. G. R. (1971). Unpublished data.
Varma, M. G. R., and Pudney, M. (1969). *J. Med. Entomol.* **6**, 432.
Varma, M. G. R., and Pudney, M. (1971). Unpublished data.
Varma, M. G. R., and Pudney, M. (1971). *Trans. Roy. Soc. Trop. Med. Hyg.* **65**, 102.
Vaughn, J. L. (1965). *J. Invertebr. Pathol.* **7**, 524.
Vaughn, J. L. (1970). Personal communication.
Vaughn, J. L. (1971). Unpublished data.
Vaughn, J. L., and Faulkner, P. (1963). *Virology* **20**, 484.
Vaughn, J. L., and Stanley, M. S. M. (1970). *J. Invertebr. Pathol.* **16**, 357.
Vaughn, J. L., Louloudes, S. J., and Dougherty, K. (1971). *Curr. Top. Microbiol. Immunol.* **55**, 92.
Wallis, R. C., and Lite, S. W. (1970). *Ann. Entomol. Soc. Amer.* **63**, 1788.
Wang, C. M., Matsumura, F., and Boush, G. M. (1970). *J. Insect Physiol.* **16**, 1283.
Weiss, E., ed. (1971). "Arthropod Cell Cultures and Their Application to the Study of Viruses," Springer-Verlag, Berlin and New York.
White, J. F. (1971). *Curr. Top. Microbiol. Immunol.* **55**, 102.
Willmer, E. N. (1965). *In* "Cells and Tissues in Culture" (E. N. Willmer, ed.), Vol. 1, pp. 143–176. Academic Press, New York.
Wyatt, G. R., Lougheed, T. C., and Wyatt, S. S. (1956). *J. Gen. Physiol.* **39**, 853.
Wyatt, S. S. (1956). *J. Gen. Physiol.* **39**, 841.
Wyatt, S. S., and Wyatt, G. R. (1971). *Gen. Comp. Endocrinol.* **16**, 369.
Yagi, S., Kondo, E., and Fukaya, M. (1969). *Appl. Entomol. Zool.* **4**, 70.
Yunker, C. E. (1971). Personal communication.
Yunker, C. E. (1971). *Curr. Top. Microbiol. Immunol.* **55**, 113.
Yunker, C. E., and Cory, J. (1968). *Amer. J. Trop. Med. Hyg.* **17**, 889.
Yunker, C. E., and Cory, J. (1969). *J. Virol.* **3**, 631.
Yunker, C. E., an Cory, J. (1972). *Proc. Int. Colloq. Invertebr. Tissue Cult., 3rd, 1971* (in press).
Yunker, C. E., Vaughn, J. L., and Cory, J. (1967). *Science* **155**, 1565.
Yunker, C. E., Ormsbee, R. A., Cory, J., and Peacock, M. G. (1970). *Acta Virol. (Prague), Engl. Ed.* **14**, 383.
Zielinska, Z. M., and Saska, J. (1972). *Proc. Int. Colloq. Invertebr. Tissue Cult., 3rd, 1971* (in press).

Metabolites from Animal and Plant Cell Culture

IRVING S. JOHNSON AND GEORGE B. BODER

The Lilly Research Laboratories,
Eli Lilly and Company,
Indianapolis, Indiana

I.	Introduction	215
II.	Macromolecules	216
	A. Nucleic Acids	216
	B. Proteins	216
	C. Biopolymers and Related Materials	218
III.	Hormones	219
	A. Animal Hormones	219
	B. Plant Hormones	220
IV.	Alkaloids, Glycosides, Steroidal Glycosides	220
V.	Lipids, Sterols, and Steroids	221
	A. Animal Cells	221
	B. Plant Cells	222
VI.	Miscellaneous Metabolites	223
	A. Antimicrobial Substances	223
	B. Pigments	224
	C. Polyamines and Other Miscellaneous Metabolites	225
VII.	Future Applications	225
	References	227

I. Introduction

The original intent of this review was to compare transformation of organic compounds in cell cultures of higher plant and animal origin. While there is a steadily increasing literature of this type in the plant cell culture field, there is an almost total lack of such studies using animal cells in culture. It would have been interesting in a teleological sense to make a direct comparison between the two kingdoms. The apparent recent increased interest in the use of animal cells in culture for drug metabolism studies should serve to increase the animal cell literature in this field.

Consequently, the scope of this review was broadened to include reports of any metabolities, large or small, which have been produced by cell cultures from both kingdoms with the exception of cells from insects, other invertebrates, and cold-blooded vertebrates. By expanding the literature reviewed, along with some speculation concerning the potentialities of animal and plant cells for chemical transformation, we hope to stimulate additional work in this area.

Although animal cells presumably contain all of the genetic material necessary for toti-potentiality, a marked difference between animal and plant cells is noted when we attempt to demonstrate this phenomenon in culture. In recent years newer techniques have increased the ability of animal cell cultures to maintain functional states similar to the tissue of origin. Historically, however, there has been a problem of "dedifferentiation" and loss of function. This is distinct from differentiation of embryonic tissues which can be demonstrated *in vitro*.

In contrast even somatic cells of plants, for example, callus cells which are produced in response to injury in intact plants, possess the capability of developing into differentiated plantlets (Hildebrandt, 1970). It seems reasonable to predict from this increased toti-potentiality *in vitro* that the synthetic abilities of cultured plant cells represent relatively undeveloped sources of viable systems with biosynthetic potential. The biosynthetic capacity of cultured plant cells, whether grown as callus pieces on semisolid agar medium or as cellular suspension cultures, appears to be limited by the resemblance of these cells to each other, more than to the tissue of origin. The common feature of actively growing plant cells is lack of storage materials such as starch (Krikorian and Steward, 1969). Since these similarities may be due to common methods of culture rather than to actual dedifferentiation of the cells, continued development of culture techniques to favor "functional" cells appears to be necessary. In spite of these limitations a wide variety of materials have been isolated or identified as products of plant cell cultures. In plants as in animals, organ-specific functions are influenced by the activities, functions, and products of other cells through open and closed circulatory systems.

II. Macromolecules

A. Nucleic Acids

While there is little evidence for secretion of nucleic acids by either plant or animal cells, there is evidence that exogenous DNA (deoxyribonucleic acid) may alter synthesis of animal cellular products by induction (Glick, 1969). By inference with bacterial transformation *in vitro*, similar results might be expected with plant cells.

B. Proteins

A large number of proteins have been produced by many types of both plant and animal cells. In animal cells, these include interferon and many of the serum proteins including albumin and gamma globulins.

1. Serum Components and Products of Lymphoid Cultures

Serum albumin has been reported from liver cultures of mouse, rat, and human origin (Richardson et al., 1969; Bissel and Tilles, 1971). On occasion other cell types have also been reported as synthesizing serumlike proteins. Embryonic chick fibroblasts (Halpern and Rubin, 1970) have been reported as synthesizing pre-albumin as well as other unidentified proteins sharing antigenic determinants with serum proteins. Immunoglobulins have been secreted by cultures of peripheral lymphoid origin from both normal and malignant conditions (Matsuoka et al., 1968). Immunoglobulins reported include κ type IgG, IgA, and free κ type light chains. In one study (Finegold et al., 1967), 19 of 27 lymphoid cell lines produced immunoglobulins which ranged from *only* light chain types to 2 heavy chains and 2 light chain types. These were linked to form 7S or 18S immunoglobulin molecules. In a series of clones of lymphoid cells (Hinuma and Grace, 1967) some clones produced single heavy chains and others 2 components. One clone produced only IgM and another both IgM and IgG.

A number of other products have been reported from a variety of lymphoid cultures. These include the following: A dialyzable transfer factor (DTF) (Laurence, 1971) which is a small nonantigenic dialyzable molecule of less than 10,000 mol wt, which converts a normal host to a specific antigenic responsive state. A migration inhibitory factor (MIF) which inhibits migration of macrophages (Rocklin and David, 1971). A proliferation inhibitory factor (PIF) which is a macromolecule inhibiting cloning of some malignant cells in culture (Cooperbund and Green, 1971). Interferon, an inhibitor of viral replication, which is produced as a normal host response to some viral infections (Green and Cooperbund, 1971).

A number of other "soluble factors" have also been reported (Hartzman and Bach, 1971) including a blastogenic factor, a factor inhibiting the lymphocytic response, a factor stimulating lymphocyte responses, and a conditioned medium reconstituting factor (CMRF) which appears to be necessary to allow lymphocytes to respond, in the mixed leucocyte culture (MLC) test. In addition, a number of skin reactive factors (SRF) associated with induction of the inflammatory response and cellular immunity (Pick et al., 1971), and an inhibitor of DNA synthesis (Bauscher et al., 1971) have been reported. Transplantation antigens (HL-A) have been reported from mass cultures of human lymphocytic cells from a normal donor (Reisfeld et al., 1970).

A number of other biologically active materials which can be found in serum or plasma have been produced in tissue culture. These include renin (A. L. Robertson et al., 1965), a plasminogen activator (Bernick and Kwaan, 1967), a factor VIII-like activity (Zacharski et al., 1969), and the ninth component of complement (Tashjian et al., 1970).

2. Enzymes

Enzymes may be found both intracellularly and extracellularly. While both are important for biosynthesis and bioconversion, it is beyond our task to tabulate all of the intracellular enzymes which have been demonstrated in plant and animal cell cultures. The extracellular enzymes of animal cells have not been well documented beyond the few proteolytic ones already described. Collagenolytic and some proteolytic enzymes are capable of being induced in animal cells by anti-inflammatory drugs (Houck and Sharma, 1968). A number of other enzymes are also capable of induction *in vitro*. One interesting example is tryosine amino transferase, optimally induced in the C_2 clone of HTC rat hepatoma cells by cortisol, dexamethasone, corticosterone, and to lesser degrees by 11β-hydroxyprogesterone, 11-deoxycortisol, and deoxycorticosterone (Samuels and Tomkins, 1970).

A number of extracellular enzymes from plant cell cultures have been described. Among those found in filtrates from cultures of a large variety of plants, are α-amylase, amylase, arginine-degrading enzyme, IAA (indole acetic acid) oxidase (Krikorian and Steward, 1969), and peroxidase (Veliky et al., 1969). Other workers have also found β-galactosidase, pectin methylesterase, acid phosphatase, ribonuclease, and ascorbate oxidase (Yamaoka et al., 1969), phenol oxidase (Berlin et al., 1971), and isomerase (Miura and Mills, 1971).

C. BIOPOLYMERS AND RELATED MATERIALS

A number of polysaccharides and glycoproteins as well as other high molecular weight materials are produced in culture by both plant and animal cells. These include the hydroxyproline-containing proteins such as collagen. One line of fibroblasts (3T6) appears to produce a procollagen molecule from proline with a molecular weight of 500,000–600,000 (Church et al., 1971). Some animal cells capable of synthesizing amino acids from gluconolactone (Priest and Bublitz, 1967) form hydroxyproline when grown in a medium devoid of amino acids. Proline is converted to hydroxyproline with subsequent collagen synthesis by cultures of avian osteoblasts (Smith and Jackson, 1957). A number of hydroxyproline-containing proteins are produced by cultures of fibroblasts containing both $\alpha1$ and $\alpha2$ chains (Layman *et al.*, 1971). In the plant kingdom sycamore cells in suspension culture also synthesize a glycoprotein rich in hydroxyproline (Dashek, 1970).

Chondrocytes in culture synthesize a sulfated polysaccharide indistinguishable from chondroitin sulfate (Nameroff and Holtzer, 1967). Chondroitin sulfate synthesis may be influenced by the presence of cortisol or cortisone in the medium (Calcagno *et al.*, 1970). Glial cells in cultures are reported as producing hyaluronic acid, chondroitin sulfate, and heparin sulfate (Dorfman and Ho, 1970). Chinese hamster ovary cells (CHO) also produce heparin and related glycosaminoglycans or N-sulfated derivatives (Kraemer, 1968). Chondroitin sulfate and hyaluronic acid were produced by mass cultures of human fetal skin and bone, bovine fetal skin, and rat subcutaneous tissue (Grossfeld *et al.*, 1957). A series of murine teratocarcinomas have been adapted to continuous culture. These produce large amounts of basement membrane material (Pierce and Verney, 1961).

III. Hormones

A. Animal Hormones

There is increasing evidence to suggest that many, if not all currently known mammalian hormones could be produced in continuous culture (Tashjian, 1969). To date these include growth hormone, prolactin (Bancroft and Tashjian, 1970), and adrenocorticotropic hormone (ACTH) from pituitary cultures (Yasumura *et al.*, 1966). Thyrocalcitonin has been reported from a culture derived from a medullary carcinoma of the thyroid (Tashjian, 1969) as well as thyroglobulin (Pavlovic-Hournac *et al.*, 1971). Thyroglobulin and thyroxine (T_4) have also been produced by nonmalignant cells derived from thyroid tissue (Tong, 1969). Parathyroid hormone has been produced in cultures initiated from parathyroid adenomas (Tashjian, 1969). Cultures derived from adrenal cortex tumors have produced a number of adrenal hormones (Kowal, 1970). In prolonged primary submerged cultures of pancreatic islets, insulin, glucagon, and a number of unidentified proteins have been produced (Boder *et al.*, 1969). Insulin has also been reported from cultures derived from β-cell adenocarcinomas. Clonal testicular Leydig-cell cultures have produced Δ^4-3-ketosteroids (Yasumura *et al.*, 1966), and a trophoblastic cell (BeWo) is documented as producing chorionic gonadotropin (Pattillo and Gey, 1968). In addition, cultures derived from embryonic gonadal tissue of mice and chicks have been shown to secrete dihydroepiandrosterone, testosterone (Haffen and Cedard, 1967), androstenedione (Weniger *et al.*, 1967), estrone and estradiol (Weniger *et al.*, 1967). Little attention has been directed by mammalian cell culturists toward providing mammalian cell cultures with the appropriate hormonal milieu. To those interested in "functional" cultures, this is to be deplored.

B. Plant Hormones

Plant hormones are generally simpler molecules than those of animals, i.e., nonproteinaceous, and have not been described as found in filtrates of plant cell cultures. Many plant cell cultures, however, appear to require them in the media for growth and/or differentiation. Although the exact physiological function of ethylene in the plant is subject to dispute, its role as a gaseous plant hormone is generally accepted. Plant cell culture studies have contributed to an understanding of ethylene production and its regulation of growth (Larue and Gamborg, 1971).

IV. Alkaloids, Glycosides, Steroidal Glycosides

In plant cultures especially, work has been stimulated by the number of pharmaceutically useful alkaloids and glycosides which have been isolated from intact plants. The *Catharanthus*, *Atropa*, *Belladonna*, *Nicotiana*, *Rauwolfia*, and *Datura* alkaloid-containing plants have received particular attention.

The *Catharanthus* alkaloids are of particular interest since at least four of the total number isolated from the plant have been shown to have oncolytic activity. Many attempts have been made to demonstrate synthetic ability of both the cultured plant callus and crown gall tissue in both agar and suspension cultures. *Catharanthus* root, stem, leaf, and petiole cultures were established (Krikorian and Steward, 1965). Alkaloids found were different from those identified in the plant. In later experiments (Krikorian and Steward, 1969) a change in alkaloid production was induced by medium manipulation. Ajmalicine, a major alkaloid of the stem, was found. A strain of *Catharanthus* crown gall which had been in culture for more than 20 years synthesized vindoline in suspension and agar cultures (Boder *et al.*, 1964); moreover, these cultures were capable of metabolizing vindoline added to the cultures. Later other investigators (Richter *et al.*, 1965) reported detection of vindoline and vindolinine in *Catharanthus* stem and leaf callus cultures. A large number of alkaloids were produced by *Catharanthus* seedling callus and leaf callus in suspension culture, but neither vindoline nor vindolinine was detected, which may reflect a large number of variables in strains and methodology (Patterson and Carew, 1969).

Tropane alkaloids were produced by *Datura* callus and suspension cultures (Chan and Staba, 1965). Several reports (West and Mika, 1957; Thomas and Street, 1970) had shown that the root of *Belladonna* was the primary site of atropine synthesis. As early as 1942 (Dawson, 1942), it was shown that nicotine was synthesized in excised tobacco roots. Indole alkaloid production was established (Staba and Laursen, 1966) and later factors involved with alkaloid production and growth were studied (Dobberstein and Staba, 1969) with indications that the main precursors of indole alkaloids are mevalonic acid and tryptophan. Identification of nicotine and anatabine was established (Furuya et al., 1966, 1967). A more recent report (Furuya et al., 1971b) showed that 2,4-D treated tobacco callus produced no alkaloids while callus treated with indole acetic acid (IAA) produced nicotine, anatabine, and anabasine. Transfer of 2,4-D treated callus to IAA medium and IAA treated callus to 2,4-D medium demonstrated that nicotine biosynthesis in the callus tissue is activated by IAA and suppressed by 2,4-D. A report (Sairam and Khanna, 1971) showed that choline and scopolamine were produced in *Datura* callus cultures. Using another species of *Datura* (Stohs, 1969) it was shown that hyoscyamine and and scopolamine were produced by *Datura* suspension cultures. Caffeine was produced by tea callus cultures (Ogutuga and Northcote, 1970). In this study, it was learned that caffeine arose from purines released from breakdown of nucleic acids rather than directly from a purine pool. Cell cultures of *Ruta graveolens* produced seven different coumarins and four alkaloids (Steck et al., 1971). All the alkaloids were monomeric but one of the coumarins was 6,7-dimethoxy-3-(1,1-dimethyl allyl) coumarin, a new natural product.

Cultures of various *Digitalis* species were established by several different groups of investigators to examine cardiac glycoside content (Staba, 1962, Büchner and Staba, 1963).

Early work (Sargent and Skoog, 1960, 1961) documented scopolin in tobacco tissue cultures. Hormonal effects on glycoside production were established (Skoog and Montaldi, 1961).

V. Lipids, Sterols, and Steroids

A. ANIMAL CELLS

A strain of hamster kidney cells (NIL 2) were shown to incorporate ^{14}C palmitate into sphingomyelin, phosphatidyl choline, phosphatidyl ethanolamine, and phosphatidyl inositol as well as producing ceramide which provided for the generation of ceramide dihexoside, hematoside, and ceramide di-, tri-, and tetrahexoside (Robbins and Macpherson, 1971). A line (N1 S1-67) of Novikoff rat hepatoma cells produced membrane-associated phosphorylcholine and phosphatidylcholine from choline. By using chromatographic methods another line of rat hepatoma cells (MH_1C_1) were shown to take up bilirubin and excrete materials found in rat bile.

Many continuous animal cell lines require cholesterol in the media. Appreciable amounts of lipids contained in the serum-supplement of most media are taken up by animal cells through pinocytosis. Cholesterol may be excreted in a nonspecific way at rates several times that of accumulation in the cell (J. M. Bailey et al., 1959). It has been suggested that β-lipoproteins regulate uptake or adsorption of free sterols and α-lipoproteins influence excretion or desorption of free cholesterol (Rothblatt and Kritchevsky, 1968). However, the human diploid cell (WI-38) and SV_{40} transformed "sister" line actively synthesize cholesterol (Rothblatt et al., 1971). Aortic endothelial cells in culture are reported to synthesize cholesterol from sodium acetate and mevalonate with squalene concentration increasing with decreased cholesterol synthesis (Pollack, 1969).

Fed appropriate substrates, many mammalian cells convert steroids to other metabolites in culture. Cells from fetal adrenal glands convert progesterone to 11-deoxycortisol and cortisol as well as cholesterol to isocaproic acid (Milner and Villee, 1970). The human trophoblastic cell line (BeWo) will convert pregnenolone-^{3}H to progesterone and a number of unidentified products (Huang et al., 1969). Murine tumors of the adrenal cortex have been established as continuous cell lines. Cloned liver cell cultures of various hepatomas differ in some degree, in their steroidogenic output, but were fairly uniform in numbers of products formed. When incubated with cholesterol, adrenal cells produce 11β-hydroxy-20α-dihydroprogesterone either via pregnenolone → progesterone → 11β-hydroxyprogesterone or pregnenolone → 20α-dihydropregnenolone → 20α-dihydroprogesterone. The 11β-hydroxy-20α-dihydroprogesterone is then converted to 11-keto-20α-dihydroprogesterone (Kowal, 1970).

B. Plant Cells

Lipids have not been reported from filtrates of plant cell cultures. Sterols, terpenes, and related compounds are widespread in various plants but little definitive work has been reported in biosynthesis in culture or in the intact plants. Technical difficulties in isolation and structure determination have contributed to this problem. Progesterone has been isolated from the leaves of certain plants and cholesterol has been detected in intact plants. Pregnenolone was converted to progesterone by a variety of plant cultures (Graves and Smith, 1967). These authors pointed out that culture techniques are advantageous in carrying out these conversions. Later studies (Furuya et al., 1971a) concerned metabolism of progesterone by *Nicotiana* and *Sophora* cultures.

Visnagin was produced in suspension cultures of *Ammi visnagi* (Kaul and Staba, 1965). Incorporation studies showed that 2-^{14}C-acetate served as a precursor of visnagin (Chen et al., 1969). Cholesterol was stimulatory to visnagin production if it was added on the tenth day of culture but inhibitory if added upon culture initiation (Kaul et al., 1969). Yeast extract increased the amount of disogenin per cell but was inhibitory to cell growth. Steroidal production was demonstrated both by specific location reagents with thin-layer chromatography and by biological activity (Medora et al., 1967). Recent reports (Heble et al., 1968; Vagujfalvi et al., 1971) dealt with diosogenin and β-sitosterol production by *Solanum* cultures.

Recently, it was found that three new sesquiterpene lactones, but not the diterpenes of the intact plant, were produced by *Andrographis* cultures (Butcher and Connolly, 1971). Although plant material is a major source of volatile oils, none have been found in cultured tissue in spite of many attempts at production (Krikorian and Steward, 1969). However, identification of squalene and farnesol as products of mevalonic metabolism in a cell-free system from tobacco tissue grown *in vitro* was reported (Benveniste et al., 1970). This publication had been preceded by a series of papers concerning the production of various sterols by cultured tobacco plants (Benveniste, 1968; Benveniste et al., 1964a,b,c, 1966). The major triterpene in *Solanum* cultures was lupeol (Heble et al., 1971).

VI. Miscellaneous Metabolites

A. Antimicrobial Substances

The secretion of antimicrobial substances by cultured plant materials has been described (Campbell et al., 1965), by testing media used to support the growth of isolated lettuce and cauliflower tissue for antimicrobial activity. By means of a rapid screening method, secretion of antimicrobial materials by cultured root, stem, and leaf tissue from various plants has been investigated (Mathes, 1967). Test organisms were *Bacillus subtilis* and *Salmonella typhimurium*. Eleven of the thirty cultured plants inhibited *B. subtilis* but none inhibited *S. typhimurium*. Antimicrobial activity derived from submerged avocado cultures has been demonstrated (Nickell, 1962). An evaluation of the antimicrobial activity of twenty-four cultured plant tissues against several organisms has been reported with significant antimicrobial activity found in 80% of the tissue cultures (Khanna and Staba, 1968).

Other workers (Khanna *et al.*, 1971) have also recently reported significant antimicrobial activity in secretions of a variety of plant tissues. It has been shown that potato tuber resistance to fungal infection was due to the production of phenolic compounds by the plant (N. F. Robertson *et al.*, 1968). Although it may be difficult at the present time to produce large amounts of these substances by tissue culture procedures, this logistic problem also occurs in antibiotic production. Moreover, identification of new chemical structures with antimicrobial activity is of interest to the natural product chemist.

Among the potentially most interesting antimicrobials produced by higher plants are the phytoalexins. Phytoalexins are compounds specifically produced by various plants in response to infectious challenge. In function they are analogous to antibodies or interferon. They appear to be specific for the type of plant which makes them rather than the inciting agent. They are not considered a normal metabolite of the plant cell. Axenic cultured pea callus cells produced the phytoalexin, pisatin (J. A. Bailey, 1970).

Certain mammalian cell lines produce a material which is probably an α-ketoaldehyde capable of inhibiting the growth of several species of bacteria (Kenny and Sparkes, 1968). When separate lines of two of these cells were contaminated with PPLO (pleuropneumonia-like organisms), no inhibitor was produced. The authors suggested some similarity to the "retine" of Szent-Györgyi.

B. Pigments

Many pigmented strains of plants have been noted in culture. Chlorophyll synthesis has been widely examined in a variety of cultures. Recent work (Schneider, 1971) showed that purified enzyme extracts from cultured tobacco tissue which had been exposed to white light showed an increase in porphobilinogen deaminase/uroporphyrinogen cosynthetase activities which are in the pathway to chlorophyll synthesis. Anthocyanin production has been studied in stem callus culture (Harborne *et al.*, 1970). Two acyanic species (carrot and Jerusalem artichoke) were shown (Ibrahim *et al.*, 1971) to produce cyanin in culture. Anthocyanin was also formed in flax and rose cultures. A report (Goldstein et al., 1962) dealt with leucoanthocyanins in *Acer* cultures and discussed conditions which appeared to alter the amounts of pigment produced. Lowered O_2 tension inhibited pigment production. Cultured lemon tissue produced a material in the medium which reacted with ammonia to produce a yellow color characteristic of flavonoids (Kordan and Morgenstern, 1962). A new anthroquinone (4-hydroxydigitolutein) was isolated from callus tissue of *Digitalis lanata* (Furuya and Kojima, 1971).

Many mammalian melanomas are known to produce melanin in culture. Cultured mast cells also make melanin (Okun, 1967). Production was enhanced by corticotropin but corticotropin was not necessary for synthesis. Dopa oxidase was present in the cells as was tyrosinase. It was suggested that the same factors which prevent melanization of albino melanocytes *in vivo* but allow their melanization *in vitro* apply to mast cells.

C. Polyamines and Other Miscellaneous Metabolites

Several polyamine derivatives (*p*-coumaroyl, caffeoyl, and feruloyl putrescine) were isolated from tobacco callus cultures (Mizusaki et al., 1971). There is considerable evidence that polyamine synthesis is a factor paralleling and possibly controlling RNA synthesis rates in rapidly growing tissue and may be involved with growth regulatory mechanisms, although the physiological roles of putrescine and polyamines are not completely understood.

It was shown (Nilsson et al., 1970) that coenzyme Q_{10} synthesized from *p*-hydroxybenzoic acid by beating heart cell cultures was released into the medium. In the heart, coenzyme Q_{10} is essential for maintenance of function. Before coenzyme Q_{10} was discovered, it was known that *p*-hydroxybenzoic acid is essential for growth of certain microorganisms.

VII. Future Applications

Possibilities for future application of plant and animal cell cultures have barely been tapped. A great deal of the work with animal cells has been directed toward establishment of long-term continuous cell strains. This has led inevitably to the selection of those cells capable of rapid growth with characteristics which are considered as "transformed" or "malignant." Much of this work and its interpretation is derived from the use of mouse cells which appear much less genetically stable than those of other species but which are more available to most laboratory workers.

The establishment of these long-term continuous cell strains has been of great benefit to virologists, experimental oncologists, molecular biologists, and biochemists, but of little use to those interested in specific cell function. What work has been performed in relation to specific function has generally been in organ culture. Cell biology has now reached such a state of scientific sophistication and technical development that this should no longer be so.

Cells from a number of tissues can now be isolated and cloned. There are techniques available to isolate cells of a particular tissue on a preparative scale with separation into different populations without serious injury to the cells.

Many of these approaches may take the form of "prolonged primary" cultures of normal cells capable of limited growth and indefinite function. Examples of this approach have been published recently in regard to heart cells and islets of Langerhans.

In the past several years, functional cells derived from most of the mammalian endocrine system have been established with animal cells. These have frequently been established by inducing tumors of endocrine glands in laboratory animals and passing these tumors back and forth under *in vitro* and *in vivo* conditions. While use of materials derived from this type of culture may raise the specter of contamination with the malignant genome, isolation techniques and appropriate tests should eliminate any such concern.

It is far past time, however, for mammalian cell culturists to start adding appropriate amounts of the circulating hormones to the various media. Circulating levels of almost all hormones have now been determined. Massive cultures of normal cells of the hematopoietic system are now available. With appropriate applications of more sophisticated instrumentation, recycling of used media with addition of depleted metabolites, and use of available knowledge concerning the hormonal milieu, etc., it now seems possible to culture almost any functional cell of interest.

The use of mammalian cells for transformation of organic compounds is a totally unexplored field. Massive cultures of functional hepatic parenchymal cells would be of obvious but not exclusive interest. Cells from the plant kingdom have similar capabilities. Cytochrome p_{450} is a component of many mixed functional oxidase systems particularly in hydroxylation of steroids, carcinogens, insecticides, and other drugs. It is a hemoprotein containing protoheme and is found in association with lipid in microsomes of almost every mammalian tissue including adrenal and liver. Cytochrome p_{450} has also been described in certain bacterial systems. Recently, it has been demonstrated in *Claviceps purpurea* cultures where it can be converted to cytochrome p_{420}. Stimulation by phenobarbital stimulates total alkaloid production in this organism (Ambike *et al.*, 1970; Ambike and Baxter, 1970). Besides transformation products however, there exists the possibility of producing transplantation antigens, functional antibodies, cells for replacement of marrow, maturation factors, etc. The use of substrate enzyme induction, feeding of known peptide fragments, and other similar techniques in mammalian cell culture are also virtually unexplored.

In a similar fashion the field of plant tissue culture also seems ripe for industrial application and the techniques for manipulating functions are even more developed than those for mammalian cells. Cells of most plants can be grown in fermentation as well as in shaker flasks and on solid substrates. Their chief drawback might be a slower rate of growth than bacteria and fungi. A second drawback is an apparent low level of many metabolites secreted into the medium. In plant cultures, cloning techniques have not been extensively investigated to increase yields of metabolites. In both plant and animal cell culture, enrichment techniques coupled with strain selection have not been widely employed.

In spite of this, however, the possibilities of new structural entities with antimicrobial activity and of the plant-specific phytoalexins seem achievable for production of metabolites of medical and agricultural application. If these factors do not yield to production procedures, they may yield to the natural product chemist. Use of phytoalexins for control of plant disease is particularly attractive from an ecological viewpoint. With no dramatic solution to the population problem and increasing ecological pressure, plant cultures may also be of industrial interest as a source of plants. A single flask of carrot cells can give rise to 10^6 plantlets. Pathogen-free starch can be produced, etc. While we have not included them in the scope of this review, cell cultures from insects also appear to be a potentially useful source of agents and materials for use in the agricultural field. We have the strong impression that the future of both plant and animal cell culture is ripe for industrial exploitation and development. Success is almost entirely dependent upon the imagination of the innovative scientist and the magnitude and capability of his support.

References

Ambike, S. H., and Baxter, R. M. (1970). *Phytochemistry* **9**, 1959.
Ambike, S. H., Baxter, R. M., and Zahid, N. D. (1970). *Phytochemistry* **9**, 1953.
Bailey, J. A. (1970). *J. Gen. Microbiol.* **61**, 409.

Bailey, J. M., Gey, G. O., and Gey, M. K. (1959). *Proc. Soc. Biol. Med.* **100**, 686.
Bancroft, F. C., and Tashjian, A. H. (1970). *In Vitro* **6**, 180.
Bauscher, J. A. C., Adler, W. H., and Smith, R. T. (1971). In *"In Vitro* Methods in Cell-Mediated Immunity" (B. R. Bloom and P. R. Glade, eds.), pp. 475–480. Academic Press, New York.
Benveniste, P. (1968). *Phytochemistry* **7**, 951.
Benveniste, P., Hirth, L., and Ourisson, G. (1964a). *C. R. Acad. Sci.* **258**, 5515.
Benveniste, P., Durr, A., Hirth, L., and Ourisson, G. (1964b). *C. R. Acad. Sci.* **259**, 2005.
Benveniste, P., Hirth, L., and Ourisson, G. (1964c). *C. R. Acad. Sci.* **259**, 2284.
Benveniste, P., Hirth, L., and Ourisson, G. (1966). *Phytochemistry* **5**, 45.
Benveniste, P., Ourisson, G., and Hirth, L. (1970). *Phytochemistry* **9**, 1073.
Berlin, J., Barz, W., Harms, H., and Haider, K. (1971). *FEBS Lett.* **16**, 141.
Bernick, M. B., and Kwaan, H. C. (1967). *J. Lab. Clin. Med.* **70**, 650.
Bissel, D. M., and Tilles, J. G. (1971). *J. Cell Biol.* **50**, 222.
Boder, G. B., Gorman, M., Johnson, I. S., and Simpson, P. J. (1964). *Lloydia* **27**, 328.
Boder, G. B., Root, M. A., Chance, R. E., and Johnson, I. S. (1969). *Proc. Soc. Exp. Biol. Med.* **131**, 507.
Büchner, S. A., and Staba, E. J. (1963). *Lloydia* **6**, 208.
Butcher, D. N., and Connolly, J. D. (1971). *J. Exp. Bot.* **22**, 314.
Calcagno, M., Goyena, H., Arrambide, E., and de Urse, C. A. (1970). *Exp. Cell Res.* **63**, 131.
Campbell, G., Chan, E. C. S., and Barker, W. G. (1965). *Can. J. Microbiol.* **11**, 785.
Chan, W. N., and Staba, E. J. (1965). *Lloydia* **28**, 55.
Chen, M., Stohs, S. J., and Staba, E. J. (1969). *Lloydia* **32**, 339.
Church, R. L., Pfeiffer, S. E., and Tanzer, M. L. (1971). *Proc. Nat. Acad. Sci. U.S.* **68**, 2638.
Cooperbund, S. R., and Green, J. A. (1971). In *"In Vitro* Methods in Cell-Mediated Immunity" (B. R. Bloom and P. R. Glade, eds.), pp. 381–400. Academic Press, New York.
Dashek, W. V. (1970). *Plant Physiol.* **46**, 831.
Dawson, R. F. (1942). *Amer. J. Bot.* **48**, 813.
Dobberstein, R. H., and Staba, E. J. (1969). *Lloydia* **32**, 141.
Dorfman, A., and Ho, P. L. (1970). *Proc. Nat. Acad. Sci. U.S.* **66**, 495.
Finegold, I., Fahey, J. L., and Granger, H. (1967). *J. Immunol.* **99**, 839.
Furuya, T., and Kojima, H. (1971). *Phytochemistry* **10**, 1607.
Furuya, T., Kojima, H., and Syono, K. (1966). *Chem. Pharm. Bull.* **14**, 1189.
Furuya, T., Kojima, H., and Syono, K. (1967). *Chem. Pharm. Bull.* **15**, 901.
Furuya, T., Hirotani, M., and Kawaguchi, K. (1971a). *Phytochemistry* **10**, 1013.
Furuya, T., Kojima, H., and Syono, K. (1971b). *Phytochemistry* **10**, 1529.
Glick, J. L. (1969). In "Axenic Mammalian Cell Reactions" (G. L. Tritsch, ed.), pp. 117–150. Dekker, New York.
Goldstein, J. L., Swain, T., and Tjhio, K. H. (1962). *Arch. Biochem. Biophys.* **98**, 176.
Graves, J. M. H., and Smith, W. K. (1967). *Nature (London)* **214**, 1248.
Green, J. A., and Cooperbund, S. R. (1971). In *"In Vitro* Methods in Cell-Mediated Immunity" (B. R. Bloom and P. R. Glade, eds.), pp. 501–513. Academic Press, New York.
Grossfeld, H., Meyer, K., Godman, G., and Linker, A. (1957). *J. Biophys. Biochem. Cytol.* **3**, 391.
Haffen, K., and Cedard, L. (1967). *C. R. Acad. Sci.* **264**, 1923.
Halpern, M., and Rubin, H. (1970). *Exp. Cell Res.* **60**, 86.

Harborne, J. B., Arditti, J., and Ball, E. A. (1970). *Amer. J. Bot.* **57**, 754.
Hartzman, R. J., and Bach, F. H. (1971). *In* "*In Vitro* Methods in Cell-Mediated Immunity" (B. R. Bloom and P. R. Glade, eds.), pp. 463–474. Academic Press, New York.
Heble, M. R., Narayanaswami, S., and Chadha, M. S. (1968). *Science* **161**, 1145.
Heble, M. R., Narayanaswami, S., and Chadha, M. S. (1971). *Phytochemistry* **10**, 783.
Hildebrandt, A. C. (1970). *In* "Control Mechanisms in the Expression of Cellular Phenotypes" (H. A. Pardykula, ed.), pp. 147–161. Academic Press, New York.
Hinuma, Y., and Grace, J., Jr. (1967). *Proc. Soc. Exp. Biol. Med.* **124**, 107.
Houck, J. C., and Sharma, V. K. (1968). *Science* **161**, 1361.
Huang, W. Y., Pattillo, R. A., Delfs, E., and Mattingly, R. F. (1969). *Steroids* **14**, 755.
Ibrahim, R. K., Thakur, M. L., and Permanand, B. (1971). *Lloydia* **34**, 175.
Kaul, B., and Staba, E. J. (1965). *Science* **150**, 1731.
Kaul, B., Stohs, S. J., and Staba, E. J. (1969). *Lloydia* **32**, 347.
Kenny, C. P., and Sparkes, B. G. (1968). *Science* **161**, 1344.
Khanna, P., and Staba, E. J. (1968). *Lloydia* **31**, 180.
Khanna, P., Mohan, S., and Nag, T. N. (1971). *Lloydia* **34**, 168.
Kordan, H. A., and Morgenstern, L. (1962). *Nature (London)* **195**, 163.
Kowal, J. (1970). *Recent Progr. Horm. Res.* **26**, 623–687.
Kraemer, P. M. (1968). *J. Cell. Physiol.* **71**, 109.
Krikorian, A. D., and Steward, F. C. (1965). *Plant Physiol.* **40**, Suppl., v–vi.
Krikorian, A. D., and Steward, F. C. (1969). *In* "Plant Physiology" (F. C. Steward, ed.), Vol. 5B, pp. 227–328. Academic Press, New York.
Larue, T. A. G., and Gamborg, O. L. (1971). *Plant Physiol.* **48**, 394.
Laurence, H. S. (1971). *In* "*In Vitro* Methods in Cell-Mediated Immunity" (B. R. Bloom and P. R. Glade, eds.), pp. 95–150. Academic Press, New York.
Layman, D. L., McGoodwin, E. B., and Martin, G. R. (1971). *Proc. Nat. Acad. Sci. U.S.* **68**, 454.
Mathes, M. C. (1967). *Lloydia* **30**, 177.
Matsuoka, Y., Takahashi, M., Yagi, Y., Moore, G. E., and Pressman, D. (1968). *J. Immunol.* **101**, 1111.
Medora, R., Kosegarten, D., Tsao, D. P. N., and De Feo, J. J. (1967). *J. Pharm. Sci.* **56**, 540.
Milner, A. J., and Villee, D. B. (1970). *Endocrinology* **87**, 596.
Miura, G. A., and Mills, S. E. (1971). *Plant Physiol.* **47**, 483.
Mizusaki, S., Tanabe, Y., Noguchi, M., and Tamaki, E. (1971). *Phytochemistry* **10**, 1347.
Nameroff, M., and Holtzer, H. (1967). *Develop. Biol.* **16**, 250.
Nickell, L. G. (1962). *Advan. Appl. Microbiol.* **4**, 213–236.
Nilsson, J. L. G., Nilsson, I., Scholler, J., and Folkers, K. (1970). *Int. J. Vitamin Res.* **40**, 374.
Ogutuga, D. B. A., and Northcote, D. H. (1970). *Biochem. J.* **117**, 715.
Okun, M. R. (1967). *J. Invest. Dermatol.* **48**, 424.
Patterson, B. D., and Carew, D. P. (1969). *Lloydia* **32**, 131.
Pattillo, R. A., and Gey, G. O. (1968). *Cancer Res.* **28**, 1231.
Pavlovic-Hournac, M., Rappaport, L., and Nunez, J. (1971). *Exp. Cell Res.* **68**, 332.
Pick, E., Krejci, J., and Turk, J. L. (1971). *In* "*In Vitro* Methods in Cell-Mediated Immunity" (B. R. Bloom and P. R. Glade, eds.), pp. 515–530. Academic Press, New York.
Pierce, G. B., Jr., and Verney, E. L. (1961). *Cancer* **14**, 1017.
Pollack, O. J. (1969). *In* "Monographs on Atherosclerosis (Tissue Cultures)" (H. S. Simms, J. E. Kirk, and O. J. Pollack, eds.), Vol. 1, p. 82. Williams & Wilkins, Baltimore, Maryland.

Priest, R., and Bublitz, E. (1967). *Lab. Invest.* **17**, 371.
Reisfeld, R. A., Pellegrino, M., Papermaster, B. W., and Kahan, B. D. (1970). *J. Immunol.* **104**, 560.
Richardson, U. I., Tashjian, A. H., and Levine, L. (1969). *J. Cell Biol.* **40**, 236.
Richter, I., Stolle, K., Groger, D., and Mothes, K. (1965). *Naturwissenschaften* **52**, 305.
Robbins, P. W., and Macpherson, I. (1971). *Nature (London)* **229**, 569.
Robertson, A. L., Smeby, R. R., Bumpus, F. M., and Page, I. H. (1965). *Science* **149**, 650.
Robertson, N. F., Friend, I., Aveyard, M., Brown, J., Huffee, M., and Homas, A. L. (1968). *J. Gen. Microbiol.* **54**, 261.
Rocklin, R. E., and David, J. R. (1971). *In* "*In Vitro* Methods in Cell-Mediated Immunity" (B. R. Bloom and P. R. Glade, eds.), pp. 281–288. Academic Press, New York.
Rothblatt, G. H., and Kritchevsky, D. (1968). *Exp. Mol. Pathol.* **8**, 314.
Rothblatt, G. H., Boyd, R., and Deal, C. (1971). *Exp. Cell Res.* **67**, 436.
Sairam, T. V., and Khanna, P. (1971). *Lloydia* **34**, 170.
Samuels, H. H., and Tomkins, G. M. (1970). *J. Mol. Biol.* **52**, 57.
Sargent, J. A., and Skoog, F. (1960). *Plant Physiol.* **35**, 934.
Sargent, J. A., and Skoog, F. (1961). *Physiol. Plant.* **14**, 504.
Schneider, H. A. W. (1971). *Phytochemistry* **10**, 319.
Skoog, F., and Montaldi, E. (1961). *Proc. Nat. Acad. Sci. U.S.* **47**, 36.
Smith, R. H., and Jackson, S. F. (1957). *J. Biophys, Biochem. Cytol.* **3**, 913.
Staba, E. J. (1962). *J. Pharm. Sci.* **51**, 249.
Staba, E. J., and Laursen, P. (1966). *J. Pharm. Sci.* **55**, 1099.
Steck, W., Bailey, B. K., Shyluk, J. P., and Gamborg, O. L. (1971). *Phytochemistry* **10**, 191.
Stohs, S. J. (1969). *J. Pharm. Sci.* **58**, 703.
Tashjian, A. H. (1969). *Biotechnol. Bioeng.* **11**, 109.
Tashjian, A. H., Bancroft, F. C., Richardson, U. I., Goldlust, M. B., Rommel, F. A., and Ofner, P. (1970). *In Vitro* **6**, 32.
Thomas, E., and Street, H. E. (1970). *Ann. Bot. (London)* [N.S.] **34**, 657.
Tong, W. (1969). *In* "Axenic Mammalian Cell Reactions" (G. L. Tritsch, ed.), pp. 261–305. Dekker, New York.
Vagujfalul, D., Maroti, M., and Tetenyi, P. (1971). *Phytochemistry* **10**, 1389.
Veliky, I., Sandquist, A., and Martin, S. M. (1969). *Biotechnol. Bioeng.* **11**, 1247.
Weniger, J. P., Ehrhardt, J. D., and Fritig, B. (1967). *C. R. Acad. Sci.* **264**, 1069.
West, F. R., Jr., and Mika, E. S. (1957). *Bot. Gaz.* **119**, 50.
Yamaoka, T., Hayashi, T., and Sato, S. (1969). *J. Fac. Sci., Univ. Tokyo, Sect. 3* **10**, 117.
Yasumura, Y., Tashjian, A. H., and Sato, G. H. (1966). *Science* **154**, 1186.
Zacharski, L. R., Bowie, E. J. W., Titus, J. L., and Owen, C. A. (1969). *Mayo Clin. Proc.* **44**, 784.

Structure-Activity Relationships in Coumermycins

JOHN C. GODFREY AND KENNETH E. PRICE

Bristol Laboratories, Division of Bristol-Myers Company, Syracuse, New York

I.	Introduction	231
II.	The Natural Coumermycins	232
	A. Structures and Activities *in Vitro*	232
	B. The Producing Organisms	236
	C. Coumermycin Production and Isolation	239
	D. Laboratory and Clinical Studies with Coumermycin A_1	241
III.	Semisynthetic Coumermycins	250
	A. Alteration at the 4-Hydroxyl	251
	B. Alteration at the 2'-Hydroxyl	252
	C. Alterations at the 3'-Ester	253
	D. Alterations at the 3-Nitrogen	253
	E. Antibacterial Activity of Various Acylated *PNC*-Amines	258
	F. Pharmacological Properties of Selected Semisynthetic Coumermycins	278
	G. Laboratory and Human Studies with BL-C43	285
	References	294

I. Introduction

The family of antibiotics which is now known as the coumermycins was discovered about 1960 by workers at the Hoffmann-LaRoche Laboratories (Berger *et al.*, 1966), and about the same time by the Bristol-Banyu Research Institute, Ltd. (Kawaguchi *et al.*, 1965a). The naturally occurring coumermycins are produced in fermentation broths of several different species of *Streptomyces*. Kawaguchi *et al.* (1965a) reported that several bioactive components, including coumermycin A_1 (known to the Bristol-Banyu group as No. 620, BU-620, and notomycin), were elaborated by a new species which they designated "*Streptomyces rishiriensis*," isolated from a soil sample collected in Rishiri Island, Hokkaido, Japan. Another *Streptomyces* isolate, originally designated as species X-7763 by the Hoffmann-LaRoche group, was obtained from a soil sample collected in Gaspé, Canada. It was also found to produce multiple antibiotics, each containing a coumarin-like moiety. The principal antibiotic in this mixture, first called sugordomycin, was soon recognized to be coumermycin A_1 (Berger *et al.*, 1966).

Chromatographic and degradative studies quickly revealed to both groups the fact that the antibacterial activity of the culture filtrates was attributable to a number of closely related substances. Both groups recognized that one of the major components, coumermycin A_1, was much more active than the others, and methodology was developed which allowed it to be fermented and isolated as almost the sole product (Kawaguchi et al., 1965b). Although coumermycin A_1 has appreciable activity *in vitro* against a representative group of gram-negative organisms (minimum inhibitory concentrations of about $1-12$ μg/ml; Kawaguchi et al., 1965a), it is about 20 times more active against gram-positive organisms. It shows exceptional activity against a wide variety of staphylococci (Kawaguchi et al., 1965a; Grunberg and Bennett, 1966) both *in vitro* and *in vivo*, and these observations provided the basis for interest in coumermycin A_1 and its derivatives as potential therapeutic agents. Unfortunately, it was soon found in laboratory studies and subsequently confirmed in the clinic that coumermycin A_1 was poorly absorbed by the oral route and had high irritation liability when administered parenterally. Because of this, a long series of semisynthetic coumermycin A_1 derivatives was prepared and their biological properties investigated. As the accompanying report shows, considerable success in overcoming some of coumermycin A_1's liabilities was realized. However, BL-C43, the most promising candidate in the semisynthetic series, has given indications in the course of a clinical tolerance study that it causes undesirable, albeit transient and reversible, effects on liver function.

II. The Natural Coumermycins

A. STRUCTURES AND ACTIVITIES *in Vitro*

1. *Coumermycins A_1 and A_2 and Their Close Relatives*

Degradative studies at Bristol-Banyu Research Institute established the structures of coumermycins A_1 and A_2 (Kawaguchi et al., 1965b,c,d). Almost simultaneously, the Hoffmann-LaRoche group reported the total synthesis of coumermycin A_1 (Hoffmann-LaRoche, 1965; Berger et al., 1966; Furlenmeier et al., 1969) with the indicated structure showing identity to that found by degradative procedures. In one of the Hoffmann-LaRoche reports (Berger et al., 1966), it was proposed that all the compounds whose structures are presented in Fig. 1 were being produced in coumermycin fermentations. However, because of their close structural similarity, the in-

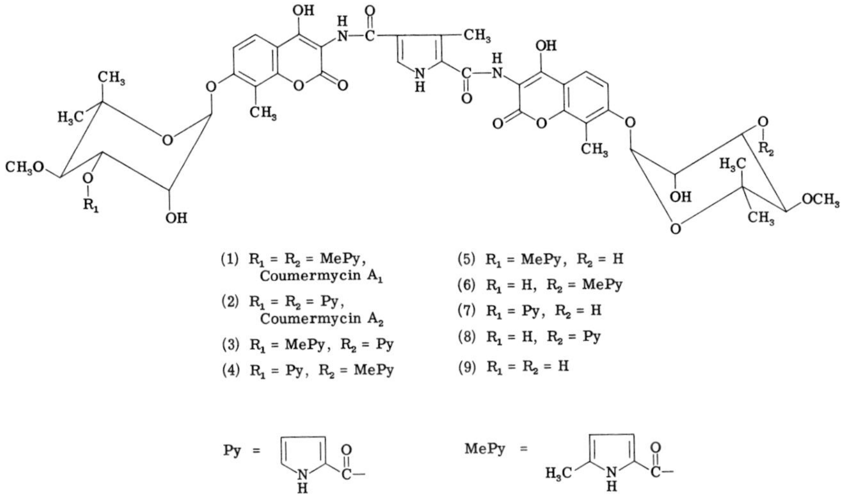

FIG. 1. Chemical structures of members of coumermycin complex.

(1) $R_1 = R_2 =$ MePy, Coumermycin A_1
(2) $R_1 = R_2 =$ Py, Coumermycin A_2
(3) $R_1 =$ MePy, $R_2 =$ Py
(4) $R_1 =$ Py, $R_2 =$ MePy
(5) $R_1 =$ MePy, $R_2 =$ H
(6) $R_1 =$ H, $R_2 =$ MePy
(7) $R_1 =$ Py, $R_2 =$ H
(8) $R_1 =$ H, $R_2 =$ Py
(9) $R_1 = R_2 =$ H

dividual members of pairs (3) and (4), (5) and (6), and (7) and (8) were never completely separated from each other. The comparative biopotency of this group of compounds as estimated by these workers utilizing a *Staphylococcus aureus* strain as the test organism, is presented in Table I. The results clearly show the superiority of coumermycin A_1's antistaphylococcal activity over that of other coumermycins. Evidence that the greater potency of coumermycin A_1 is not restricted to this one type of microorganism is given in Table II where the antimicrobial spectra of coumermycins A_1 and A_2 are directly compared.

TABLE I
COMPARATIVE *in Vitro* ACTIVITY OF COMPONENTS OF THE COUMERMYCIN COMPLEX[a]

Component	In vitro activity (units/mg)
1, Coumermycin A_1	5500
2, Coumermycin A_2	200
3 + 4 (mixture)	1400
5 + 6 (mixture)	1100
7 + 8 (mixture)	90
9 (NCDCN)	0.7

[a]Taken from Berger et al., 1966.

TABLE II
Coumermycin Antimicrobial Spectra (MIC's in µg/ml, Heart Infusion Broth)[a]

Organism	Coumermycin A_1	Coumermycin A_2
Staphylococcus aureus Smith	0.004	0.25
Staphylococcus aureus Smith + serum	1.6	50
Streptococcus pyogenes	0.062	0.5
Bacillus subtilis	6.3	12.5
Escherichia coli	12.5	> 100
Klebsiella pneumoniae	6.3	12.5
Proteus morganii	6.2	25
Proteus mirabilis	12.5	> 100
Proteus vulgaris	3.1	> 100
Pseudomonas aeruginosa 8206A	12.5	> 100
Pseudomonas aeruginosa Yale	25	> 100
Salmonella enteritidis	12.5	50
Salmonella typhi	12.5	> 100

[a]Taken from Cron *et al.*, 1970.

Here it is apparent that a markedly broader range of microorganisms is inhibited by coumermycin A_1. Thus it appears that among the naturally produced coumermycins, the degree of methylation of the terminal pyrrole moieties and the presence or absence of this heterocyclic function correlates positively with intrinsic biopotency and very probably with breadth of antimicrobial spectrum.

2. *Relationship to Novobiocin*

From its physical, chemical, and biological properties, it was quickly apparent to the workers in this field that the coumermycin complex was closely related to novobiocin (**10**), and there is little question that this observation facilitated the elucidation of the struc-

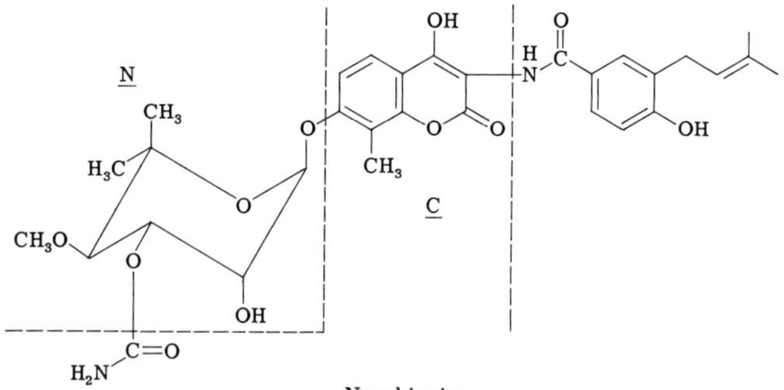

Novobiocin

(**10**)

tures of the coumermycins. Although their antimicrobial spectra were similar, coumermycin A_1 proved to be a considerably more potent antibiotic. Comparative data illustrating this conclusion are shown in Table III.

TABLE III
COMPARISON OF COUMERMYCIN A_1 WITH NOVOBIOCIN in Vitro[a]

	MIC (μg/ml)[b]	
Organism	Coumermycin A_1	Novobiocin
Bacillus cereus (Yale)	6.2	6.2
Staphylococcus aureus Smith	0.1	0.4
S. aureus M2614	0.1	0.8
S. aureus M1845	0.1	1.6
S. albus NOTC 7292	0.4	100
Diplococcus pneumoniae M1638	0.4	1.6
Streptococcus faecalis ATCC 8043	6.2	1.6
S. pyogenes Digonnet #7	0.8	1.6
Klebsiella pneumoniae (Yale)	12.5	25

[a]Taken from Keil *et al.*, 1971.
[b]The minimal inhibitory concentrations (MIC) were determined by the 2-fold agar dilution method. In this procedure, antibiotics were incorporated into Trypticase Soy Agar (BBL) with 2% defibrinated sheep blood added. The minimal inhibitory concentrations were determined after 18 hr of incubation at 37°C.

3. Related Natural Analog(s)

The only natural analog of the coumermycins which has been reported is compound (**11**), called 18,631 R.P., from the laboratories of Rhone-Poulenc S.A. (Mancy *et al.*, 1969, 1970). Although the positions

(11)

of the chlorine atom and the pyrrole methyl are unspecified, it seems clear that this antibiotic is a hybrid of the novobiocin and coumermycin structures, which has the additional feature of a chlorine atom instead of a methyl group in the coumarin nucleus. Its *in vitro* antimicrobial properties (Table IV) appear to be very similar to those of coumermycin A_1 and to those of some of the more active semisynthetic coumermycins (see below).

TABLE IV
ANTIBACTERIAL ACTIVITY OF 18,631 R.P. *in Vitro*[a]

Bacterial organisms tested	Minimum bacteriostatic concentrations (μg/ml)
Staphylococcus aureus, 209 P strain – ATCC 6538 P	0.005
S. aureus, Smith strain	0.003
Sarcina lutea – ATCC 9341	0.02
Streptococcus faecalis – ATCC 9790	0.04
Streptococcus viridans (Institut Pasteur)	3
Streptococcus pyogenes hemolyticus (Dig 7 strain, Institut Pasteur)	0.05
Diplococcus pneumoniae (Til strain, Institut Pasteur)	0.03
Neisseria catarrhalis (A 152 – Institut Pasteur)	0.005
Neisseria meningitidis (5813 – Institut Pasteur)	0.05
Neisseria gonorrhoeae (A 50 – Institut Pasteur)	0.4
Lactobacillus casei – ATCC 7469	1
Bacillus subtilis – ATCC 6633	0.6
Bacillus cereus – ATCC 6630	0.5
Mycobacterium species – ATCC 607	10
Escherichia coli – ATCC 9637	10
Proteus vulgaris	13
Klebsiella pneumoniae – ATCC 10,031	1
Pseudomonas aeruginosa (Bass strain – Institut Pasteur)	4
Brucella bronchiseptica (CN 387 – Wellcome Institute)	0.3
Brucella abortus bovis B 19 (52,135 – Institut Pasteur)	0.1
Pasteurella multocida (A 125 – Institut Pasteur)	7
Reiter's treponema	12

[a]Taken from Mancy *et al.*, 1970.

B. THE PRODUCING ORGANISMS

1. *Streptomyces rishiriensis*

This soil microorganism was characterized as having aerial mycelia in the form of tangled and branched hyphae with sporophores occurring as sinistrose spirals. Electron micrographs revealed that the spores are oval to elliptical in shape and that they possess a very smooth surface (Kawaguchi et al., 1965a).

2. Streptomyces hazeliensis var. hazeliensis

The morphology of this organism was similar to that of *Streptomyces rishiriensis* with sporophores described as being "straight to flexuous, ending sometimes in a wide hook or loop, or in loose, open, characteristically wide (diameter 7 μ) spirals with 1 to 5 loops, the spirals being quite irregular in shape and distance between the loops." Spores were described as oval to cylindrical, while the nature of their surface was not indicated. Hoffmann-LaRoche workers considered the microorganism to be a new species and gave it the designation "*Streptomyces hazeliensis* var. *hazeliensis*" (Hoffmann-LaRoche, 1968a). The *S. rishiriensis* and *S. hazeliensis* var. *hazeliensis* cultures were compared under identical conditions and found to be indistinguishable (Table V). Since *S. rishiriensis* was adequately described and named prior to *S. hazeliensis* var. *hazeliensis*, the former designation is preferred for the coumermycin A_1-producing culture.

3. Streptomyces spinicoumarensis and Streptomyces spinichromogenes

The discovery of two new *Streptomyces* species that produce coumermycin A_1 has been recently reported (Institute of Microbial Chemistry, 1971). The *S. spinicoumarensis* culture was isolated from a soil specimen collected in Chiba City, while the *S. spinichromogenes*-bearing specimen was found in Toyooka City, Japan. The cultures were deposited in the Fermentation Research Institute, MITI, Japan, with the former culture receiving the designation FERM-P371, the latter FERM-P372. It is assumed that these microorganisms represent valid new species.

4. The 18,631 R.P.-Producing Organisms

The organisms which produce 18,631 R.P. are *Streptomyces hygroscopicus*, strain DS 9,751 – NRRL 3418; *S. albocinerescens*, DS 21,647 – NRRL 3419; and *S. roseochromogenes* var. *oscitans*, strain DS 12,976 – NRRL 3504 (Mancy et al., 1970). The morphological characteristics of these organisms are consistent with those previously described for these species and are clearly different from those of *S. rishiriensis* and *S. hazeliensis* var. *hazeliensis*. The *S. roseochromogenes* culture did vary from the type species in several minor characteristics and therefore was considered to be a special

TABLE V
MORPHOLOGICAL CHARACTERISTICS AND BIOCHEMICAL PROPERTIES OF CULTURES PRODUCING COUMERMYCINS

Characteristic	Streptomyces rishiriensis ATCC 14812	S. hazeliensis var. hazeliensis NRRL 2938	S. hygroscopicus NRRL 3418	S. albocinerescens NRRL 3419	S. roseochromogenes var. oscitans NRRL 3504
Color of the aerial mycelium	White to gray, becoming gray-brown	White to gray, becoming gray-brown	Light gray to deep gray, with small hygroscopic patches	White to light gray	Light grayish pink (poor sporulation)
hygroscopic					
Spore chain morphology	Flexuous to hooks and spirals	Flexuous to hooks and spirals	Tight spirals of 1 to 5 turns	Long, straight to flexuous	Long, ending in spirals of 1 to 2 turns
Spore surface	Smooth	Smooth	Not given	Not given	Not given
Spore shape and size	Oval to elliptical	Oval to cylindrical	Cylindrical	Oval with rounded ends	Oval
	$0.7–1.0 \times 1.0–2.5\,\mu$	$1.0–1.2 \times 1.4–2.4\,\mu$	$0.6–0.9 \times 0.9–1.2\,\mu$	$0.5–0.7 \times 1.0–1.2\,\mu$	$0.3–0.4 \times 0.6–0.8\,\mu$
Melanin production	Positive	Positive	Negative	Negative	Positive
Carbon utilization	Positive Negative Glucose Mannitol Arabinose Sucrose Xylose Inositol Fructose Rhamnose Raffinose	Positive Negative Same as S. rishiriensis ATCC 14812	Positive Negative Glucose Arabinose Sucrose Xylose Inositol Fructose Rhamnose Raffinose Mannitol	Positive Negative Glucose Xylose Arabinose Inositol Sucrose Mannitol Fructose Rhamnose Raffinose	Positive Negative Glucose Rhamnose Arabinose Sucrose Xylose Inositol Mannitol Fructose Raffinose
Antibiotics produced	Coumermycins	Coumermycins (sugordomycins)	18631 R.P.	18631 R.P.	18631 R.P.

variety of the species. For this reason it was designated as *S. roseochromogenes* var. *oscitans*. The major morphological and biochemical properties of these organisms are summarized in Table V.

C. Coumermycin Production and Isolation

Coumermycin A_1 is produced by *Streptomyces rishiriensis* in aerated submerged culture (Kawaguchi *et al.*, 1965a). The producing organism grows luxuriantly at 26–30°C in various media of which the following could be considered representative: Hydrolyzed corn starch (4%), pharmamedia (3.5%), glycerin (0.4%), yeast extract (0.2%), $CaCO_3$ (0.8%), KH_2PO_4 (1%), and KCl (0.2%). Peak yields of coumermycin A_1 are attained after 6 to 8 days of incubation. The antibiotic is found in both the mycelium and the broth filtrate, but principally in the former. Subsequent reports by Kawaguchi and his co-workers (1965b,c,d) revealed that *S. rishiriensis* fermentations also contained coumermycin A_2 and three bioactive factors designated as components B, C, and D.

A report by other Japanese workers (Institute of Microbial Chemistry, 1971) indicates that coumermycin A_1 can also be produced in agitated submerged fermentation media containing carbon sources such as glycerin, glucose, and starch and with various nitrogen sources, e.g., soybean meal and corn steep liquor. The *S. spinicoumarensis* and *S. spinichromogenes* cultures grown at 27° C to 32° C produce maximum quantities of coumermycin A_1 after a 6- to 8-day fermentation.

Berger *et al.* (1966) found that the coumermycin A_1-producing culture (*S. rishiriensis*) grew well on a variety of complex nitrogen and carbohydrate sources. Furthermore, reasonable growth could be attained in a simple synthetic medium composed of 0.1% L-proline, 1% soluble starch, and inorganic salts (Pridham and Gottlieb, 1948). They observed excellent antibiotic yields in media containing yellow split peas as a nitrogen source and starch, dextrose, glucose, or mannitol as a source of carbon. Antibiotic levels, however, were markedly suppressed when 0.25% concentrations of citric, fumaric, malic, acetic, lactic, gluconic, succinic, or glutamic acids were present in the fermentation medium. Since use of various media resulted in different antitiotic activity ratios in tests against selected sensitive organisms, it was apparent that multiple antibiotic moieties were being produced during the fermentations. Additional investigation allowed the Hoffmann-LaRoche workers to identify 9 different components (Fig. 1).

The identification of these coumermycin components (Fig. 1) as compounds (**1**), (**2**), (**9**), and the pairs (**3**) and (**4**), (**5**) and (**6**), (**7**) and (**8**) was made possible by the development of quantitative analytical methods for the determination of the terminal pyrrole moieties. Using a combination of pyrolysis-vapor phase chromatography and ultraviolet spectral measurements, it was possible to not only demonstrate the compositions of these compounds and pairs of compounds, but in the case of mixtures to give semiquantitative estimates of the compositions. These estimates were based upon the known behavior of the pure components of the antibiotic complex which had been isolated by Craig countercurrent distribution (Scannell, 1968).

Since the A_1 material is by far the most potent antibiotic in the mixture, it was a stroke of good fortune to find that minute amounts of cobalt ion directed the biosynthesis exclusively to coumermycin A_1 (Claridge et al., 1966; Claridge and Gourevitch, 1968). It is probable that the producing organism has a requirement for vitamin B_{12} in order to carry out the methylation of the precursor to the pyrrole-2-carboxylic acid moiety, and that the trace of cobalt ion insures the presence of enough vitamin B_{12} to complete that sequence. Further information as to the nature of the biosynthetic pathway by which coumermycin A_1 is produced has been contributed by Scannell and Kong (1970). These investigators found through use of radio-labeled precursors that glucose is incorporated into the nonpyrrole portions of the molecule and that methionine serves as a precursor for both pyrrolic and nonpyrrolic moieties. Their observation that proline was incorporated into the pyrrole groups to a 10-fold greater extent than δ-aminolevulinic acid suggested to them that the coumermycin A_1 pyrrole formation mechanism differed from that occurring in porphyrin formation.

The coumermycin complex, or coumermycin A_1 in the case where the biosynthesis has been directed mainly to the production of coumermycin A_1, is readily isolated. The fermentation broth is filtered at its ambient pH, usually about pH 8, and the filtrate is adjusted to pH 6 and extracted with methylisobutyl ketone (MIBK). The damp filter cake is extracted with acetone and the acetone is removed from the resulting filtrate under vacuum. The resulting aqueous concentrate is adjusted to pH 6 and extracted with MIBK. Crude coumermycin A_1 may be recovered from the combined MIBK extracts by concentration and precipitation with petroleum ether, or it may be further purified by transfer back and forth one or more times between solvent and water at pH's 10 and 6. A very similar procedure is described for the isolation of coumermycin A_1 from the fermentation carried out with *Streptomyces spinichromogenes* or *Streptomyces spinicoumarensis* (Institute of Microbial Chemistry, 1971). Coumermycin A_1 may be obtained as analytically pure material by countercurrent distribution, as described by Kawaguchi et al. (1965a,b,c). Pure, crystalline coumermycin A_1 decomposes at 258–260°C, has the molecular composition $C_{55}H_{59}N_5O_2$, and possesses characteristic infrared and ultraviolet spectra.

Antibiotic 18,631 R.P., as has been found with coumermycin A_1, can be produced in fermentation broths that are comprised of a wide variety of assimilable sources of carbon and nitrogen. Useful carbon sources include glucose, sucrose, lactose, starch, molasses, and various organic acids. Certain oils such as lard, soybean, and cottonseed are also acceptable. Nitrogen sources range from simple chemical compounds, e.g., nitrates, ammonium salts, urea, and amino acids to complex proteins such as casein, lactalbumen, gluten, fish meal, and yeast extract. Improved yields result from the addition of certain organic salts containing alkali metals, carbonates, or phosphates, and particularly chlorides. Fermentations can be carried out aerobically or under submerged conditions at pH's ranging from 6.0 to 7.8. Temperature requirements are wide, ranging from 23° C to 35° C but more optimally 25° to 28° C. Maximum antibiotic yields are obtained after 4 to 7 days, depending upon the nature of the production medium, and are reported to be as high as 288 μg/ml.

Antibiotic 18,631 R.P. is isolated by acidification of the fermentation broth to pH 5 and separation of the mycelium, which contains most of the activity. The antibiotic is extracted therefrom with aqueous methanol at pH 7. Concentration to a small volume and acidification to pH 5 allows the antibiotic to be transferred into a relatively large volume of n-butanol from which it is recovered as the sodium salt by concentration and neutralization with sodium methoxide. At this stage it is approximately 40% pure. The product is purified by successive column chromatography over Dowex 1X2 (chloride form) in water, alumina in ethyl acetate, and finally crystallization as the free acid from a mixture of acetone-dioxane-water, final melting point 206°C.

D. Laboratory and Clinical Studies with Coumermycin A_1

The *in vitro* antimicrobial spectrum of coumermycin A_1 has been examined by a number of investigators (Price, 1965; Kawaguchi et al., 1965a; Grunberg and Bennett, 1966; Fedorko et al., 1969; Michaeli et al., 1970). A compilation of results obtained with this antibiotic in tube dilution MIC (minimal inhibitory concentrations) tests against various gram-positive bacteria are presented in Table VI.

It is apparent from these data that coumermycin A_1 is not only an extremely potent antistaphylococcal agent, but also has marked in-

TABLE VI
In Vitro Activity of Coumermycin A_1 Against Various Gram-Positive Bacteria

Reference[a]	Geometric mean MIC in $\mu g/ml$[b]					
	Diplococcus pneumoniae	Streptococcus pyogenes	Enterococci	Staphylococcus aureus	Clostridium sp.	Bacillus anthracis
1	0.44 (4)	0.54 (6)	3.1 (1)	0.004 (6)	0.1 (4)	—
2	0.78 (2)	0.78 (—)	—	0.002 (7)	—	0.1 (1)
3	0.09 (1)	0.09 (1)	—	0.0016 (126)	—	—
4	0.62 (11)	0.62 (3)	25.0 (4)	0.004 (100)	—	—
5	0.22 (2)	2.89 (17)	4.05 (15)	0.26 (30)	—	—

[a] 1, Price, 1965; 2, Kawaguchi et al., 1965a; 3, Grunberg and Bennett, 1966; 4, Fedorko et al., 1969; 5, Michaeli et al., 1970.
[b] Numbers in parentheses indicate the number of strains.

TABLE VII
In Vitro Activity of Coumermycin A_1 Against Various Gram-Negative Bacteria

Reference[a]	Geometric mean MIC in $\mu g/ml$[b]							
	Pseudomonas aeruginosa	Escherichia coli	Shigella sp.	Klebsiella-Enterobacter sp.	Proteus mirabilis	Proteus sp. (indol +)	Serratia marcescens	Salmonella sp.	Haemophilus influenzae
1	25.0 (1)	12.5 (2)	6.25 (4)	12.5 (2)	3.1 (1)	3.9 (3)	—	12.5 (5)	0.08 (3)
2	12.5 (1)	7.4 (4)	3.9 (3)	0.78 (1)	—	—	—	3.1 (2)	—
3	—	12.5 (1)	—	—	—	—	—	12.5 (1)	—
4	15.9 (26)	16.6 (12)	—	41.0 (21)	3.55 (32)	13.4 (20)	—	—	—
5	13.2 (17)	11.16 (14)	25.0 (2)	31.25 (10)	—	—	10.64 (8)	>25.0 (7)	0.84 (6)

[a] 1, Price, 1965; 2, Kawaguchi et al., 1965a; 3, Grunberg and Bennett, 1966; 4, Fedorko et al., 1969; 5, Michaeli et al., 1970.
[b] Numbers in parentheses indicate the number of strains.

hibitory effects against other gram-positive bacteria of clinical significance. Despite the variety of test conditions utilized by the different investigators, there was in general very good agreement among mean MIC values reported for the *Staphylococcus aureus, Streptococcus pyogenes,* and *Diplococcus pneumoniae* strains.

Additional confirmation of coumermycin A_1's potent antistaphylococcal activity has been provided by Hoeprich (1968) who reported that when a number of antibiotics were tested *in vitro* against clinical strains of staphylococci (sixty-two), coumermycin A_1 was among the most active of these agents. Michaeli *et al.* (1969) found that all 13 methicillin- and cephalothin-resistant staphylococcal isolates examined were susceptible to <0.08 μg/ml of coumermycin A_1. Members of another gram-positive species, *Mycobacterium tuberculosis*, have also been shown to be quite susceptible to coumermycin A_1. Kawaguchi *et al.* (1965a) tested a single strain (H37Rv) and indicated that it was inhibited by 0.78 μg/ml. This indication of high susceptibility was confirmed by Duma and Warner (1969), who reported that all 8 strains of *M. tuberculosis* var. *hominis* tested by them were inhibited at coumermycin A_1 concentrations below 2.5 μg/ml. Only 1 of these 8 strains was found to be comprised of a population that contained a significant proportion of coumermycin A_1-resistant cells.

Coumermycin A_1 also has inhibitory activity against some gram-negative bacterial species. A summary of data obtained from the indicated sources is presented in Table VII.

Results tabulated here show that coumermycin possesses moderate activity against a wide variety of gram-negative bacteria and has particularly potent inhibitory effects against *Haemophilus influenzae* and *Proteus mirabilis* strains. As was the case with gram-positive organisms, there was remarkably good agreement between MIC values reported by the different investigators. The only exceptions of significance were the low MIC reported by Kawaguchi *et al.* (1965a) for the single *Klebsiella* isolate they tested and the high mean MIC value found for *Salmonella* strains by Michaeli *et al.* (1970).

In addition to the group of susceptible gram-negative species listed in Table VII, strains of *Neisseria meningitidis* are also reportedly inhibited by very low concentrations of coumermycin A_1. Devine and Hagerman (1970) found that the highest MIC obtained for any of the 40 strains tested was 0.00048 μg/ml. Furthermore, the geometric mean MIC was 0.0001 μg/ml, about 200 times lower than that of rifampin, the next most active agent among the 49 antimicrobials evaluated.

Recent laboratory studies by Cleeland et al. (1970) revealed that coumermycin A_1's antibacterial activity against selected strains of *Pseudomonas aeruginosa, Escherichia coli, Klebsiella pneumoniae,* and *Proteus mirabilis* could be enhanced some 8- to 40-fold when EDTA (ethylenediamine tetraacetic acid) was incorporated in a Mg^{++}-free synthetic test medium (Brock and Brock, 1959). The response of novobiocin was similar but of somewhat lesser magnitude. The enhancing effect could be virtually eliminated by addition of Mg^{++} to the synthetic medium and did not occur to an appreciable extent in complex medium. The authors suggested that the outer lipopolysaccharide complexes of the gram-negative organisms were interfering with the entry of coumermycin A_1 into the cells but they were unable to arrive at any conclusion as to the antibiotic's mode of action. A significant contribution to this area, however, was made by Michaeli et al. (1971) who studied the effects of coumermycin A_1 on the growth and metabolism of a *Staphylococcus aureus* strain. Their results suggest that the antibiotic's primary action is not directed at suppression of cell-wall formation or antagonism of cellular protein synthesis, but to a rapid antagonism of nucleic acid synthesis, with DNA (deoxyribonucleic acid) showing particularly high, but not exclusive, susceptibility.

The effect of inoculum size on the *in vitro* antimicrobial activity of coumermycin A_1 has been examined by several groups of investigators (Price, 1965; Kawaguchi et al., 1965a; Fedorko et al., 1969). There is general agreement that the antibiotic is not appreciably influenced by the size of the inoculum employed. However, the MIC did increase somewhat when the number of cells in the inoculum exceeded 10^6/ml.

The same investigators along with Michaeli et al. (1970) reported that the minimum bactericidal concentration (MBC) of coumermycin A_1 was significantly higher than the MIC for some microorganisms whereas the two values were identical for others. Studies in the laboratories of the present authors indicate that organisms in the first category usually have a fairly high incidence of resistant cells in their populations, while no evidence of such heteroresistance is apparent in those strains for which similar MIC and MBC values are found. The present authors also observed that the antibiotic has bactericidal effects against both proliferating and nonproliferating cells.

Attempts were made by all of the above-cited authors to determine the rate at which resistance to coumermycin A_1 emerges when microorganisms are repeatedly transferred in sublethal concentrations of the drug. The consensus was that resistance will develop in a progressive fashion under this condition but at a much slower rate than is usually found with streptomycin. Cross-resistance occurred exclusively and completely with novobiocin.

Another factor that dramatically influences the antimicrobial activity of coumermycin A_1 is the pH of the test medium. The present authors and Kawaguchi et al. (1965a) observed that the compound's antistaphylococcal activity at pH 6.0 was 100- to 1000-fold greater than its activity at pH 8.0. In the range of pH 7.0–7.2, the antibiotic was 10- to 20-fold less active than at pH 6.0.

Kawaguchi et al. (1965a), Fedorko et al. (1969), and Michaeli et al. (1970) all noted that the antimicrobial potency of coumermycin A_1 was markedly reduced when studies were conducted in media containing human serum or plasma. The present authors, on the basis of results obtained with both an equilibrium dialysis method and a procedure described by Scholtan and Schmid (1962) estimated that coumermycin A_1 is bound by fresh human serum to the extent of 98–99%. Despite this high degree of binding, coumermycin A_1, as is demonstrated by the data in Tables VIII and IX, gives significant therapeutic effects in experimental infections of mice.

Table VIII summarizes results obtained by several different groups of investigators when coumermycin A_1 was administered by the oral or parenteral route to mice experimentally infected with gram-positive organisms. *Staphylococcus aureus* Smith infections responded to both oral and parenteral therapy. However, CD_{50}'s obtained by the oral route were 10 to 30 times higher than those found after subcutaneous administration of coumermycin A_1. This, of course, suggests that the compound has a relatively low degree of oral absorbability. *Diplococcus pneumoniae* and *Streptococcus pyogenes* infections also responded to parenteral coumermycin A_1 therapy but at much higher doses than were required following staphylococcal challenge. No evidence of oral effectiveness was observed in infections caused by the diplococcal and streptococcal strains. Finally, the one *Mycobacterium tuberculosis*-induced infection studied did not respond to intramuscular administration of 20 mg/kg of coumermycin A_1.

Table IX summarizes results obtained by Hoffmann-LaRoche (Grunberg and Bennett, 1966; Grunberg et al., 1967) and Bristol Laboratories workers when they evaluated the efficacy of coumermycin A_1 in experimental infections of mice produced by intraperitoneal challenge with various gram-negative organisms.

Grunberg et al. (1967) found the antibiotic to be quite effective in an experimental *Neisseria meningitidis* infection and, much as was the case with gram-positive infections, the oral CD_{50} was considerably

TABLE VIII

THERAPEUTIC ACTIVITY OF COUMERMYCIN A_1 IN EXPERIMENTAL INFECTIONS[a] OF MICE CAUSED BY GRAM-POSITIVE ORGANISMS

CD_{50} (curative dose, 50%) in mg/kg (total)

Reference[b]	Staphylococcus aureus Smith		Diplococcus pneumoniae		Streptococcus pyogenes		Mycobacterium tuberculosis	
	Oral	Parenteral	Oral	Parenteral	Oral	Parenteral	Oral	Parenteral
1	—	0.37[c]	—	64[c]	—	66[c]	—	—
2	4.3	0.13	—	—	—	—	—	—
3	3.0	0.30	>500	28	>500	17	—	>20
4	16.0	—	>400	—	>400	—	—	—
5	17.0	—	—	—	—	—	—	—

[a] Mice challenged by the intraperitoneal route with lethal doses of the infecting organism.
[b] 1, Price, 1965; 2, Kawaguchi et al., 1965a; 3, Grunberg and Bennett, 1966; 4, Cron et al., 1970; 5, Price et al., 1970a.
[c] Intramuscular route employed, all other parenteral treatments administered subcutaneously.

TABLE IX

THERAPEUTIC ACTIVITY OF COUMERMYCIN A_1 IN EXPERIMENTAL INFECTIONS[a] OF MICE CAUSED BY GRAM-NEGATIVE ORGANISMS

CD_{50} (curative dose, 50%) in mg/kg (total)

Reference[b]	Neisseria meningitidis		Klebsiella pneumoniae		Proteus vulgaris		Escherichia coli		Salmonella typhi	
	Oral	Parenteral	Oral	Parenteral	Oral	Parenteral	Oral	Parenteral	Oral	Parenteral
1	—	—	—	>100[c]	—	88[c]	—	—	—	—
2	—	—	229	11	>500	>50	>500	>50	>500	>50
3	22	2	—	—	—	—	—	—	—	—
4	—	—	—	260[c]	—	—	—	—	—	—

[a] Mice challenged by the intraperitoneal route with lethal doses of the infecting organism.
[b] 1, Price, 1965; 2, Grunberg and Bennett, 1966; 3, Grunberg et al., 1967; 4, Price et al., 1970a.
[c] Intramuscular route employed, all other parenteral treatments administered subcutaneously.

higher than that obtained parenterally. The *Klebsiella pneumoniae* infection responded to both oral and parenteral coumermycin A_1 therapy although results were somewhat variable. Mice infected with *Proteus vulgaris* were protected only by parenteral treatment, while no treatment route proved effective in *Escherichia coli* and *Salmonella typhi* infections. Grunberg and Bennett (1966) reported that coumermycin A_1 also failed to protect mice challenged with *Pseudomonas aeruginosa* and *Salmonella schottmuelleri*. Thus, on the basis of these results and those shown in Table IX, it appears that coumermycin A_1 has limited efficacy in infections in mice caused by gram-negative organisms, particularly when administered by the oral route.

A somewhat surprising observation regarding the antimicrobial spectrum of coumermycin A_1 was made by Grunberg *et al.* (1967). These authors found that the antibiotic was active in an infection of mice produced by a member of the psittacosis-lymphogranuloma group of viruses, namely, the murine meningopneumonitis agent. The compound's activity following subcutaneous administration was greater ($CD_{50} = 14$ mg/kg) than that of antibiotics such as tetracycline ($CD_{50} = 79$ mg/kg) and chloramphenicol ($CD_{50} = 325$ mg/kg) when they were given by the same route. No effect against the virus was found when coumermycin A_1 was given at oral doses up to 500 mg/kg.

Table X shows that coumermycin A_1 has low acute toxicity for mice. The high ratio of oral to parenteral toxicity (about 10:1) offers additional evidence that the antibiotic is poorly absorbed by the oral route in mice. This suggestion was confirmed by the present authors who found the peak coumermycin A_1 blood level in mice after a single oral dose of 25 mg/kg to be only 0.9 μg/ml, a concentration some 20-fold lower than that obtained after administration of a com-

TABLE X
ACUTE TOXICITY OF COUMERMYCIN A_1 FOR MICE

Reference[a]	Acute LD_{50} (lethal dose, 50%) in mg/kg			
	Intraperitoneal	Subcutaneous	Intramuscular	Oral
1	—	380	—	>2000
2	—	—	500[b]	—
3	183	—	—	—
4	230	—	—	2400

[a] 1, Kawaguchi *et al.*, 1965a; 2, Kawaguchi *et al.*, 1965b; 3, Newmark and Berger, 1970; 4, Price *et al.*, 1970a.
[b] Value for coumermycin complex.

parable dose by the intramuscular (IM) route. Kawaguchi et al. (1967) found that mice receiving single 100 mg/kg doses by either the oral or IM route had relatively high coumermycin A_1 levels in the stomach, intestine, lung, and liver, and low, but detectable, concentrations in the brain and kidney.

Coumermycin A_1 also proved to be well tolerated in terms of survival and weight gain when administered chronically to rats by the IM or the oral route. Animals that received the drug at 25 mg/kg per day IM or 100 mg/kg per day orally for 60 days had normal growth rates and displayed no gross evidence of drug-related pathology (Kawaguchi et al., 1965a).

Since animal studies clearly showed that coumermycin A_1 administered by the IM route is extremely irritating (Price et al., 1970a) and that there is appreciable "retardation of movement of the drug" from the injection site (Newmark et al., 1970), further efforts to utilize parenteral dosage forms of this drug have been abandoned.

Early trials in humans (Kawaguchi et al., 1967) showed that oral absorbability of coumermycin A_1, either as the free acid or as a simple salt, was also poor in this species. The best absorbed dosage form was a Carbowax preparation which gave an average peak level of about 2 μg/ml after administration of a 500-mg dose. Hoffmann-LaRoche workers also studied various dosage forms, one of which was a combination of 1 part coumermycin A_1 and 4 parts N-methylglucamine. In both dogs and humans, this mixture was found to enhance oral absorption of coumermycin A_1 by 5- to 15-fold (Newmark and Berger, 1970). Utilizing this preparation Newmark et al. (1970) estimated that the "post-absorptive" half-life in man is about 10 hr. These same authors confirmed the earlier report of Kawaguchi et al. (1967) that little or no bioactivity is detectable in the urine of humans receiving coumermycin A_1.

Others (Fedorko et al., 1969; Devine et al., 1970; Hoeprich, 1971; Michaeli et al., 1970; Newmark, 1970) have also determined serum levels in humans following oral administration of mixtures with N-methylglucamine that contained from 50 to about 150 mg of coumermycin A_1. Average peak serum concentrations of the antibiotic in these studies ranged from 1 to 4 μg/ml. Kaplan (1970) reported that 3 subjects receiving a single 100-mg dose of coumermycin A_1 as an N-methylglucamine mixture each absorbed 14% of the administered drug, while a fourth subject absorbed 28%. He estimated the antibiotic's half-life to be in the range of 15 to 46 hr.

Michaeli et al. (1970) gave repeated doses of the mixture containing 100 mg coumermycin A_1 to 8 volunteers. Each one was dosed at

8-hr intervals for 4 consecutive days. Eight hours after drug ingestion on the fourth day, the average serum concentration was 1.28 µg/ml as compared to a mean of 1.07 found 8 hr after dosing on day 1. Thus, there appeared to be little or no accumulation of drug occurring at this dose level. The major side effect observed was gastrointestinal distress. Seven of the 8 subjects complained of abdominal pains, "heartburn," nausea, and vomiting. Although symptoms were suppressed by use of antacids and sedatives, 6 of the 8 subjects had diarrhea. No other untoward reactions were noted.

Kawaguchi et al. (1967) also found a high incidence of gastrointestinal disturbances when the disodium salt of coumermycin A_1 was used (1 to 3 grams/day) to treat 64 patients having a variety of clinical conditions. Included were patients with furuncle, phlegmon, infected atheroma, lymphadenitis, mastitis, bronchitis, and tonsilitis. According to the authors, satisfactory results were obtained in 49 cases (76.6%). There were 12 failures (18.8%) and 3 cases that could not be evaluated. The cure rate for staphylococcal infections was about 81% (18 of 22 cases). Also responding well were infections caused by α- or β-hemolytic streptococci and corynebacteria. Daily 2-gram doses appeared to be more effective than 1-gram doses.

Another human study with coumermycin A_1 was conducted by Devine et al. (1970) who attempted to eliminate the meningococcal carrier state in normal individuals. Their study was prompted by a previous observation (Devine and Hagerman, 1970) that coumermycin A_1 has greater potency against *Neisseria meningitidis* than any of 48 other antimicrobials tested. Subjects received 100 mg of coumermycin A_1 as the N-methylglucamine mixture every 12 hr for 7 days. A control group of comparable size (81 individuals) received a placebo that was similar in appearance. The antibiotic, as determined by culture techniques, failed to have any influence on the incidence of *N. meningitidis* in the nasopharyngeal tissues. These disappointing results were attributed to the absence of significant drug concentrations in the saliva of treated subjects. The drug's failure, when used for this particular application, was predicted by Hoeprich (1971), who showed that only a minimal or nondetectable concentration of coumermycin A_1 was present in the saliva and tears of the normal subjects he examined, despite the fact that serum concentrations in these individuals averaged about 2.3 µg/ml.

In an earlier clinical study, Colmore et al. (1968) reported that orally-administered coumermycin A_1, as one could readily predict on the basis of its failure to appear in urine, was generally ineffective in controlling urinary tract infections. A "success" rate of 24% was found for the 21 patients treated. This compared very unfavorably with the 68% success rate achieved with tetracycline in the same study.

Thus, coumermycin A_1 has essentially been eliminated as a candidate for an orally-active chemotherapeutic agent. Its low degree of absorbability, its poor distribution characteristics, its innate potential for causing gastrointestinal distress, and finally, its unimpressive record of therapeutic successes, have all contributed significantly to its demise.

III. Semisynthetic Coumermycins

Early recognition that coumermycin A_1 had too much irritation liability to be used parenterally, that it is remarkedly susceptible to serum binding and possible inactivation, and finally, that it is very poorly absorbed following oral administration, made it a prime candidate for chemical modification.

It was planned to direct structural changes toward (1) preservation of coumermycin A_1's broad spectrum and potent antimicrobial activity, (2) reduction of its susceptibility to serum binding, (3) lowering its irritation liability, and (4) increasing its oral absorbability.

An analytical appraisal of the coumermycin A_1 molecule with an eye to possible modifications suggested the changes shown in structure (12). For convenience, the coumermycin subunits are herein referred to as P, 5-methylpyrrole; N, noviose; C, 4-hydroxy-8-methyl-coumarin; and *D*, 2,4-dicarboxamido-3-methylpyrrole. No nondestructive method for electrophilic substitution on the coumarin nucleus has been reported, so the remarks which follow are confined to the other four possibilities.

A. ALTERATION AT THE 4-HYDROXYL

1. Ethers

Treatment of the di-2′-tetrahydropyranyl ether of coumermycin A_1 (see below) in tetrahydrofuran with an excess of ethereal diazomethane produced compound (**13**). Similar treatment of coumermycin A_1 produced compound (**14**) (Keil, 1966). Both (**13**) and (**14**) were inactive.

2. 4-Deoxy Analogs

Catalytic reduction (PtO_2, H_2) of oxazoles (**15**) derived from coumermycin A_1 produced 4-deoxy analogs (**16**) of the semisynthetic coumermycins. All analogs with structure (**16**) were inactive, as were the oxazole precursors (**15**) (Keil, 1966).

B. ALTERATION AT THE 2'-HYDROXYL

1. Esters

If coumermycin A_1 is acylated under mild conditions (e.g., in pyridine with one equivalent of acid chloride or anhydride at 0–25°C) the result is simple esterification of the noviose hydroxyls to give compounds (**17**). The coumarin-4-hydroxyl is phenolic and does not acylate under these conditions (or if it does acylate to some extent, the acyl group is washed off in the course of isolation of the product).

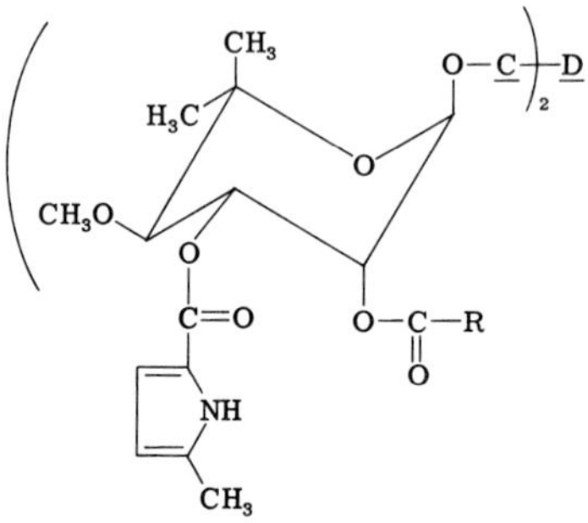

(17)

Most such derivatives (**17**) are inactive, and it has been shown by chromatography with bioautograph that the slight activity sometimes observed is due to a small degree of hydrolysis back to the parent compound (Claridge, 1967). The analogs (**18**) were prepared as intermediates in the total synthesis of coumermycins A_1 and A_2 (Batcho et al., 1969; Whitaker, 1968), but no claims were made for their bioactivity.

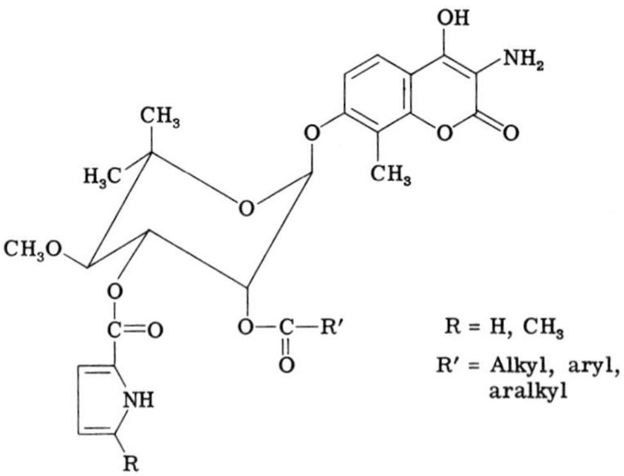

R = H, CH_3
R' = Alkyl, aryl, aralkyl

(18)

2. Ethers and the Tetrahydropyranyl Acetal

The formation of simple ethers of coumermycins at the 2'-hydroxyls has not been reported. On the basis of the lack of activity of the 2'-esters and the tetrahydropyranyl acetal, it is to be expected that 2'-ethers would be inactive. Preparation of the bis-2'-tetrahydropyranyl acetals of coumermycins A_1 and A_2 is described in a U.S. patent (Nettleton, 1968). Although they are biologically inactive, these derivatives proved to be exceptionally useful intermediates for the preparation of semisynthetic coumermycins. They form readily and in high yield, are stable, crystalline solids, and are readily reconvertible to coumermycins A_1 and A_2 on treatment with toluenesulfonic acid in methanol. They are stable under conditions used for displacement of the D moiety, but are readily cleaved from the resulting intermediates to provide fully active, semisynthetic coumermycins. In short, the bis-THP derivatives were found to be near-ideal intermediates for the preparation of a series of new, active compounds.

C. Alterations at the 3'-Ester

Unlike novobiocin and dihydronovobiocin, wherein the carbamoyl group readily equilibrates between the 3' and 2' positions under mildly basic conditions (Hinman et al., 1957b; Birlova and Trakhtenberg, 1966), no such migration of the 2-pyrrolecarbonyl function has been observed in coumermycins A_1 and A_2, or any of the semisynthetic analogs. As noted in Section II, the 5-methyl group on the pyrrole ester is necessary for high activity, and removal of a methyl group from even one of the terminal pyrroles reduces the activity to about one-fourth that of coumermycin A_1 (see mixture **3+4** in Table I). If one of the terminal esters is absent, further reductions in activity result (mixtures **5+6** and **7+8**, Table I) and if both are removed then the product, NCDCN (**9**), is virtually inactive. In the course of attempts to find hydrolytic conditions specific for removal of the D moiety of coumermycin A_1, it was discovered that warming the coumermycin or its bis-tetrahydropyranyl derivative in 95–100% hydrazine gave very high yields of compound **9** or the corresponding bis-THP analog (Cron and Nettleton, 1965). No use was made of these observations because it was felt that the 5-methylpyrrole probably represents the ultimate in desirable R-groups for the ester.

D. Alterations at the 3-Nitrogen

This was the area which proved to be most fruitful, in terms of both interesting chemistry and production of active antibiotics.

1. Oxazole Formation

Hinman et al. (1956, 1957a) had shown that hot acetic anhydride cleaves novobiocin according to the reaction

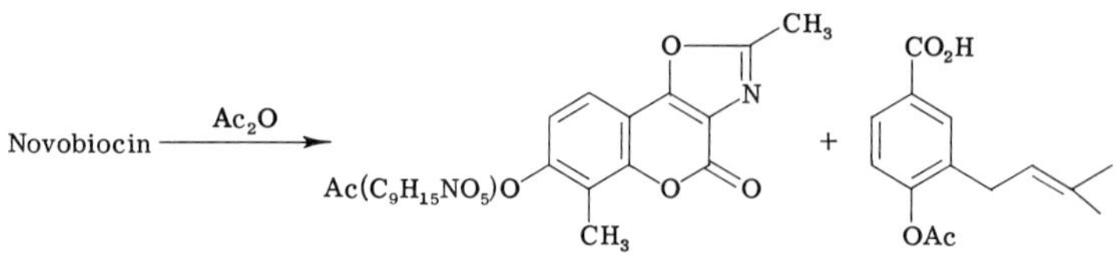

$$\text{Novobiocin} \xrightarrow{Ac_2O} Ac(C_9H_{15}NO_5)O\text{-[oxazole-coumarin]} + \text{[acetoxybenzene with CO}_2\text{H and prenyl]}$$

This reaction applied to coumermycin A₁ produced (**19**), while the bis-THP derivative yielded (**20**) (Keil et al., 1971) which could be cleaved to (**21**).

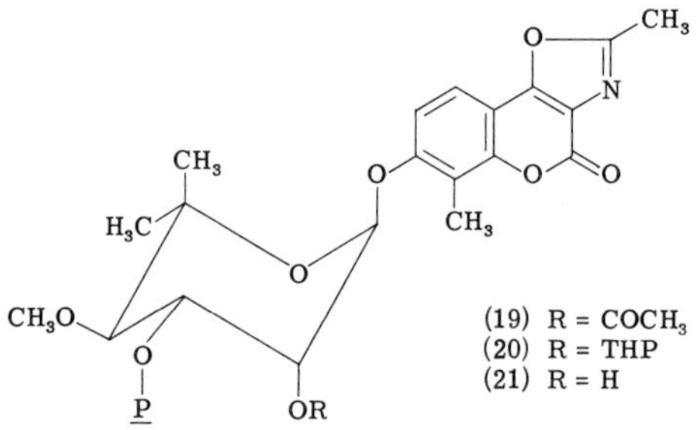

(**19**) R = COCH₃
(**20**) R = THP
(**21**) R = H

Although variation of the acid anhydride or chloride gave a variety of analogs of (**21**) with alkyl, aryl, aralkyl, and other groups in place of the oxazole methyl, none of these derivatives were active against bacteria. It was noted, however, that unless a considerable excess of acylating agent and quite vigorous reaction conditions (e.g., boiling pyridine) were employed, the oxazole was always accompanied by chromatographically detectable amounts of antibacterial substances (after removal of the THP group) not identical with coumermycin A₁. Furthermore, the amounts of these active byproducts were inversely proportional to the severity of reaction conditions. These observations led to the discovery of the transacylation reaction of coumermycins.

2. The Transacylation Reaction

The following sequence illustrates this very useful reaction.

The intermediate (**22**) has not been isolated because of its instability which results in rapid rearrangement to (**23**), examples of which have been isolated and characterized (Keil et al., 1969, 1971).

That (**22**) is in fact a necessary intermediate is supported by the observations that the 4-methoxy (**13**) and 4-deoxy (**16**) analogs of (**14**) do not undergo the reaction (Keil, 1966), and that acyl O to N migrations in similar 2-aminophenols are well known (Bell, 1931). A number of the THP-precursors (**24**) of the semisynthetic coumermycins

(27) have also been isolated and fully characterized, as has the by-product (25) (Keil et al., 1969). The 3-methylpyrrole-2,4-dicarboxylic acid was not isolated in pure form, although chromatographic evidence of its presence was occasionally seen.

a. Aromatic Amides

One hundred and six derivatives with structure (27), where R is aromatic, heteroaromatic, or β-phenylethylene (derived from cinnamic acids) have been reported (Keil et al., 1969; Schmitz et al., 1968; Schmitz and Godfrey, 1970; Schmitz and DeVault, 1970; Cron, 1970).

3. PNC-Amine

It was recognized very early in the coumermycin program that degradation of coumermycin A_1 to PNC-amine would provide an intermediate of great flexibility for the preparation of semisynthetic derivatives. However, because of the relatively greater sensitivity of the ester and coumarin lactone functions to solvolytic cleavage, all such attempts to specifically hydrolyze the amide linkages were frustrated. When the transacylation reaction was discovered, early attempts to carry it out with carbobenzoxy chloride also failed. It was only after a painstaking investigation of reaction conditions that the reaction sequence shown below was brought to fruition (Keil et al., 1970, 1971; Keil and Hooper, 1969). As in the original transacylation sequence, intermediates (28) and (29) were isolated and characterized. Compound (29) proved to have antibacterial activity in vitro and in vivo, but was not orally absorbed. PNC-amine (30), was very weakly active (MIC = 4.0 μg/ml vs. Streptococcus pyogenes, 12.5 μg/ml against staphylococci). The successful preparation of PNC-amine made possible the preparation of a variety of semisynthetic coumermycins derived from aliphatic and olefinic acids (which formed not at all or in very poor yield via the transacylation route), isocyanates, isothiocyanates, sulfonyl halides, and alkyl halides. None of the latter four types of derivatives could be made by the transacylation procedure. [This observation may be taken as secondary confirmation of the intermediacy of compounds (22) in the transacylation sequence, since none of the four reagents can give intermediates (22) which are likely to rearrange to analogs of (23).]

a. Aliphatic and Olefinic Amides

Thirty-six semisynthetic coumermycins (27) wherein R is aliphatic or olefinic were prepared from the corresponding acid halides, anhydrides, mixed anhydrides, or from the free acid plus dicyclohexylcarbodiimide (Keil et al., 1970; Keil and Hooper, 1970).

b. Ureides and Thioureides

These derivatives formed readily on refluxing of a pyridine solution of *PNC*-amine with isocyanates or isothiocyanates. At best, they were weakly active *in vitro* and inactive *in vivo* (Keil et al., 1970).

c. Sulfonamides

Sulfonyl halides condense readily with *PNC*-amine under mild conditions to produce the expected sulfonamides. The few examples prepared (only the *p*-toluenesulfonamide was purified for analysis) all showed very poor activity, so this area was quickly abandoned (Keil *et al.*, 1970).

d. Aliphatic Secondary Amine

Since several examples proved that this class of derivatives is virtually inactive, only the *n*-propyl analog, *PNC*-NHCH$_2$CH$_2$CH$_3$, was carried through to analytical purity (Keil *et al.*, 1970).

E. ANTIBACTERIAL ACTIVITY OF VARIOUS ACYLATED *PNC*-AMINES

More than 160 derivatives of coumermycin A$_1$ were synthesized by direct transacylation of the antibiotic itself or by acylation of the free coumarin amino group possessed by *PNC*-amine. Minimal inhibitory concentrations (MIC) of the *PNC*-derivatives were determined by the standard 2-fold broth dilution technique (Gourevitch *et al.*, 1961) utilizing Antibiotic Assay Broth or the same medium containing the indicated amount of human serum. All derivatives were tested against 1 strain of each of 3 gram-positive species (*Diplococcus pneumoniae, Streptococcus pyogenes,* and *Staphylococcus aureus*), as well as against one strain of each of 7 gram-negative species (*Proteus morganii, P. mirabilis, Escherichia coli, Salmonella typhosa, S. enteritidis, Klebsiella pneumoniae,* and *Pseudomonas aeruginosa*). It was observed that any structural modification in the test compounds that affected their activity against the highly sensitive *S. aureus* Smith strain appeared to induce a corresponding and generally proportional change in activity against the other 2 gram-positive species. Similarly, the degree of inhibitory effect of the compounds against the *K. pneumoniae* strain, the most sensitive of the gram-negative species, was found to be representative of their activity against the other 6 members of this group of organisms. Consequently, inhibition values obtained in tests with these 2 strains are considered to be accurate indicators of the extent of a given compound's overall antimicrobial potency. The higher MIC values obtained in the medium containing human serum (25% by volume) as compared to those found in the serum-free medium were attributed to binding of the test compounds by serum proteins.

The median curative dose (CD$_{50}$) of each of the *PNC* derivatives was determined in an experimental *S. aureus* Smith infection of mice. The challenge, which contained sufficient cells to kill all mice within 24 hr, was administered in 5% mucin by the intraperitoneal (IP) route. Animals were treated once by the oral route, immediately after challenge. The number of mice surviving at 96 hr was recorded and the CD_{50} (the dose in mg/kg required to cure 50% of the infected mice) estimated by means of a log-probit plot.

Klebsiella pneumoniae infections of mice were induced by IP-administration of the bacterial cells in mucin-free saline. Oral or IM (intramuscular) treatment was given both at the time of challenge and 4 hr later. The CD_{50} was determined as described above.

1. Variation with Type of Amide

Although the bulk of compounds synthesized were *PNC*-carboxamido derivatives, a few members of two other classes (*PNC*-sulfonamido and *PNC*-ureido derivatives) were prepared. Table XI shows comparative MIC and CD_{50} values for coumermycin A_1 and representative derivatives from the three classes of amides.

The marked activity of coumermycin A_1 against *Staphylococcus aureus* Smith in *in vitro* tests was considerably reduced when the medium was supplemented with 25% human serum. Its MIC against *Klebsiella pneumoniae* in serum-free medium, while still low (3.1 µg/ml), was more than 1000 times higher than that found for the *S. aureus* Smith culture. The *in vitro* antibacterial activity of all the *PNC* derivatives was much lower than that of coumermycin A_1, with the poorest activity being produced by the nonacylated *PNC*-amine. Although all remaining *PNC* compounds were more active than the free amine derivative, only four inhibited *S. aureus* Smith at 1.0 µg/ml or less, while none displayed significant activity against the *K. pneumoniae* strain. The four compounds having the greatest inhibitory effect on the *S. aureus* strain were representatives of the aryl-, alkyl-, and aralkylcarboxamido and arylsulfonamido classes. The lowest MIC observed among this group of semisynthetic coumermycins was that of the phenylcarboxamido. This compound also proved to be the most efficacious in the *S. aureus* Smith mouse protection test. The other two carboxamido derivatives (compounds **33** and **34**) had poorer *in vivo* antistaphylococcal activity, while the sulfonamido derivative (compound **35**) was inactive. The failure of this compound to display *in vivo* activity and the high CD_{50} obtained with the aralkylcarboxamido representative can very likely be attributed to their relatively great susceptibility to serum binding. A high degree of binding for these compounds is suggested by the fact that their inhibitory concentrations in 25% serum are at least 50 times higher than in serum-free medium. The poor intrinsic activity of the aryl- and alkylureido derivatives probably is responsible for their lack of *in vivo* efficacy.

TABLE XI
ANTIBACTERIAL ACTIVITY OF PNC^a DERIVATIVES HAVING VARIOUS SUBSTITUENTS IN THE 3-POSITION OF THE COUMARIN MOIETY

Compound number	R Group		Minimum inhibitory concentration (μg/ml)			$CD_{50}{}^b$ (mg/kg)
	Class	Structure	Staphylococcus aureus Smith		Klebsiella pneumoniae	S. aureus Smith
			No serum	25% Serum	No serum	
(1)	Coumermycin A₁	(see Fig. 1)	0.0008	0.016	3.1	17
(31)	Amino	—N(H)—H	12.5	50	100	>100

(**32**)	Arylcarboxamido	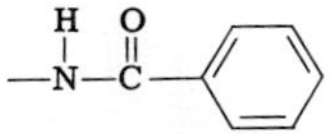	0.063	0.6	50	5.5
(**33**)	Alkylcarboxamido	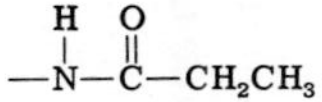	1	1.8	100	20
(**34**)	Aralkylcarboxamido	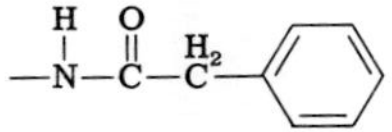	0.4	50	100	68
(**35**)	Arylsulfonamido	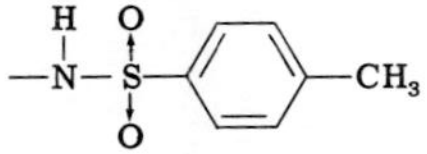	0.8	50	>100	>100
(**36**)	Arylureido	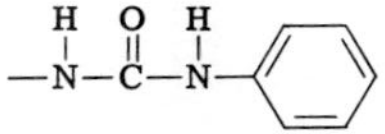	>1	12.5	>100	>100
(**37**)	Alkylureido	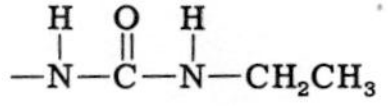	3.1	12.5	100	>100

[a]4-hydroxy-8-methyl-7 [3-0-(5-methyl-2-pyrrolylcarbonyl) noviosyloxy] coumarin.
[b]Compounds were administered by oral route.

On the basis of results shown in Table XI, further structure-activity relationship studies were confined to those PNC derivatives possessing carboxamido groups.

2. Amides of Aromatic Acids

a. Monosubstituted on the Aromatic Ring

The PNC derivative with the unsubstituted benzamido side chain (Table XI) gave a lower oral *S. aureus* Smith CD_{50} than coumermycin A_1 despite having an MIC about 80 times higher in serum-free medium and 20 times higher in serum-containing medium. Substitution of a chlorine or methyl group into the *ortho-*, *meta-*, or *para*-position of the benzene ring did not have a favorable effect on its *in vitro* or *in vivo* activity. Regardless of the substituent's position, MIC and CD_{50} values were 2- to 4-fold higher than those obtained with the unsubstituted parent compound. Similar results were obtained with the *p*-iodo, *p*-bromo, and *p*-fluorobenzamido derivatives.

Several other *PNC*-benzamido compounds having *p*-substituents with varying electron-withdrawing capacities were subjected to *in vitro* and *in vivo* evaluation. Results of these studies are summarized in Table XII.

It can be seen that there were no appreciable differences between compounds in regard to their *in vitro* activity versus *S. aureus* Smith and *K. pneumoniae* when tests were conducted in serum-free medium. However, addition of 25% human serum to the test medium resulted in an increase in MIC that paralleled to some extent the increase in electron-withdrawing capacity of the *para*-substituent. *In vivo* efficacy also appeared to be affected since there was a corresponding increase in the *S. aureus* Smith CD_{50}. The most active members of this series had *in vivo* activity about equal to that of the unsubstituted benzamido derivative.

b. Disubstituted on the Aromatic Ring

In most cases introduction of 2 chlorine atoms into the benzene ring had a depressing effect on *in vivo* activity. However, there were profound differences in the compounds' *in vivo* efficacy depending upon the location of the substituents. Data obtained with several such derivatives are shown in Table XIII.

As evidenced by the CD_{50} obtained with the 2,6-dichlorobenzoyl derivative, disubstitution in the *ortho*-position had a negligible effect on *in vivo* efficacy. This was not the case, however, with the 3,4-dichloro derivative which failed to give a measurable protective effect in the experimental mouse infection. Measurable but poor *in*

TABLE XII

ANTIBACTERIAL ACTIVITY OF *PNC*-BENZAMIDO[a] DERIVATIVES HAVING SUBSTITUENTS WITH VARYING ELECTRON-WITHDRAWING CAPACITIES IN THE PARA-POSITION OF THE BENZENE RING

Compound number	R Group		Minimum inhibitory concentration (μg/ml)				CD_{50}[b] (mg/kg)
			Staphylococcus aureus Smith		Klebsiella pneumoniae		S. aureus Smith
	Name	Structure	No serum	25% Serum	No serum	No serum	
(38)	Amino	—NH$_2$	0.08	0.8	50		8
(39)	Hydroxy	—OH	0.16	0.8	>100		8
(40)	Methoxy	—OCH$_3$	0.08	3.2	>100		12
(41)	Nitro	—NO$_2$	0.08	50	50		22

[a] 3-benzamido-4-hydroxy-8-methyl-7 [3-0-(5-methyl-2-pyrrolylcarbonyl) noviosyloxy] coumarin.
[b] Compounds were administered by the oral route.

TABLE XIII
Antibacterial Activity of $PNC\text{-}NH_2{}^a$ Derivatives Having Side Chains Prepared from Dichlorobenzoic Acids

Compound number	R Group		Minimum inhibitory concentration (μg/ml)			$CD_{50}{}^a$ (mg/kg)
			Staphylococcus aureus Smith		Klebsiella pneumoniae	S. aureus Smith
	Name	Structure	No serum	25% Serum	No serum	
(32)	Benzoyl	—C(=O)—C₆H₅	0.063	0.6	100	5.5

No.	R				
(42)	2,6-Dichlorobenzoyl	0.031	0.31	50	6
(43)	2,5-Dichlorobenzoyl	0.031	2.5	>100	50
(44)	3,5-Dichlorobenzoyl	0.063	50	>100	70
(45)	3,4-Dichlorobenzoyl	0.16	50	50	>100

[a] 3-amino-4-hydroxy-8-methyl-7 [3-0-(5-methyl-2-pyrrolylcarbonyl) noviosyloxy] coumarin.
[b] Compounds were administered by the oral route.

TABLE XIV
ANTIBACTERIAL ACTIVITY OF PNC-NH$_2$[a] DERIVATIVES HAVING SIDE CHAINS PREPARED FROM METHYL AND HYDROXYL SUBSTITUTED BENZOIC ACIDS

Compound number	R Group		Minimum inhibitory concentration (μg/ml)			CD$_{50}$[b] (mg/kg)
			Staphylococcus aureus Smith		Klebsiella pneumoniae	S. aureus Smith
	Name	Structure	No serum	25% Serum	No serum	
(32)	Benzoyl	–C(O)–C$_6$H$_5$	0.063	0.6	100	9.5

(46)	4-Hydroxy-3-methyl-benzoyl	![structure with CH3 and OH]	0.032	0.4	25	7
(47)	3-Hydroxy-4-methyl-benzoyl	![structure with OH and CH3]	0.25	3.1	50	50
(48)	2-Hydroxy-3-methyl-benzoyl	![structure with HO and CH3]	0.06	50	50	45

[a] 3-amino-4-hydroxy-8-methyl-7 [3-0-(5-methyl-2-pyrrolylcarbonyl) noviosyloxy] coumarin.
[b] Compounds were administered by the oral route.

vivo activity was found for the 2,5- and 3,5-dichlorobenzoyl compounds. The vast range of activity displayed by this group of chlorine-substituted derivatives could not have been predicted from MIC values in serum-free medium, but did correlate well with those obtained in serum-supplemented medium.

The MIC in serum and the *S. aureus* Smith CD_{50} of the 3,5-dinitro- and 3,5-diaminobenzoyl derivatives were no better than those of the 3,5-dichlorobenzoyl compound. However, poor efficacy is not a general property of 3,5-disubstituted compounds since the dimethylbenzoyl analog was only slightly less active in *in vitro* and *in vivo* tests than the unsubstituted parent compound.

c. Special Case of 3-Substituted-4-hydroxybenzamides

Among the semisynthetic coumermycins with disubstituted benzoic acid side chains examined, there were several that contained both hydroxyl and methyl substituents in the benzene ring. Data collected with these compounds are shown in Table XIV.

The 4-hydroxy-3-methylbenzoyl derivative was by far the most potent of the three compounds. It had good *in vivo* activity and its MIC values for *Staphylococcus aureus* Smith and *Klebsiella pneumoniae* were somewhat lower than those of the unsubstituted benzoyl. The excellent activity observed with this disubstituted compound was not completely unexpected in view of the fact that the closely related antibiotic, novobiocin, also contains a 3-alkyl-4-hydroxybenzamido group at the 3-position of the coumarin moiety.

Other derivatives having up to 6 carbons in a straight chain in the 3-position were subjected to *in vitro* and *in vivo* studies. *In vitro* antistaphylococcal activities of compounds having 3, 4, 5, or 6 carbons were up to 20 times better than that of the 3-methyl homolog. The gram-negative activity of the derivatives, based on the *K. pneumoniae* MIC, was also considerably affected by the size of the *n*-alkyl side chain. Figure 2 shows the antibacterial activity in serum-free medium of this series as a function of the number of carbons in the 3-alkyl group.

The striking increase in activity achieved by lengthening the *n*-alkyl chain was, with the possible exception of the *n*-hexyl derivative, almost inapparent in medium containing 25% serum. The 6-carbon homolog had an MIC in serum about one-tenth that of the other compounds. This derivative gave the lowest *S. aureus* Smith CD_{50}, although all members of the series had a median curative dose that fell within a very narrow range (6 to 14 mg/kg).

Much the same activity pattern was observed with branched-chain

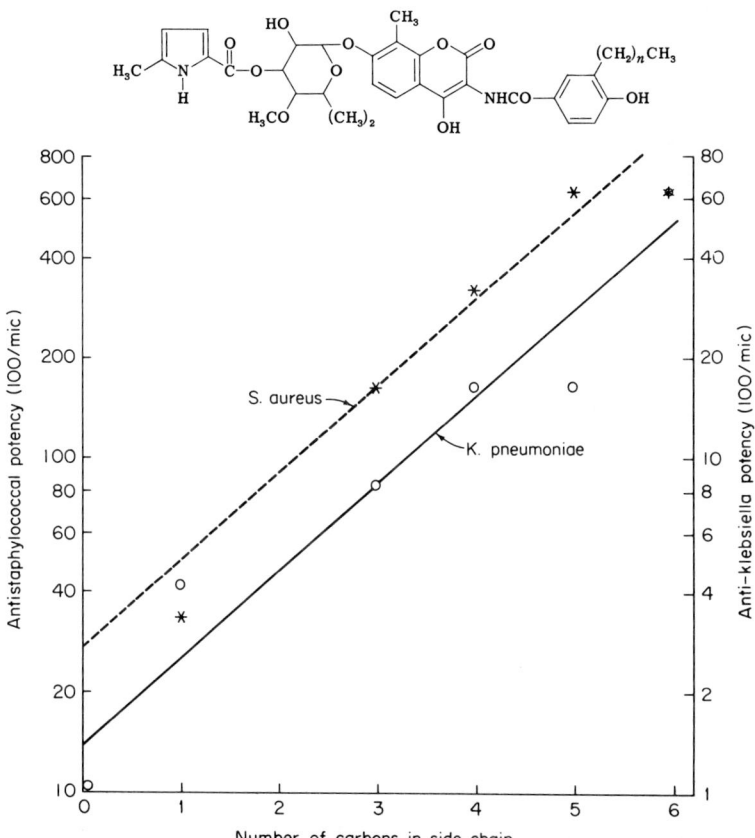

FIG. 2. Relative *in vitro* antibacterial potency against *Staphylococcus aureus* (✱) and *Klebsiella pneumoniae* (o) of 4-hydroxy-8-methyl-7-[3-O-(5-methyl-2-pyrrolylcarbonyl)noviosyloxy] coumarins substituted in the 3-position of the coumarin moiety with various 3-*n*-alkyl-4-hydroxybenzamido groups.

derivatives. Compounds with 3, 4, or 5 carbons in the 3-position, such as the *i*-propyl, 1-methylpropyl, *t*-butyl, 3-methylbutyl, and the 1,2-dimethylpropyl, displayed good *in vitro* activity against both *S. aureus* Smith and *K. pneumoniae*. The CD_{50} obtained in the *S. aureus* Smith infection ranged from 4 to 10 mg/kg for these compounds.

Several unsaturated 3-alkyl side chains were also examined. These were the allyl, the 1-methylallyl, and the novobiocin side chain, the 3-methyl-2-butenyl. Despite having comparable *in vitro* activity, both the allyl and 1-methylallyl were much less effective *in vivo* (CD_{50} values of 23.5 and 25 mg/kg, respectively) than their corresponding saturated analogs. However, the 3-methyl-2-butenyl derivative was as active as the saturated compound since their respective median curative doses were 5.0 and 8.6 mg/kg.

Introduction of a benzyl, a phenethyl, or a cyclohexyl group into the 3-position of the 4-hydroxybenzamido side chain resulted in compounds that have *in vitro* and *in vivo* antibacterial effects similar to those found with the longer chain *n*-alkyl analogs. Table XV shows antibacterial data for one of these compounds (cyclohexyl) as well as for outstanding representatives among the derivatives having an *n*-alkyl or branched-chain alkyl group.

All of the 3-substituted derivatives just discussed had better *in vitro* antistaphylococcal and anti-*Klebsiella pneumoniae* activity in serum-free medium than did the unsubstituted compound. However, in the presence of 25% serum, only compound **50**, the *n*-hexyl analog, had a *Staphylococcus aureus* Smith MIC appreciably lower than that of the compound having no 3-substituent. All of the derivatives were effective in *S. aureus* Smith infections with oral CD_{50} values ranging from 6 to 12.5 mg/kg. This level of *in vivo* activity compares very favorably with that of coumermycin A_1. The derivative with the greatest activity in the *S. aureus* Smith infection, the 3-*n*-hexyl analog, also proved to be as effective as coumermycin A_1 in an experimental *K. pneumoniae* infection of mice when the compounds were administered by the IM route. Neither was active against this infection when treatment was given orally at doses up to 300 mg/kg × 2. None of the other compounds shown in Table XV had measurable *in vivo* activity against *K. pneumoniae*.

d. Trisubstituted on the Aromatic Ring

Acylation of the coumarin amine with trisubstituted benzoic acids produced compounds having a higher degree of serum binding and poorer *in vivo* efficacy than the unsubstituted benzoic acid derivative. Among derivatives examined were those having 2,4,6-trimethylbenzamido, 3,5-dimethyl-4-hydroxybenzamido, and 3,5-dipropyl-4-hydroxybenzamido side chains.

e. Amides of Aromatic and Heteroaromatic Acids

The relative antibacterial efficacy of some representative derivatives in the aromatic and heteroaromatic series is considered in Table XVI. Good *in vitro* and *in vivo* activity was demonstrated by all compounds except the fused ring derivative (1-naphthoyl) and the phenyl-substituted isoxazole (compound **57**). It seems probable that the bulky nature of these acyl groups increased their degree of binding to serum with the net result that their *in vivo* effectiveness was decreased.

TABLE XV
ANTIBACTERIAL ACTIVITY OF PNC-4-HYDROXYBENZAMIDO[a] DERIVATIVES HAVING VARIOUS SUBSTITUENTS IN THE 3-POSITION OF THE BENZENE RING

Compound number	R Group Name	R Group Structure	Minimum inhibitory concentration (μg/ml)				CD_{50}^{b} (mg/kg)	
			Staphylococcus aureus Smith		K. pneumoniae		S. aureus Smith	K. pneumoniae
			No serum	25% Serum	No serum			
(49)	Hydrogen	—H	0.16	0.8	>100		14.3	—
(50)	n-Hexyl	—(CH$_2$)$_5$—CH$_3$	0.0016	0.02	1.6		6	140
(51)	3-Methylbutyl	—(CH$_2$)$_2$—CH—CH$_3$ CH$_3$	0.0032	0.2	6.2		8.6	>300
(52)	Cyclohexyl		0.0063	0.8	3.1		12.5	>200
(I)	Coumermycin A$_1$	(see Fig. 1)	0.0008	0.016	3.1		17	130

[a] 3(4-hydroxybenzamido)-4-hydroxy-8-methyl-7 [3-0-(5-methyl-2-pyrrolylcarbonyl) noviosyloxy] coumarin.
[b] Compounds administered by oral route (IX) for S. aureus Smith; by IM (2×) for K. pneumoniae.

TABLE XVI
ANTIBACTERIAL ACTIVITY OF PNC-$NH_2{}^a$ DERIVATIVES HAVING SIDE CHAINS PREPARED FROM AROMATIC AND HETEROCYCLIC ACIDS

Compound number	Name	R Group Structure	Minimum inhibitory concentration (μg/ml)			$CD_{50}{}^b$ (mg/kg)
			Staphylococcus aureus Smith		Klebsiella pneumoniae	S. aureus Smith
			No serum	25% Serum	No serum	
(32)	Benzoyl		0.063	0.4	100	14
(53)	2-Picolinoyl		0.16	0.8	100	17

(54)	1-Naphthoyl		0.063	50	>100	23
(55)	Thiazole-4-carbonyl		0.16	0.8	100	13
(56)	3,5-Dimethylisoxazole-4-carbonyl		0.32	3.2	>100	13
(57)	5-methyl-3-phenyl-isoxazole-4-carbonyl		0.32	50	>100	70

[a] 3-amino-4-hydroxy-8-methyl-7 [3-0-(5-methyl-2-pyrrolylcarbonyl) noviosyloxy] coumarin.
[b] Compounds were administered by oral route.

TABLE XVII
ANTIBACTERIAL ACTIVITY OF $PNC\text{-}NH_2{}^a$ DERIVATIVES HAVING SIDE CHAINS PREPARED FROM CINNAMIC AND ACRYLIC ACIDS

Compound number	Name	R Group Structure	Minimum inhibitory concentration (µg/ml)			$CD_{50}{}^b$ (mg/kg)
			Staphylococcus aureus Smith		Klebsiella pneumoniae	S. aureus Smith
			No serum	25% Serum	No serum	
(58)	Cinnamoyl	–C(O)–CH=CH–C₆H₅	0.125	1.6	50	9.0

(59)	4-Methoxycinnamoyl	−C(=O)−CH=C(H)−C₆H₄−OCH₃	0.032	1.6	25	8.0
(60)	2-Hydroxycinnamoyl	−C(=O)−CH=C(H)−C₆H₄−OH (2-OH)	0.125	0.8	25	20
(61)	α-Chlorocinnamoyl	−C(=O)−C(Cl)=C(H)−C₆H₅	0.125	3.2	100	>50
(62)	β-(2-Thienyl) Acryloyl	−C(=O)−CH=C(H)−(2-thienyl)	0.125	>3.2	100	35

[a] 3-amino-4-hydroxy-8-methyl-7 [3-0-(5-pyrrolylcarbonyl) noviosyloxy] coumarin.
[b] Compounds were administered by the oral route.

f. Amides of Cinnamic Acids

Some members of the cinnamic acid group of derivatives, as is apparent from the data in Table XVII, possessed excellent *in vitro* and *in vivo* antistaphylococcal activity while others, possibly as a result of their greater susceptibility to serum binding, were relatively ineffective *in vivo*. Somewhat surprising was the marked difference in *in vivo* activity found for the compound with the cinnamic acid side chain and its close structural relative, the β-(2-thienyl) acrylic acid derivative.

g. Amides of Aralkylcarboxylic Acids

A series of aralkyl compounds including the 2-phenylacetamido, 3-phenylpropionamido, 3-cyclohexylpropionamido, and a number of related derivatives with various substituents in the ring or on the α-carbon were examined for their antibacterial properties. Although most had a low MIC against *S. aureus* Smith in serum-free medium, addition of 25% serum resulted in a marked and essentially uniform decrease in antistaphylococcal activity. The lowest *S. aureus* Smith CD_{50} in this series (21 mg/kg) was obtained with the 2-phenylbutyramido derivative.

3. Amides of Aliphatic Acids

A number of PNC derivatives were prepared by acylation of the coumarin amine with aliphatic acid chlorides. Chain length of this series of homologs varied from 2 to 22 carbons. Figure 3 shows their *in vitro* potency against *S. aureus* Smith as a function of the number of carbons in the acyl group.

It is apparent that chain length has a profound effect on antistaphylococcal activity since potency of the derivatives varied over a greater than 1000-fold range (inhibitory concentrations ranged from 0.08 to 100 μg/ml). An increase in chain length starting from the 2-carbon acetyl group resulted in an increase in potency (about 10- to 20-fold) that peaked with the 10-carbon acyl radical. Activity then fell off rather precipitously as the number of carbons in the chain was increased still further. Compounds containing 21- and 22-carbon acyl radicals had about one-hundredth the activity of the 2-carbon derivative. Unfortunately, in serum-containing medium, an *S. aureus* Smith MIC of 50 μg/ml or greater was obtained with all the alkylcarboxamido compounds except the acetamido and propionamido which had inhibitory concentrations of 0.4 and 1.8 μg/ml, respectively. *In vivo* antistaphylococcal activity was also rather poor for this series since the most active compound (propionamido) had a CD_{50} of 30 mg/kg. None of the members of the series displayed significant *in vitro* activity against *K. pneumoniae*.

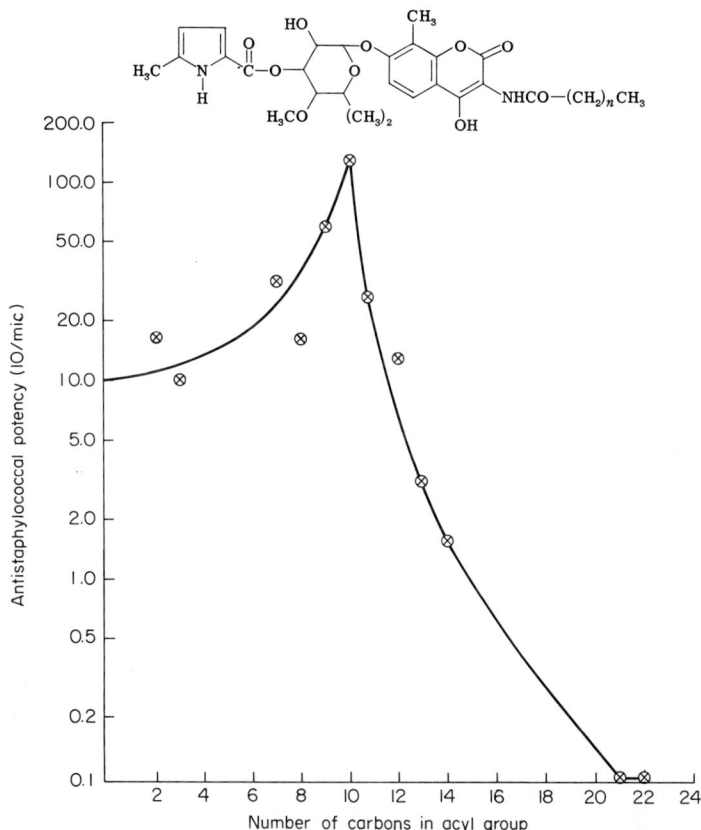

FIG. 3. Relative *in vitro* antibacterial potency against *Staphylococcus aureus* Smith of 4-hydroxy-8-methyl-7-[3-0-(5-methyl-2-pyrrolylcarbonyl)noviosyloxy] coumarins substituted in the 3-position of the coumarin moiety with various 3-*n* alkylamido groups.

In addition to the *n*-alkylamido compounds discussed above, a number of semisynthetic coumermycins having branched-chain acyl groups were examined. In contrast to the straight-chain analogs, several of the compounds were quite active *in vivo*. *Staphylococcus aureus* Smith CD_{50} values falling within an 8 to 13 mg/kg range were obtained with the 2-methylpropionamido, 3-methylbutyramido, and 4-methylpentanamido derivatives. Median curative doses obtained with the 2,2-dimethylpropionamido, 3,3-dimethylbutyramido, 2-ethylbutyramido, and 2-*n*-propylpentanamido were higher, ranging from 23 to 40 mg/kg.

F. PHARMACOLOGICAL PROPERTIES OF SELECTED SEMISYNTHETIC COUMERMYCINS

1. Oral Absorbability

Coumermycin A_1 and four semisynthetic derivatives that had demonstrated good *in vivo* efficacy in the *Staphylococcus aureus* Smith infection were selected for absorption studies. Blood concentrations of each of the compounds were determined at various time intervals following oral administration to mice of a single 25 mg/kg dose. Results obtained with coumermycin A_1 and these PNC derivatives are presented in Fig. 4.

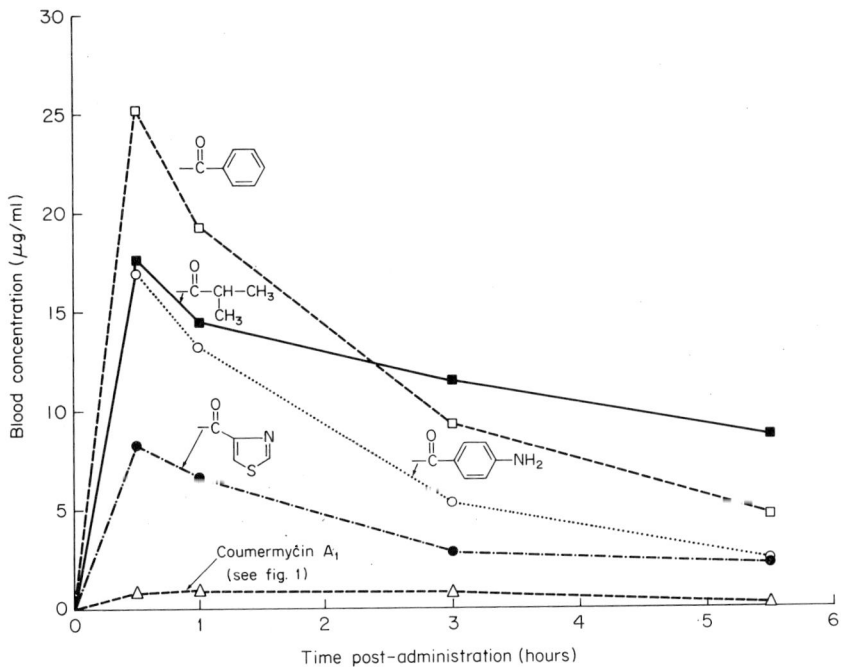

FIG. 4. Concentration of coumermycin A_1 and several amide derivatives of 3-amino-4-hydroxy-8-methyl-7-[3-0-(5-methyl-2-pyrrolylcarbonyl)noviosyloxy] coumarin in mouse blood after oral administration of a single 25 mg/kg dose. Each point is the average value for 12 mice.

Coumermycin A_1 displayed a low blood level peak (0.9 µg/ml at 1 hr post-treatment) but still had a measurable blood concentration (0.5 µg/ml) 5½ hr after drug administration. In sharp contrast were the blood levels produced by the same dose of the PNC derivatives. Peak concentrations of these compounds, which in all cases occurred at ½ hr post-administration, were considerably higher than the peak concentration of coumermycin A_1. Despite a rather wide range (8.4 to 24.9 µg/ml) of peak levels, all *PNC* derivatives with aromatic carboxamido side chains had similar blood "disappearance" curves. This was not the case, however, for the derivative with the 2-methylpropionamido side chain. Its blood concentration appeared to decrease at a significantly slower rate than that of the aromatic carboxamido compounds. The blood level of this derivative at 5½ hr was only 50% below the peak level as compared to a minimum 75% decrease from the peak for each of the other *PNC* derivatives.

Three additional *PNC* compounds were subjected to oral absorption studies in the mouse. These were benzamido analogs and had peak blood levels ranging from 14.6 to 26.7 µg/ml. In all cases peak blood levels occurred at ½ hr and serum disappearance curves were similar to those found for the other aromatic carboxamido derivatives.

The relationship between the *S. aureus* Smith MIC (in 25% serum), peak mouse blood levels, and therapeutic effectiveness in the Smith infection for coumermycin A_1 and the seven *PNC* derivatives that had been subjected to absorption studies, is shown in Table XVIII.

It is apparent that all of the *PNC* derivatives were quite effective in controlling the *S. aureus* Smith infection despite the fact that in most instances their antibacterial potency was inferior to that of coumermycin A_1. It must be presumed therefore that their relatively high efficacy in the Smith infection was due to their greater oral absorbability. When both antibacterial potency ($A = 1/MIC$) and peak antibiotic concentrations in the blood (P) are simultaneously considered, a coefficient showing the relative efficacy of the test compound (t) to coumermycin A_1 (c) can be obtained as follows:

$$\frac{A_t}{A_c} \times \frac{P_c}{P_t} = \text{"efficacy" coefficient}$$

Although the CD_{50} values obtained vary only within a narrow range (6 to 17 mg/kg), there does appear to be some relationship between the efficacy coefficient and therapeutic effectiveness. As a general rule, the higher the coefficient, the lower the CD_{50}. *PNC* derivatives having a coefficient of 2 or higher were at least twice as effective in the *S. aureus* Smith infection as coumermycin A_1 (coefficient of 1).

2. Irritation Studies

Coumermycin A_1 and several semisynthetic derivatives that had displayed a high level of effectiveness in an experimental mouse infection were tested to determine the extent of their irritating

TABLE XVIII
RELATIONSHIP BETWEEN MIC, PEAK BLOOD LEVEL, AND CD_{50} FOR COUMERMYCIN A_1 AND 8 PNC-NH_2[a] DERIVATIVES

Name	R group Structure	In vitro activity[b] ratio (1) Coumermycin A_1 compound	Peak blood level[c] ratio (2) Compound coumermycin A_1	Efficacy coefficient $(1) \times (2)$	Staphylococcus aureus Smith CD_{50}[d]
Coumermycin A_1	(see Fig. 1)	1	1	1	17
2-Methylpropionyl	—C(=O)—CH(CH₃)—CH₃	0.02	19.7	0.4	13
Thiazole-4-carbonyl	—C(=O)—(thiazole)	0.04	9.3	0.4	13

Substituent	MIC[b]	Blood level[c]	CD_{50}[d]	(value)
Benzoyl	0.05	27.7	1.4	9.5
2,6-Dichlorobenzoyl	0.1	21	2.1	6
4-Hydroxy-3-methyl-benzoyl	0.08	29.7	2.4	7
4-Hydroxy-3(3-methyl-butyl)-benzoyl	0.16	27.6	4.4	8
3-Hexyl-4-hydroxy-benzoyl	0.4	16.2	6.5	6

[a] 3-amino-4-hydroxy-8-methyl-7 [3-0-(5-methyl-2-pyrrolylcarbonyl) noviosyloxy] coumarin.
[b] *In vitro* activity based on the compound's minimum inhibitory concentration (MIC) for *Staphylococcus aureus* Smith in antibiotic assay broth (BBL) containing 25% human serum.
[c] Blood levels were determined after oral administration of a 25 mg/kg dose to mice.
[d] CD_{50} (Median curative dose) determined for *S. aureus* Smith infection (challenge given by IP route). A single oral treatment was given immediately after challenge.

properties for the rat foot-pad when injected as 1 or 5% solutions or suspensions. To be considered irritating, a compound must produce a swelling or edema volume of 0.25 ml or greater for at least 4 hr. The degree of edema obtained at representative time periods following administration of the test compounds is shown in Fig. 5.

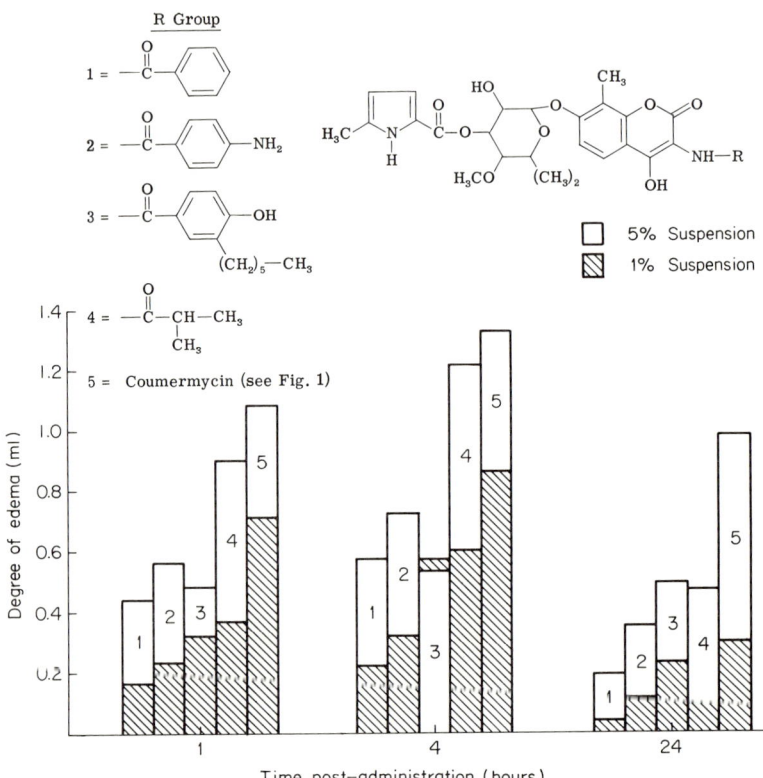

FIG. 5. Relative irritability to the rat foot-pad of coumermycin A_1 and four derivatives of 3-amino-4-hydroxy-8-methyl-7-[3-0-(5-methyl-2-pyrrolylcarbonyl)noviosyloxy] coumarin. One and 5% solutions or suspensions were injected in 0.1-ml volume into the plantar surface (foot-pad) of five rats per dose level. The degree of edema was measured by the volume displacement method.

Coumermycin A_1 proved to be extremely irritating at both 1 and 5% concentrations. Although the PNC derivatives had a lesser effect, all were irritating at the 5% dosage level and only two could be considered to be free of irritating properties at the 1% concentration level. The nonirritating derivatives were the benzoyl (compound **1**) and the *p*-aminobenzoyl (compound **2**). In most cases, the degree of edema reached a maximum at 4 hr post-injection and appeared to be directly related to the dosage administered. The only exception to this was noted with the 3-n-hexyl-4-hydroxybenzoyl analog (compound 3). This derivative, possibly because of its extremely low degree of solubility, was only slightly more irritating at the 5% than at the 1% level.

It is of interest that the swelling induced by all of the *PNC* compounds receded much more rapidly than that caused by coumermycin A_1. By 48 hr post-injection (not shown in Fig. 5), no swelling of significance was found for any of the *PNC* derivatives, whereas the 5% coumermycin suspension still produced an average foot-pad edema volume of >0.5 ml.

No evidence of pain on injection was noted for coumermycin A_1 or any of the *PNC* derivatives.

3. Acute Toxicity Studies

Intraperitoneal (IP) and oral (PO) acute LD_{50} (lethal dose, 50%) values in the mouse were determined for coumermycin A_1 and three *PNC* derivatives. Results are summarized in Table XIX.

The acute IP LD_{50} values of *PNC* derivatives other than the 2-6-dichlorobenzoyl were about the same as or higher than that of coumermycin A_1. The suggestion that the chlorine-containing compound has somewhat greater toxicity for mice is reinforced by the finding that its oral LD_{50} was also lower than the values of the other *PNC* compounds. On the whole, however, all of the derivatives displayed a surprisingly low level of oral toxicity relative to coumermycin A_1 considering that their degree of absorbability by this route is some 15 to 30 times greater.

4. Assessment of the Effects of Structural Modification on Biological Properties

The primary objectives of the structural modification program were to maintain or enhance the desirable properties of coumermycin A_1 and to reduce or eliminate some of its undesirable characteristics. There is little question but that a certain degree of success was achieved. The result for some of the more active compounds was a net 2- to 3-fold improvement in *in vivo* antistaphylococcal therapeutic effect. Furthermore, in experimental *Diplococcus pneumoniae* and *Streptococcus pyogenes* mouse infections, the oral efficacy of derivatives such as the 4-aminobenzoyl, the 3-benzyl-4-hydroxybenzoyl, the 3-hydroxy-4-(3-methylbutyryl)benzoyl, and the 2-methylpropionyl exceeded that of coumermycin A_1 by more than 4- to 8-fold. It is

TABLE XIX
ACUTE TOXICITY OF COUMERMYCIN A₁ AND SEVERAL PNC-NH₂[a] DERIVATIVES FOR MICE

[Structure diagram of coumermycin-type molecule with pyrrole-carbonyl-noviose-coumarin-NH—R backbone]

Name	R Group Structure	Intraperitoneal	Oral
		(Route of administration)	
Coumermycin A₁	(see Fig. 1)	230[b]	2400
Benzoyl	—C(=O)—C₆H₅	370	1980
2,6-Dichlorobenzoyl	—C(=O)—C₆H₃Cl₂	130	940
2-Methylpropionyl	—C(=O)—CH(CH₃)—CH₃	350	1400

[a] 3-amino-4-hydroxy-8-methyl-7-[3-0-(5-methyl-2-pyrrolylcarbonyl) noviosyloxy] coumarin.
[b] LD_{50} (lethal dose, 50%) in mg/kg.

noteworthy that the improved therapeutic effect of the semisynthetic antibiotics was achieved in most cases without a corresponding increase in toxicity to the host.

Compounds having activity of the order described above against gram-positive bacteria were usually obtained only when the 3-amino group of the coumarin moiety was acylated with aromatic or heteroaromatic carboxylic acids, either without, or with mono- or di-substituents on the ring. While several compounds prepared from branched-chain aliphatic acids had comparable *in vivo* efficacy, those derived from n-aliphatic or aryl-substituted aliphatic acids, as well as those prepared from tri-substituted benzoic acids, were much less effective.

The structural requirements for inhibitory activity against gram-negative bacteria would appear to be much more stringent since only those derivatives prepared by acylation with 3-substituted-4-hydroxybenzoic acids had significant activity. However, it is likely that such inhibitory effects are merely a reflection of these compounds' relatively high potency and cannot be attributed to a specifically broadened antibacterial spectrum. The most active representative of this class (the 3-n-hexyl-4-hydroxybenzoic acid derivative) and coumermycin A_1 had comparable efficacy when administered parenterally in an experimental *Klebsiella pneumoniae* infection of mice. However, neither proved to be effective when given by the oral route.

Many of the new semisynthetic compounds were found to possess markedly greater oral absorbability in mice. This improvement over the parent compound was of such magnitude (in some cases, >25-fold) that it more than compensated for the lower *in vitro* antibacterial potency possessed by most of the derivatives.

In addition to their superior oral absorbability, most of the semisynthetic derivatives tested proved to be significantly less irritating to local tissues following parenteral administration. Several derivatives, specifically those prepared by acylation with benzoic acid or 4-aminobenzoic acid, were less irritating when given as a 5% suspension than was coumermycin A_1 when given as a 1% suspension. Even the swelling produced by the most irritating derivative receded much more rapidly than that induced by coumermycin A_1.

The ultimate choice of a candidate for clinical evaluation from among the more active compounds was the 2-methylpropionyl or the 3-isobutyramido-4-hydroxy-8-methyl-7-[3-0-(5-methyl-2-pyrrolylcarbonyl)noviosyloxy]coumarin, the sodium salt of which will hereafter be referred to as BL-C43. The selection of this derivative was based on the fact that its antimicrobial activity was not inferior to that of other semisynthetic coumermycins, and that relative to the other compounds under consideration, it was less susceptible to serum binding (78% for BL-C43 vs. 90% or more for the others) and was better absorbed by the dog following oral administration.

G. Laboratory and Human Studies with BL-C43

Since the antibacterial spectrum of BL-C43 closely resembles the spectra of erythromycin, novobiocin, lincomycin, and oleandomycin, these antibiotics were included for reference purposes in the present studies.

The *in vitro* antibacterial spectra of BL-C43 and several other antibiotics as determined by tube dilution tests are shown in Table XX.

TABLE XX
ANTIMICROBIAL SPECTRUM OF BL-C43 AND SEVERAL COMMERCIAL ANTIBIOTICS

Organism	No. of strains tested	MIC[a]				
		BL-C43	Novobiocin	Lincomycin	Oleandomycin	Erythromycin
Staphylococcus aureus (penase −)[b]	2	0.5	<0.01	1.6	3.2	0.18
S. aureus (penase +)[b]	4	0.9	0.18	2	6.3	2
Streptococcus pyogenes	4	2	0.5	0.5	4	0.05
S. faecalis	4	16	1	8	8	0.18
Diplococcus pneumoniae	4	3.2	6.3	0.5	1.8	0.05
Listeria monocytogenes	2	6.3	0.25	8	8	0.18
Haemophilus influenzae	2	2	8	32	>32	1
Neisseria gonorrhoeae	2	2	0.35	>16	8	1
Pasteurella multocida	2	0.25	56	22.6	32	1.4
Escherichia coli	3	180	32	>250	400	28
Klebsiella pneumoniae	1	63	250	>500	>500	63
Enterobacter aerogenes	2	>500	16	>500	>500	125
Proteus mirabilis	2	63	32	>500	>500	250
Proteus sp. (indol +)	8	70	80	>500	>500	250
Pseudomonas aeruginosa	4	300		>500	>500	300

[a] MIC = geometric mean in μg/ml.
[b] Penase − = non-penicillinase-producer, penase + = penicillinase-producer.

It is apparent that BL-C43 is active against both gram-positive and gram-negative bacteria. However, the minimum inhibitory concentration (MIC) values obtained with the latter group of organisms are probably too high to be of significance from the clinical standpoint. Thus BL-C43 can be considered of interest primarily because of its gram-positive spectrum. It should be noted, however, that a few gram-negative species such as *Haemophilus influenzae, Neisseria gonorrhoeae,* and *Pasteurella multocida,* which are susceptible to many penicillins, are also quite sensitive to BL-C43. Lincomycin and oleandomycin have activity against gram-positive bacteria only, with their level of potency against these organisms being generally comparable to that of BL-C43. Erythromycin, on the other hand, possesses the same broad spectrum of antibacterial activity found for BL-C43 and novobiocin. However, it is considerably more inhibitory for gram-positive bacteria than are either of these antibiotics.

The distribution of MIC values of BL-C43 and the commercial antibiotics for a large number of strains of selected gram-positive and gram-negative bacterial species is presented in Tables XXI and XXII. These strains, which were predominantly of clinical origin, were tested by agar dilution methods utilizing the multiple inoculator apparatus of Steers *et al.* (1959).

Data in Table XXI show that novobiocin is by far the most potent antistaphylococcal agent with about 98% of the strains being inhibited at a concentration of 0.25 μg/ml or less. BL-C43 and erythromycin were almost as active, with 85 and 91% of the strains, respectively, inhibited at a concentration of 0.5 μg/ml, Lincomycin and oleandomycin were only slightly less inhibitory than the latter antibiotics.

Results obtained with 17 strains of *Streptococcus pyogenes* are also presented in Table XXI. Here erythromycin inhibited all strains at a concentration of 0.125 μg/ml. Lincomycin was only slightly less inhibitory since 100% of the cultures were sensitive at 0.5 μg/ml. The other antibiotics were less effective, requiring 2 to 4 μg/ml to inhibit all 17 strains.

A similar distribution of MIC values was found in tests against 17 strains of *Diplococcus pneumoniae,* where erythromycin was again the most active antibiotic. BL-C43, although displaying the least activity, did inhibit all strains at a concentration of 4 μg/ml. Lincomycin, novobiocin, and oleandomycin each inhibited 100% of the strains at a concentration of 1 or 2 μg/ml.

Table XXII summarizes results obtained with 18 strains of *Haemophilus influenzae.* Outstandingly active was novobiocin which inhibited all strains at 1 μg/ml. BL-C43 and erythromycin were effective

TABLE XXI

Relative Inhibitory Activity of BL-C43 and Several Commercial Antibiotics for *Staphylococcus aureus*, *Streptococcus pyogenes*, and *Diplococcus pneumoniae*

Compound	MIC (μg/ml)									
	0.016	0.032	0.063	0.125	0.25	0.5	1	2	4	>4
Staphylococcus aureus – 88 strains										
BL-C43	–	–	–	–	7[a]	85	98	–	–	100
Novobiocin	–	–	9	66	98	–	–	–	–	100
Lincomycin	–	–	–	–	–	9	91	100	–	–
Oleandomycin	–	–	–	–	–	6	74	98	–	100
Erythromycin	–	–	–	21	89	91	–	92	100	–
Streptococcus pyogenes – 17 strains										
BL-C43	–	–	–	–	–	–	–	–	100	–
Novobiocin	–	–	–	–	–	12	71	100	–	–
Lincomycin	–	–	–	–	18	100	–	–	–	–
Oleandomycin	–	–	–	–	–	–	–	100	–	–
Erythromycin	–	18	59	100	–	–	–	–	–	–
Diplococcus pneumoniae – 17 strains										
BL-C43	–	–	–	–	–	12	41	65	100	–
Novobiocin	–	–	6	–	35	77	94	100	–	–
Lincomycin	–	–	–	12	77	94	100	–	–	–
Oleandomycin	–	–	–	–	6	35	71	100	–	–
Erythromycin	71	100	–	–	–	–	–	–	–	–

[a]Numbers in body of table indicate total percentages of strains inhibited at indicated concentration.

TABLE XXII
Relative Inhibitory Activity of BL-C43 and Several Commercial Antibiotics for *Haemophilus influenzae* and *Neisseria gonorrhoeae*

Compound	MIC (μg/ml)									
	0.06	0.13	0.25	0.5	1	2	4	8	16	>16
Haemophilus influenzae — 18 strains										
BL-C43	—	—	—	—	50[a]	83	89	100	—	—
Novobiocin	—	61	83	89	100	—	—	—	—	—
Lincomycin	—	—	—	—	—	—	—	—	28	100
Oleandomycin	—	—	—	—	—	—	—	—	—	100
Erythromycin	—	—	—	11	50	67	100	—	—	—
Neisseria gonorrhoeae — 14 strains										
BL-C43	—	7	—	—	14	93	100	—	—	—
Novobiocin	—	—	—	7	—	—	29	93	100	—
Lincomycin	—	—	—	—	—	—	—	—	7	100
Oleandomycin	—	14	—	—	—	—	43	100	—	—
Erythromycin	14	—	—	36	100	—	—	—	—	—

[a] Numbers in body of table indicate total percentages of strains inhibited at indicated concentration.

at somewhat higher concentrations, while no measurable activity was found for lincomycin or oleandomycin.

Erythromycin was the most active antigonococcal antibiotic since a concentration of 1 μg/ml inhibited all 14 strains (see Table XXII). BL-C43, the next most active compound, was only slightly more effective than oleandomycin and novobiocin. Lincomycin inhibited only one of the 14 strains at the highest concentration tested (16 μg/ml).

MIC values of BL-C43 were found to be essentially unaffected by inoculum size. A 10,000-fold increase in the numbers of *Staphylococcus aureus* cells in the inoculum required only a doubling of the concentration of BL-C43 in order to inhibit growth of the organism in standard tube dilution tests.

A study of the relative bacteriostatic and bactericidal concentrations of BL-C43 for 2 strains of staphylococci revealed that BL-C43 was capable of sterilizing both cultures at concentrations only 2- to 4-fold higher than the MIC. As would be predicted, the ability of the staphylococcal strain to produce penicillinase had no significant effect on its sensitivity to BL-C43.

The rate at which *S. aureus* strain A9497 develops resistance to BL-C43, novobiocin, erythromycin, and lincomycin was determined by transferring cultures repeatedly in the presence of sublethal concentrations of the antibiotic. A similar stepwise development of resistance was noted in all cases and all 4 of the antibiotics could be considered to induce development of resistance at a comparable rate. BL-C43 was cross-resistant with novobiocin but not with the other antibiotics. It was noted that there was a considerable amount of cross-resistance between lincomycin and erythromycin.

BL-C43 was found to be remarkably stable to low pH. Its half-life or the time required for destruction of one-half the original 100 μg/ml concentration at pH 2.0 was estimated to be in excess of 10 hr. This indicates that the compound would not undergo significant degradation as a result of exposure to highly acidic gastric juices following its oral administration.

The relative effectiveness of BL-C43 and the other antibiotics in experimental bacterial infections of mice was determined. CD_{50} (curative dose, 50%) values obtained in these studies are shown in Table XXIII.

As can be observed, BL-C43 had excellent activity against all 4 strains of *S. aureus*. Novobiocin, lincomycin, and oleandomycin were also active against all 4 staphylococcal strains, but considerably less so than BL-C43. Erythromycin had a very low level of activity against these strains. As expected, the ability to produce penicillinase did not influence responses of the test organisms to these agents.

BL-C43 was found to be considerably more active than novobiocin, lincomycin, oleandomycin, and erythromycin in tests against *Streptococcus pyogenes* and *Diplococcus pneumoniae*. Interestingly, in the case of BL-C43, novobiocin, and erythromycin, there seemed to be little if any advantage in treating four times rather than twice. However, there was some indication that lincomycin and oleandomycin were more effective when four doses were administered.

The infection caused by *Pasteurella multocida*, the one gram-negative species tested, responded very well to BL-C43 therapy, but was quite refractory to treatment with the other antibiotics.

Comparative oral blood levels, in mice, of BL-C43, sodium novobiocin, linocomycin hydrochloride, erythromycin estolate, and triacetyloleandomycin after administration of a 25 mg/kg dose of each are shown in Fig. 6.

The blood level of BL-C43 reached its peak at 1 hr post-administration with a level in excess of 11 μg/ml being attained. The antibiotic tended to persist in the blood, since a 3 μg/ml concentration was still

TABLE XXIII
In Vivo ANTIBACTERIAL ACTIVITY OF BL-C43 AND SEVERAL OTHER ANTIBIOTICS
FOLLOWING THEIR ORAL ADMINISTRATION TO EXPERIMENTALLY INFECTED MICE

Challenge organism	No. of treatments	BL-C43	Novobiocin (Na)	Lincomycin (HCl)	Oleandomycin-triacetyl	Erythromycin estolate
Staphylococcus aureus A9537 (penase −)[a]	2	8[b]	26	76	66	124
	4	4	24	108	88	80
S. aureus A9497 (penase −)[a]	2	10	88	140	92	320
S. aureus A9606 (penase +)[a]	2	10	38	54	22	230
S. aureus A9631 (penase +)[a]	2	40	120	104	114	260
Streptococcus pyogenes A9604	2	196	>500	1000	560	280
	4	140	−	140	176	216
Diplococcus pneumoniae A9585	2	180	800	>1000	>1000	580
	4	160	600	880	680	376
Pasteurella multocida A20139	4	92	560	>1000	1000	800

[a]Penase − = non-penicillinase-producer, penase + = penicillinase-producer.
[b]Numerals listed under names of antibiotics indicate total dose in mg/kg required to protect 50% of the mice.

present at 12 hr post-administration. Novobiocin also peaked at a high level (about 9 µg/ml) and persisted for at least 9 hr. Of the remaining antibiotics, the highest peak occurred with triacetyloleandomycin (1.3 µg/ml). This compound could not be detected at 2 hr, while erythromycin and lincomycin were not found in the 6-hr samples.

The marked absorbability of BL-C43 after oral administration in the mouse also occurs following its administration by that route to the dog and to man. This is indicated in Figs. 7 and 8 where this antibiotic's concentrations in serum following single oral doses have been compared with those of coumermycin A_1.

In addition to the strikingly high peak serum levels produced by BL-C43 in both species, it is particularly noteworthy that the drug

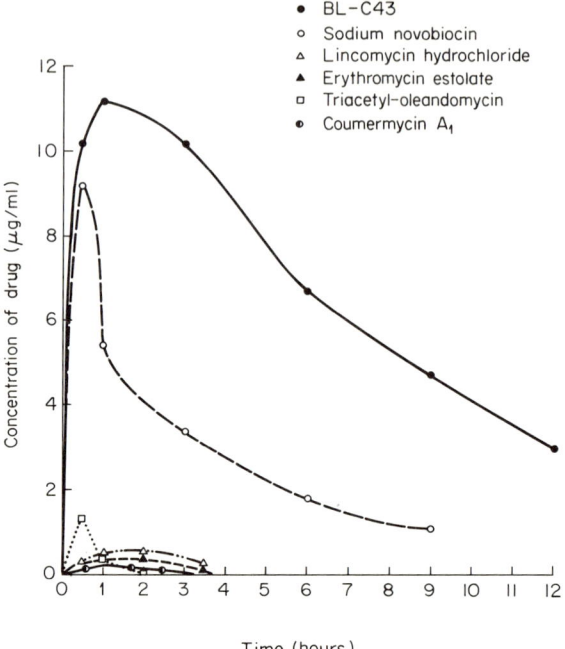

FIG. 6. Concentrations of BL-C43 and several other drugs in the blood of mice (eight mice/compound) after oral administration of a 25 mg/kg dose.

tends to persist in the serum for many hours after ingestion of the oral dose.

Price *et al.* (1970b), who determined BL-C43 concentrations in the tissues and organs of mice following intravenous (IV) administration of a single 20 mg/kg dose, hypothesized that the primary route of excretion of BL-C43 is via biliary excretion. Less than 1% of the IV-administered dose was recovered in the urine, and there was no evidence that the kidney, heart, or lung tissues concentrate the drug. The antibiotic appears to be accumulated by the liver, concentrated in bile, and then released into the small intestine where it is probably reabsorbed to a considerable extent. They found in cases where very high doses (50 to 200 mg/kg) were administered by the oral route, that the absorption rate may approach or exceed the excretion rate.

BL-C43 underwent toxicological examination in rats and dogs and

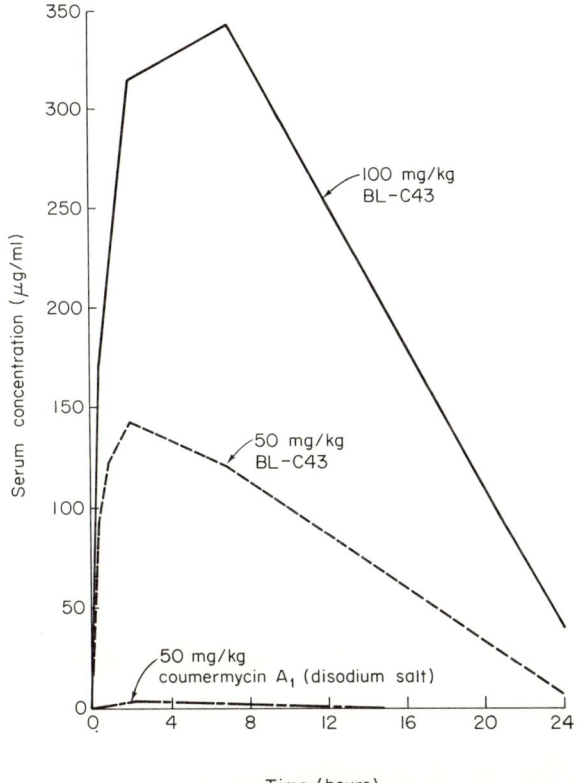

FIG. 7. Blood serum levels of BL-C43 Na in dogs after a single oral dose.

was found to cause no gross or microscopic pathological changes at the maximum oral doses administered (100 mg/kg per day for 39 days to rats, 100 mg/kg per day for 36 days to dogs).

Plans were then made to conduct a 14-day BL-C43 tolerance study in man. The compound was given to each of 25 volunteers as 250-mg capsules t.i.d. On the eighth day, one individual developed a rash and appeared jaundiced. Seventeen of the remaining 24 subjects also showed evidence of mild jaundice. One subject on each of days 9 and 10 developed signs of congestive heart failure. At this point, the study was terminated. Following withdrawal of the drug, all volunteers made an uneventful recovery.

On the basis of these results, no further plans to evaluate semisynthetic coumermycin derivatives in man have been contemplated.

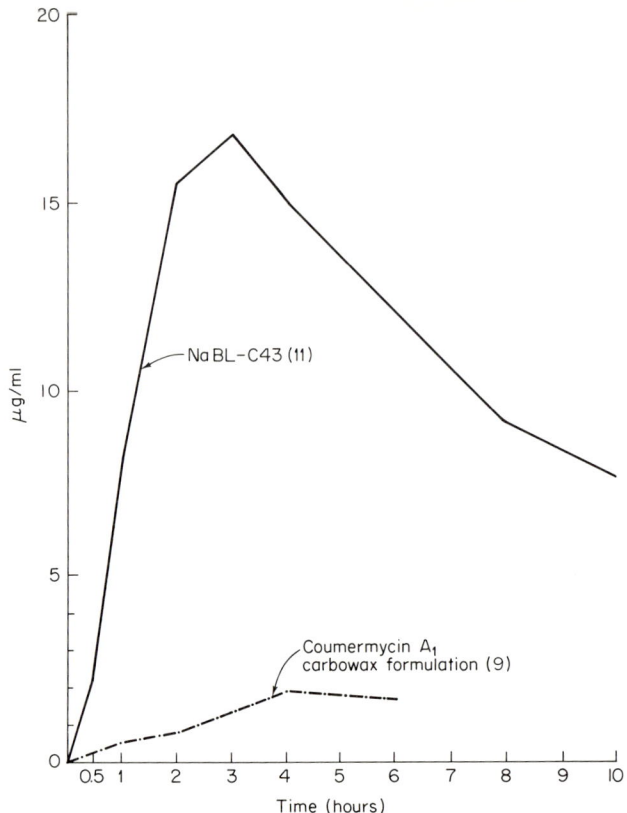

FIG. 8. Human oral blood levels after a 500-mg single dose. BL-C43 data from eleven subjects, coumermycin A_1 data from nine subjects.

REFERENCES

Batcho, A. D., Berger, J., Furlenmeier, A., Keller, O., Pecherer, B., Schocher, A. J., Spiegelberg, H., and Vaterlaus, B. P. (1969). Swiss Pat. 472,428.
Bell, F. (1931). *J. Chem. Soc., London* p. 2962.
Berger, J., Schocher, A. J., Batcho, A. D., Pecherer, B., Keller, O., Maricz, J., Karr, A. E., Vaterlaus, B. P., Furlenmeier, A., and Spiegelberg, H. (1966). *Antimicrob. Ag. Chemother.* pp. 778–785.
Birlova, L., and Trakhtenberg, D. M. (1966). *Antibiotiki (Moscow)* **11**, 395.
Brock, T. D., and Brock, M. L. (1959). *Arch. Biochem. Biophys.* **85**, 176–185.
Claridge, C. A. (1967). Private communication.
Claridge, C. A., and Gourevitch, A. (1968). U.S. Pat. 3,403,078.
Claridge, C. A., Rossomano, V. Z., Buono, N. S., Gourevitch, A., and Lein, J. (1966). *Appl. Microbiol.* **14**, 280–283.
Cleeland, R., Beskid, G., and Grunberg, E. (1970). *Infec. Immunity* **2**, 371–375.
Colmore, J. P., Braden, B., and Wilkerson, R. (1968). *Abstr. Intersci. Conf. Antimicrob. Ag. Chemother., 8th, 1968* p. 36.

Cron, M. J. (1970). U.S. Pat. 3,547,903.
Cron, M. J., and Nettleton, D. E., Jr. (1965). Unpublished results.
Cron, M. J., Godfrey, J. C., Hooper, I. R., Keil, J. G., Nettleton, D. E., Jr., Price, K. E., and Schmitz, H. (1970). In "Progress in Antimicrobial and Anticancer Chemotherapy," Proc. VIth, pp. 1069–1082. Amer. Soc. Microbiol., Bethesda, Maryland.
Devine, L. F., and Hagerman, C. R. (1970). Appl. Microbiol. **19**, 329–334.
Devine, L. F., Johnson, D. P., Hagerman, C. R., Pierce, W. E., Rhode, S. L., III, and Peckinpaugh, R. O. (1970). Amer. J. Med. Sci. **260**, 165–170.
Duma, R. J., and Warner, J. F. (1969). Appl. Microbiol. **18**, 404–405.
Fedorko, J., Katz, S., and Allnoch, H. (1969). Appl. Microbiol. **18**, 869–872.
Furlenmeier, A., Schocher, A. J., Spiegelberg, H., Vaterlaus, B. P., Batcho, A. D., Berger, J., Keller, O., and Pecherer, B. (1969). Swiss Pat. 467,799.
Gourevitch, A., Hunt, G. A., Luttinger, J. R., Carmack, C. C., and Lein, J. (1961). Proc. Soc. Exp. Biol. Med. **107**, 455–458.
Grunberg, E., and Bennett, M. (1966). Antimicrob. Ag. Chemother. pp. 786–788.
Grunberg, E., Cleeland, R., and Titsworth, E. (1967). Antimicrob. Ag. Chemother. pp. 397–398.
Hinman, J. W., Hoeksema, H., Caron, E. L., and Jackson, W. G. (1956). J. Amer. Chem. Soc. **78**, 1072.
Hinman, J. W., Caron, E. L., and Hoeksema, H. (1957a). J. Amer. Chem. Soc. **79**, 3789.
Hinman, J. W., Caron, E. L., and Hoeksema, H. (1957b). J. Amer. Chem. Soc. **79**, 5321.
Hoeprich, P. D. (1968). Antimicrob. Ag. Chemother. pp. 697–704.
Hoeprich, P. D. (1971). J. Infec. Dis. **123**, 125–133.
Hoffmann-LaRoche (1965). Belg. Pat. 665,237.
Hoffmann-LaRoche (1967a). S. Afr. Pat. 66/5644.
Hoffmann-LaRoche (1967b). Neth. Pat. Appl. 6,613,927.
Hoffmann-LaRoche (1968a). Brit. Pat. 1,111,511.
Hoffmann-LaRoche (1968b). Brit. Pat. 1,132,287.
Institute of Microbial Chemistry. (1971). Jap. Pat. 15675/1971.
Kaplan, S. A. (1970). J. Pharm. Sci. **59**, 309–313.
Kawaguchi, H., Tsukiura, H., Okanishi, M., Miyaki, T., Ohmori, T., Fujisawa, K., and Koshiyama, H. (1965a). J. Antibiot., Ser. A **18**, 1–10.
Kawaguchi, H., Tsukiura, H., Okanishi, M., and Miyaki, T. (1965b). U.S. Pat. 3,201,386.
Kawaguchi, H., Naito, T., and Tsukiura, H. (1965c). J. Antibiot., Ser. A **18**, 11–25.
Kawaguchi, H., Miyaki, T., and Tsukiura, H. (1965d). J. Antibiot., Ser. A **18**, 220–222.
Kawaguchi, H., Ueda, H., and Ishiyama, S. (1967). Proc. Int. Congr. Chemother., 5th, 1967 pp. 1015–1017.
Keil, J. G. (1966). Unpublished results.
Keil, J. G., and Hooper, I. R. (1969). U.S. Pat. 3,454,548.
Keil, J. G., and Hooper, I. R. (1970). U.S. Pat. 3,494,914.
Keil, J. G., Hooper, I. R., Cron, M. J., Fardig, O. B., Nettleton, D. E., Jr., O'Herron, F. A., Ragan, E. A., Rousche, M. A., Schmitz, H., Schreiber, R. H., and Godfrey, J. C. (1969). Antimicrob. Ag. Chemother. p. 120.
Keil, J. G., Hooper, I. R., Schreiber, R. H., Swanson, C. L., and Godfrey, J. C. (1970). Antimicrob. Ag. Chemother. p. 200.
Keil, J. G., Godfrey, J. C., Cron, M. J., Hooper, I. R., Nettleton, D. E., Jr., Price, K. E., and Schmitz, H. (1971). Abstr., Symp. Antibiot., 1971, see also Pure Appl. Chem. **28**, 571 (1971).
Mancy, D., Ninet, L., and Preud'Homme, J. (1969). Ger. Pat. 1,907,556.
Mancy, D., Ninet, L., and Preud'Homme, J. (1970). Brit. Pat. 1,208,842.
Michaeli, D., Meyers, B., and Weinstein, L. (1969). J. Infec. Dis. **120**, 488–490.

Michaeli, D., Meyers, B. R., and Weinstein, L. (1970). *Antimicrob. Ag. Chemother.* pp. 463–467.
Michaeli, D., Molavi, A., Mirelman, D., Hanoch, A., and Weinstein, L. (1971). *Antimicrob. Ag. Chemother.* pp. 95–99.
Nettleton, D. E., Jr. (1968). U.S. Pat. 3,380,994.
Newmark, H. L. (1970). U.S. Pat. 3,519,712.
Newmark, H. L., and Berger, J. (1970). *J. Pharm. Sci.* **59,** 1246–1248.
Newmark, H. L., Berger, J., and Carstensen, J. T. (1970). *J. Pharm. Sci.* **59,** 1249–1251.
Price, K. E. (1965). Unpublished data.
Price, K. E., Chisholm, D. R., Godfrey, J. C., Misiek, M., and Gourevitch, A. (1970a). *Appl. Microbiol.* **19,** 14–26.
Price, K. E., Chisholm, D. R., Leitner, F., and Misiek, M. (1970b). *Antimicrob. Ag. Chemother.* pp. 209–218.
Pridham, T. G., and Gottlieb, D. (1948). *J. Bacteriol.* **56,** 107–114.
Scannell, J. (1968). *Antimicrob. Ag. Chemother.* pp. 470–474.
Scannell, J., and Kong, Y. L. (1970). *Antimicrob. Ag. Chemother.* pp. 139–143.
Schmitz, H., and DeVault, R. L. (1970). U.S. Pat. 3,547,902.
Schmitz, H., and Godfrey, J. C. (1970). *J. Antibiot.* **23,** 497.
Schmitz, H., DeVault, R. L., McDonnell, C. D., and Godfrey, J. C. (1968). *J. Antibiot.* **21,** 603.
Scholtan, W., and Schmid, J. (1962). *Arzneim.-Forsch.* **12,** 741–750.
Steers, E., Foltz, E. L., and Graves, B. S. (1959). *Antibiot. Chemother. (Washington, D.C.)* **9,** 307–311.
Whitaker, W. D. (1968). Brit. Pat. 1,114,468.

Chloramphenicol

VEDPAL S. MALIK

*Department of Nutrition and Food Science,
Massachusetts Institute of Technology,
Cambridge, Massachusetts*

I.	Introduction	297
II.	Chloramphenicol as a Growth Inhibitor	299
III.	Effect on Morphology	300
IV.	Mode of Action	300
V.	Structure-Activity Relationships	303
VI.	Differential Effects on Proteins and Protein Synthesis	305
	A. Esterases	305
	B. Regulatory Proteins	305
	C. Nitrate Reductase	306
	D. Membrane-Bound Enzymes	306
	E. Chloramphenicol Particles	307
VII.	Effect on RNA Metabolism	307
VIII.	Effect on Energy-Producing Systems	308
IX.	Effect on Cell Wall Biosynthesis	310
X.	Effect on Biosynthesis of Aromatic Compounds	311
XI.	Effect on Antibiotic Biosynthesis	312
XII.	Resistance to Chloramphenicol	312
	A. Physiological Immunity	312
	B. Drug Inactivation	313
	C. Permeability Barrier	317
XIII.	Chloramphenicol as a Mutagen	318
XIV.	Chloramphenicol as a Therapeutic Agent	318
XV.	Chloramphenicol as a Secondary Metabolite	319
XVI.	Biosynthesis of Chloramphenicol	321
XVII.	Catabolism of Chloramphenicol by the Producing Organism	323
XVIII.	Effect of Chloramphenicol on the Producing Organism	323
	A. Adaptation to Chloramphenicol	324
	B. Protein Synthesis and Development of Resistance	325
	C. Role of Antibiotic Metabolism in Resistance	326
	D. Chloramphenicol Hydrolase and Resistance	326
XIX.	Regulation of Chloramphenicol Biosynthesis	328
XX.	Effect of p-Nitrophenylserinol on Chloramphenicol Biosynthesis	330
	References	331

I. Introduction

Chloramphenicol was the first broad-spectrum antibiotic introduced into medicinal use to inhibit the growth of bacteria, rickettsiae, and organisms of the psittacosis-lymphogranuloma group (Brock, 1964; Hahn, 1967). It was the first antibiotic to be chemically synthesized (Controulis et al., 1949) and is still the only antibiotic industrially produced in this way. Because of its relatively simple chemical structure, a large number of derivatives have been prepared. Studies of these compounds have led to useful conclusions about the relationship between chemical structure and biological activity (Hahn et al., 1956).

Chloramphenicol interests the organic chemist because it was the first natural product found to contain nitro and chloro groups. It appears to be derived from an unusual diversion of intermediates in the shikimic acid pathway of aromatic biosynthesis (Vining et al., 1968). In recent years, chloramphenicol has become a tool of molecular biologists working on the synthesis and function of nucleic acids. At low concentrations (e.g., 10 μg/ml) it enables the investigator to specifically block protein synthesis in order to investigate other processes.

The discovery of chloramphenicol as the antibiotic in cultures of *Streptomyces venezuelae* was made independently in two laboratories (Ehrlich et al., 1947; Gottlieb et al., 1948). Rebstock et al. (1949) showed it to have the structure shown in Fig. 1. Several systematic

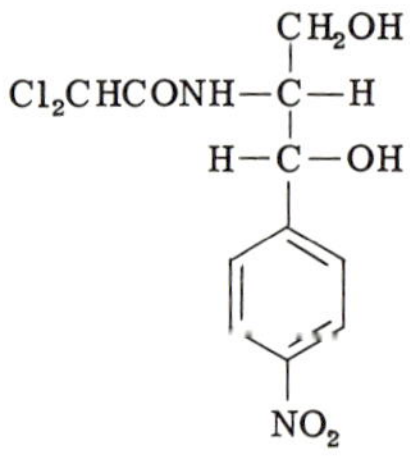

$$Cl_2CHCONH-\underset{\underset{\displaystyle H-C-OH}{|}}{\overset{\overset{\displaystyle CH_2OH}{|}}{C}}-H$$

(1)

FIG. 1. Chloramphenicol.

names have been used. *Chemical Abstracts* prefers D-(−)-*threo*-2,2'-dichloro-N-[β-hydroxy-α-(hydroxymethyl)-p-nitrophenylethyl]-acetamide, but the name more commonly encountered is D-(−)-*threo*-2-dichloroacetamide-p-nitrophenyl-1,3-propanediol (Rebstock et al., 1949). Since the molecule has two asymmetric carbon atoms, four stereoisomers are possible. The two *erythro*-isomers are biologically inactive. The L-(+)-*threo*-isomer is less than 0.5% as active as the natural D-(−)-*threo*-isomer which is synthesized and sold as Chloromycetin by Parke Davis and Company of Detroit, and as Chlorocol by J. Webster Laboratories, Ltd., Toronto.

Several strains of chloramphenicol-producing *Streptomyces* have been described (Ehrlich *et al.*, 1947; Gottlieb *et al.*, 1948; Umezawa *et al.*, 1948; C. G. Smith, 1958). In a complex medium, growth and chloramphenicol production are dissociated, but in a defined medium, antibiotic production and growth are closely linked (Malik and Vining, 1970), and for most of the growth phase cells are exposed to concentrations much higher than the 10 μg/ml which inhibits the growth of many bacteria (Brock, 1961).

Knowledge of the mechanism by which *Streptomyces* escapes the harmful effects of its own toxic metabolite might have useful application. Strains highly resistant to chloramphenicol might produce high yields of antibiotic, possibly enabling fermentation methods to displace chemical synthesis for industrial production. However, the nature of the relationship between growth and antibiotic production is not well understood. Mechanisms by which the producing organism may regulate antibiotic synthesis have been discussed (Bu'Lock, 1967; Demain, 1968), and the possible roles of these and other secondary metabolites in the life and survival of microorganisms in nature have been repeatedly mentioned (Schaeffer, 1969; Dhar and Khan, 1971), but convincing evidence bearing on these questions is badly needed.

Because so much information is available on its production, biosynthesis, regulation, and mode of action, chloramphenicol appears to be a suitable compound for studies of this type.

II. Chloramphenicol as a Growth Inhibitor

Chloramphenicol is predominantly a bacteriostatic agent, inhibiting the growth of blue-green algae, bacteria, and related procaryotic organisms (Brock, 1961). Fassin *et al.* (1955) reported exceptions to this rule in that *Shigella flexneri* and an unidentified gram-positive spore-forming organism were rapidly killed by the antibiotic. They found, however, that *Salmonella typhosa* was completely inhibited by 7.5 μg/ml but not killed by concentrations as high as 1500 μg/ml. Ciak and Hahn (1958) found a progressive decrease in the growth rate of *Escherichia coli* as the concentration of chloramphenicol was increased, until complete inhibition was achieved at 10 μg/ml.

Allison *et al.* (1962) have carried out a precise study of the changes in viable as well as total cell numbers in cultures of *E. coli* exposed to growth-inhibitory concentrations of chloramphenicol. Total cell count continued to increase by increments of almost 50% for 14 generation times; viable count increased for only 1 or 2 generation times, and then decreased to approximately 25% of the highest value. When chloramphenicol (500 µg/ml) was added to an exponentially growing culture of *Tetrahymena pyriformis* growth was completely inhibited within a few generations (Turner and Lloyd, 1971). These studies have shown that the view of chloramphenicol as a "bacteriostatic" agent, in the literal sense, is an oversimplification. The more complex situation is not apparent from conventional turbidimetric inhibition analysis or routine plate counting with a fairly high standard deviation.

III. Effect on Morphology

Cells growing in chloramphenicol concentrations too low to completely inhibit growth may assume abnormal shapes (Pulvertaft, 1952). Although these have been called L-forms, they do not continue to divide and grow and are apparently not analogous to the L-forms induced by penicillin (Dienes *et al.*, 1950). Changes in the nuclear bodies have also been observed; Bergerson (1953), using acid Giemsa staining in *E. coli*, first reported that the nuclear material became arranged in long, irregular bars. Hahn *et al.* (1957) found that nuclear bodies of *E. coli* grew larger after prolonged incubation in 10 µg/ml of chloramphenicol and returned to normal when the antibiotic was removed. The increase in nuclear size correlated with increased DNA (deoxyribonucleic acid) content. DeLamater *et al.* (1955) observed very similar changes in *Bacillus megatherium*, although these were not specific for chloramphenicol, occurring also with tetracyclines, erythromycin, carbomycin, and streptomycin.

Morgan and associates (1967) examined thin sections of *E. coli* by electron microscopy at intervals after addition and removal of chloramphenicol. The first changes, after 1 hr, were disappearance of the ribosomes and aggregation of the nuclear material toward the center. At 2 hr, aggregates of abnormal cytoplasmic granules appeared and subsequently grew larger. By 23 hr, amorphous, electron-dense material had accumulated within and at the periphery of the nuclear matrix. When chloramphenicol was removed, the bacteria became normal in appearance, passing through a series of stages that were sequential but not synchronous. After 145 minutes the bacteria were seen undergoing abnormal division.

IV. Mode of Action

The mode of action of chloramphenicol has been the subject of numerous reviews (Brock, 1961, 1964; B. D. Davis and Feingold, 1962; Gale, 1963; Hahn, 1964, 1967; Goldberg, 1965; Newton, 1965;

Vazquez, 1966; Weisblum and Davies, 1968). In bacteria, formation of many specific proteins is inhibited by concentrations of 10 μg/ml and above. Only the D-*threo* isomer is effective. Since these same concentrations inhibit growth, it is assumed that growth inhibition is due to inhibited protein biosynthesis (Brock, 1964). Bacteriostatic concentrations do not prevent the formation of nucleic acids (Gale and Folkes, 1953), cell walls (Hancock and Park, 1958; Mandelstam and Rogers, 1958, 1959), or polysaccharides (Hopps *et al.*, 1954).

In the cells of higher animals much greater concentrations of chloramphenicol (over 300 μg/ml) are usually required to inhibit growth and protein synthesis (Fusillo *et al.*, 1952; Pomerat and Leake, 1954; Lallier, 1962; Farese, 1964), although a few cases have been reported in which the antibiotic was effective at bacteriostatic levels (Djordjevic and Szybalski, 1960; Ambrose and Coons, 1963; Amos, 1964). At 300–3000 μg/ml it inhibits protein synthesis in reticulocytes (Borshook *et al.*, 1957; Allen and Schweet, 1962) and incorporation of ^{59}Fe into heme in bone marrow cells (Vas *et al.*, 1962). It does not inhibit the synthesis of protein from endogenous templates in cell-free systems from liver (Rendi, 1959) or reticulocytes (Allen and Schweet, 1962; Weisberger *et al.*, 1963; Weisberger and Wolfe, 1964) at concentrations which affect intact cells. These results suggest that the mechanism of chloramphenicol's action in mammalian cells differs from that in bacteria.

It will not be possible to locate the exact site of action of chloramphenicol until the genetics of the ribosome and the biochemical details of protein synthesis are completely known. However, it is clear that the antibiotic does not inhibit the first step, the activation of amino acids (DeMoss and Novelli, 1955). Neither does it affect the transfer of activated amino acids to their tRNAs (transfer ribonucleic acids) since aminoacyl tRNAs accumulate in bacteria treated with chloramphenicol (Lacks and Gros, 1959).

Chloramphenicol does not alter the ribosomal dissociation, promoted by the dissociation factor, which could interfere with the process of protein synthesis at the ribosome cycle level (Garcia-Patrone *et al.*, 1971). The peptidyl transfer reaction, inhibited by increasing additions of chloramphenicol, is barely affected by erythromycin (Teraoka, 1970). Thus, the effect of chloramphenicol on the activity of *E. coli* ribosomes is clearly distinct from that of erythromycin (Olenick *et al.*, 1968; Cerna *et al.*, 1971; Vogel *et al.*, 1971).

It is generally agreed that the ribosome is the sensitive target. Chloramphenicol binds to the 50S subunit of the procaryotic ribosome and alters its structure (Schweet and Heintz, 1966; Vazquez,

1966; Hurwitz and Braun, 1967, 1968). It may prevent peptide chain elongation by competing with the carboxyl-terminal amino acid of the growing peptide chain for a ribosomal site (Das et al., 1966) or perhaps its conformation resembles that of aminoacyl-RNA or peptidyl-RNA and thus interferes with a reaction in which one of these substrates participates (Coutsogeorgopoulos, 1967; Pestka, 1969).

The effect of chloramphenicol on amino acid incorporation into peptides *in vitro* depends on the messenger RNA used (Kuccan and Lipmann, 1964). Peptide synthesis directed by synthetic polynucleotides is relatively insensitive, though sensitivity with homopolynucleotides varies inversely with their ability to stimulate amino acid incorporation. Incorporation directed by natural messenger RNA (mRNA) is quite sensitive to chloramphenicol, suggesting that in synthetic messenger-directed incorporation the chloramphenicol-sensitive step can be bypassed. This step could not be the peptidyl transfer, since assays measured incorporation of radioactive amino acid into acid-precipitable material and were therefore based on the peptidyl transfer (Weber and DeMoss, 1969).

Bresler et al. (1968) found that addition of 200 μg/ml of chloramphenicol together with ^{32}P-tRNA to a cell-free system did not interfere with the formation of labeled peptidyl tRNA. Presumably the antibiotic does not inhibit the first transfer of peptide to bound aminoacyl tRNA, but does prevent the formation of subsequent peptide bonds; this would explain why chloramphenicol inhibits protein synthesis in cell-free systems by only 85–90%. These results are supported by experiments conducted by Julian (1965) which showed that addition of chloramphenicol during polylysine synthesis directed by a polyadenylic acid template does not substantially change the number of chains synthesized. However, the chains consist only of di- and tripeptides. According to Gurgo et al. (1969) a new kind of monosome is formed in the presence of chloramphenicol; it contains mRNA, two aminoacyl tRNAs, a 30S and a 50S ribosome subunit, but no peptidyl tRNA at the peptidyl site.

Peptidyl chains on washed ribosomes prepared from chloramphenicol-inhibited *Escherichia coli* were released slowly when puromycin was added. If chloramphenicol was added too, no release occurred (Weber and DeMoss, 1969). This evidence that chloramphenicol inhibits puromycin-induced release of peptides has been substantiated by many workers, and most recently by Pestka (1969) whose results indicate that chloramphenicol competitively inhibits the action of puromycin. Thus the site of inhibition is limited to either of two reactions, aminoacyl tRNA binding or peptidyl transferase, since puromycin is known to substitute for aminoacyl tRNA (Dagley et al., 1962). Despite earlier evidence by Nakamoto et al. (1963), Jardetsky and Julian (1964), and Suarez and Nathans (1965), that net aminoacyl tRNA binding is not reduced by chloramphenicol, Pestka concluded this to be the site of action. His evidence is based upon an assay believed to be specific for aminoacyl tRNA binding reaction. Chloramphenicol caused inhibition at concentrations as low as 2 µg/ml.

Celma et al. (1971) also found that chloramphenicol decreased binding of UACCA-leucine to 70S ribosomes and 50S ribosomal subunits. However, this inhibition does not prove that the aminoacyl end of tRNA and the chloramphenicol bind at the same site. The two sites might be allosterically linked in such a way that binding of chloramphenicol prevents participation of aminoacyl ends of tRNA in a reaction catalyzed by peptidyl transferase instead of inhibiting the activity of peptidyl transferase.

Although this conclusion cannot be accepted unreservedly until it is confirmed by other methods and the conflicting results of earlier workers are explained, it is attractive. Competitive inhibition by chloramphenicol of aminoacyl tRNA binding to a specific ribosomal site (as distinct from codon-anticodon interaction with mRNA) seems likely because of the stereochemical resemblance between chloramphenicol and a pyrimidine nucleotide (Dunitz, 1952; Jardetsky, 1963).

V. Structure-Activity Relationships

Hahn et al. (1956), Shemyakin et al. (1956), and Shemyakin (1961) have derived rules for predicting the effect of structure on biological activity in the chloramphenicol series (Fig. 2).

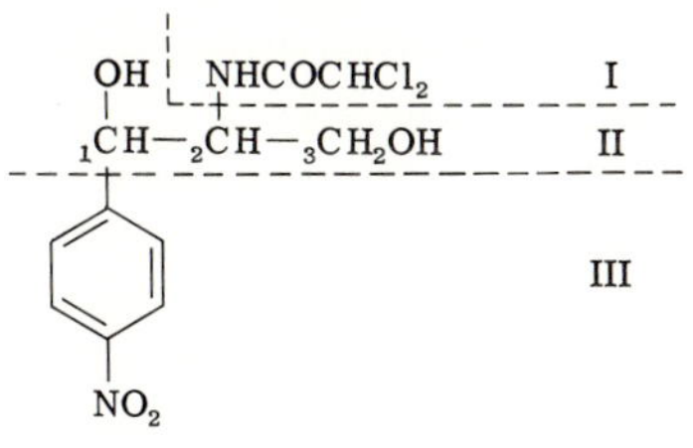

$$\text{OH} \quad \text{NHCOCHCl}_2 \qquad \text{I}$$
$$_1\text{CH}-_2\text{CH}-_3\text{CH}_2\text{OH} \qquad \text{II}$$
$$\bigcirc \qquad \text{III}$$
$$\text{NO}_2$$

FIG. 2. Chloramphenicol: structure-activity relationships.

In moiety I the dichloroacetamide side chain can be varied with only moderate loss of activity provided the strongly electronegative character of the acyl residue is maintained. The N-acetyl analogue is only 14% as active as chloramphenicol. Although there is no absolute requirement for the chlorine atoms, it is important that the size of the dichloroacetyl group is not far exceeded. If the dichloroacetamide side chain is deleted, the remaining base has only 1–2% of the original activity. Coutsogeorgopoulos (1966) modified chloramphenicol by replacing the dichloroacetyl with an α-aminoacyl group and by reducing the nitro to an amino group. The L-leucyl, L-p-methoxyphenylalanyl, and reduced L-leucyl analogues were as effective as chloramphenicol in inhibiting polyuridylate-directed incorporation of ^{14}C-L-phenylalanine. None were as effective in inhibiting polyuridylate-cytidylate-directed incorporation of ^{14}C-L-phenylalanine, although the glycyl L-leucyl, L-phenylalanyl, and L-p-methoxyphenylalanyl analogues were 60–80% as effective. Apparently the nature of the mRNA is important.

In moiety II (Fig. 2) the propanediol structure with the correct stereochemical configuration of substituents at positions 1 and 2 is essential for antibacterial activity. Some diesters of chloramphenicol are active but may actually require restoration of the antibiotic molecule by hydrolysis. If either hydroxyl group is replaced by hydrogen all activity is lost. The propane chain cannot be extended without loss of activity, nor can the hydrogen at position 2 be replaced by a methyl group. The hydroxyl groups in moiety II form an alicyclic ring through hydrogen bonding (Jardetsky, 1963), and presumably this is essential for the action of chloramphenicol.

In moiety III the nitro group is not essential for activity. Shemyakin *et al.* (1956) consider that the most important feature of this group is its polarizing ability, the geometry having little effect. The acyl and nitrophenyl parts of the molecule appear to play a role in the molecule's penetration into the bacterial cell (Telesnina *et al.*, 1967). Hurwitz and Braun (1967) showed that racemic mixtures in which the nitro group is replaced by methylsulfonyl or methylthio groups are, respectively, 31% and 48% as effective as chloramphenicol in inhibiting induced synthesis of β-galactosidase in *E. coli* B.

However, *in vivo* activity of an antibiotic is also determined by its ability to enter cells. If the p-methylthio analog enters cells more easily than chloramphenicol, it would be a better therapeutic agent against the pathogens whose permeability barrier makes them resistant to chloramphenicol (Malik, 1970).

The structural requirements are similar for the inhibition of protein synthesis in mitochondria and bacterial extracts by chloramphenicol isomers and analogues (Freeman, 1970).

VI. Differential Effects on Proteins and Protein Synthesis

A. ESTERASES

Early in the biochemical investigation of chloramphenicol, G. N. Smith *et al.* (1949) tested a wide range of enzymes and found that the activities of all except esterases were insensitive to the antibiotic. Liver and bacterial esterases were inhibited by low concentrations, and a correlation was obtained between inhibition of bacterial esterase and of the producing bacterium's growth. The possibility therefore exists that chloramphenicol affects ester metabolism.

B. REGULATORY PROTEINS

Sypherd and DeMoss (1963) have presented several lines of evidence that chloramphenicol preferentially inhibits synthesis of repressor-controlled enzymes. A specific mutation in the regulator gene can eliminate the enzyme's susceptibility, and it is believed that chloramphenicol-promoted repression of enzyme synthesis is not a manifestation of catabolic repression but the result of increased levels of specific repressor molecules (Magasanik, 1961; Sypherd and DeMoss, 1963).

Levine and Sinsheimer (1968) consider chloramphenicol resistance to be a quantitative rather than a qualitative property. They describe a protein as chloramphenicol-resistant if it can be synthesized in 30 μg/ml chloramphenicol. In *Escherichia coli* the replicator protein required for initiation of DNA synthesis and the *lac* repressor are of this type (Pardee and Prestidge, 1959; C. Lark and Lark, 1964; K. G. Lark and Lark, 1966). Using a double-label technique, Levine and Sinsheimer (1968) have isolated from *E. coli* infected with ϕX174 bacteriophage a chloramphenicol-resistant viral-directed protein involved in synthesis of progeny replicative forms. It was shown to be associated with the bacterial membrane (Levine and Sinsheimer, 1969). The protein isolated from cells infected with λ-bacteriophage (Levine and Sinsheimer, 1968) and a host protein involved in lysogenization (Naha, 1969) appear to be of the same type. Several *cis*-acting proteins of bacteriophage are resistant to chloramphenicol and interact with the membrane (Ray, 1970).

All of these proteins are believed to interact with nucleic acid to control DNA replication or to regulate transcription, and may represent a distinctive class; Levine and Sinsheimer (1968) have also suggested that chloramphenicol decreases the overall rate at which proteins are synthesized instead of completely inhibiting most proteins while allowing a few to be made at their normal rates. Those proteins which, because of repression, are ordinarily made in small amounts and of which only very few molecules are needed for observable function might then appear to be chloramphenicol resistant.

C. Nitrate Reductase

Since in *Escherichia coli* chloramphenicol permits exponential growth and synthesis of constitutive enzymes while inhibiting the formation of inducible enzymes (Sypherd *et al.*, 1962), Ramsey (1966) was surprised to find that *Staphylococcus aureus* increased its production of nitrate reductase 3-fold when 50 µg/ml chloramphenicol was added. Growth was completely inhibited and total protein synthesis was reduced by approximately 45%. Chloramphenicol alone did not induce nitrate reductase. Nitrate was required for enzyme formation proportional to chloramphenicol concentration within the 5–50 µg/ml range. Under these conditions chloramphenicol did not induce the formation of chloramphenicol reductase or chloramphenicolase (G. N. Smith and Worrell, 1949).

In contrast, Egami *et al.* (1950) have reported that high concentrations of chloramphenicol inhibited the nitrate reductase of *Staphylococcus hemolyticus*. Chloramphenicol and nitrate ions compete for the active site of this enzyme which acts on both organic and inorganic nitro compounds. Schrader *et al.* (1967) showed differential effects of chloramphenicol on the induction of nitrate and nitrite reductases in green leaf tissue. They concluded that nitrate reductase is synthesized by a cytoplasmic, and the nitrite reductase by a chloroplastic, ribosomal system. The average stimulation of nitrate reductase formation with 1.6 mg/ml chloramphenicol was 74%. In contrast, nitrite reductase was repressed. Neither enzyme was inhibited by chloramphenicol *in vitro*.

D. Membrane-Bound Enzymes

Vambutas and Salton (1970) observed differential inhibitory effects of chloramphenicol on the synthesis of cytoplasmic enzymes and membrane ATPase of *Micrococcus lysodeikticus*. When chloramphenicol (100 µg/ml) was added to cultures for about half a mean generation time, total protein and the levels of cytoplasmic enzymes (polynucleotide phosphorylase, adenosine deaminase, and glucose 6-phosphate dehydrogenase) were approximately 75% of the amounts found in the untreated cultures. Under similar conditions the formation of membrane-bound Ca^+-dependent ATPase was unaffected by chloramphenicol. Since puromycin inhibits to the same extent the syntheses of both ATPase and total protein, the existence of a pool of inactive subunits and continued assembly in the presence of chloramphenicol are doubtful. It remains unclear whether these observations can be explained by qualitative or quantitative differences in the chloramphenicol binding site on ribosomes or by accessibility of different classes of ribosomes to the antibiotic.

E. Chloramphenicol Particles

Addition of chloramphenicol to exponentially growing *Escherichia coli* cultures can reduce the rate of protein synthesis is more than 100-fold with little effect, initially, on RNA synthesis. The RNA content of the cell is approximately doubled during the first 90 minutes, and much of the RNA synthesized associates with protein in particles that sediment at 18–25S. RNA extracted from these "chloramphenicol particles" sediments at 16S and 23S, like that from mature ribosomes. It had been assumed that chloramphenicol particles were incomplete ribosomes, the protein being drawn from a pool of free ribosomal protein existing in the cell at the time the antibiotic was added (Hosokawa and Nomura, 1965). However Schleif (1968) has found that most of the protein is nonribosomal. On removal of chloramphenicol, it dissociates from the RNA which combines with newly formed ribosome protein to produce mature ribosomes. F. C. Davis and Sells (1969) have provided evidence that at least four proteins are preferentially synthesized and assembled into 50S ribosomal subunits during recovery from chloramphenicol treatment. Possibly the information for these has accumulated as messenger RNA during chloramphenicol inhibition or has been synthesized early during the recovery phase.

In certain mutants of *E. coli* (Lewandowski and Brownstein, 1966) and in cells treated with low concentrations of chloramphenicol (Otaka *et al.*, 1967), 43S particles accumulate and may later be converted to mature ribosomes. Otaka *et al.* (1967) have demonstrated that the 43S particles which accumulate in the presence of low concentrations of chloramphenicol lack four of the proteins found on the 50S subunit. These may be synthesized maximally during early recovery from chloramphenicol treatment. Chloramphenicol-treated *E. coli* contain defective ribosomes, and initiation factor F_2 is substantially decreased (Young and Nakada, 1971).

VII. Effect on RNA Metabolism

Concentrations of chloramphenicol which blocked protein synthesis in *Escherichia coli* also markedly reduced the rate of ribosomal RNA synthesis (Kurland and Maaløe, 1962). Synthesis of functional tRNA and a low molecular weight RNA, however, continued (Ezekiel and Valulis, 1965; Jordan *et al.*, 1971), although some had altered chromatographic properties. This was not the result of under-methylation or degradation. It has been suggested that these isoaccepting species might be intermediates in the maturation of tRNA. If so, they may prove useful in elucidating the steps in tRNA biogenesis (Adesnik and Levinthal, 1969).

In minimal medium, degradation of RNA follows removal of chloramphenicol and constitutes, in physiological terms, a recovery period that precedes resumption of balanced growth and cell division (Neidhardt and Gros, 1957). When the recovery medium is enriched with a full complement of amino acids, no RNA breakdown is observed. Protein synthesis resumes immediately, at first without concomitant RNA synthesis but soon accompanied by the nucleic acid synthesis characteristic of balanced growth (Aronson and Spiegelman, 1961). The situation is similar to that in bacteria which are "shifted down" to a poorer growth medium (Maaløe and Kjeldgaard, 1966) and temporarily possess more RNA than is needed for balanced growth (Neidhardt, 1964).

Stimulation of RNA synthesis by chloramphenicol has been reported by Midgley and Gray (1971). In cultures growing slowly in glucose-salts or lactate-salts media, chloramphenicol caused an immediate acceleration of 2- to 3-fold in the overall rate of RNA synthesis. Some of this apparent acceleration could be due to the stabilization of the mRNA fraction formed after the addition of chloramphenicol. This would be in agreement with the observation of Varmus *et al.* (1971) that chloramphenicol stabilizes preexisting *lac* mRNA but inhibits further accumulation by allowing rapid degradation of nascent message.

By an unknown mechanism, chloramphenicol arrests ribosome movement and induces artificial polarity. This effect is relieved by the presence of the SuA allele which codes for a defective nuclease (Morse, 1970, 1971; Morse and Guertin, 1971). The data are consistent with a mechanism for polarity based on mRNA degradation.

VIII. Effect on Energy-Producing Systems

Chloramphenicol, but not its L-isomer, severely inhibits growth of the fungus *Pythium ultimum* at a concentration of 100 µg/ml (Marchant and Smith, 1968). Since concentrations above 100 µg/ml have little effect on the small amount of residual growth, D. G. Smith and

Marchant (1968) concluded that there are two mechanisms for energy acquisition in *P. ultimum:* one highly sensitive to chloramphenicol and another which is insensitive but maintains only a low growth rate. Since chloramphenicol-treated and normal mycelium showed the same specific oxygen uptake, the alternative mechanism of terminal oxidation must be relatively inefficient. In mycelium exposed to the antibiotic, mitochondria lacked cytochromes a, a_3, and b which are necessary for oxidative phosphorylation, but cytochrome c increased in the cytoplasm. Negatively stained mitochondrial preparations failed to show stalked particles on the cristae, and respiration was insensitive to antimycin A, which blocks electron transfer from cytochrome b to c (Potter and Reif, 1952). Thus the chloramphenicol-insensitive energy metabolism appears neither to be mediated by normal cytochromes, nor to depend on glycolysis because *P. ultimum* does not grow anaerobically.

Chlormaphenicol does not inhibit cytochrome synthesis directly in *Tetrahymena* but dissociates growth and division of mitochondria from growth and division of the organism (Turner and Lloyd, 1971); organisms from cholamphenicol-inhibited cultures have a greater proportion of small mitochondria (Poole *et al.*, 1971).

Kroon and Jansen (1968) and DeVries and Kroon (1970) showed that cultures of beating heart cells from newborn rats had reduced cytochrome c oxidase activity after exposure for 4 days to 50–100 μg/ml chloramphenicol. The effect was specific for the D-*threo*-isomer, and was not caused by overall inhibition of cell growth. At comparatively high concentrations (100–150 μg/ml) chloramphenicol inhibits respiration of HeLa cells as well as their isolated mitochondria (Firkin and Linnane, 1968).

Yeast and regenerating rat liver cells grown in the presence of 10–20 μg/ml of the antibiotic no longer synthesize cytochromes a, a_3, b, and c_1, but in nondividing cells the mitochondrial cytochromes are not affected (Firkin and Linnane, 1968, 1969). The four insoluble cytochromes of the terminal electron transport system are believed, therefore, to be synthesized by the mitochondria. As a second consequence of chloramphenicol administration, the endoplasmic reticulum of some regenerating rat liver cells was extensively dilated and the mitochondria were swollen and appeared to contain less cristae. Direct inhibition of cellular respiration occurred only at relatively high concentrations of 75–180 μg/ml.

Hanson and Hodges (1963) found that chloramphenicol at 300 μg/ml inhibited oxidative phosphorylation in isolated maize mitochondria by acting as an uncoupling agent. Inhibition of salt accumulation in plant tissue by high concentrations of chloramphenicol also appears to result from the uncoupling of oxidative phosphorylation rather than from specific inhibition of protein synthesis (Stoner et al., 1964). At the high level of 1 mg/ml, chloramphenicol appeared to interfere with ATP (adenosine triphosphate) formation in intact rabbit reticulocytes (Godchaux and Herbert, 1966).

Chloramphenicol inhibits 2-amino isobutyric acid influx into *Streptomyces hydrogenans* by 50%, but only after an exposure of at least 20 minutes (Gross and King, 1969). These experimental data can best be explained by assuming that some protein which is involved in carrier-mediated transport has a rapid turnover, and when its amount is reduced by chloramphenicol, it becomes rate limiting.

Freeman and Halder (1967) noted that concentration of chloramphenicol above 30 µg/ml inhibits $NADH_2$ oxidase (nicotinamide-adenine dinucleotide, reduced form) activity (but not the succinate oxidase activity) of isolated rat liver mitochondria. More recently, Mackler and Haynes (1970) described the inhibitory effects of chloramphenicol on purified preparations of electron transport particles and $NADH_2$ dehydrogenase from heart muscle, and electron transport particles from various yeasts. The site of inhibition in heart preparations is between the flavin component (FMN) of $NADH_2$ oxidase and cytochrome b; Freeman (1970), Lloyd et al. (1970), and Turner and Lloyd (1971) have also reported the inhibitory effect of chloramphenicol on the $NADH_2$ oxidase activity of mammalian mitochondria.

IX. Effect on Cell Wall Biosynthesis

The incorporation of amino acids into bacterial cell wall polymers is not inhibited by chloramphenicol (Hancock and Park, 1958; Mandelstam and Rogers, 1959). An effect on the incorporation of diaminopimelic acid into sporulating *Bacillus cereus* has been attributed to inhibition of induced enzyme synthesis (Vinter, 1963).

Shockman (1965) described thickening of cell walls in *Streptococcus faecium* when protein synthesis was inhibited by chloramphenicol. Similar results have been obtained with *Rhodotorula glutinis* by D. G. Smith and Marchant (1968). Cells grown in a high concentration (500 µg/ml) of chloramphenicol showed a significant increase in size, but the increase in wall material was probably a simple accretion and not caused by a disturbance of wall synthesis. Rates of wall synthesis in the presence and absence of chloramphenicol do not differ significantly and the abnormal appearance can probably be attributed to restricted synthesis of proteins and cell constituents other than wall material. Elaborations of the plasmalemma observed in the chloramphenicol-inhibited cells may be a result of unbalanced membrane synthesis leading to convolution and vesiculation.

Anraku and Landman (1968) have reported that, in *Bacillus subtilis*, chloramphenicol inhibits a late stage in the reversion of protoplasts to the osmotically stable bacillary form and simultaneously inhibits synthesis of a phosphorylated wall polymer believed to be a teichoic acid. It was suggested that inhibition of synthesis of wall polymers, including the teichoic acid, was an indirect result of the inhibited synthesis of appropriate enzyme proteins. However, Stow *et al.* (1971) have since reported that chloramphenicol powerfully inhibits the biogenesis of a wall teichoic acid in a cell-free system of fragmented cytoplasmic membrane from *Bacillus licheniformis*. This occurs through direct action on the system synthesizing teichoic acid and is unrelated to protein synthesis since neither streptomycin nor pyromycin is effective. Stow *et al.* conclude that, although chloramphenicol inhibits the growth of most microbial cells primarily through its effect on protein synthesis, the direct inhibition of the transfer of glucose to teichoic acid may be a contributory factor in some microbes. The effect is apparently specific to glucose and is not observed with the transfer of other residues from nucleotides. The site of inhibition seems to be at the stage of transfer of glucose from nucleotide precursor to lipid carrier; in the cases studied there was no inhibition of glucose transfer in syntheses where lipid carriers were not involved. Although this may provide an alternative explanation of the effect observed by Anraku and Landman (1968), a more detailed study of their system is required.

X. Effect on Biosynthesis of Aromatic Compounds

Woolley (1950) detected reversal of chloramphenicol inhibition with various aromatic compounds and postulated that the antibiotic was an analogue of phenylalanine. However the large number of aromatic compounds which were effective and the small degree of reversal indicated some nonspecific mechanism. A similar explanation may apply to the observation by Mulherkar *et al.* (1967) that the action of chloramphenicol on chick morphogenesis can be reversed with equimolar concentrations of phenylalanine, tryosine, *o*-aminobenzoic acid, and phenyllactic acid.

Truhaut *et al.* (1951) suggested that chloramphenicol interfered with the biosynthesis of tryptophan. Conversion of anthranilic acid to indole was thought to be the sensitive reaction because growth inhibition by chloramphenicol was partially reversed when essential metabolites were added to bacterial auxotrophs which required anthranilic acid, indole, or tryptophan (Bergmann and Sicher, 1952). In wild-type *Escherichia coli,* however, tryptophan fails to reverse the effect of chloramphenicol on growth (Hopps *et al.,* 1956) and it is unlikely that the action of chloramphenicol is induction of tryptophan deficiency.

Chloramphenicol (but not its inactive stereoisomers) does inhibit the biosynthesis of indole in a mutant of *E. coli* that accumulates indole as the result of a block in the conversion of indole to tryptophan (Gibson *et al.,* 1955, 1956; McDougall and Gibson, 1958). The effect was also observed with nongrowing suspensions of bacteria in which indole synthesis had been proceeding for hours and which apparently possessed a complete set of the necessary enzymes. Gibson and McDougall (1961) extended the study to auxotrophs accumulating indole-3-glycerol, anthranilic acid, or 5-dehydroshikimic acid and concluded that chloramphenicol interfered with a reaction preceding 5-dehydroshikimic acid. Tetracyclines produce the same effect, so the activity is not specific for chloramphenicol. However, the possible relationship between the inhibition of indole and protein syntheses warrants further study.

XI. Effect on Antibiotic Biosynthesis

Inhibitors of protein synthesis stimulate biosynthesis of some antibiotics (Light, 1967). Increased incorporation of amino acids into actinomycin by *Streptomyces antibioticus* is believed due to the increased endogenous pool when incorporation into protein has been inhibited (E. Katz *et al.,* 1965). Paulus and Gray (1964) found that chloramphenicol slightly stimulated incorporation of ^{14}C-threonine into polymyxin by *Bacillus polymyxa.* In both examples rapid antibiotic synthesis followed growth and active protein synthesis.

XII. Resistance to Chloramphenicol

Organisms resist the effect of toxic agents by three general mechanisms: (1) Physiological immunity. The organism does not possess, or depend on, the system which the agent inhibits. (2) Drug inactivation. The organism produces enzymes capable of detoxifying the agent. (3) Permeability barrier. The organism possesses a wall or membrane which protects sensitive areas.

A. Physiological Immunity

Chloramphenicol binds to 70S but not to 80S ribosomes. All organisms and systems sensitive to low concentrations of it (bacteria, blue-green algae, mitochondria, and chloroplasts) have 70S ribosomes whereas those resistant to chloramphenicol (yeast, fungi, mammalian cell cytoplasm, pea seedling cytoplasm, and protozoa) contain 80S ribosomes (Spirin and Gavrilova, 1969). So far chloramphenicol resistance arising from modification of the ribosomal site of action in bacteria has not been demonstrated.

B. Drug Inactivation

G. N. Smith and Worrell (1949, 1950, 1953) observed that chloramphenicol was decomposed by susceptible cultures of *Escherichia coli*, *Proteus vulgaris*, *Bacillus mycoides*, and *B. subtilis* if it was incubated with heavy cell suspensions for several days. All functional groups of chloramphenicol were vulnerable; the possible degradative pathways are summarized in Fig. 3. Both *B. subtilis* and *P. vulgaris* produced considerable quantities of chloramphenicolase (G. N. Smith and Worrell, 1950), the enzyme responsible for hydrolyzing the amide linkage. As a result p-nitrophenylserinol (**2**, Fig. 3) accumulated, but could be rapidly converted to p-aminophenylserinol (**3**). The rate at which p-aminophenylserinol accumulated was approximately twice the rate at which chloramphenicol was reduced to 2-N-dichloroacetyl-p-aminophenylserinol (**4**).

None of the degradation products are appreciably antibacterial and some may even stimulate growth (G. N. Smith and Worrell, 1952; Rannenberg and Arnold, 1968). Resistant strains did not seem to degrade the antibiotic to a greater extent than sensitive ones, and the overall rate was quite low.

Some saprophytes which use chloramphenicol as a nutrient have special pathways for its decomposition (Lingens and Oltmanns, 1966). The various products shown in Fig. 4 were identified in cultures of a *Flavobacterium* (Lingens et al., 1966). No chloramphenicol-destroying enzyme could be demonstrated in the culture filtrate.

Malek et al. (1961) have described a *Streptomyces* utilizing chloramphenicol as a source of carbon and nitrogen. The organism grew in concentrations of up to 600 μg/ml and completely degraded the antibiotic in 13 days. Neither growth nor destruction of antibiotic was detected in a medium containing 1 mg/ml. Chloramphenicol at first inhibited growth of a mutant of *Chlamydomonas reinhardi* requiring p-aminobenzoic acid, but after a lag the alga grew. Chloramphenicol was then able to replace p-aminobenzoic acid, presumably because catabolism had produced the growth factor (Rannenberg and Arnold, 1968).

Escherichia coli cells adapted to chloramphenicol became more aerobic than normal cells (Molho-Lacroix and Molho, 1952). Cells

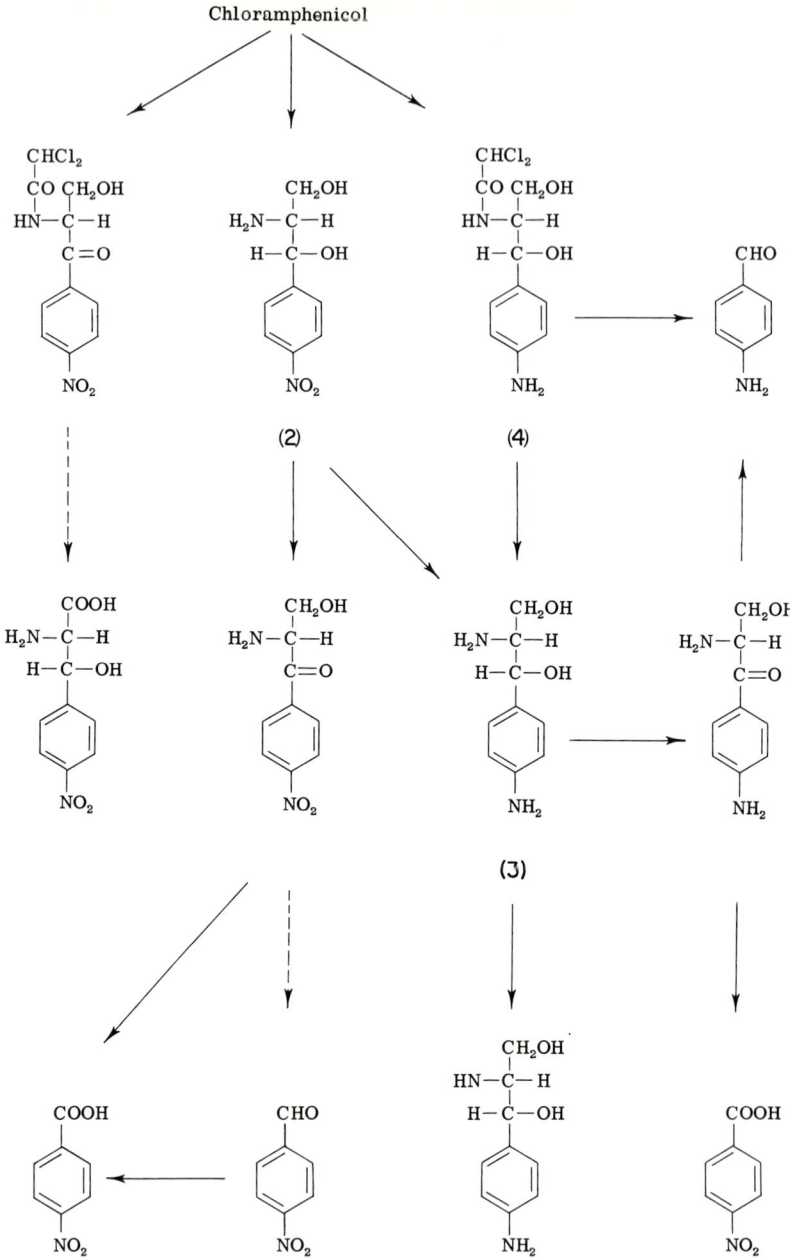

FIG. 3. Products isolated from bacterial metabolism of chloramphenicol and pathways postulated (Smith and Worrell, 1950).

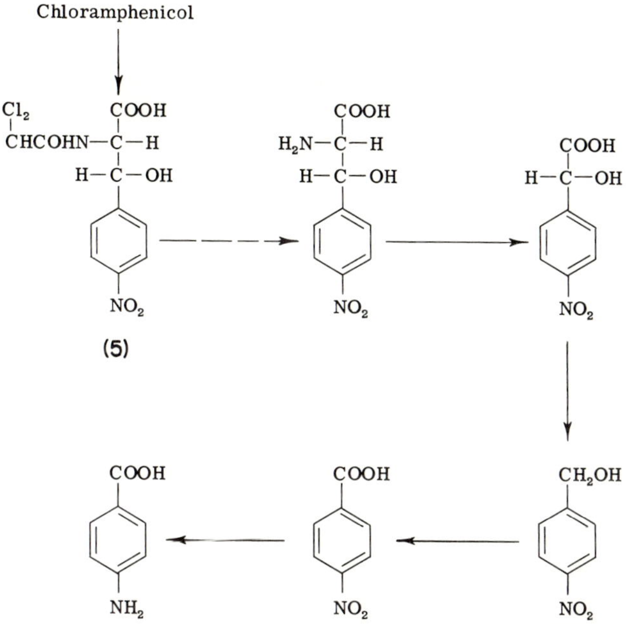

FIG. 4. Products isolated from bacterial metabolism of chloramphenicol and pathway postulated (Lingens et al., 1966).

adapted to N-dichloroacetyl-p-nitrophenylserine (5, Fig. 4) had a shorter lag phase when subsequently exposed to chloramphenicol than unadapted cells or cells grown in the presence of N-dichloroacetyl-p-nitrophenylalanine, N-dichloroacetylphenylalanine, N-dichloroacetyl-p-nitrophenol, phenylalanine, p-nitrophenylalanine, or phenylserine. Bacteria adapted to chloramphenicol hydrolyzed the dichloroacetyl group from N-dichloroacetylthienylalanine, but from none of the above compounds.

Merkel and Steers (1953) claimed that resistance in a mutant of *E. coli* B was related to chloramphenicol reductase activity. A relationship between resistance and inactivation was noted by Miyamura (1964) in clinical isolates of *Shigella*, *Escherichia*, and *Staphylococcus* but not *Pseudomonas* strains.

Shaw and Brodsky (1967) have shown that clinical isolates of enteric bacteria with an R factor for chloramphenicol resistance contain chloramphenicol acetyltransferase. This enzyme catalyzes acetylation by acetyl coenzyme A to the 3-acetoxy and 1,3-diacetoxy derivatives. In *E. coli* a single enzyme catalyzes the acetylation of both hydroxyl groups of chloramphenicol (Mise and Suzuki, 1968). Propionoxy-chloramphenicol accumulates in the presence of propionate (Shaw, 1967). The acetylated products are biologically inactive.

Resistant strains of *E. coli* grow in the presence of chloramphenicol after an augmented lag period approximately proportional to the concentration of chloramphenicol (Shaw and Brodsky, 1967). This lag reflects the time required for inactivation of the antibiotic and depends on the amount of acetylating enzyme. After drug inactivation, resistant strains grow at the same rate as cultures free of chloramphenicol.

Chloramphenicol acetyl transferases from *Staphylococcus aureus* and the enteric bacteria are similar in their pH optimum and molecular weight but differ in their affinity for chloramphenicol, heat stability, electrophoretic mobility, and immunological reactivity (Shaw *et al.*, 1970; Shaw, 1971). An important biological distinction is that the *S. aureus* enzyme is inducible whereas the enzyme in *E. coli* is constitutive (Shaw, 1967). The induction has been studied in some detail because it is unusual for the inducer also to be a potent inhibitor of protein synthesis (Shaw and Brodsky, 1968). Addition of the antibiotic was followed promptly by a modest decrease in the exponential growth rate. Synthesis of chloramphenicol acetyltransferase began within 10 minutes, but increased at a linear rate as chloramphenicol disappeared from the medium. A specific activity plot showed that chloramphenicol acetyltransferase was synthesized much faster than total protein.

Induction occurred only with compounds stereochemically and structurally related to chloramphenicol (Kono *et al.*, 1971). The minimum structural requirements defined by Shaw and Winshell (1968) are (1) a D-*threo* configuration of the 2-amino 1 propanol moiety, (2) electronegative acyl substitution of the 2-amino group, and (3) an unsubstituted hydroxyl group at C-1. The substituent on the phenyl ring does not appear to be important, nor does the presence of the 3-hydroxyl group which is the site of primary acetylation. The analogue (Fig. 5) is a potent inducer of chloramphenicol acetyl-

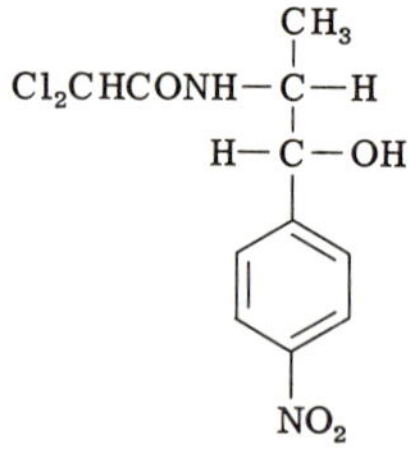

FIG. 5. Gratuitous inducer of chloramphenicol acetyltransferase.

transferase, and, by virtue of its inertness as a substrate and its low antibacterial activity, is useful for analysis of induction under gratuitous conditions.

Okamoto et al. (1967) surveyed gram-negative bacteria for chloramphenicol acetyltransferase. Activity was detected in almost all strains of the *Proteus* genus, with resistant strains showing higher levels. A few chloramphenicol-sensitive strains lacked the enzyme. Only sensitive strains containing the enzyme readily yielded resistant mutants with higher enzyme activity. This parallels the observation that mutants producing penicillinase are never obtained from parents which are penicillinase negative (Pollock, 1960). All *Pseudomonas aeruginosa* strains were resistant to chloramphenicol. Most contained low levels of enzyme, but not enough to explain their resistance. Other bacterial groups examined (except one strain of *Enterobacter cloacae*) lacked the enzyme, although most strains of the *Streptococcus marcescens* and *Klebsiella-Aerobacter* groups were resistant to chloramphenicol. Sompolinsky and Samra (1968) obtained *E. coli* B mutants which, through one or two mutational events, were resistant to high levels of chloramphenicol. All mutants inactivated the drug. Growth depended on inoculum size, medium composition, and chloramphenicol concentration. No growth was observed with lactose as sole energy source unless the organisms had been previously exposed to a β-galactosidase inducer.

C. Permeability Barrier

Decreased permeability to chloramphenicol in *E. coli* mutants was noted by Vazquez (1964) and by Unowsky and Rachmeler (1966). Vazquez (1966) demonstrated a striking correlation between the antibacterial activity of chloramphenicol isomers and analogues and the extent to which these compounds accumulated within the cells. Inactive analogues showed little affinity for bacterial ribosomes and did not inhibit protein synthesis. Shaw and Unowsky (1968) have postulated that the apparent impermeability of chloramphenicol in *Escherichia coli* strains carrying a resistance factor is due to failure of acetylated chloramphenicol to bind to ribosomes (Unowsky and Rachmeler, 1966). As a result no radioactivity accumulates in the cells when they are treated with ^{14}C-chloramphenicol.

It was once thought that the R factor conferring resistance to streptomycin, tetracycline, chloramphenicol, sulfonamide, kanamycin, and neomycin brought about a single biochemical change that caused a broad alteration in membrane properties to exclude all six antibiotics (Lebek, 1963). However, it is now clear that R factors carrying many resistance markers may dissociate to form smaller R factors conferring resistance to fewer antibiotics (Novick, 1969), still by modifying the permeability properties of the cell (Watanabe, 1963, 1971). However, loss of resistance to some but not all antibiotics argues against a single biochemical mechanism, and now Okamoto and Mizuno (1964) have shown that *E. coli* with an R factor conferring resistance to tetracycline, sulfonamide, and chloramphenicol were impermeable to these antibiotics, but resistance to each depended on distinct biochemical processes.

Only after obtaining several mutations were Sompolinsky and Samra (1968) able to obtain cultures of *E. coli* K-12 with a high level of resistance to chloramphenicol. None of these mutants inactivated the antibiotic. Growth was extremely slow, even in the absence of the drug, and resistance was considered due to a nonspecific decrease in permeability.

XIII. Chloramphenicol as a Mutagen

Carnevali *et al.* (1971) reported the induction of petites in yeast by erythromycin and tetracycline but not by chloramphenicol. However, Weislogel and Butow (1970) have shown that chloramphenicol induces petites in a particular strain of *Saccharomyces cerevisiae,* and Williamson *et al.* (1971) induced cytoplasmic petite mutations with erythromycin and chloramphenicol. Whatever the mechanism, the primary event appears to be the blockage of protein synthesis in the mitochondria. A reasonable hypothesis is that a protein necessary for correct replication of mitochondrial DNA is built on the mitochondrial ribosomal system; disturbance of this protein synthesis inhibits the synthesis of mitochondrial DNA, and petite mutations result.

XIV. Chloramphenicol as a Therapeutic Agent

Despite the description of hematologic toxicity and the occurrence of the so-called gray baby syndrome there has been no curtailment in the use of chloramphenicol. A recent survey of antibiotics used in hospitals revealed that 1.51 lb of semisynthetic penicillins were prescribed during a period when 1.89 lb of tetracycline and 1.84 lb of chloramphenicol were dispensed (Ingall and Sherman, 1970).

Chloramphenicol U.S.P. is sold in capsules (50, 100, and 250 mg) as a cream (1%) for topical use, as an ointment (1%), as a powder (25 mg), and as an oral solution (0.5%). A powder (1 gm) is supplied for intramuscular administration and a 2-ml vial (500 mg) is produced for intravenous therapy. Chloramphenicol palmitate solution is provided as an oral suspension (155 mg/5 ml). The currently popular injectable derivative is sodium chloramphenicol succinate, which is dispensed in 1-gm doses of dry powder. Chloramphenicol is given locally, orally, intramuscularly, or intravenously. Intravenous administration should be slow (over several hours); the drug may also be given subcutaneously (with hyaluronidase), intrathecally, rectally, intraperitoneally, and intra-articularly.

Several side effects have been ascribed to chloramphenicol, e.g., gastrointestinal reactions (nausea, vomiting, diarrhea pruritusani), allergic manifestations (fever, rashes, and Herxheimer reactions), dose-related phenomena (gray baby syndrome, reversible anemia), superinfection and toxic reactions (aplastic anemia, thrombocytopenia, optic neuritis, and digital paresthesias). Considering the notoriety given the toxicity of chloramphenicol, the drug should be used with greater caution while the etiology of these reactions is critically evaluated.

XV. Chloramphenicol as a Secondary Metabolite

Chloramphenicol production is a variable activity of *Streptomyces* strain 13S. Cultures grow rapidly in complex media and produce chloramphenicol, in relatively low yield (25 mg/liter), during the stationary phase. In the defined medium they grow slowly but produce larger amounts of chloramphenicol (150 mg/liter) during the exponential phase (Malik and Vining, 1970). Generally similar results have been obtained previously by Umezawa *et al.* (1948), Oyaas *et al.* (1950), and Gottlieb *et al.* (1956).

Bu'Lock (1961) has pointed out that the appearance of a secondary metabolite in parallel with growth is exceptional. Such fermentations often do not involve any extensive biosynthetic activity; several are, in fact, rather clear examples of partial utilization of substrates. More typically the synthesis of a secondary metabolite is suppressed while the cells are actively multiplying and begins as the culture enters the stationary phase. Thus if one considers only the sequence of events in defined medium (Fig. 6), chloramphenicol is not a characteristic secondary metabolite associated with idiophase (Bu'Lock, 1967). Its production reflects the immediate environment, not the previous history of the organism. However, Brar *et al.* (1968) have observed that the pattern of ergot alkaloid production by *Claviceps paspali* is strongly influenced by the growth rate of the producing culture.

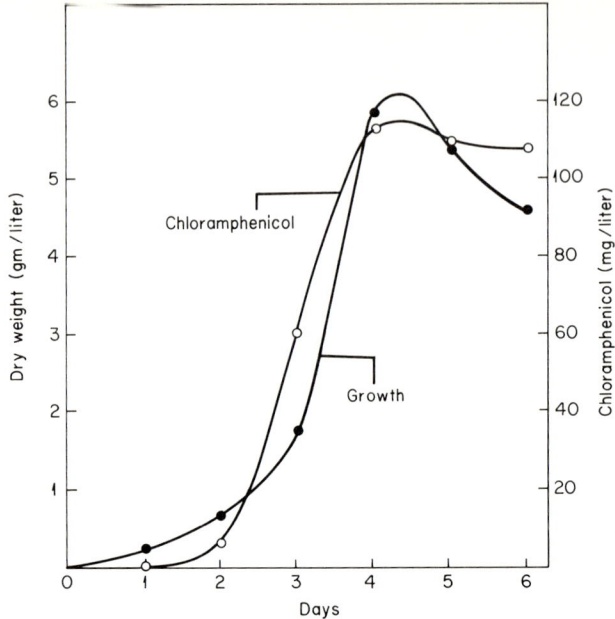

FIG. 6. Growth and chloramphenicol production: *Streptomyces* strain 13S in defined (GSL) medium.

Alkaloid was formed during the trophophase if the medium supported slow growth, but in the idiophase if the medium supported rapid growth. Their results parallel those obtained for chloramphenicol and suggest that any classification of secondary metabolites into growth-linked and idiophase products is meaningless.

The constitution of a microbial cell is determined to a considerable extent by the growth medium (Maaløe and Kjeldgaard, 1966). In the efficient and regular pattern of balanced growth, synthesis of secondary metabolites is suppressed while the physiological processes of the cell regulate each other. However, if control mechanisms fail to synchronize and modulate reactions, changes in cell composition may induce or repress specific enzymes and initiate secondary metabolism. Unbalanced growth, leading to stress in the regulatory systems, can occur during cell proliferation in some media, or after it in others. Present knowledge of the pathway of chloramphenicol biosynthesis indicates that chloramphenicol might be formed as a result of increased levels of the normal intermediates of aromatic biosynthesis when other pathways requiring these intermediates are partly or wholly closed.

XVI. Biosynthesis of Chloramphenicol

Streptomyces 13S produces chloramphenicol on a chemically defined medium with glycerol as the main carbon source, DL-serine as the nitrogen source, and sodium lactate as the auxiliary carbon source (GSL medium). This *Streptomyces* grows on quinic acid, shikimic acid, glutamic acid, pyruvate, glucose, phenylalanine, acetate, lactose, tyrosine, and sucrose.

The chloramphenicol-producing streptomycete displays and develops resistance to a variety of inhibitory compounds, such as chloramphenicol, *p*-methylthio and *p*-methylsulfo analogues of chloramphenicol, *p*-fluoro-DL-phenylalanine, β-2-thienyl-DL-alanine, norleucine, 5-methyltryptophan, and DL-4-fluorotryptophan. However, sensitivity to growth inhibition by some of the analogs results from manipulation of the nutritional environment. For example, when the chloramphenicol-producing streptomycete is grown with glutamic acid as a source of carbon and energy, the cells become sensitive to inhibition by β-2-thienylalanine and DL-4-fluorotryptophan. This increased sensitivity could be due to a decreased rate of synthesis of aromatic compounds when glutamic acid is the carbon source. Resistant mutants are obtained when DL-4-fluorotryptophan crystals are placed on streptomycete-seeded minimal agar plates that have glutamic acid as a carbon source. Most such mutants are tryptophan excretors (Malik, 1972).

Phenylalanine, tyrosine, *p*-aminobenzoic acid, and *p*-nitrophenylserinol are not incorporated directly into chloramphenicol. The stimulation of antibiotic production by these aromatics is presumably due to an indirect effect on the regulatory systems of streptomycetes. The shikimic acid pathway is used in the biosynthesis of chloramphenicol (Vining *et al.*, 1968). The route of chloramphenicol synthesis has at least one intermediate in which the shikimic and phosphoenol pyruvate precursors are linked. This route diverges from that of phenylpropanoid amino acids between 3-enolpyruvyl-5-phosphoshikimic and prephenic acids (Fig. 7). The nitro groups of chloramphenicol is not introduced by a biological nitration employing nitrate ions. Both nitrogen atoms of the antibiotic are derived from a common pool in which nitrogen from different sources is equilibrated.

When the chloride concentration is limiting, *Streptomyces venezuelae* can synthesize analogues of chloramphenicol (C. G. Smith, 1958) and derivatives of *p*-nitrophenylserinol (N-acetyl, N-propionyl, N-butyryl, and probably N-pentanoyl and N-hexanoyl derivatives). When bromide is added to a halogen-free medium, N-bromochloroacetyl and N-dibromoacetyl derivatives are formed. The presence

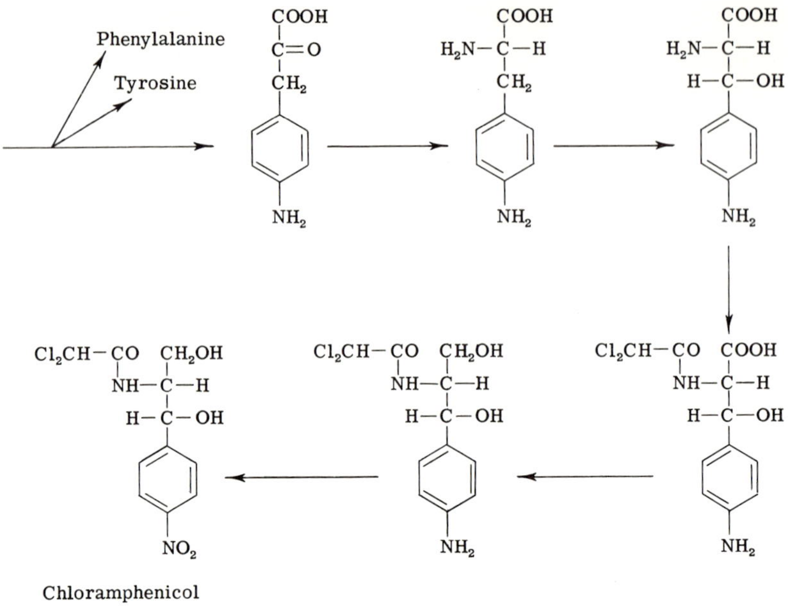

FIG. 7. Biosynthesis of chloramphenicol.

of acyl analogues in chloramphenicol fermentations has also been reported by Stratton and Rebstock (1963), who also found D-*threo*-N-dichloroacetyl-*p*-aminophenylserinol, the *p*-amino analogue of chloramphenicol.

Cultures grown in many different media can be used as seed; the course of growth and chloramphenicol production in GSL medium is not influenced significantly by the prior history of the inoculum. Addition of yeast extract (50 mg/liter) stimulates early growth, indicating that the extended lag phase which occurs in this medium is probably not caused by a toxic substance. The effect is similar to that observed in preutilized medium which may likewise contain growth factors and nutrients excreted during the lag phase. A toxic substance does appear to be formed if GSL medium is autoclaved with nutrient broth, and a similar toxicity in autoclaved preincubated medium is further evidence for excretion of nutrients during the lag phase.

The importance of lactate to chloramphenicol production is well established, but the mechanism is still obscure. The suggestion that it forms a protective complex with zinc during autoclaving (Gallicchio and Gottlieb, 1958) seems inadequate since variations in the method of preparation and sterilization of media did not affect antibiotic production (Malik, 1970). Lactate may act by controlling the pH of the medium; another possibility is that it acts by a feedback mechanism, altering the balance between the glycolytic and pentose phosphate pathways in favor of the latter, thereby promoting the flow of precursors into the shikimic acid pathway.

XVII. Catabolism of Chloramphenicol by the Producing Organism

In fermentations the concentration of a metabolite declines after reaching a peak (Gottlieb and Shaw, 1967). Legator and Gottlieb (1953), investigating chloramphenicol accumulation in cultures of *Streptomyces venezuelae*, discovered that chloramphenicol, added in amounts greater than normally produced, disappeared from the medium. If *p*-nitrophenylserinol was added it was metabolized to the N-acetyl derivative, and some *p*-nitrobenzoic acid was formed (Gottlieb *et al.*, 1956). During studies on chloramphenicol biogenesis Siddiqueullah *et al.* (1968) isolated *p*-nitrobenzoic acid and *p*-nitrobenzyl alcohol and suggested that the latter might be formed by catabolism of chloramphenicol.

Another chloramphenicol producer, *Streptomyces*, hydrolyzed chloramphenicol to *p*-nitrophenylserinol, but this was quickly acetylated to N-acetyl-*p*-nitrophenylserinol (Malik and Vining, 1970). Small amounts were metabolized to *p*-nitrobenzoic acid and *p*-nitrobenzyl alcohol, as well as some unidentified products detected by the use of ^{14}C-labeled chloramphenicol. *p*-Nitrophenylserinol and N-acetyl-*p*-nitrophenylserinol are less toxic than chloramphenicol, but the extent of chloramphenicol inactivation was insufficient to account for development of resistance during lag phase.

A constitutive enzyme, chloramphenicol hydrolase, was detected bound to particulate cell fraction in *Streptomyces*. The enzyme hydrolyzed chloramphenicol to *p*-nitrophenylserinol. Its specific activity in the mycelium did not vary in response to exogenous chloramphenicol concentration (Malik and Vining, 1971). No specific recovery protein could be detected in the protein synthesized during or immediately after the lag phase (when cultures become resistant to chloramphenicol). As soon as chloramphenicol starts to be biosynthesized, it stops being catabolized.

XVIII. Effect of Chloramphenicol on the Producing Organism

Microorganisms producing an antibiotic are often challenged by their product which accumulates and inhibits growth. In certain cases, such as the formation of penicillin by *Penicillium chrysogenum*, the producer is immune to the effects of its product; it lacks the site of antibiotic action. However, many examples are known of producing cultures that are inhibited markedly by the antibiotics they make (Dulaney, 1951; McDaniel and Hodges, 1951; Niri *et al.*, 1963; Woodruff, 1966). Only if the antibiotic accumulates during the stationary phase of growth can broth levels become significantly greater than those to which the growing organism is sensitive (Woodruff, 1966).

Waksman *et al.* (1947) have claimed that antibiotics in moderate amounts are not toxic to their producing organism. More recently Okami *et al.* (1967) have tested the reactions of a wide range of *Streptomyces* to 22 antibiotics; they showed that producers generally resist their own antibiotics and exhibit a characteristic spectrum of sensitivity. Similarly, Legator and Gottlieb (1953) observed normal growth of *Streptomyces venezuelae* in the presence of chloramphenicol. But Malik and Vining (1970) found a growing *Streptomyces* very sensitive to its own antibiotic. It later acquired resistance to chloramphenicol, however, so perhaps the discrepancy between these results and those of Waksman, Okami, and their collaborators lies in the time, i.e., the phase of growth, at which resistance was determined.

Streptomyces 13S, then, shifts from a state of chloramphenicol sensitivity when no antibiotic is being formed to chloramphenicol resistance during the production phase (Malik, 1970); similar observations have been made by Niri *et al.* (1963) with oxytetracycline-producing cultures of *Streptomyces rimosus*, by Yoshida *et al.* (1966) with an actinomycin-producing *Streptomyces antibioticus*, and by Al'-Nuri and Egorov (1968) with the novobiocin producer, *Actinomyces spheroides*. The fungus *Cordyceps militaris* behaves differently. Cordycepin is not taken up when added to producing cultures at the time of inoculation (Chassy and Suhadolnik, 1969). Once formed and excreted into the media, it does not equilibrate with the intracellular pool, which is very small.

A. Adaptation to Chloramphenicol

One of the most interesting aspects of chloramphenicol biosynthesis is the ability of the producing organism to become resistant to concentrations which were originally inhibitory. During the first incubation in the presence of the drug, long filamentous cells are formed, perhaps as a result of the inhibited cell division reported by Pulvertaft (1952) and Dienes *et al.* (1950). On subculture, in the presence of chloramphenicol, the tendency to abnormal clones vanishes. The lag also disappears, but the growth rate remains below normal perhaps because chloramphenicol has impaired the energy-producing system or decreased permeability to nutrients.

The manner in which resistance to increasing chloramphenicol concentrations develops, and the equally rapid loss of resistance when the organism is grown without antibiotic, is convincing evidence that the adaptation is quite distinct from genetic change, which involves selection of preexisting variants (Malik, 1970). There is no permanent extension of the range of drug tolerance and the level of resistance is nicely adjusted to the concentration of chloramphenicol at which the cells have been trained. Selection can account for this result only if extremely improbable systems of polygenes are postulated (Cavalli and Maccacaro, 1952).

B. Protein Synthesis and Development of Resistance

In vivo as well as *in vitro* experiments show that chloramphenicol inhibits protein synthesis in *Streptomyces* 13S (Malik, 1970). Such evidence is consistent with measurements showing *Streptomyces* ribosomes to be of the 70S prokaryotic type. Moreover, these ribosomes bind chloramphenicol to almost the same extent as do those of a chloramphenicol-sensitive culture of *Escherichia coli.* When chloramphenicol was added to cultures growing in complex medium, protein synthesis halted but resumed after a lag. In GSL medium, too, protein synthesis was initially sensitive to chloramphenicol, but resistance developed as the culture began to produce the antibiotic. Since protein synthesis always stayed sensitive to higher concentrations, resistance is unlikely to have been due to development of a chloramphenicol-insensitive ribosomal system. Decreased accessibility of the protein-synthesizing machinery is a more likely explanation, and changed membrane permeability seems probable.

Analysis of doubly labeled cell protein indicates that development of resistance does not require *de novo* synthesis of a specific protein. Recovery from inhibition may involve a relatively trivial modification of a preformed molecule, such as a change in hydrogen bonding, sulfur-sulfur linkages, or the gain or loss of only a few building blocks. Any change in lipid composition during development of resistance should be investigated since decreased membrane permeability may prevent the entrance of antibiotic into the cell (Dunnick and O'Leary, 1970).

C. ROLE OF ANTIBIOTIC METABOLISM IN RESISTANCE

The initial step in metabolism of chloramphenicol by *Streptomyces* 13S is hydrolytic removal of the N-dichloroacetyl group (Malik and Vining, 1971). *p*-Nitrophenylserinol formed in this reaction inhibits growth less effectively than chloramphenicol, and is reacylated to N-acetyl-*p*-nitrophenylserinol which has only weak antibiotic activity (Gottlieb *et al.*, 1956). The N-acetyl-*p*-nitrophenylserinol does not prolong lag, but affects the growth rate (Malik, 1970).

N-Acetylation of *p*-nitrophenylserinol has also been reported by Gottlieb *et al.* (1956). This reaction may be similar to the acetylation of amino acids by extracts of *E. coli* (Maas *et al.*, 1953), the acetylation of amino acids and aliphatic amines by *Clostridium kluyveri* (J. Katz *et al.*, 1953), or bacterial N-acetylation of the L-lysine antagonist, S(β-aminoethyl)-L-cysteine (Soda *et al.*, 1969).

p-Nitrobenzoic acid and *p*-nitrobenzyl alcohol were found labeled in the experiment with ^{14}C-*p*-nitrophenylserinol, indicating that *p*-nitrophenylserinol is not only acetylated but can be further metabolized. Similar degradation of chloramphenicol has been observed in other microbes (G. N. Smith and Worrel, 1950; Lingens *et al.*, 1966). Metabolism of excreted metabolites by their producing organisms has been reported by various investigators (Gourevitch *et al.*, 1961; Kojima *et al.*, 1969; Miller and Walker, 1969; Chassy and Suhadolnik, 1969), but the physiological importance of this phenomenon is uncertain. The possibility remains that nonspecific enzymes metabolize any portion of the product which reenters the cells.

D. CHLORAMPHENICOL HYDROLASE AND RESISTANCE

Antibiotic resistance is often accompanied by increased production of an enzyme capable of destroying or inactivating the drug. An example is the penicillinase formed in certain types of penicillin-resistant bacteria (Pollock, 1960). Chloramphenicol resistance in *Staphylococcus aureus* and enteric bacteria involves the enzyme chloramphenicol acetyltransferase (Shaw, 1971). However, destruction or inactivation of chloramphenicol cannot solely account for the resistance of *Streptomyces* 13S, since the concentration of chloramphenicol in the medium never falls below inhibitory levels. Chloramphenicol hydrolase appears to be a constitutive enzyme and the presence of chloramphenicol in the medium does not result in elevated levels in the mycelium (Malik and Vining, 1971).

Metabolism of chloramphenicol increases during the log phase; the activity of chloramphenicol hydrolase also increases during this period and then falls off. Perhaps one of the amidases, required in biosynthesis of amino acids, degrades chloramphenicol. To gain further insight into the role of chloramphenicol hydrolase, its specificity must be established.

Decomposition of chloramphenicol in *Streptomyces* 13S cultures is proportional to the exogenous concentration of the antibiotic and may depend on the rate of penetration into the cell. Presumably, intracellular chloramphenicol is fairly rapidly destroyed by the hydrolytic enzyme, but a permeability barrier between the substrate and the enzyme would limit activity to the rate at which substrate could penetrate it. The most obvious barrier to free access of chloramphenicol would be the cell wall or membrane, and access might be mediated by a specific permease (Cohen and Monod, 1957). At present, however, there is no evidence for such a specific permease. That cellular membranes can, by virtue of their structure, limit random diffusion is generally accepted. Diffusion does not appear to be completely prevented, however, and apolar substances penetrate more readily than polar ones. Thus the greater ease with which the *p*-methylthio analogue of chloramphenicol would be expected to diffuse into the cell might explain its greater toxicity.

A completely different explanation for the development of chloramphenicol resistance is that exposure of *Streptomyces* 13S cells to chloramphenicol activates a biochemical pump which eliminates the antibiotic from the cell. Bu'Lock (1967) observed that when he added 6-methylsalicylate to *Penicillium urticae* mycelium the supplement was removed from the medium very rapidly, but after a period varying between 5 minutes and 1 hr, it reappeared. This occurred with washed trophophase and idiophase mycelia, irrespective of their capacity to make 6-methylsalicylate *de novo,* and only changed with late idiophase mycelium which rapidly converted 6-methylsalicylate into patulin. Bu'Lock (1967) thought the excretion effect might be due to a change in intracellular pH.

After the onset of chloramphenicol biosynthesis, *Streptomyces* 13S acquires resistance by what appears to be a change in permeability of the cells, and metabolism of chloramphenicol stops. The permeability relationship may be quite complex, since the organism is also excreting the antibiotic against a concentration gradient. This pumping out action itself may well reduce net uptake.

Where chloramphenicol is formed in the cells, how it reaches the outside, and how it is released are still unknown. It is obvious that the membrane must play some role in this process and the following possibilities are suggested: (1) Chloramphenicol is synthesized and released into the cytoplasm in a free state. It subsequently passes through the membrane by free diffusion, or (2) is pumped out by a permease. (3) It is formed either within the membrane or on its inner surface and reaches the exterior without being released in the cytoplasm. (4) Synthesis occurs on the outher side of the membrane.

XIX. Regulation of Chloramphenicol Biosynthesis

Demain (1968) has marshaled the evidence indicating that the biosynthesis of secondary metabolites may be controlled by feedback mechanisms. One of the more cogent examples is given by Legator and Gottlieb (1953) who reported that addition of chloramphenicol to cultures of *Streptomyces venezuelae* at various intervals during growth did not appreciably alter the final concentration in the medium. They suggested that chloramphenicol exerted an inhibitory effect on its own biosynthesis.

In the experiments described by Legator and Gottlieb, production of chloramphenicol by *S. venezuelae* did not parallel growth but lagged considerably behind. The fastest rate of antibiotic synthesis occurred when growth had almost ceased. In no instance were large amounts of chloramphenicol found in mycelium.

If chloramphenicol was added to the culture medium normal growth was observed, but the amount added appeared to influence the amount produced endogenously. The concentration at the time of harvest was invariably less than the total expected if the organism had produced as much as in unsupplemented cultures. If the maximum quantity of chloramphenicol normally produced was added to media at the beginning of the fermentation, no increase in antibiotic titer occurred. Smaller initial supplements allowed production of a total concentration equivalent to the normal yield. Addition of the antibiotic at any time during the growth phase exerted a limiting action on further synthesis, but the productivity of cells exposed to high concentrations was not permanently injured. When washed cells were suspended in high concentrations of chloramphenicol no decomposition of the antibiotic was detected (Gottlieb and Shaw, 1967).

Measurements of ^{3}H-chloramphenicol production in cultures of *Streptomyces* grown on a medium containing 6-^{3}H-D-glucose and 3-^{14}C-chloramphenicol also showed that chloramphenicol inhibits its own biosynthesis (Malik, 1970). Similar results were obtained in cultures supplemented with the antibacterial *p*-methylthio analogue of chloramphenicol. Here synthesis of the antibiotic was completely suppressed until the concentration of analogue had been reduced by inactivating enzymes. In contrast, the L-*threo* and *p*-methylsulfonyl analogues did not delay growth of the organism and had little effect on chloramphenicol biosynthesis. However, like chloramphenicol and its *p*-methylthio analogue, both compounds were degraded.

Degradation of chloramphenicol and its p-methylsulfonyl analogue ceased when endogenously produced antibiotic reached a concentration of 10–30 mg/liter, suggesting that changes in cell permeability are associated with the onset of chloramphenicol synthesis.

Since chloramphenicol, aromatic amino acids, or shikimic acid pathway intermediates did not repress or inhibit 3-deoxy-D-arabino-heptulosonate 7-phosphate synthetase (Lowe and Westlake, 1971), the site of metabolic control of chloramphenicol biogenesis remains a mystery. The absence of the antibiotic inside the producing streptomycete (Legator and Gottlieb, 1953) does not exclude chloramphenicol as a controlling agent.

The possibility that chloramphenicol inhibits its own excretion and regulates general permeability of the microbe to the aromatics should be examined. The results of Malik (1970) show that some inhibition of chloramphenicol biosynthesis occurs. However, the apparent constancy of yield is due, in part, to catabolism of the added antibiotic. Even when chloramphenicol was added in concentrations higher than those normally attained by endogenous synthesis, biosynthesis was not completely inhibited.

The effect of chloramphenicol upon *Streptomyces* 13S is hard to evaluate because of the difficulty of dissociating its specific roles in resistance development and regulation from nonspecific effects due to its action on protein biosynthesis. It would be worthwhile to search for a gratuitous inducer of resistance which would not affect the metabolic processes. The limitation of biosynthesis by exogenous chloramphenicol may be entirely due to a general toxic action, or it may be due to more specific but indirect effects on the antibiotic-synthesizing machinery.

An additional complication in understanding the effect on cultures of varied antibiotic concentrations is that antibiotic catabolism ceases at a certain point after endogenous synthesis has begun. This, of course, accounts for the results from gas-liquid chromatographic analysis of cultures grown in GSL medium which showed only one strong peak corresponding to chloramphenicol. No appreciable concentration of N-acetyl-p-nitrophenylserinol was present, as would be expected if endogenously produced antibiotic were also degraded. Furthermore, little degradation of the methylthio and methylsulfonyl analogues of chloramphenicol occurs when cells are actively engaged in antibiotic synthesis (Malik, 1970).

Examination of the plot of growth and chloramphenicol production in GSL medium (Fig. 6) shows that the most rapid rate of antibiotic synthesis occurs in the early growth phase. Toward the end of the growth phase, the rate decreases, This may be because the antibiotic is inhibiting its own biosynthesis, but it might also be caused by an antibiotic-promoted decrease in cell permeability restricting uptake of nutrients.

XX. Effect of p-Nitrophenylserinol on Chloramphenicol Biosynthesis

The new metabolite 2(S)-dichloroacetamido-3-(p-acetamido phenyl)propan-1-ol, isolated from cultures growing in complex medium to which p-nitrophenylserinol has been added, appears to be the result of inhibited chloramphenicol biosynthesis (Wat et al., 1971). Inspection of the structure suggests that, like chloramphenicol, it is a biosynthetic product of p-aminophenylalanine. Since it retains

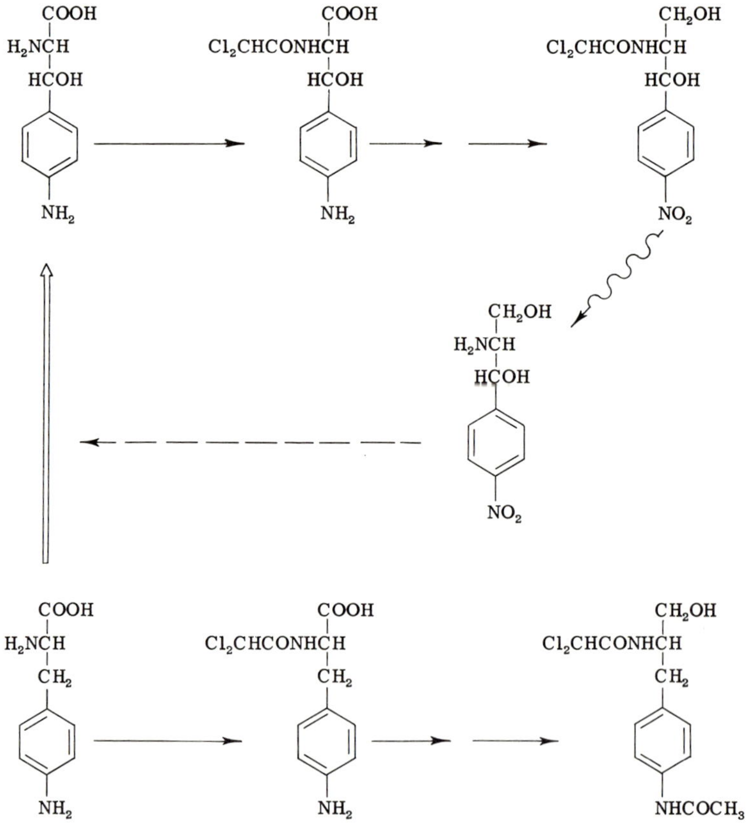

FIG. 8. Postulated scheme for inhibition of chloramphenicol biosynthesis by p-nitrophenylserinol or one of its metabolites.

the methylene group present in the amino acid, the site of inhibition is probably at the enzyme which hydroxylates p-aminophenylalanine to p-amino-*threo*-phenylserine (Fig. 8). Analogy with the subsequent reactions proposed for chloramphenicol biosynthesis suggests that the initial steps are N-dichloroacetylation and carboxyl reduction. Presumably the product is not a substrate for the enzyme which oxidizes the p-amino to a nitro group, but is readily acetylated. However, the timing of the acetylation is unknown. Biosynthetic studies with isotopically labeled precursors are required to provide definitive information.

Acknowledgment

I am grateful to Professors A. L. Demain and L. C. Vining and to Mrs. R. B. Kittredge for their help during preparation of this manuscript.

References

Adesnik, M., and Levinthal, C. (1969). *J. Mol. Biol.* **46**, 281.
Allen, E. H., and Schweet, R. (1962). *J. Biol. Chem.* **237**, 760.
Allison, J. L., Hartman, R. E., Hartman, R. S., Wolfe, A. D., Ciak, J., and Hahn, F. E. (1962). *J. Bacteriol.* **83**, 609.
Al'-Nuri, M. A., and Egorov, N. S. (1968). *Mikrobiologiya* **37**, 413.
Ambrose, C. T., and Coons, A. H. (1963). *J. Exp. Med.* **117**, 1075.
Amos, H. (1964). *Biochim. Biophys. Acta* **80**, 269.
Anraku, N., and Landman, O. E. (1968). *J. Bacteriol.* **95**, 1813.
Aronson, A. I., and Spiegelman, S. (1961). *Biochim. Biophys. Acta* **53**, 84.
Bergerson, F. J. (1953). *J. Gen. Microbiol.* **9**, 353.
Bergmann, E. D., and Sicher, S. (1952). *Nature (London)* **170**, 931.
Borshook, H., Fischer, E. H., and Keighley, G. (1957). *J. Biol. Chem.* **229**, 1059.
Brar, S. H., Giam, C. S., and Taber, W. A. (1968). *Mycologia* **60**, 806.
Bresler, S., Grajevskaja, R., Kirilov, S., and Saminski, E. (1968). *Biochim. Biophys. Acta* **155**, 465.
Brock, T. D. (1961). *Bacteriol. Rev.* **25**, 32.
Brock, T. D. (1964). *Exp. Chemother.* **3**, 119–169.
Bu'Lock, J. D. (1961). *Advan. Appl. Microbiol.* **3**, 293.
Bu'Lock, J. D. (1967). "Essays in Biosynthesis and Microbial Development." Wiley, New York.
Carnevali, F., Leoni, L., Morpurgo, G., and Conti, G. (1971). *Mutat. Res.* **12**, 357.
Cavalli, L. L., and Maccacaro, G. A. (1952). *Heredity* **6**, 311.
Celma, M. L., Monro, R. E., and Vazquez, D. (1971). *FEBS Lett.* **13**, 247.
Cerna, J., Jonak, J., and Rychlik, I. (1971). *Biochim. Biophys. Acta* **240**, 109.
Chassy, B. M., and Suhadolnik, R. J. (1969). *Biochim. Biophys. Acta* **182**, 307.
Ciak, J., and Hahn, F. E. (1958). *J. Bacteriol.* **75**, 125.
Cohen, G. N., and Monod, J. (1957). *Bacteriol. Rev.* **21**, 169.
Controulis, J., Rebstock, M. C., and Crooks, H. M., Jr. (1949). *J. Amer. Chem. Soc.* **71**, 2463.
Coutsogeorgopoulos, C. (1966). *Biochim. Biophys. Acta* **129**, 214.
Coutsogeorgopoulos, C. (1967). *Proc. Int. Congr. Chemother., 5th,* Vol. 4, p. 371.

Dagley, S., White, A. E., and Wild, D. G. (1962). *Nature (London)* **194**, 25.
Das, H. K., Goldstein, A., and Kanner, L. C. (1966). *J. Mol. Pharmacol.* **2**, 158.
Davis, B. D., and Feingold, D. S. (1962). *In* "The Bacteria" (I. C. Gunsalus and R. Y. Stanier, eds.), Vol. 4, p. 343. Academic Press, New York.
Davis, F. C., and Sells, B. H. (1969). *J. Mol. Biol.* **39**, 503.
DeLamater, E. D., Hunter, M. E., Szybalski, W., and Bryson, V. (1955). *J. Gen. Microbiol.* **12**, 203.
Demain, A. L. (1968). *Lloydia* **31**, 395.
DeMoss, J. A., and Novelli, G. D. (1955). *Biochim. Biophys. Acta* **18**, 592.
DeVries, H., and Kroon, A. M. (1970). *Biochim. Biophys. Acta* **204**, 531.
Dhar, M. M., and Kahn, A. W. (1971). *Nature (London)* **233**, 182.
Dienes, L., Weinberger, H. J., and Madoff, S. (1950). *J. Bacteriol.* **59**, 755.
Djordjevic, B., and Szybalski, W. (1960). *J. Exp. Med.* **112**, 509.
Dulaney, E. L. (1951). U.S. Pat. No. 2,545,572.
Dunitz, J. D. (1952). *J. Amer. Chem. Soc.* **74**, 995.
Dunnick, J. K., and O'Leary, W. M. (1970). *J. Bacteriol.* **101**, 892.
Egami, F., Ebata, M., and Sato, R. (1950). *Nature (London)* **167**, 118.
Ehrlich, J. E., Bartz, Q. R., Smith, R. M., Joslyn, D. A., and Burkholder, P. R. (1947). *Science* **106**, 417.
Ezekiel, D. H., and Valulis, B. (1965). *Biochim. Biophys. Acta* **108**, 135.
Farese, R. V. (1964). *Biochim. Biophys. Acta* **87**, 699.
Fassin, W., Hengel, R., and Klein, U. P. (1955). *Z. Hyg. Infektionskrankh.* **141**, 363.
Firkin, F. C., and Linnane, A. W. (1968). *Biochem. Biophys. Res. Commun.* **32**, 398.
Firkin, F. C., and Linnane, A. W. (1969). *Exp. Cell Res.* **55**, 68.
Freeman, K. B. (1970). *Can. J. Biochem.* **48**, 469.
Freeman, K. B., and Halder, D. (1967). *Biochem. Biophys. Res. Commun.* **28**, 8.
Fusillo, M. H., Metzger, J. F., and Kuhns, D. M. (1952). *Proc. Soc. Exp. Biol. Med.* **79**, 376.
Gale, E. F. (1963). *Pharmacol. Rev.* **15**, 481.
Gale, E. F., and Folkes, J. P. (1953). *Biochem. J.* **53**, 493.
Gallicchio, V., and Gottlieb, D. (1958). *Mycologia* **50**, 490.
Garcia-Patrone, M., Perazzolo, C. A., Baralle, F., Gonzalez, N. S., and Algranati, I. D. (1971). *Biochim. Biophys. Acta* **246**, 291.
Gibson, F., and McDougall, B. (1961). *Aust. J. Exp. Biol. Med. Sci.* **39**, 171.
Gibson, F., Jones, M. J., and Teltscher, H. (1955). *Nature (London)* **176**, 164.
Gibson, F., McDougall, B., Jones, M. J., and Teltscher, H. (1956). *J. Gen. Microbiol.* **15**, 446.
Godchaux, W., and Herbert, E. (1966). *J. Mol. Biol.* 21, 537.
Goldberg, I. H. (1965). *Amer. J. Med.* **39**, 722.
Gottlieb, D., and Shaw, P. D., eds. (1967). "Antibiotics," Vol. II. Springer-Verlag, Berlin and New York.
Gottlieb, D., Bhattacharya, P. K., Anderson, H. W., and Carter, H. E. (1948). *J. Bacteriol.* **55**, 409.
Gottlieb, D., Robbins, P. W., and Carter, H. E. (1956). *J. Bacteriol.* **72**, 153.
Gourevitch, A., Pursiano, T. A., and Lein, J. (1961). *Arch. Biochem. Biophys.* **93**, 283.
Gross, W., and King, K. (1969). *FEBS Lett.* **4**, 319.
Gurgo, C., Apirion, D., and Schlessinger, D. (1969). *J. Mol. Biol.* **45**, 205.
Hahn, F. E. (1964). *Proc. Int. Congr. Chemother.*, 3rd, Vol. 42, p. 215.
Hahn, F. E. (1967). *In* "Antibiotics," Vol. I. (D. Gottlieb and P. D. Shaw, eds.), pp. 308–330, Springer-Verlag, Berlin and New York.

Hahn, F. E., Hayes, J. E., Wisseman, C. L., Jr., Hopps, H. E., and Smadel, J. E. (1956). *Antibiot. Chemother. (Washington, D.C.)* **6**, 531.
Hahn, F. E., Schaechter, M., Ceglowski, W. S., Hopps, H. E., and Ciak, J. (1957). *Biochim. Biophys. Acta* **26**, 269.
Hancock, R., and Park, J. T. (1958). *Nature (London)* **181**, 1050.
Hanson, J. B., and Hodges, T. K. (1963). *Nature (London)* **200**, 1009.
Hopps, H. E., Wisseman, C. L., Jr., and Hahn, F. F. (1954). *Antibiot. Chemother. (Washington, D.C.)* **4**, 857.
Hopps, H. E., Wisseman, C. L., Jr., Hahn, F. E., Smadel, J. E., and Ho, R. (1956). *J. Bacteriol.* **72**, 561.
Hosokawa, K., and Nomura, M. (1965). *J. Mol. Biol.* **12**, 225.
Hurwitz, C., and Braun, C. (1967). *J. Bacteriol.* **93**, 1671.
Hurwitz, C., and Braun, C. (1968). *Biochim. Biophys. Acta* **157**, 392.
Ingall, D., and Sherman, J. D. (1970). In "Antimicrobial Therapy" (B. Kagan, ed.), p. 61. Saunders, Philadelphia, Pennsylvania.
Jardetsky, O. (1963). *J. Biol. Chem.* **238**, 2498.
Jardetsky, O., and Julian, G. R. (1964). *Nature (London)* **201**, 396.
Jordan, B. R., Forget, B. G., and Monier, R. (1971). *J. Mol. Biol.* **55**, 407.
Julian, G. R. (1965). *J. Mol. Biol.* **12**, 9.
Katz, E., Wise, M., and Weissbach, H. (1965). *J. Biol. Chem.* **240**, 3071.
Katz, J., Lieberman, I., and Barker, H. A. (1953). *J. Biol. Chem.* **200**, 417.
Kojima, M., Yamada, Y., and Umezawa, H. (1969). *Agr. Biol. Chem.* **33**, 1181.
Kono, M., Hara, O., Nagawa, K., and Mitsuhashi, S. (1971). *Jap. J. Microbiol.* **15**, 219.
Kroon, A. M., and Jansen, R. J. (1968). *Biochim. Biophys. Acta* **155**, 629.
Kuccan, Z., and Lipmann, F. (1964). *J. Biol. Chem.* **239**, 516.
Kurland, C. G., and Maaløe, O. (1962). *J. Mol. Biol.* **4**, 193.
Lacks, S., and Gros, F. (1959). *J. Mol. Biol.* **1**, 301.
Lallier, R. (1962). *J. Embryol. Exp. Morphol.* **10**, 563.
Lark, C., and Lark, K. G. (1964). *J. Mol. Biol.* **10**, 120.
Lark, K. G. (1966). *Bacteriol. Rev.* **30**, 3.
Lark, K. G., and Lark, C. (1966). *J. Mol. Biol.* **20**, 9.
Lebek, G. (1963). *Zentralbl. Bakteriol., Parasitenk., Infektimskr. Hyg., Abt. 1: Orig.* **188**, 944.
Legator, M., and Gottlieb, D. (1953). *Antibiot. Chemother. (Washington, D.C.)* **3**, 809.
Levine, A. J., and Sinsheimer, R. L. (1968). *J. Mol. Biol.* **32**, 567.
Levine, A. J., and Sinsheimer, R. L. (1969). *J. Mol. Biol.* **39**, 655.
Lewandowski, L. J., and Brownstein, B. L. (1966). *Biochem. Biophys. Res. Commun.* **25**, 554.
Light, R. J. (1967). *Arch. Biochem. Biophys.* **122**, 494.
Lingens, F., and Oltmanns, O. (1966). *Biochim. Biophys. Acta* **130**, 336.
Lingens, F., Eberhardt, H., and Oltmanns, O. (1966). *Biochim. Biophys. Acta* **130**, 345.
Lloyd, D., Evans, D. A., and Venables, S. A. (1970). *J. Gen. Microbiol.* **61**, 33.
Lowe, D. A., and Westlake, D. W. S. (1971). *Can. J. Biochem.* **49**, 448.
Maaløe, O., and Kjeldgaard, N. O. (1966). "Control of Macromolecular Synthesis." Benjamin, New York.
Maas, W. K., Novelli, G. D., and Lipmann, F. (1953). *Proc. Nat. Acad. Sci. U.S.* **39**, 1004.
McDaniel, L. E., and Hodges, A. B. (1951). U.S. Pat. No. 2,545,554.
McDougall, B., and Gibson, F. (1958). *Aust. J. Exp. Biol.* **36**, 245.
Mackler, B., and Haynes, B. (1970). *Biochim. Biophys. Acta* **197**, 317.
Magasanik, B. (1961). *Cold Spring Harbor Symp. Quant. Biol.* **26**, 249.

Malek, Y. A., Monib, M., and Hazem, A. (1961). *Nature (London)* **189**, 775.
Malik, V. S. (1970). Ph.D. Thesis, Dalhousie University, Halifax (N.S.), Canada.
Malik, V. S. (1972). Unpublished data.
Malik, V. S., and Vining, L. C. (1970). *Can. J. Microbiol.* **16**, 173.
Malik, V. S., and Vining, L. C. (1971). *Can. J. Microbiol.* **17**, 1287.
Mandelstam, J., and Rogers, H. J. (1958). *Nature (London)* **181**, 956.
Mandelstam, J., and Rogers, H. J. (1959). *Biochem. J.* **72**, 654.
Marchant, R., and Smith, D. G. (1968). *J. Gen. Microbiol.* **50**, 391.
Merkel, J. R., and Steers, E. (1953). *J. Bacteriol.* **66**, 389.
Midgley, J. E. M., and Gray, W. J. H. (1971). *Biochem. J.* **122**, 149.
Miller, A. L., and Walker, J. B. (1969). *J. Bacteriol.* **99**, 401.
Mise, K., and Suzuki, Y. (1968). *J. Bacteriol.* **95**, 2124.
Miyamura, S. (1964). *J. Pharm. Sci.* **53**, 604.
Molho-Lacroix, L., and Molho, D. (1952). *Bull. Soc. Chim. Biol.* **34**, 93.
Morgan, C., Rosenkranz, H. S., Carr, H. S., and Rose, H. M. (1967). *J. Bacteriol.* **93**, 1987.
Morse, D. E. (1970). *Cold Spring Harbor Symp. Quant. Biol.* **35**, 495.
Morse, D. E. (1971). *J. Mol. Biol.* **55**, 113.
Morse, D. E., and Guertin, M. (1971). *Nature (London)* **232**, 165.
Mulherkar, L., Joshi, P. N., and Diwan, B. N. (1967). *Experientia* **23**, 901.
Naha, P. M. (1969). *Biochem. Biophys. Res. Commun.* **35**, 920.
Nakamoto, T., Conway, T. W., Allende, J. E., Spyrides, G. J., and Lipmann, F. (1963). *Cold Spring Harbor Symp. Quant. Biol.* **28**, 227.
Neidhardt, F. C. (1964). *Progr. Nucl. Acid Res. Mol. Biol.* **3**, 145–173.
Neidhardt, F. C., and Gros, F. (1957). *Biochim. Biophys. Acta* **25**, 513.
Newton, B. A. (1965). *Annu. Rev. Microbiol.* **19**, 209.
Niri, L., Lengyl, Z. L., and Erdelyi, A. (1963). *J. Antibiot., Ser. A* **16**, 80.
Novick, R. P. (1969). *Bacteriol. Rev.* **33**, 210.
Okami, Y., Hashimoto, T., and Suzuki, M. (1967). *J. Antibiot.* **13**, 223.
Okamoto, S., and Mizuno, D. (1964). *J. Gen. Microbiol.* **35**, 125.
Okamoto, S., Suzuki, Y., Mise, K., and Nakaya, R. (1967). *J. Bacteriol.* **94**, 1616.
Oleinick, N. L., Wilhelm, J. M., and Corcoran, J. W. (1968). *Biochim. Biophys. Acta* **155**, 290.
Otaka, E., Itoh, T., and Osawa, S. (1967). *Science* **157**, 1452.
Oyaas, J. E., Ehrlich, J. E., and Smith, R. M. (1950). *Ind. Eng. Chem.* **42**, 1775.
Pardee, A. B., and Prestidge, L. S. (1959). *Biochim. Biophys. Acta* **36**, 545.
Paulus, H., and Gray, E. (1964). *J. Biol. Chem.* **239**, 865.
Pestka, S. (1969). *Biochem. Biphys. Res. Commun.* **36**, 589.
Pollock, M. R. (1960). *Brit. Med. Bull.* **16**, 16.
Pomerat, C. M., and Leake, C. D. (1954). *Ann. N.Y. Acad. Sci.* **58**, 1110.
Poole, R. K., Nicholl, W. G., Turner, G., Roach, G. I., and Lloyd, D. (1971). *J. Gen. Microbiol.* **67**, 161.
Potter, V. R., and Reif, R. E. (1952). *J. Biol. Chem.* **194**, 287.
Pulvertaft, R. J. V. (1952). *J. Pathol. Bacteriol.* **64**, 75.
Ramsey, H. H. (1966). *Biochem. Biophys. Res. Commun.* **23**, 353.
Rannenberg, H., and Arnold, C. G. (1968). *Z. Pflanzenphysiol.* **59**, 7.
Ray, D. S. (1970). *J. Mol. Biol.* **53**, 239.
Rebstock, M. C., Crooks, H. M., Jr., Controulis, J., and Bartz, Q. R. (1949). *J. Amer. Chem. Soc.* **71**, 2458.
Rendi, R. (1959). *Exp. Cell Res.* **18**, 187.

Schaeffer, P. (1969). *Bacteriol. Rev.* **33**, 48.
Schleif, R. F. (1968). *J. Mol. Biol.* **37**, 119.
Schrader, L. E., Beevers, L., and Hageman, R. H. (1967). *Biochem. Biophys. Res. Commun.* **26**, 14.
Schweet, R., and Heinz, R. (1966). *Annu. Rev. Biochem.* **35**, 723.
Shaw, W. V. (1967). *J. Biol. Chem.* **242**, 687.
Shaw, W. V. (1971). *Ann. N.Y. Acad. Sci.* **182**, 234.
Shaw, W. V., and Brodsky, R. F. (1967). *Antimicrob. Ag. Chemother.* p. 257.
Shaw, W. V., and Brodsky, R. F. (1968). *J. Bacteriol.* **95**, 28.
Shaw, W. V., and Unowsky, J. (1968). *J. Bacteriol.* **95**, 1976.
Shaw, W. V., and Winshell, E. (1968). *Antimicrob. Ag. Chemother.* p. 7.
Shaw, W. V., Bentley, D. W., and Sands, L. (1970). *J. Bacteriol.* **104**, 1095.
Shemyakin, M. M. (1961). "Khimia Antibiotikov," Vol. I. Acad. Sci. USSR, Moscow.
Shemyakin, M. M., Kolosov, M. N., Levitov, M. M., Germanova, K. I., Karapetian, M. G., Shvetsov, Iu. B., and Bamdas, E. M. (1956). *Zh. Obshch. Khim.* **26**, 773.
Shockman, G. D. (1965). *Bacteriol. Rev.* **29**, 345.
Siddiqueullah, M., McGrath, R., Vining, L. C., Sala, F., and Westlake, D. W. S. (1968). *Can. J. Biochem.* **46**, 9.
Smith, C. G. (1958). *J. Bacteriol.* **75**, 577.
Smith, D. G., and Marchant, R. (1968). *Arch. Mikrobiol.* **60**, 262.
Smith, G. N., and Worrell, C. S. (1949). *Arch. Biochem.* **24**, 216.
Smith, G. N., and Worrell, C. S. (1950). *Arch. Biochem.* **28**, 232.
Smith, G. N., and Worrell, C. S. (1952). *Arch. Biochem. Biophys.* **40**, 314.
Smith, G. N., and Worrell, C. S. (1953). *J. Bacteriol.* **65**, 313.
Smith, G. N., Worrell, C. S., and Swanson, A. L. (1949). *J. Bacteriol.* **58**, 803.
Soda, K., Tanaka, H., and Yamamoto, T. (1969). *Arch. Biochem. Biophys.* **130**, 610.
Sompolinsky, D., and Samra, Z. (1968). *J. Gen. Microbiol.* **50**, 55.
Spirin, A. S., and Gavrilova, L. P. (1969). "The Ribosome." Springer-Verlag, Berlin and New York.
Stoner, C. D., Hodges, T. K., and Hanson, J. B. (1964). *Nature (London)* **203**, 258.
Stow, M., Starkey, B. J., Hancock, I. C., and Baddiley, J. (1971). *Nature (London)* **229**, 56.
Stratton, C. D., and Rebstock, M. C. (1963). *Arch. Biochem. Biophys.* **103**, 159.
Suarez, G., and Nathans, D. (1965). *Biochem. Biophys. Res. Commun.* **18**, 743.
Sypherd, P. S., and DeMoss, J. A. (1963). *Biochim. Biophys. Acta* **76**, 589.
Sypherd, P. S., Strauss, N., and Treffers, H. P. (1962). *Biochem. Biophys. Res. Commun.* **7**, 477.
Telesnina, G. N., Novikova, M. A., Zhdanov, G. D., Kolosov, M. N., and Shemyakin, M. M. (1967). *Experientia* **23**, 427.
Teraoka, H. (1970). *Biochim. Biophys. Acta* **213**, 535.
Truhaut, R., Lambin, S., and Boyer, M. (1951). *Bull. Soc. Chim. Biol.* **33**, 387.
Turner, G., and Lloyd, D. (1971). *J. Gen. Microbiol.* **67**, 175.
Umezawa, H. T., Kametani, R., Osati, T., Takeda, K., Kanari, H., and Kawahara, O. (1948). *Jap. Med. J.* **1**, 790.
Unowsky, J., and Rachmeler, M. (1966). *J. Bacteriol.* **92**, 358.
Vambutas, V. K., and Salton, M. R. (1970). *Biochim. Biophys. Acta* **203**, 94.
Varmus, H. E., Perlman, R. L., and Pastan, I. (1971). *Nature (London)* **230**, 41.
Vas, M. R., Bain, B., and Lowenstein, L. (1962). *Nature (London)* **204**, 1100.
Vazquez, D. (1964). *Nature (London)* **203**, 257.
Vazquez, D. (1966). *Symp. Soc. Gen. Microbiol.* **16**, 1969.

Vining, L. C., Malik, V. S., and Westlake, D. W. S. (1968). *Lloydia* **31**, 355.
Vinter, V. (1963). *Experientia* **19**, 307.
Vogel, Z., Vogel, T., Zamir, A., and Elson, D. (1971). *J. Mol. Biol.* **60**, 339.
Waksman, S. A., Reilly, H. C., and Harris, D. A. (1947). *Proc. Soc. Exp. Biol. Med.* **66**, 617.
Wat, C., Malik, V. S., and Vining, L. C. (1971). *Can. J. Chem.* **49**, 3653.
Watanabe, T. (1963). *Bacteriol. Rev.* **27**, 87.
Watanabe, T. (1971). *Curr. Top. Microbiol. Immunol.* **56**, 43.
Weber, M. J., and DeMoss, J. A. (1969). *J. Bacteriol.* **97**, 1099.
Weisberger, A. S., and Wolfe, S. (1964). *Fed. Proc., Fed. Amer. Soc. Exp. Biol.* **23**, 976.
Weisberger, A. S., Armentrout, S., and Wolfe, S. (1963). *Proc. Nat. Acad. Sci. U.S.* **50**, 86.
Weisblum, B., and Davies, J. (1968). *Bacteriol. Rev.* **32**, 493.
Weislogel, P. O., and Butow, R. A. (1970). *Proc. Nat. Acad. Sci. U.S.* **67**, 52.
Williamson, D. H., Maroudas, N. G., and Wilkie, D. (1971). *Mol. Gen. Genet.* **111**, 209.
Woodruff, H. B. (1966). *In* "Biochemical Studies of Antimicrobial Drugs," pp. 22–46. Cambridge Univ. Press, London and New York.
Woolley, D. W. (1950). *J. Biol. Chem.* **185**, 293.
Yeigan, C. D., and VanderSlice, R. W. (1971). *J. Bacteriol.* **108**, 849.
Yoshida, T., Weissbach, H., and Katz, E. (1966). *Arch. Biochem. Biophys.* **114**, 252.
Young, R. M., and Nakada, D. (1971). *J. Mol. Biol.* **57**, 457.

Microbial Utilization of Methanol [1]

CHARLES L. COONEY AND DAVID W. LEVINE

*Department of Nutrition and Food Science,
Massachusetts Institute of Technology, Cambridge, Massachusetts*

I.	Introduction	337
II.	Aerobic Utilization of Methanol	338
	A. Microbiology	338
	B. Biochemistry	339
	C. Synthesis of Cellular Material	346
III.	Anaerobic Utilization of Methanol	350
	A. Microbiology	350
	B. Biochemistry	352
	C. Energetics	354
IV.	Single-Cell Protein from Methanol	355
	A. Methanol vs. Methane and Other Hydrocarbons	355
	B. Methanol Economics	356
	C. Status of SCP from Methanol	357
	D. Composition of Methanol-Grown Cells	358
	E. Engineering Aspects of Methanol-Derived SCP	360
V.	Fermentation Products from Methanol	362
VI.	Summary and Conclusions	362
	References	363

I. Introduction

Interest in the microbial utilization of C_1 compounds began when Söhngen first reported, in 1906, the isolation of bacteria capable of growing on methane. Only in the last 20 years, however, with the works of Hutton and ZoBell (1949), Dworkin and Foster (1956), and Leadbetter and Foster (1958) has the interest been more than peripheral. In the case of aerobic methanol utilization, the initial concern had been simply a matter of positive or negative utilization; it represented an interesting characteristic of organisms being studied for different reasons. Initial studies on the anaerobic utilization of methanol were primarily concerned with the mechanism of methane formation from other C_1 compounds. These concerns are understandable considering the similarity between methane and methanol, and the likelihood that a pathway for methane oxidation or methane formation would proceed via methanol. More recently, interest in methanol as a fermentation substrate has increased. As a consequence, the number of organisms isolated for growth on methanol and the accumulation of quantitative data have increased so that now there are

---

[1] Contribution number 1093 from the Department of Nutrition and Food Science.

essentially two approaches to methanol utilization. One is the concern with methanol as a convenient substrate for use in studies on mechanisms of oxidation of C_1 compounds. The other is consideration of methanol as a substrate for industrial fermentations. Each of these areas of interest is documented by its own literature. At this time, it seems appropriate to review the available literature and assess the state-of-the-art in the microbial utilization of methanol, attempting to point out gaps in our knowledge and define directions for the future. Herein lies the motivation for this review. Consideration will be given to the microbiology and biochemistry of both aerobic and anaerobic assimilation of methanol, and the status of single-cell protein from methanol will be examined.

II. Aerobic Utilization of Methanol

A. Microbiology

Söhngen isolated a methane-oxidizing bacterium in 1906; Bassalik observed growth of *Bacillus extorquens* on methanol in 1914; yet until after the work of Dworkin and Foster (1956) little effort was devoted toward the elucidation of the mechanism of aerobic C_1 metabolism. Early work was directed toward isolation of methane utilizers and taxonomic characterization of these cultures; growth on methanol was incidental to the objectives of the work. In recent years, however, the need for inexpensive carbon-energy sources for the production of single-cell protein has stimulated interest in methane and methanol metabolism. Coty (1969), Whittenbury et al. (1970), and Naguib and Overbeck (1970) have reviewed the literature on microbial methane utilization, and Ribbons et al. (1970) have comprehensively reviewed the metabolism of C_1 compounds. It is the objective of this portion of the review to complement these works by examining the aerobic metabolism of methanol.

A compilation of recent work on the aerobic microbial utilization of methanol is presented in Table I. Two aspects of this table are striking, the preponderance of gram-negative bacterial isolates and the lack of quantitative data. The high incidence of gram-negative bacteria is partially a result of the techniques used for enrichment and isolation and is not totally a reflection of the genetic profile of C_1 utilizers. Most enrichments were carried out at pH values of 6 to 7 and temperatures of 25 to 30°C in simple mineral-salts media. The interesting exceptions to the gram-negative isolates are the gram-positive isolates of Akiba et al. (1970) and the various yeast isolates. Ogata *et al.* (1969, 1970a,b) isolated several species of yeast in a vitamin-containing medium at pH 6, however, he has only reported detailed data on *Kloeckera* sp. No. 2201. Cooney *et al.* (1971) used a continuous enrichment culture with a vitamin-free, mineral-salts medium at a pH of 3.5 to 4.0. This isolation was carried out at 37°C and through the use of a continuous flow device, was designed to select for only those yeasts capable of dividing in less than 10 hr in the enrichment medium. The low pH provided preferential selection of yeast over bacteria. Asthana *et al.* (1971) added tetracycline to the isolation medium and lowered the pH to 4.5 to effect an enrichment of yeast over bacteria. The procedures of Cooney *et al.* (1971) and Asthana *et al.* (1971) should also be useful for isolating fungal species of methanol utilizers. These examples demonstrate that manipulation of parameters in the enrichment medium, e.g., pH, temperature, vitamin or selective inhibitor addition, provides the investigator with a tool for obtaining a wide range of methanol utilizers, thus not restricting the biochemists, microbiologists, or biochemical engineers interested in methanol metabolism to *Pseudomonas*-like organisms.

As the interest in aerobic methanol utilization increases, the availability of quantitative data describing microbial growth on methanol increases. Cellular yields are found to vary from 0.19 to 0.45 gm dry cell weight per gram of methanol with most cultures having yields of about 0.4. To place these data into perspective, typical cellular yields for glucose, normal alkanes, and methane are 0.5, 1.0, and 0.6 gm of cell per gram of substrate consumed. In single-cell protein production, high yields are desired while in waste treatment processes, low yields are preferred.

Specific growth rates or mass doubling times vary tremendously among the various isolates depending on the growth environment and the methanol concentration. Higher concentrations, e.g., greater than 1 to 2% methanol, frequently inhibit culture growth (Kaneda and Roxburgh, 1959a; Peel and Quayle, 1961; Häggström, 1969; Kirikova, 1970; Ogata *et al.*, 1969; Asthana *et al.*, 1971). Methanol utilizers are capable of doubling their mass in as little as 2 hr (Mateles and Chalfan, 1970; Tannahill and Finn, 1970). However, from a practical point of view, fast growth rates are not always desirable since it may become difficult to meet the oxygen demand of the culture; a point to be discussed in more detail later.

B. BIOCHEMISTRY

The pathway of aerobic methanol oxidation has been shown to

TABLE I
MICROBIAL CULTURES CAPABLE OF AEROBIC GROWTH ON METHANOL

Organism	Doubling time (hours)	pH	Temperature (°C)	Growth on methane	Yield (gm cells/ gm MeOH)	Comments	Reference
BACTERIA							
Pseudomonas methanica (*Methanomonas methanica*)	—	6.0–6.6 opt.	30	+	—	Motile. Growth factor required.[a,b,c] Pink pigment.	Dworkin and Foster (1956)
	3.5	—	25	—	—	Nonmotile. Pink pigment. No capsule. Lipid inclusion.[c]	Harrington and Kallio (1960)
		5.8–7.4 range 6.6–6.8 opt.	30	+	0.4	Motile. Ocher-yellow and pink pigments. Forms capsule.[a,c]	Whittenbury et al. (1970)
Pseudomonas PRL-W4	7.2	5.0–8.0 range 7.0 opt.	30 opt.	—	—	Motile. Red pigment. Inhibited by > 1% methanol. Requires biotin.	Kaneda and Roxburgh (1959a)
Pseudomonas AM-1	4–10	6–8.5 range 7.0 opt.	25–37 range 30 opt.	—	0.2	Motile. Pink pigment. Isolated on methylamine. 6% of dry cell weight is poly-β-hydroxybutyrate. 1% methanol inhibitory.	Peel and Quayle (1961)
Strain L-8	—	—	25	+		Gram-negative rods. L-47 pink pigmented. L-49 yellow pigmented.	Elizarova (1963)
Strain L-47	—	—	—	+			
Strain L-49	—	—	—	+			

Organism					Notes	Reference
Pseudomonas M27	—	7.0	30	—	Motile. Pink pigment. Lipid inclusions.	Anthony and Zatman (1964a)
Methanomonas methanooxidans	—	6.1 opt.	30 opt.	+	Gram-negative rods, motile.[a,b,c]	Brown et al. (1964); Stocks and McClesky (1964b)
Hypomicrobium vulgare	14	7.2	30	—	Filamentous bacteria. Accumulates poly-β-hydroxybutyrate. Light-sensitive cultures.	Hirsch and Conti (1964)
Protaminobacter ruber	—	7.0	30	—	Isolated by den Dooren de Jong (1927)	Johnson and Quayle (1964)
Bacillus extorquens (*Pseudomonas extorquens*)	—	7.0	30	—	Isolated by Bassalik (1914)	Johnson and Quayle (1964)
Vibrio extorquens	3.8	—	—	—	Gram-negative rods, motile. Isolated as contaminant of methane-grown culture. Pink pigment. Accumulates polyhydroxybutyrate.[d]	Stocks and McClesky (1964a); Hamer and Norris (1971)

(Continued)

TABLE I (*Continued*)

Organism	Doubling time (hours)	pH	Temperature (°C)	Growth on methane	Yield (gm cells/ gm MeOH)	Comments	Reference
Methylococcus capsulatus	3.5–5	7.0	55 range 33 opt.	+	—	Gram-negative coccus. Forms capsule.[a,b,c]	Foster and Davis (1966)
Strain HR (Mixed culture)	2.1	6.50 opt.	30 opt.	+	0.33	Gram-negative rods. Yellow and greenish pigments on yeast extract. Lipid inclusions.	Vary and Johnson (1967)
Strain TM20 (Mixed culture)	3.2	6.0–6.30 opt.	31 opt.	—	0.32	Gram-negative rods. Optimum methanol concentration 0.1–0.2%.	Häggström (1969)
Bacillus cereus	39.5	—	15–45 range 25–37 opt.	?	0.38	Motile.	Akiba *et al.* (1970)
Arthrobacter rufescens B-62-1	—	—	10–45 range 25–37 opt.	—	—	Nonmotile. Red pigment.	Akiba *et al.* (1970)
Pseudomonas, Strain 2	—	7.0 opt.	30 opt.	—	4.0×10^{11} cells/gm methanol at 0.1% methanol	Motile. Pink pigment. Optimum methanol concentration is 0.1–0.5%.	Kirikova (1970)

Organism		pH	Temp			Notes	Reference
Pseudomonas sp.	2	—	37	?	0.31	Optimum methanol concentration <2%. Polysaccharide formed.	Mateles and Chalfan (1970)
Pseudomonas or *Xanthomonas*	2	—	—	?	0.4	Yellow pigment. Polysaccharide production.	Tannahill and Finn (1970)
Methylomonas albus	3	6.8 opt.	30	+	0.4	Gram-negative rods. Motile. *M. rubrum* forms capsule and red pigment.[a]	Whittenbury et al. (1970)
Methylomonas agile	3.5	6.8 opt.	30	+	0.4	—	Whittenbury et al. (1970)
Methylomonas rubrum	4	6.8 opt.	30	+	0.4	—	Whittenbury et al. (1970)
Strain TM10 (Mixed culture)	4.36	4.9–7.0 range	31	+	—	Gram-negative rods. Forms microcolonies. Lipid inclusion. Pink pigmented. Polyhydroxybutyrate accumulates.[a,c]	Bewersdorff and Dostalek (1971)
Hyphomicrobium sp. Strains WC, B522	—	7.2	25	?	—	Both aerobic and anaerobic growth. Pale pink to white growth.	Sperl and Hoare (1971)

(*Continued*)

TABLE I (*Continued*)

Organism	Doubling time (hours)	pH	Temperature (°C)	Growth on methane	Yield (gm cells/ gm MeOH)	Comments	Reference
YEASTS							
Kloeckera sp. No. 2201	9	2–9 range 5–6 opt.	5–45 range 30 opt.	?	0.29	Asporogenous, bipolar budding. Requires thiamine. Optimum methanol concentration 1%.	Ogata *et al.* (1969, 1970a,b)
Candida silvicola DL-1	4	5.5–6.0 opt.	37 opt.	?	0.37	Asporogenous, multilateral budding. Grows with biotin and thiamine.	Cooney *et al.* (1971)
Torulopsis glabrata	8	3.0–8.0 range	30 opt.	?	0.45	Greater than 3% methanol inhibitory	Asthana *et al.* (1971)

[a] Organism isolated with methane.
[b] Data presented were obtained with methane as carbon source.
[c] Obligate C_1 compound utilizers.
[d] Proposed to be identical to: *P. methanica* (Harrington and Kallio, 1960), P. AM-1, PRL-W4, *Protaminobacter ruber*, *Bacillus* (*Pseudomonas*) *extorquens*.

proceed in the manner first proposed by Dworkin and Foster (1956) for methane oxidation by *Pseudomonas methanica*:

$$CH_4 \to CH_3OH \to HCHO \to HCOOH \to CO_2 \quad \text{(I)}$$

This pathway was proposed on the basis of experiments with methane- and methanol-grown cells, and formaldehyde and formate oxidation by resting cell suspensions. Further support for this pathway is provided by the work of Brown and Strawinski (1957, 1958), and Brown *et al.* (1964), with the methane utilizer *Methanomonas methanooxidans*. These workers, using sodium sulfite as a trapping agent, were able to show the accumulation as formaldehyde of 60 to 70% of the methane or methanol utilized. The use of iodoacetate, to inhibit alcohol dehydrogenase activity, resulted in the accumulation of 70% of utilized methane as methanol. Finally, they reported a low level of formate accumulation in the growth media of cells oxidizing either methane, methanol, or formaldehyde.

To date, three different alcohol dehydrogenases catalyzing the oxidation of methanol have been described. Kaneda and Roxburgh (1959b) working with the organism PRL-W4, reported a nicotinamide-adenine dinucleotide (NAD)-linked methanol dehydrogenase activity in cell-free extracts. Harrington and Kallio (1960) working with crude enzyme extracts of *Pseudomonas methanica* showed methanol oxidation to be dependent on the availability of hydrogen peroxide. They postulated that *in vivo*, methanol oxidation occurred via a peroxidase, with methanol as the hydrogen donor. The most definitive work on methanol dehydrogenase activity has been carried out, however, by Anthony and Zatman (1964a,b, 1965, 1967a,b) on their own isolate *Pseudomonas* M27. They have isolated and purified an enzyme which catalyzes the oxidation of methanol and is NAD-independent. Early *in vitro* work employed phenazine methosulfate as the hydrogen acceptor. Free ammonia served as an enzyme activator. Johnson and Quayle (1964) have reported the presence of a similar enzyme in extracts of *Pseudomonas* AM-1, *P. methanica, P. extroquens,* and *Protaminobacter ruber,* and Anthony (1970) has reported the dehydrogenase of *Pseudomonas* AM-1 to be identical to that of *Pseudomonas* M27. Anthony and Zatman (1967a,b) believe this enzyme to represent a totally new alcohol dehydrogenase. Work to date indicates that the prosthetic group of the enzyme is probably a pteridine derivative. The enzyme is not specific for methanol, but it is specific for primary alcohols in general. Enzymatic activity has been demonstrated *in vitro* with primary alcohols as the substrate, except for the cases of amino alcohols, their N-substituted derivatives, and insoluble solid alcohols (Anthony and Zatman, 1965).

More recently, Anthony (1970) has reported the presence of cytochromes a, b, and c in *Pseudomonas* AM-1. From whole cell studies he has shown that cytochrome c is reduced in the presence of methanol, formaldehyde, and formate, provided the cells have been grown on C_1 compounds, thus causing the formation of the necessary dehydrogenases. Mutants incapable of forming the methanol dehydrogenase do not reduce cytochrome c in the presence of methanol. Using cytochrome c negative mutants of *Pseudomonas* AM-1, Anthony (1970) and Anthony and Dunstan (1971) have further shown that these cells cannot oxidize methanol to formaldehyde, thereby adding additional evidence for the involvement of cytochrome c in methanol oxidation. These mutants, however, are still able to oxidize formaldehyde and formate.

To distinguish this methanol dehydrogenase system from the peroxidative system proposed by Harrington and Kallio (1960), Anthony and Zatman (1964a) have grown cells of *Pseudomonas* M27 on methanol in the presence of a catalase inhibitor, as have Johnson and Quayle (1964) for *Pseudomonas* AM-1.

Two different enzymes catalyzing the oxidation of formaldehyde have been described for cells grown on methanol. Harrington and Kallio (1960) working with *Pseudomonas methanica*, described a specific formaldehyde dehydrogenase which functions *in vitro* in the presence of NAD and glutathione. Johnson and Quayle (1964) described the existence of a similar enzyme in extracts of *P. methanica*, *P. extorquens*, and *Protoaminobacter ruber*. On the other hand, in *Pseudomonas* AM-1, specific formaldehyde oxidation activity was lacking. However, *Pseudomonas* AM-1 does have nonspecific aldehyde dehydrogenase activity, as do the other three organisms.

Johnson and Quale (1964) also reported NAD-linked formate dehydrogenase activity for *Protaminobacter ruber*, *Pseudomonas* AM-1, *P. methanica*, and *P. extorquens*. This enzyme was specific for formate.

C. Synthesis of Cellular Material

Two different pathways for carbon incorporation into cellular material have been postulated by Quayle and his co-workers. These paths have been both suggested and supported by the results of ^{14}C-labeling studies carried out with various organisms grown aerobically on labeled C_1 compounds. The pathways involve either direct incorporation of formaldehyde into a phosphorylated sugar—the Allulose Pathway (Kemp and Quayle, 1966, 1967; Lawrence *et al.*,

1970), or a condensation of formaldehyde with glycine to form serine—the Serine Pathway (Large et al., 1962a,b; Large and Quayle, 1963).

C_1 units are incorporated into the Allulose Pathway by condensation of three formaldehydes with three ribose 5-phosphate molecules to form the six-carbon, allulose 6-P as shown in Fig. 1. The allulose 6-P is epimerized to fructose 6-P. One fructose 6-P is phosphorylated to fructose 1,6-diphosphate and then split to glyceraldehyde-P and dihydroxyacetone-P. The dihydroxyacetone-P can then enter the glycolytic pathway, and the glyceraldehyde-P can react with the remaining two fructose 6-P molecules to regenerate three ribose 5-P. Regeneration of the ribose units follows reactions similar to those of the Calvin cycle for photosynthetic CO_2 fixation. The Allulose Pathway has been worked out for two obligate C_1 utilizers, *P. methanica* (Kemp and Quayle, 1966, 1967), and *Methylococcus capsulatus* (Lawrence et al., 1970). Cell-free extracts of *P. methanica* and *M. capsulatus*, have been demonstrated to be capable of catalyzing the condensation of formaldehyde and ribose 5-P, however, there is little information on the nature of the cofactors or the specific enzyme system involved.

The Serine Pathway has been examined in detail in *Pseudomonas* AM-1 by Quayle and co-workers, and it has been shown to exist in *Methanomonas methanooxidans* (Lawrence and Quayle, 1970). The importance of serine and its probable involvement in C_1 compound incorporation was first noted by Kaneda and Roxburgh (1959b). ^{14}C incorporation data accumulated by Quayle and co-workers led to the proposed pathway shown in Fig. 2. C_1 compounds are incorporated at the oxidation level of formaldehyde by a condensation with glycine to form a molecule of serine. The enzyme catalyzing the condensation, serine hydroxymethyl transferase, has been found in cell-free extracts. The C_1 compound prior to transfer to glycine, is bound in the form of 5,10-methylenetetrahydrofolate. The proposed scheme requires the presence of one molecule of glycine for every carbon incorporated; however, the method of glycine formation or regeneration has not yet been determined. Tracer studies indicate that the carboxyl carbon is derived from carbon dioxide and the methylene carbon from methanol (Large et al., 1961). The two hypothesized paths for glycine formation involve either a condensation of carbon dioxide and the C_1 compound to form glycine or a carboxylation of phosphoenolpyruvate to form oxaloacetate. Oxaloacetate is then converted to an unknown C_4 compound which splits into two C_2 compounds, one of which is incorporated into cellular material and the other serving to regenerate

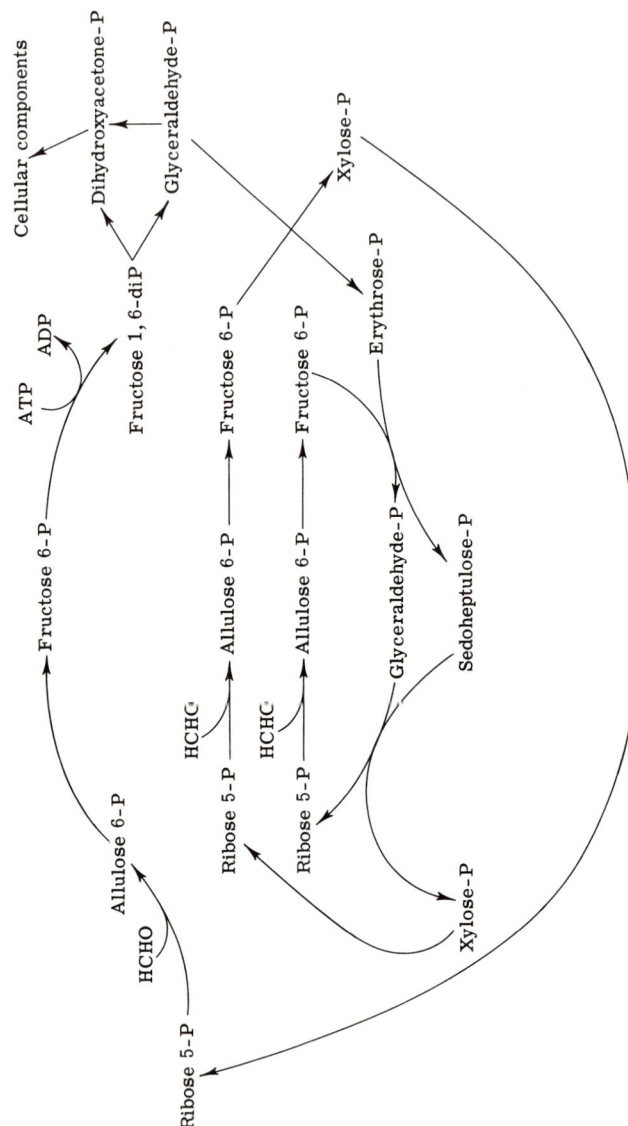

Fig. 1. The Allulose Pathway of C_1-compound incorporation. From Ribbons et al. (1970).

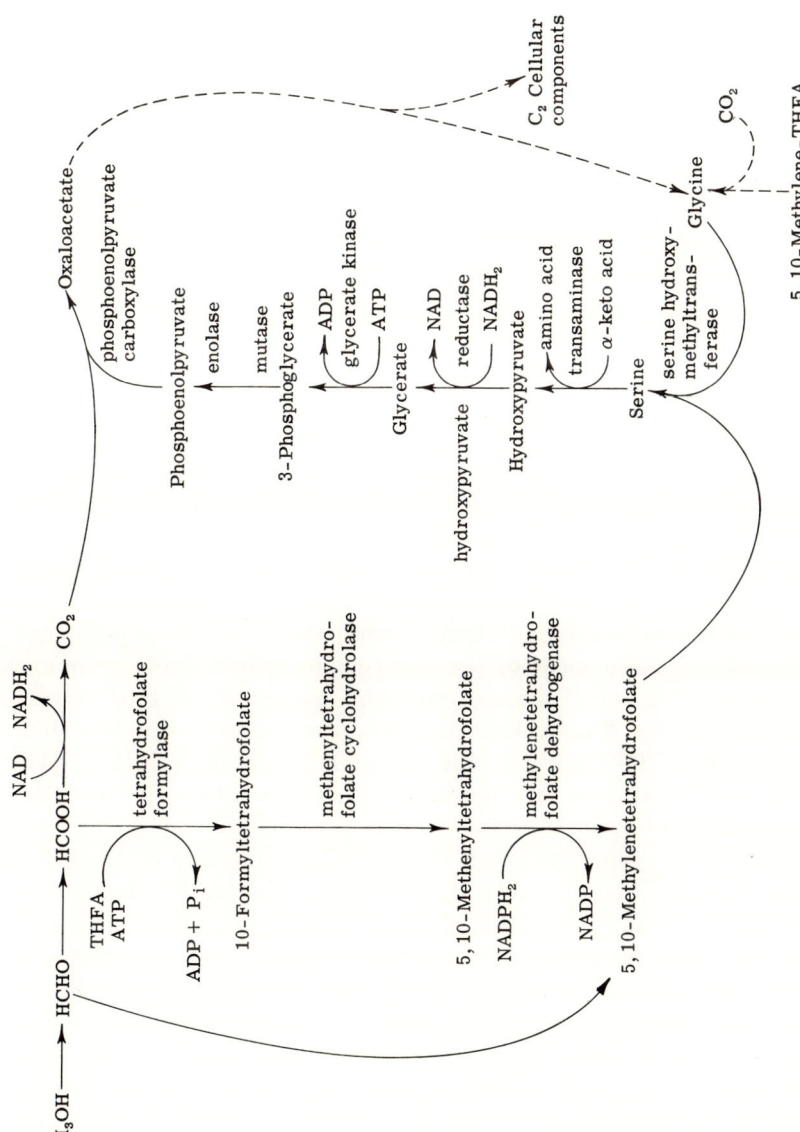

FIG. 2. The Serine Pathway of C_1-compound incorporation. From Kemp and Quayle (1967).

glycine. The most recent work with blocked pathway mutants indicates that glyoxalate is a probable precursor of glycine, however, further work is required to determine the nature of glyoxalate synthesis. Enzymatic studies indicate that it is not formed by way of the glyoxalate cycle.

Lawrence and Quayle (1970) have examined a number of methane and methanol utilizers for hexose phosphate synthetase activity, indicative of the Allulose Pathway, and for hydroxypyruvate reductase activity indicative of the Serine Pathway. Interestingly enough, they find a correlation between the presence of either of these enzymes with one of two characteristic membrane systems described by Davies and Whittenbury (1970). Further work is required, however, to determine the biochemical significance of these different membrane systems.

III. Anaerobic Utilization of Methanol

A. MICROBIOLOGY

Anaerobic growth on C_1 compounds, resulting in the formation of methane, was first observed by Söhngen in 1910 (Stephenson and Stickland, 1933). Anaerobic growth on methanol, however, was not reported until 1933, when Stephenson and Stickland isolated an unknown organism(s) for its ability to grow on formic acid and demonstrated its ability to grow on other C_1 compounds including methanol. These authors, like most of the workers to follow, were primarily interested in the mechanism of methane formation. Stadtman (1967) has reviewed the literature on methane fermentation through 1966. More recently, Toerien and Hattingh (1969), and Kirsch and Sykes (1971), in describing the anaerobic digestion process, have also reviewed this literature.

Despite the long-term interest in the biochemistry of anaerobic metabolism of C_1 compounds, there is little quantitative data on the microbial utilization of methanol. Reasons for this state of affairs are associated with the difficulty in growing these obligate anaerobes and in isolating pure cultures of the methanogenic organisms. A summary of the cultures capable of anaerobically utilizing methanol is provided in Table II. Care is needed, however, in interpreting this summary, since all of the described cultures are not pure. As recently as 1967 Bryant *et al.* found the culture of *Methanobacterium (Methanobacillus) omelianskii* to consist of two species of bacteria, only one of which was methanogenic. Other cultures may also exist as symbiotic relationships between two species.

TABLE II
Microbial Cultures Capable of Anaerobic Growth on Methanol

Culture	Reference
Unknown isolate	Stephenson and Stickland (1933)
Methanosarcina barkerii	Kluyver and Schnellen (1947)
Methanobacterium omelianskii (*Methanobacillus omelianskii*)	Kluyver and Schnellen (1947)
Methanosarcina methanica	Barker *et al.* (1940)
Methanosarcina sp.	Stadtman and Barker (1951)
Methanobacterium suboxydans	Cited by Toerien and Hattingh (1969)
Methanobacterium propionicum	Cited by Toerien and Hattingh (1969)
Hyphomicrobium sp.	Sperl and Hoare (1971)

The only quantitative data on methanol metabolism by pure methanogenic cultures are the results of Stadtman (1967). She observed a yield of 0.1 gm of cell per gram of methanol in a 200-liter batch fermentation of *Methanosarcina barkerii*. More recently, however, workers have become more aware of the need for quantitative data to facilitate the design of waste treatment processes, and some data are available for mixed populations. Data cited by Andrews *et al.* (1964) show a cell yield on methanol of 0.229 gm of cell per gram of methanol. Moore and Schroeder (1970) found the cell yield on methanol to vary with reactor residence time in a continuous denitrification process; the yield ranged from 0.19 to 0.28 gm of cells per gram of methanol at retention times of 2 and 6 days respectively in a continuous flow digestor. Increasing the retention time further had little effect on the yield. In these studies, nitrate was the growth-limiting nutrient at the longer retention times and we suspect that under such conditions there is probably an accumulation of carbon-energy reserve material. Sperl and Hoare (1971) have shown that pure cultures of *Hyphomicrobium* growing anaerobically on methanol and nitrate do accumulate poly-β-hydroxybutyric acid. At the shorter retention times, methanol may be the limiting nutrient; thus, the lower yield value is probably more representative of the yield on methanol. These yields are also consistent with the results of McCarty (as cited by Moore and Schroeder, 1970). McCarty *et al.* (1969) have derived the following material balance for denitrification with methanol:

$$NO_3^- + 1.08\ CH_3OH + H^+ \longrightarrow 0.065\ C_5H_7O_2N + 0.47\ N_2 + 0.76\ CO_2 + 2.44\ H_2O \quad \textbf{(II)}$$

indicating that yields of 0.2 gm of cell per gram of methanol should be expected from the denitrification process.

Interest in methanol utilization during microbial denitrification has been stimulated by the need to clean up waste water from the irrigation of farm land. These waters are low in organic carbon but high in nitrate and phosphate and as a consequence accelerate eutrophication in receiving waters. Several treatment processes employing a methanol feed to supply necessary organic carbon have been developed. McCarty *et al.* (1969) examined a variety of carbon sources and concluded that methanol was the most economically attractive feed material. Sheffield (1969) came to similar conclusions. Both McCarty *et al.* (1969) and Moore and Schroeder (1970) found that 2.5 to 3.0 gm of methanol were required to remove 1 gm of nitrate nitrogen. Extensions of this approach have been applied by Barth *et al.* (1968) to the denitrification of municipal waste waters and by Parkhurst *et al.* (1967) for a denitrification process using activated carbon filters. Barth *et al.* (1968) found a methanol to nitrate consumption ratio of 4. These processes typically require a methanol feed which provides an initial methanol concentration of 15 to 75 mg/liter. Methanol is attractive in these processes because of its low cost, high solubility in water, ease of handling and storage, and ease of removal of any undesired residual by gas stripping.

Sperl and Hoare (1971) have recently examined denitrification with methanol using pure isolates of *Hyphomicrobium* species. These cultures were selected and enriched on a medium containing methanol and nitrate; both aerobic and anaerobic enrichment were successful. The isolate *Hyphomicrobium* WC would grow on methanol under aerobic or anaerobic conditions; however, only under anaerobic conditions would it denitrify. This study is an excellent example of how selective enrichment may be used to obtain cultures capable of performing a specified biochemical job.

B. Biochemistry

Well-documented descriptions of methane fermentation are provided by several recent reviews (Stadtman, 1967; Toerien and Hattingh, 1969; Kirsch and Sykes, 1971) and only those facts pertinent to methanol metabolism are presented here. Although much is known about the process, the detailed mechanism still remains obscure and, furthermore, progress has been hindered by difficulties in handling obligate anaerobes and in obtaining pure cultures.

The original observations of Stephenson and Stickland in 1933 suggested that methanol was reduced by hydrogen in the following manner:

$$CH_3OH + H_2 \rightarrow CH_4 + H_2O \tag{III}$$

Subsequent studies proved that this was not the case. With the knowledge that carbon dioxide was reduced to methane according to

$$CO_2 + 4H_2 \longrightarrow CH_4 + 2H_2O \qquad \text{(IV)}$$

and the observation of Stadtman and Barker (1951) and Pine and Vishniac (1957) that methanol was reduced directly to methane and not first oxidized to carbon dioxide, it was realized that the actual stoichiometry for methanol assimilation was

$$4CH_3OH \longrightarrow 3CH_4 + CO_2 + 2H_2O \qquad \text{(V)}$$

In Barker's generalized concept of the methane fermentation (1956, 1967), methanol was visualized as entering the pathway via a carrier

$$CH_3OH + H\text{-}X \longrightarrow CH_3 - X + H_2O \qquad \text{(VI)}$$

followed by terminal reduction to methane according to

$$CH_3 - X + 2H \longrightarrow CH_4 + H - X \qquad \text{(VII)}$$

thereby regenerating the carrier. Much effort has been expended to substantiate this mechanism, yet the details are not totally worked out. The carrier appears to be vitamin B_{12} or some closely related derivative (Stadtman, 1967; Barker, 1967). This conclusion is based on the formation of methane from methyl cobalmine in cell-free extracts of *Methanosarcina barkerii* (Blaylock and Stadtman, 1964), the inhibition of methane formation by alkylhalides followed by photoreactivation (Stadtman, 1967), and the natural occurrence of corrinoid compounds in C_1-metabolizing cells (Barker, 1967). The factors required for the reduction of methanol include a corrinoid-containing protein, a ferredoxin, an unknown protein fraction, an unidentified cofactor, ATP (adenosine triphosphate), and magnesium (Stadtman, 1967). Most of the work leading to the present state of knowledge has been done using whole cells and cell-free extracts of *Methanoscarcina barkerii* and *Methanobacterium omelianskii*. With these tools an elucidation of the roles of all of the components can be hoped for in the future.

The preceding discussion accounts for approximately 3 of each 4 moles of methanol consumed; consideration still must be given to the methanol oxidized to carbon dioxide and the methanol incorporated into cellular material. Both of these points have received little attention in the literature. Methanol oxidation probably occurs in a manner analogous to that in aerobic metabolism (see Section II,B).

5,10-Methyltetrahydrofolate has been implicated in methane formation from carbon dioxide (Stadtman, 1967) and may also function in the oxidative pathway. However, the actual cofactors involved in coupling the oxidation-reduction reactions do not appear to be known. Furthermore, the mechanism of ATP generation via substrate-level phosphorylation in this pathway is not fully substantiated. Stadtman (1967) has hypothesized on the basis of free energy changes that sufficient energy is available in the oxidation of formaldehyde to a formyl derivative to generate one ATP. This point will be considered in more detail in Section III, C. Clearly, this phase of methanol metabolism will require more work before it is completely understood.

The incorporation of methanol into cellular material is yet another vague area of our knowledge of anaerobic methanol assimilation. Very likely, the incorporation of methanol carbon proceeds in the same basic manner in both aerobic (see Section II, C) and anaerobic growth. Methanogenic bacteria are able to use the β-carbon of serine for methane formation, thus producing a glycine molecule (Barker, 1956). This pathway may be reversible, thereby permitting the incorporation of a C_1 unit into a three-carbon unit. One major unknown, however, is the mechanism for glycine regeneration. This same problem exists in the aerobic Serine Pathway (Section II, C). Clearly, this point needs to be clarified and both the Serine (see Fig. 2) and Allulose (see Fig. 1) Pathways examined in detail to elucidate their significance in methanol-assimilating anaerobes.

C. Energetics

Cellular yields in the anaerobic methane fermentation are low relative to most aerobic fermentations. The reason for this is that relatively little free energy of oxidation is available per mole of substrate consumed. Aerobic oxidation of methanol to carbon dioxide and water yields 347 kcal per mole of methanol, whereas the anaerobic conversion of 4 moles of methanol to 3 moles of methane and 1 mole of carbon dioxide yields only 18.8 kcal per mole of methanol (Stadtman, 1967). This latter calculation is based on the following free energy changes: methanol to methane, -26 kcal; methanol to formaldehyde, $+10$ kcal; formaldehyde to formate, -8 kcal; and formate to carbon dioxide, 0 to -2 kcal. If in a coupled oxidation-reduction system, one ATP is derived by substrate-level phosphorylation in the step from a formaldehyde to a formyl derivative, then the cell yield can be estimated from the general rule that 10 gm of cells are produced for each mole of ATP generated (Bauchop and Elsden, 1960). Stadtman (1967) found that *Methanosarcina barkerii* yielded 280 gm of cells from 85 moles of methanol in a 200-liter batch fermentation. This is better than the 210 gm predicted from the preceding calculation thereby suggesting some inaccuracies in the estimated free energy changes. From these data it is also possible to estimate the amount of energy wasted as heat. From 85 moles of methanol, the total available energy is about -1600 kcal. The 280 gm of cells produced have a heat of combustion of about 1000 kcal (Guenther, 1965), therefore, assuming the entropy change to be small relative to the enthalpy change, about 600 kcal are released as heat in the process. This corresponds to an efficiency of energy utilization of about 60% which is slightly higher than the average of 50% observed by Cooney et al (1968). Knowledge of the energetics of methanol metabolism is lacking and more quantitative data are required.

IV. Single-Cell Protein from Methanol

A. Methanol vs. Methane and Other Hydrocarbons

On the premise that there is and will continue to be a world protein shortage, many university and industrial research teams have put forth a large effort exploring the potential of single-cell protein (SCP) as a novel protein source. In the course of this work, a wide variety of carbon-energy sources have been considered for use in protein production. The cost for the carbon substrate represents a major fraction of the protein production cost (Wang, 1968). Most of the attention has been devoted to the use of hydrocarbons and in particular normal alkanes including methane.

More recently, methanol has been of increasing interest as a carbon source for SCP production. The reasons for this are best understood by examining the advantages and disadvantages of hydrocarbons and methane as compared with methanol. All these are readily available, easy to transport and store, and available at a low cost relative to other carbon sources. Typical data are shown in Table III. Methane and methanol are available as highly pure substrates,

TABLE III
Cost Data for Various Carbon Substrates

Substrate	Cost (cents/lb)
Molasses	2.0
n-Alkanes	2.0–3.0
Methane	0.4
Methanol	1.5–1.8

while normal alkanes or gas oil may have impurities which require subsequent removal from the final product. Methanol is usually sold as >99.85% pure (Mehta and Pan, 1971). Methane and other hydrocarbons are only slightly soluble in water, and as a consequence, the rate of utilization of these substrates is limited by their mass transfer rate. Wang and Ochoa (1972) have shown growth on normal alkanes to be limited by the surface area for mass transfer. Sheehan (1970) has measured mass transfer coefficients for methane and found them to be on the same order as for oxygen. The low water solubility of hydrocarbons may also necessitate a solvent wash for hydrocarbon-grown cells in order to remove residual substrate. Methanol, on the other hand, is completely miscible with water. The assimilation of all three of these substrates requires substantial amounts of oxygen; however, when compared on the basis of constant productivity, cells utilizing methanol have the lowest oxygen demand as shown in Table IV. Since heat evolution is directly proportional to the oxygen demand, methanol cultures also have a lower cooling water requirement when compared on the basis of constant productivity (Table IV). Summarizing these points we note that the present and future cost structure, the solubility and ease of handling, and the purity are all factors which make methanol an attractive raw material for SCP production.

TABLE IV
OXYGEN DEMAND AND HEATS OF FERMENTATION FOR BACTERIA GROWN ON VARIOUS CARBON SOURCES COMPARED AT CONSTANT PRODUCTIVITY[a]

Carbon source	Cell yield on substrate (gm cell/gm substrate)	Cell yield on oxygen[c] (gm cell/gm O_2)	Oxygen demand (mmole/liter/hr)	Heat of fermentation[b] (kcal/liter/hr)
Glucose	0.5	2.1	53	6.4
Methanol	0.5	0.7	150	18.0
Methane	0.6	0.2	560	67.0
n-Alkane	1.0	0.53	210	25.0

[a] Productivity = 3.5 gm cell/liter/hr.
[b] Heat of fermentation calculated from oxygen demand according to correlation of Cooney et al. (1968).
[c] Calculated from equation of Mateles (1971).

B. METHANOL ECONOMICS

The increasing demand for methanol as a commodity has continued to stimulate advancements in the technology of methanol manufacture.

As a result of advancing technology, the price of methanol has continued to decline, thus making it an economically attractive raw material for industrial fermentations. The status of methanol technology and economics has been recently summarized in a volume edited by Danner (1970) and the detailed economics of methanol production have been presented for a number of processes (Hedley et al., 1970a–d; Hiller and Marschner, 1970). Selling prices for methanol were quoted as being 1.5 to 1.8 cents per pound in April of 1971 (Prescott, 1971), and predictions for the future are as low as 1.05 cents per pound (Prescott, 1971; Kenard and Nimo, 1970). Various estimated selling prices for methanol are given in Table V, these prices are very dependent on the nature and cost of the feedstock. Most methanol is made from natural gas although in some locations naphtha or heavy fuel oil are economically attractive (Hedley et al., 1970d). Clearly, the future economics of methanol depend heavily on the economics of these feedstocks.

TABLE V
ESTIMATED SELLING PRICES FOR METHANOL MANUFACTURED BY VARIOUS PROCESSES

Process	Feedstock	Selling price (cents/lb)[a]	Reference
—	—	1.5–1.8 [Current selling price]	Prescott (1971)
Lurgi, low pressure	Natural gas[a]	1.25	Hiller and Marschner (1970)
Kellogg steam-methane	Natural gas[b]	1.05	Prescott (1971)
Low pressure[g]	Natural gas[c]	1.2	Hedley et al. (1970d)
High pressure[g]	Natural gas[c]	1.3	Hedley et al. (1970d)
Low pressure[g]	Naphtha[e]	1.7	Hedley et al. (1970d)
Low pressure[g]	Naphtha[f]	1.5	Hedley et al. (1970d)
ICI, low pressure	Natural gas	1.05	Kenard and Nimo, (1970)

[a] Gas cost, 40¢/million BTU.
[b] Gas cost, 30¢/million BTU.
[c] Gas cost, 20¢/million BTU.
[d] Gas cost, 23¢/million BTU.
[e] Naphtha cost, 6.5¢/gal.
[f] Naphtha cost, 5.0¢/gal.
[g] 800 ton/day production.

C. STATUS OF SCP FROM METHANOL

Having established the advantages of methanol over some other carbon substrates, it is now appropriate to examine the status of single-cell protein derived from methanol. It is difficult to pinpoint when the potential of methanol as a raw material for SCP was first realized. Vary and Johnson (1967) presented some of the first quantitative data on methanol utilization in a discussion of protein production from methane. Hamer (1968) alluded to the potential of methanol as a fermentation substrate in his studies on volatile organic liquids as carbon substrates for aerobic fermentations. A more intense period of activity began in 1969. Häggström (1969) reported some studies on methanol-oxidizing bacteria presenting quantitative data on growth and cell composition. Coty (1969) published a review on methane utilization, pointing out the number of cultures also able to utilize methanol. The isolation of a yeast capable of growing on methanol was also reported by Ogata and co-workers in 1969. More recently, Hamer and Norris (1971), Mateles (1971), Harrison and Hamer (1971), Humphrey (1970), and Asthana et al. (1971) have considered various aspects of SCP from methanol.

Currently both Shell Research Ltd. and Imperial Chemical Industries are reported to be actively considering methanol as a substrate for single-cell protein production. Very likely, most of the United States petroleum companies interested in SCP have also examined the use of methanol as a substrate. Other groups actively engaged in research on the microbial utilization of methanol include Heden, Molin, Dostalek, and co-workers at the Karolinska Institutet in Stockholm, Cooney and co-workers at M.I.T., Humphrey and co-workers at the University of Pennsylvania, and groups at the Institute of Biochemistry and Physiology of Microorganisms, USSR Academy of Sciences, and the All-Union Research Institute of Protein Biosynthesis, both in Moscow.

D. Composition of Methanol-Grown Cells

Cell composition is of prime concern in the production of SCP; in particular, high protein, low nucleic acid and low carbohydrate and lipid content is desired. In addition, a favorable balance of essential amino acids (e.g., relative to FAO standards) especially lysine, tryptophan, and methionine is necessary. These three amino acids are the usually deficient amino acids in plant-derived proteins with which SCP must compete economically. A summary of amino acid profiles of several methanol- and methane-grown cultures is given in Table VI. Only the *Kloeckera* species No. 2201 of yeast and the TM20 culture of Häggström (1969) were grown on methanol. While the other cultures are able to grow on methanol, the data available are for methane-grown cells. These data, however, are felt to be indicative of the amino acid profiles for growth on methanol as well.

Actual protein contents of methanol-grown cells have been reported to range from 35 to 70%. Häggström reported a crude protein of 71%

TABLE VI
AMINO ACID PROFILES OF METHANOL-ASSIMILATING ORGANISMS

Amino acid	FAO reference	Organism and reference					
		Kloeckera (Ogata et al., 1970b)	TM10[a] (Bewersdorff and Dostalek, 1971)	TM20 (Häggström, 1969)	HR[a] (Vary and Johnson, 1967)	Bacillus[a] (Wolnak et al., 1967)	M45[a] (Sheehan, 1970)
		Content (in gm amino acid/100 gm protein)					
Cystine[b]	2.0	Trace	0.28	0.32	—	—	0.35
Histidine	—	1.8	1.87	1.73	1.9	1.29	1.6
Isoleucine	4.2	5.1	4.94	3.90	6.1	—	5.0
Leucine	4.8	7.1	6.95	6.96	9.1	9.06	7.9
Lysine	4.2	7.5	5.36	5.3	5.3	3.15	4.3
Methionine	2.2	0.89	1.88	1.81	3.4	0.9	3.0
Phenylalanine	2.8	4.0	4.06	4.18	6.2	—	5.2
Threonine	2.8	5.1	4.29	4.52	4.5	—	4.9
Tryptophan	1.4	—	trace	—	—	1.15	2.7
Tyrosine[c]	—	3.3	3.51	2.91	4.1	—	3.8
Valine	4.2	5.3	5.91	5.85	8.5	—	6.2

[a]Grown on methane.
[b]Spares methionine.
[c]Spares phenylalanine.

in a mixed culture, TM20, of gram-negative rods. Bewersdorff and Dostalek (1971) found similar results in their mixed culture TM10. Both TM10 and TM20 appear very similar with respect to cell composition and amino acid profile. Vary and Johnson (1967) also working with a mixed culture, measured 60% protein in their isolate labeled HR.

Working with yeast, Ogata et al. (1970a) found *Kloeckera* sp. No. 2201 to have 45.30% crude protein and 5.5% nucleic acid. The protein content of *Candida silvicola* grown in a methanol-limited chemostat has been shown to be consistently about 50% over a range of growth rates (Cooney, 1971). In an effort to increase the protein to nucleic acid ratio of *C. silvicola,* Cooney and co-workers, using the heat shock and incubation procedure, shown by Ohta et al. (1970) to be effective for the removal of nucleic acids from yeast, were able to remove up to 88% of the nucleic acid with essentially complete retention of the cell protein.

From a nutritional point of view, it is not sufficient to speak of just protein content in that all the cell protein is not available for use. However, information on protein availability, nutritional value, or toxicology of methanol-grown cells is not yet available in the literature. The state of our knowledge on the quality of SCP from methanol is not as extensive as from n-alkanes and gas oil but this is likely to change in the near future as work in this area continues.

E. Engineering Aspects of Methanol-Derived SCP

The success or failure of single-cell protein depends on its cost per pound relative to other competing protein sources, e.g., soybean, fish protein, etc. The two predominant costs in SCP production are the carbon substrate and the processing costs. Interestingly, these two costs are not independent of each other since the physical properties of the carbon source and the pattern of its assimilation both affect the process design. For instance, insoluble substrates such as hydrocarbons or cellulose present engineering problems not found with soluble substrates such as methanol or glucose. In addition, the level of oxidation of the substrate determines the need for oxygen in the assimilation of that substrate, i.e., hydrocarbons require more oxygen per pound of protein produced than does methanol (see Mateles, 1971). As a consequence of these factors, the search for an optimum SCP process has been a complex problem.

Interest in methanol as a substrate for single-cell protein results from its low present and future cost as well as processing advantages when compared with hydrocarbons. These points have already been discussed in the first part of this section. Methanol, however, is not without its disadvantages as a carbon source for SCP. The factor limiting the productivity of a fermentor is still the ability to transfer oxygen to the growing culture. While methanol is more favorable in its oxygen demand (see Table IV), the actual demand is still high. One interesting approach to overcoming this problem is through the use of pure oxygen or oxygen-enriched aeration. Pure oxygen aeration has been shown to be economically attractive in a waste treatment process (Albertsson et al., 1970). Furthermore, the power required to transfer a pound of oxygen with pure oxygen compares favorably with the data available on large-scale fermentors transferring oxygen from air (Cooney and Wang, 1971). Very little data, however, are available on the use of oxygen aeration in SCP production. MacLennan et al., (1972) have examined the influence of dissolved oxygen on *Pseudomonas* AM-1 grown on methanol in continuous culture. These authors found the cellular yield to vary inversely with the dissolved oxygen tension (DOT) as it was increased above 100 mm of Hg. The yield decreased slowly up to a DOT of 420 mm Hg and upon further increase of DOT, the yield fell from 0.42 to 0.33. For effective utilization of oxygen in SCP protein production, the partial pressure of oxygen will need to be controlled carefully and a high efficiency of utilization for the pure oxygen feed will need to be maintained.

An additional problem with methanol-based fermentation relates to the relatively high vapor pressure of methanol at fermentation temperatures. At 40°C, for instance, the vapor pressure of methanol is 261 mm of Hg. Furthermore, as the concentration of methanol decreases, its activity coefficient increases (Hamer, 1968) and hence so does its relative volatility. The problem of stripping a volatile component from a fermentation broth has been examined theoretically and experimentally by Hamer (1965a,b). One of the advantages of continuous culture is that there is a low residual concentration of the limiting nutrient under steady state conditions. For cultures whose growth is described by the Monod model (Herbert et al., 1956), the K_S value or the half-rate saturation constant is a good indicator of residual methanol. Cooney et al. (1971) have found *Candida silvicola* to have a K_s value of 47 mg of methanol per liter. From the data of Asthana et al. (1971) a K_s value for *Torulopsis glabrata* of about 350 mg per liter can be calculated. Similar data for other organisms are not available. Also related to the concentration of methanol in the fermentation broth is the inhibition of growth by methanol. Many cultures able to grow on methanol exhibit methanol inhibition at concentrations above 1 to 2% (Asthana et al., 1971; Ogata et al., 1969).

This inhibition may be minimized by growth in continuous culture with resultant low residual methanol concentrations. The application of the Monod model to describe growth, however, may not be truly valid and the addition of an inhibition term to the model may be warranted. Further data are needed to clarify this point.

V. Fermentation Products from Methanol

As a low cost carbon source, methanol has potential application in other fermentation processes than SCP production. In a patent issued to Okumura and co-workers (1970), examples of processes for the production of L-glutamic acid, L-alanine, L-valine, L-lysine, L-threonine, α-ketoglutaric acid, citric acid, fumaric acid, and hypoxanthine are given. The product yields in these examples are low relative to industrial standards; however, the precedent has been set. An analogous case history has occurred with hydrocarbon-based fermentations. In 1965, glutamic acid yields from n-paraffins were about 10 gm per liter and today they are reported to be in excess of 100 gm per liter (Hill, 1971).

Another approach to the use of methanol as a fermentation substrate is through cooxidation of a second substrate. Leadbetter and Foster (1959) demonstrated the ability of *Pseudomonas methanica* to cooxidize ethane to acetic acid and acetaldehyde, propane to propionic acid and acetone, and n-butane to n-butyric acid and 2-butanone. The conversions were done with cells growing on methane since *P. methanica* does not use ethane, propane, or n-butane for growth. *Pseudomonas methanica* will, however, grow on methanol. Through such cooxidation methods a wide variety of methanol-based fermentations are possible yet unexplored. The versatility of this approach has been demonstrated by the work on cooxidation with hydrocarbons (Raymond *et al.*, 1971). Clearly, much work remains to be done on the microbiology of methanol utilizers capable of producing commercially important products.

VI. Summary and Conclusions

Despite the early observations on microbial utilization of methanol, only recently, in response to the need for cheap substrates in single-cell protein production and the need to understand methanol removal in waste treatment processes, have investigations begun to examine in detail the mechanisms of methanol metabolism. Most attention has been devoted to the aerobic utilization of methanol and in particular by gram-negative bacteria; although there is an increasing interest in the ability of yeast to grow on methanol. Significant progress has been made toward understanding the pathways of methanol oxidation yet much work remains before the details of its incorporation into cell mass are elucidated. Considerably less is known about the anaerobic utilization of methanol. Furthermore, there is a distinct lack of quantitative data concerning all aspects of methanol metabolism.

Along with increased recognition of the advantages of methanol as a fermentation substrate, the number of teams investigating methanol metabolism has increased. However, the level of our knowledge on methanol metabolism does not yet equal that on hydrocarbon metabolism; although the gap is rapidly closing. With respect to single-cell protein production, more data are required on the quality of protein derived from methanol utilizers and further quantitative engineering data are needed to enable a good economic assessment to be made. Lastly, there is still an untapped potential for the use of methanol as a substrate for other fermentation processes, both by direct metabolism and through cooxidation.

Acknowledgment

We gratefully acknowledge the Lewis and Rosa Strauss Foundation for supporting the work on methanol utilization performed in our laboratory.

References

Akiba, T., Veyama, H., Seki, M., and Fukimara, T. (1970). *J. Ferment. Technol.* **48**, 323.
Albertsson, J. G., McWhirter, J. R., Robinson, E. K., and Vahldieck, N. P. (1970). Federal Water Quality Administration Program No. 17050 DNW.
Andrews, J. F., Cole, R. D., and Pearson, E. A. (1964). SERL Rep. No. 64-11. School of Public Health, University of California, Berkeley.
Anthony, C. (1970). *Biochem. J.* **119**, 54p.
Anthony, C., and Dunstan, P. (1971). *Biochem. J.* **124**, 75p.
Anthony, C., and Zatman, L. J. (1964a). *Biochem. J.* **92**, 609.
Anthony, C., and Zatman, L. J. (1964b). *Biochem. J.* **92**, 614.
Anthony, C., and Zatman, L. J. (1965). *Biochem. J.* **96**, 808.
Anthony, C., and Zatman, L. J. (1967a). *Biochem. J.* **104**, 953.
Anthony, C., and Zatman, L. J. (1967b). *Biochem. J.* **104**, 960.
Asthana, H., Humphrey, A. E., and Moritz, V. (1971). *Biotechnol. Bioeng.* **13**, 923.
Barker, H. A. (1956). "Bacterial Fermentations." Wiley, New York.
Barker, H. A. (1967). *Biochem. J.* **105**, 1.
Barker, H. A., Ruben, S., and Kamen, M. D. (1940). *Proc. Nat. Acad. Sci. U.S.* **26**, 426.
Barth, E. F., Brenner, R. C., and Lewis, R. F. (1968). *J. Water Pollut. Contr. Fed.* **40**, 2040.
Bassalik, K. (1914). *Jahrb. Wiss. Bot.* **53**, 287.
Bauchop, T., and Elsden, S. R. (1960). *J. Gen. Microbiol.* **23**, 457.
Bewersdorff, M., and Dostalek, M. (1971). *Biotechnol. Bioeng.* **13**, 49.
Blaylock, B. A., and Stadtman, T. C. (1964). *Biochem. Biophys. Res. Comm.* **13**, 435.
Brown, L. R., and Strawinski, R. J. (1957). *Bacteriol. Proc.* 18.

Brown, L. R., and Strawinski, R. J. (1958). *Bacteriol. Proc.* 122
Brown, L. R., Strawinski, R. J., and McClesky, C. S. (1964). *Can. J. Microbiol.* **10**, 791.
Bryant, M. D., Wolin, E. A., Wolin, M. J., and Wolfe, R. S. (1967). *Arch Mikrobiol.* **59**, 20.
Cooney, C. L. (1971). Unpublished results.
Cooney, C. L., and Wang, D. I. C. (1971). *Biotechnol. Bioeng. Symp.* **2**, 63.
Cooney, C. L., Wang, D. I. C., and Mateles, R. I. (1968). *Biotechnol. Bioeng.* **11**, 269.
Cooney, C. L., Levine, D. W., and Wang, D. I. C. (1971). *162nd Meet. Amer. Chem. Soc., Washington, D. C., 1971.*
Coty, V. F. (1969). *Biotechnol. Bioeng. Symp.* **1**, 105.
Danner, G. A. (1970). *Chem. Eng. Progr., Symp. Ser.* **66**, No. 98.
Davies, S. L., and Whittenbury, R. (1970). *J. Gen. Microbiol.* **61**, 227.
Dworkin, M., and Foster, J. W. (1956). *J. Bacteriol.* **72**, 646.
Elizarova, T. N. (1963). *Mikrobiologiya* **32**, 1091.
Foster, J. W., and Davis, R. H. (1966). *J. Bacteriol.* **91**, 1924.
Guenther, K. R. (1965). *Biotechnol. Bioeng.* **7**, 445.
Häggström, L. (1969). *Biotechnol. Bioeng.* **11**, 1043.
Hamer, G. (1965a). *Biotechnol. Bioeng.* **7**, 199.
Hamer, G. (1965b). *Biotechnol. Bioeng.* **7**, 215.
Hamer, G. (1968). *J. Ferment. Technol.* **46**, 177.
Hamer, G., and Norris, J. R. (1971). *World Petrol. Congr., Proc., 8th, 1971.* PD 21 (2).
Harrington, A. A., and Kallio, R. E. (1960). *Can. J. Microbiol.* **6**, 1.
Harrison, D. E. F., and Hamer, G. (1971). *Biochem. J.* **124**, 78p.
Hedley, B., Powers, W., and Stobaugh, R. B. (1970a). *Hydrocarbon Process.* **49** (6), 97.
Hedley, B., Powers, W., and Stobaugh, R. B. (1970b). *Hydrocarbon Process.* **49** (7), 131.
Hedley, B., Powers, W., and Stobaugh, R. B. (1970c). *Hydrocarbon Process.* **49** (8), 117.
Hedley, B., Powers, W., and Stobaugh, R. B. (1970d). *Hydrocarbon Process.* **49** (9), 275.
Herbert, D., Elsworth, R., and Telling, R. C. (1956). *J. Gen. Microbiol.* **14**, 601.
Hill, I. D. (1971). *162nd Meet., Amer. Chem. Soc., Washington, D.C., 1971.*
Hiller, G., and Marschner, F. (1970). *Hydrocarbon Process.* **49** (9), 281.
Hirsch, P., and Conti, S. F. (1964). *Arch. Mikrobiol.* **48**, 358.
Humphrey, A. E. (1970). *Process. Biochem.* **5** (6), 19.
Hutton, W. E., and ZoBell, C. E. (1949). *J. Bacteriol.* **58**, 463.
Johnson, P. A., and Quayle, J. R. (1964). *Biochem. J.* **93**, 281.
Kaneda, T., and Roxburgh, J. M. (1959a). *Can. J. Microbiol.* **5**, 87.
Kaneda, T., and Roxburgh, J. M. (1959b). *Can. J. Microbiol.* **5**, 187.
Kemp, M. B., and Quayle, J. R. (1966). *Biochem. J.* **99**, 41.
Kemp, M. B., and Quayle, J. R. (1967). *Biochem. J.* **102**, 94.
Kenard, R. J., and Nimo, N. M. (1970). *Chem. Eng. Progr., Symp. Ser.* **66**, No. 98, 47.
Kirikova, N. N. (1970). *Mikrobiologiya* **39**, 18.
Kirsch, E. J., and Sykes, R. M. (1971). *Progr. Ind. Microbiol.* **9**, 155.
Kluyver, A. J., and Schnellen, G. T. P. (1947). *Arch. Biochem.* **14**, 57.
Large, P. J., and Quayle, J. R. (1963). *Biochem. J.* **87**, 386.
Large, P. J., Peel, D., and Quayle, J. R. (1961). *Biochem. J.* **81**, 470.
Large, P. J., Peel, D., and Quayle, J. R. (1962a). *Biochem. J.* **82**, 483.
Large, P. J., Peel, D., and Quayle, J. R. (1962b). *Biochem. J.* **85**, 243.
Lawrence, A. J., and Quayle, J. R. (1970). *J. Gen. Microbiol.* **63**, 371.
Lawrence, A. J., Kemp, M. B., and Quayle, J. R. (1970). *Biochem. J.* **116**, 631.
Leadbetter, E. R., and Foster, J. W. (1958). *Arch. Mikrobiol.* **30**, 91.
Leadbetter, E. R., and Foster, J. W. (1959). *Arch. Biochem. Biophys.* **82**, 491.

McCarty, P. L., Beck, S., and St. Amant, P. (1969). *Proc. Ind. Waste Conf.* **24**, 1271.
MacLennan, D. G., Ousby, S. C., Vasey, R. B., and Cotton, N. T. (1972). To be published.
Mateles, R. I. (1971). *Biotechnol. Bioeng.* **13**, 581.
Mateles, R. I., and Chalfan, Y. (1970). *Abstr. Int. Congr. Microbiol., 10th, 1970* Vol. 7, p. 124, Abstr. No. GA-7.
Mehta, D. D., and Pan, W. W. (1971). *Hydrocarbon Process.* **50** (2), 115.
Moore, S. F., and Schroeder, E. D. (1970). *Water Res.* **4**, 685.
Naguib, M., and Overbeck, J. (1970). *Z. Allg. Mikrobiol.* **10**, 17.
Ogata, K., Nishikawa, H., and Ohsugi, M. (1969). *Agr. Biol. Chem.* **33**, 1519.
Ogata, K., Nishikawa, H., Ohsugi, M., and Tochikura, T. (1970a). *J. Ferment. Technol.* **48**, 389.
Ogata, K., Nishikawa, H., Ohsugi, M., and Tochikura, T. (1970b). *J. Ferment. Technol.* **48**, 470.
Ohta, S., Maul, S., Sinskey, A. J., and Tannenbaum, S. R. (1970). *Appl. Microbiol.* **22**, 415.
Okumura, S., Yamanoi, A., Tsugawa, R., and Nakase, T. (1970). Brit. Pat. No. 1,210,330.
Parkhurst, J. D., Dryden, F. D., McDermott, G. N., and English, J. (1967). *J. Water Pollut. Contr. Fed.* **39**, R70.
Peel, D., and Quayle, J. R. (1961). *Biochem. J.* **81**, 465.
Pine, M. J., and Vishniac, W. (1957). *J. Bacteriol.* **73**, 736.
Prescott, J. H. (1971). *Chem. Eng.* **5** (7), 60.
Raymond, R. L., Jamison, V. W., and Hudson, J. O. (1971). *Lipids* **6**, 453.
Ribbons, D. W., Harrison, J. E., and Wadzinski, A. M. (1970). *Annu. Rev. Microbiol.* **24**, 135.
Sheehan, B. T. (1970). Ph.D. Thesis, University of Wisconsin, Madison.
Sheffield, C. W. (1969). *Proc. Ind. Waste Conf. 24th*, 620.
Söhngen, N. L. (1906). *Zentralbl. Bakteriol., Parasitenk., Infektionskr. Hyg., Abt. 2* **15**, 513.
Sperl, G. T., and Hoare, D. S. (1971). *J. Bacteriol.* **108**, 733.
Stadtman, T. C. (1967). *Annu. Rev. Microbiol.* **21**, 121.
Stadtman, T. C., and Barker, H. A. (1951). *J. Bacteriol.* **61**, 81.
Stephenson, M., and Stickland, L. H. (1933). *Biochem. J.* **27**, 1517.
Stocks, P. K., and McClesky, C. S. (1964a). *J. Bacteriol.* **88**, 1065.
Stocks, P. K., and McClesky, C. S. (1964b). *J. Bacteriol.* **88**, 1071.
Tannahill, A., and Finn, R. K. (1970). *Abstr. Pap., 160th Meet., Amer. Chem. Soc., Chicago, 1970* Abstr. No. 49.
Toerien, D. F., and Hattingh, W. H. J. (1969). *Water Res.* **3**, 385.
Vary, P. S., and Johnson, M. J. (1967). *Appl. Microbiol.* **15**, 1473.
Wang, D. I. C. (1968). *Chem. Eng.* **26** (17), 99.
Wang, D. I. C., and Ochoa, A. (1972). *Biotechnol. Bioeng.* (in press).
Whittenbury, R., Phillips, K. C., and Wilkinson, J. F. (1970). *J. Gen. Microbiol.* **61**, 205.
Wolnak, B., Andreen, B. H., Chisholm, J. A., and Saadeh, M. (1967). *Biotechnol. Bioeng.* **9**, 57.

Modeling of Growth Processes with Two Liquid Phases: A Review of Drop Phenomena, Mixing, and Growth

P. S. Shah[1], L. T. Fan, I. C. Kao, and L. E. Erickson

Department of Chemical Engineering,
Kansas State University, Manhattan, Kansas

I.	Introduction	367
II.	Dispersion and Coalescence Phenomena of Liquid Drops	368
	A. Introduction	368
	B. Simplified Mixing Models	373
	C. General Mixing Models	375
	D. Measurements of Dispersion-Coalescence Frequency and Drop-Size Distribution	387
	E. Coalescence Phenomena of Liquid Drops	400
III.	Mathematical Modeling	402
	Nomenclature	409
	References	410

I. Introduction

This work is an attempt to review recent research in the areas of drop phenomena, mixing, and growth modeling which are important to the development of mathematical models of growth processes where two liquid phases are present. Research on the production of single cell protein has been reviewed in a book by Mateles and Tannenbaum (1968) and in a number of articles such as those by Humphrey (1967, 1968, 1969), by Aiba *et al.* (1969a,b), by Mimura (1970), and by Mimura *et al.* (1971a,b). The yearly reviews in *Folia Microbiologica* by Malek and Ricica (1970) give excellent coverage of experimental research on growth processes with two liquid phases. However, the research on drop phenomena and mixing which is important to the modeling of growth processes with two liquid phases has not been as thoroughly reviewed.

A biological growth process with two liquid phases, in reality, involves at least four phases. The gas (air) phase, solid (cell) phase or phases, aqueous phase (water), and dispersed (oil) phase all contribute to the growth process. For such a process the biological growth will be affected by factors that do not exist in one liquid phase systems, e.g., size distribution and number of oil droplets, frequency of coalescence and dispersion between the droplets, fraction of cells in

---

[1] Present address: Argonne National Laboratories, Argonne, Illinois.

the continuous and dispersed phases, and mass transfer between these phases. Because of these factors, concepts such as macromixing and micromixing (Tsai *et al.*, 1969) may become more significant in modeling large-scale continuous fermentors for cultures with two liquid phases.

Some of the engineering problems which must be considered in the analysis and design of fermentors with two liquid phases include: oxygen transfer, substrate transfer, control of liquid-liquid interfacial area, coalescence and dispersion of oil drops, and cell adsorption and desorption at the surface of the oil drops. The design of the fermentor as to the size, mixing, oxygen transfer, residence time distribution of each phase, and heat removal is important. The procedures for introducing the gas phase and each liquid phase also cannot be neglected. An understanding of the fundamental processes which occur in two liquid phase systems is essential in fermentor design and operation.

Very little information is available on the mechanism of oxygen transfer in fermentations with two liquid phases. This problem, however, was considered by Mimura (1970) and Mimura *et al.* (1971a) recently. They examined the amount of oxygen required by hydrocarbon fermentations, the effect of dissolved oxygen, the problems of oxygen transfer, the optimum size of oil drop, and the critical value of substrate supply, etc. An excellent review of the literature on interfacial area of liquid-liquid processes and mass transfer has recently been written by Onda and Takeuchi (1970). The stability problems associated with two-phase systems have been extensively investigated by many investigators such as Yamazaki and Ichikawa (1970), Schmitz and Amundson (1963a,b,c,d), Luss (1966), and Luss and Amundson (1967). Comprehensive reviews on bubble and drop phenomena are also available (Brodkey, 1967; Soo, 1967; Resnick and Gal-Or, 1968; Gal-Or *et al.*, 1969; Tavlarides *et al.*, 1970).

First the literature on the dispersion and coalescence phenomena of drops will be summarized and then the literature on modeling of biological growth processes with two liquid phases and also the literature on the effect of the micromixing on growth processes with one liquid phase system will be summarized.

II. Dispersion and Coalescence Phenomena of Liquid Drops

A. INTRODUCTION

Dispersed phase systems are a common occurrence in many industrially important processes such as liquid-liquid extraction, emulsion polymerization, waste treatment, and hydrocarbon fermentation in which two liquid phases exist. Whenever two immiscible or partially miscible liquids are vigorously mixed, one of the liquids will be dispersed as liquid drops and continuous coalescence and redispersion of these drops will occur simultaneously. These coalescence and redispersion phenomena are very complex and the exact mechanism is far from clear. It is, however, known that these poorly understood phenomena will affect very greatly the process characteristics in a flow system especially where there is a residence time distribution in the dispersed drops as demonstrated by Imoto and Lee (1967, 1970). Until recently the design of dispersed reactors has been based largely on empirical methods.

As stated before, when two immiscible liquids are agitated, a dispersion will form in which continuous dispersion and coalescence of droplets will occur simultaneously. Church and Shinnar (1961) and Komasawa (1970) have pointed out that if the agitation is continued over a sufficiently long time, a local dynamical balance between breakup and coalescence will be established; that is, the average diameter of liquid drops will not change with time and it will have some steady state average value. This average size of droplet at equilibrium will then depend on the operational conditions such as the angular velocity of the impeller or stirrer (see Church and Shinnar, 1961; Shinnar and Church, 1960; Shinnar, 1961; Komasawa, 1970). It is possible to predict from theoretical considerations the influence of the turbulent velocity fluctuations on both breakup and coalescence. Kolmogoroff (1949) has put forward the hypothesis that in any turbulent flow at sufficiently high Reynolds numbers the small-scale components of the turbulent velocity fluctuations are statistically independent of the main flow and of the turbulence-generating mechanism. He has defined a length scale (Kolmogoroff length) by

$$\eta = (\nu^3/\epsilon)^{1/4} \qquad (1)$$

where

η = Kolmogoroff length
ν = kinematic viscosity for continuous phase
ϵ = local rate of energy dissipation per unit mass of fluid

The Kolmogoroff theory of isotropic turbulence provides a tool for predicting drop sizes.

The case of breakup has been treated by Kolmogoroff (1949) and Hinze (1955, 1959). In their treatment a small volume of fluid is considered in which turbulence is assumed to be locally isotropic, and also the densities and viscosities of both liquids are assumed to be similar. All of the droplets in this volume are exposed to both inertial forces due to velocity fluctuations and to viscous shear forces. It is, however, assumed that the droplets are much larger than the Kolmogoroff length, η, so that viscous forces can be neglected. Using the dimensional argument both of them independently developed the same expression for the Weber number of breakup in turbulent flow in which the diameter of a droplet is much larger than η as

$$N_{\text{We}} = \frac{\rho_c(\bar{\epsilon})^{2/3} (d_p)_{\text{max}}^{5/3}}{\sigma} = \text{constant} \quad (2)$$

where ρ_c is the density of the continuous phase, σ the interfacial surface tension, $\bar{\epsilon}$ the average local rate of turbulent energy dissipation per unit mass of the fluid, and d_p the diameter of a droplet.

It has been shown experimentally by Rushton et al. (1950) that at a high Reynolds number ($N_{\text{Re}} = N_I D^2/\nu$) the energy input of the mixing impeller per unit mass of liquid in the vessel is independent of the properties of the liquid, and a function only of the geometrical design of the agitator and its speed. Rushton et al. (1950) have shown that for a turbine agitator in a completely baffled tank the average energy dissipation per unit mass is given by

$$\bar{\epsilon} = K N_I^3 D^2 \quad (3)$$

where K is a dimensionless constant which depends on the design of the agitator, N_I is the impeller speed (rpm) and D is the diameter of the agitator. Hence, the value of the maximum stable droplet diameter can be obtained by substituting Eq. (3) into Eq. (2) as

$$(d_p)_{\text{max}} = K_1 N_I^{-6/5} D^{-4/5} \rho_c^{-3/5} \sigma^{3/5} \quad (4)$$

In the above equation, $(d_p)_{\text{max}}$ is the upper limiting breakage value of a droplet and hence whenever drops have a larger value than that of $(d_p)_{\text{max}}$, there will always be breakup.

As stated before, Eq. (4) should apply only to cases in which $(d_p)_{\text{max}}$ is much larger than Kolmogoroff length η. However, if σ is very small or ν is rather large (e.g., some emulsions), the maximum stable droplet diameter will be smaller than η. In such a case the viscous shear forces cannot be neglected; hence one can get a similar expression for this case as obtained by Shinnar (1961) and also by Sprow (1967a)

$$(d_p)_{\max} = K_1' N_I^{-3/2} D^{-1} \sigma \nu_c^{1/2} \mu_c^{-1} f\left(\frac{\mu_d}{\mu_c}\right) \qquad (5)$$

where f is a certain function, ν the kinematic viscosity, and subscripts c and d refer to the continuous phase and the dispersed phase respectively. Equation (5) should describe the breakup of droplets in emulsions, whenever $(d_p)_{\max} < \eta$. However, the data of Vermeulen et al. (1955) on gas-liquid dispersions seem to obey Eq. (5) despite the fact the drops are larger than the Kolmogoroff length, η. Sprow (1967b) has pointed out that no previous investigator has correlated experimental results from a liquid-liquid system with an expression similar to Eq. (5).

The equation for prevention of coalescence was derived by Shinnar and Church (1960). They have shown that to prevent coalescence, kinetic energy of the droplets must be larger than the energy of adhesion between them. According to Shinnar and Church (1960) the minimum drop diameter for which separation is still possible for a locally isotropic flow in a given fluid can be obtained from the relationship

$$\frac{c_1 \, \rho_c(\bar{\epsilon})^{2/3} \, (d_p)_{\min}^{8/3}}{A(h_0)} = \text{constant} \qquad (6)$$

Substitution of Eq. (3) into the above equation gives the minimum stable droplet diameter as

$$(d_p)_{\min} = K_2 \, N_I^{-3/4} \, D^{-1/2} \, \rho_c^{-3/8} \, A(h_0)^{3/8} \qquad (7)$$

where K_2 is a constant and $A(h_0)$ is the necessary energy required to separate two drops of unit radius from an initial minimum distance h_0 to infinity. In Eq. (7), $(d_p)_{\min}$ is the lower limiting coalescence value of a droplet and hence whenever drops have a smaller diameter value than that of $(d_p)_{\min}$, there will always be coalescence. Shinnar (1961) has pointed out that the results of Rodger and his co-workers (1956) for geometrically similar systems are well correlated by Eq. (7). Equation (7) can be applied only when the viscous forces are negligible, that is, only when the droplet diameter is much larger than η. Sprow (1967b) has developed a similar expression for the region of viscous shear as

$$(d_p)_{\min} = K_2' \, N_I^{-3/4} \, D^{-1/2} \, \nu_c^{1/4} \, \mu_c^{-1/2} \, A(h_0)^{1/2} \qquad (8)$$

Sprow (1967b) has also pointed out that it would be quite difficult to distinguish between Eq. (7) for the inertial region and Eq. (8) for the viscous region from available data.

Figure 1 taken from Church and Shinnar (1961) or Shinnar (1961) plots the maximum stable droplet diameter [Eq. (4)] as determined by the process of breakup and the minimum stable droplet diameter [Eq. (7)] as determined by the process of coalescence against the agitator speed. This latter quantity is proportional to $(\bar{\epsilon})^{1/3}$. It can be seen from this figure that there is a region in which neither breakup nor coalescence occurs and dispersions which fall in this region are called by Church and Shinnar (1961) "turbulence-stabilized" dispersions.

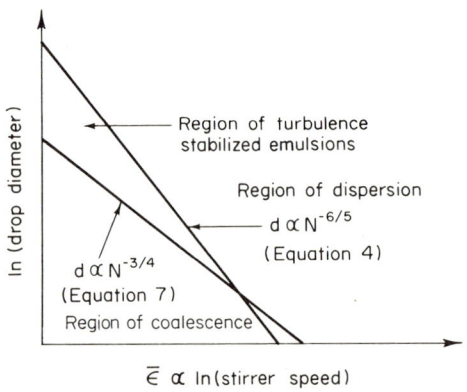

FIG. 1. Church-Shinnar plot (Church and Shinnar, 1961).

This theory of Church and Shinnar (1961) has proved its value but it has limitations as well. They have assumed that the turbulence is homogeneously distributed throughout the reactor containing the emulsion. The studies of Snell (1943), Vanderveen (1962), Weiss et al. (1962), and Groothuis and Zuiderweg (1964) have shown that this generally is not the case and that dispersion occurs in the vicinity of the impeller while coalescing occurs away from the impeller and also the coalescence occurs more frequently for the highly viscous drops.

Collins (1967) has published an excellent literature review on the dispersion of drops. A. H. Brown and Hanson (1966) have reviewed critically the mechanisms of coalescence phenomena. In their review on coalescence the effects of surface-active agents, impurities, etc. and the influence of high energy electric fields are discussed. Imoto and Lee's (1967, 1970) comprehensive reviews cover the dispersion-coalescence phenomena of liquid drops for the suspended polymerization reaction system. Belk (1965) has reported the effect of physical and chemical parameters on coalescence. Jeffreys and Hawksley (1965a,b) have also reported on the effect of physical properties on coalescence rate for hydrocarbon-water systems and theoretically analyzed these coalescence rates to verify experimental results. Furthermore, Jeffreys and Lawson (1965) have also discussed the effect of mass transfer on coalescence. Effects of various other factors such as geometry of flow path (porous media) (Jordan, 1965), surface diffusion (A. H. Miller et al., 1965), and electric charge (Waterman, 1965) on coalescence have been considered by other investigators.

Valentas et al. (1966) attempted to develop a mathematical model which relates a two-phase system to the distribution of droplet sizes, while Valentas and Amundson (1966) have developed a model to relate breakage and coalescence of droplets to steady state distribution of droplet sizes. Their model contains many mixing functions which have to be measured experimentally when the model is used to determine the effect of mixing functions on the drop-size distribution in the vessel. Measurement of each function in a separate specialized experiment would be a difficult task. However, Verhoff (1969) has showed that it is possible to simplify Valentas and Amundson's model by considering other spaces. These two works will be discussed in detail later in this work.

B. Simplified Mixing Models

Models to express the dispersion-coalescence phenomena of liquid drops are necessary. Rietema (1964) has discussed four basic models of mixing for the liquid-liquid continuously stirred flow system. In these models the dispersion-coalescence phenomena are expressed quantitatively by the dispersion-coalescence frequency, ω_i (time^{-1}), that is, the ratio of the volume of liquid drops which dispersed or coalesced per unit time to the total volume of the dispersed phase.

1. Homogeneous Mixing Model

Curl (1963) was the first to establish this type of model for chemical process with zero order reaction in a continuous stirred tank reactor. In his model, Curl (1963) has assumed that all drops have the same size. The chance of coalescence with a neighboring drop is the same for each and every drop and it is independent of time and of the concentration of the drop. Redispersion occurs immediately after coalescence of two drops to produce two equal drops of the same concentration. Furthermore, it has been assumed that coalescence and redispersion do not have any effect on mass transfer. The coalescence-redispersion frequency, ω_i, is defined by

$$\omega_i = 2\,\frac{pn_c}{N} \tag{9}$$

where p is the number of drops which coalesce per unit time [that is, collision frequency], n_c is the number of drops involved in one coalescence [that is, collision probability of the colloid liquid drops], and N is the total number of drops per unit volume.

The result of Curl has been confirmed by Veltkamp (1964) who has solved Curl's equations by means of Laplace transformation. Veltkamp and Geurts as cited by Rietema (1964) have also evaluated the case in which the order of drop conversion is 1/2. Spielman and Levenspiel (1965) calculated the influence of coalescence on the progress of reaction for this model using the Monte Carlo method on a digital computer.

2. Dead Corner Mixing Model

The walls of a stirred tank may be preferentially wetted by the dispersed phase liquid, and they may be covered with a thin stagnant layer of the dispersed phase which may act as a "dead" corner. However, a continuous coalescence of liquid drops with the stagnant layers and corners may still occur and the "dead" corners may continuously lose new drops which are taken up again in the "living" dispersed phase. In this case the dispersion-coalescence frequency, ω_i, is the ratio of the volume of the stagnant portion and volume of the liquid drops which disperse and coalesce per unit time to the total volume of the dispersed phase in the reactor. Rietema (1964) showed that this model is simpler to evaluate than the homogeneous mixing model.

3. Circulation Mixing Model

In a stirred reactor the energy dissipation coefficient is not usually homogeneous. The dissipation coefficient is larger near the impeller and it decreases with increasing distance from the impeller. Hence the diameter of the liquid drops will be smaller near the impeller and larger away from the impeller as demonstrated by Vanderveen (1962) and by Sprow (1967a). Many of the liquid drops will be dispersed in the region near the impeller while more coalescence will be found in the region away from the impeller. The drops will circulate between these two regions. In this case the dispersion-coalescence mechanism can no longer be described by one parameter but at least two parameters are needed. The dispersion-coalescence frequency, ω_i, is defined as the product of the number of circulations per unit time and the number of liquid drops which can coalesce during this circulation.

Harada et al. (1962) proposed a mixing model in which the dispersed liquid drops coalesce and redisperse as imaginary drops having the average concentration of the total dispersed drops. They further assumed that the mass transfer which continuously takes place is proportional to the average coalescence and redispersion velocity of the imaginary liquid drops.

The homogeneous mixing model has been used by many investigators to interpret their experimental results. Other models which take additional factors into consideration have not been as widely used as the homogeneous mixing model.

C. General Mixing Models

Valentas et al. (1966) developed a mathematical model which relates the breakage of droplets in a two-phase system to the distribution of droplet sizes. With the assumption of no coalescence, they showed that simultaneous mass and number balances on the completely mixed vessel lead to an integral equation of the Volterra type which involves the influent distribution function, the vessel distribution function, and a kernel describing the breakage mechanism. Valentas and Amundson (1966) extended such treatment to include both breakage and coalescence. In their extension the number of drops entering and leaving a differential mass range m to $m + dm$ in the vessel by breakage, coalescence, and flow is considered.

The droplet-size distributions in the influent and in the vessel itself are described by probability density functions as

$a(m,t)dm =$ fraction of droplets with mass between m and $m + dm$ at time t in the influent
$A(m,t)dm =$ fraction of droplets with mass between m and $m + dm$ at time t in the vessel

It is assumed here that effluent droplet-size distribution is the same as the vessel droplet distribution [assumption of "perfect mixing"]. When the droplets in the feed are distributed in size, an upper size limit, η_d, is set by the device used to disperse the feed. The effluent and the vessel droplet-size distributions are not limited by the maximum feed size, since coalescence may produce drops larger than η_d. However, a practical size limit, L, does occur and is dependent upon operating and geometrical parameters as pointed out by Kolmogoroff (1949), Hinze (1955, 1959), Shinnar and Church (1960), Shinnar (1961), and Church and Shinnar (1961).

A number balance on drops may be derived as follows:

Let
$\Phi(m,t)dm$ = rate at which droplets of mass between m and $m + dm$ are produced by breakage of larger drops
$\Lambda(m,t)dm$ = rate at which droplets of mass between m and $m + dm$ are produced by coalescence of smaller drops
$\Gamma(m,t)dm$ = rate at which droplets of mass between m and $m + dm$ are lost through breakage, coalescence, and escape
$N(t)$ = total number of droplets in the vessel at any time t
n_A = rate of incoming droplets

Then,

$$\frac{d[N(t)\,A(m,t)]}{dt} = n_A\,a(m,t) + \Phi(m,t) + \Lambda(m,t) - \Gamma(m,t) \quad (10)$$

The above equation becomes the steady state number balance equation if the left-hand side is equated to zero. Each term of Eq. (10) will be discussed separately below.

The first term on the right hand side is the number of drops which come into the mass range by flow which is determined by the number of input drops per unit time, n_A, and the input number distribution, $a(m,t)$, that is,

the number input by flow in the mass range $= n_A\,a(m,t)$ (11a)

The number output by flow from this mass range is found from the drop size distribution in the vessel, $N(t)\,A(m,t)$, and the escape frequency, $f(m)$, where

$f(m)$ = fraction of droplets of the vessel contents with mass between m and $m + dm$ flowing out per unit time

Furthermore, it is assumed that $f(m)$ is independent of all properties except droplet mass. Hence,

the number output by flow from the mass range $= N(t)\,A(m,t)\,f(m)$ (11b)

Next consider the breakage term $\Phi(m,t)$. The breakage process will not be characterized only by the breakage frequency, since it is also necessary to know the distribution of droplet sizes resulting from the breakage of a larger drop. Each droplet in the vessel has a breakage frequency, $g(m)$, which is dependent on droplet size and is defined as

$g(m)$ = fraction of droplets of the reactor contents, with mass between m and $m + dm$, disappearing through breakage per unit time

Upon breakage a droplet of mass M gives rise on the average to $\nu(m)$ smaller droplets whose sizes are distributed according to the probability density function $\beta(m:\mu)$ where

$\beta(m:\mu)d\mu$ = fractional number of droplets with mass between μ and $\mu + d\mu$ formed upon breakage of a droplet of mass m

By definition, $\beta(m:\mu) = 0$ when $m \leq \mu$. Hence,

the number input in the mass range by breakage
$= \phi(m,t)$
$$= \int_m^L \nu(m)\, g(\mu)\, \beta(\mu:m)\, N(t)\, A(\mu,t)\, d\mu \qquad (12a)$$

and

the number output from the mass range by breakage
$$= N(t)\, A(m,t)\, g(m) \qquad (12b)$$

The mathematical description of coalescence is explained in terms of the basic coalescence frequency, $F(m:\mu)$, where

$F(m,\mu)\, dm\, d\mu$ = rate at which droplets of mass between m and $m + dm$ and mass between μ and $\mu + d\mu$ are coalescing

It is assumed that the coalescence occurs only between droplet pairs and the coalescence frequency is symmetrical, that is,

$$F(m:\mu) = F(\mu:m)$$

and, to avoid counting various contributions twice, it is understood that $m \geq \mu$. The rate of formation of droplets of mass m from smaller droplets is computed from the basic coalescence frequency by considering the interaction of droplets of mass μ and $m - \mu$. Thus,

$$\Lambda(m,t)dm = \int_0^{m/2} F(m - \mu:\mu)d\mu\; d(m - \mu) \tag{13}$$

Assuming that $dm = d(m - \mu)$, one has

$$\Lambda(m,t) = \int_0^{m/2} F(m - \mu:\mu)d\mu \tag{14}$$

The upper limit of integration is dictated by symmetry, which requires that $m - \mu \geq \mu$. Similarly, the rate at which droplets of mass m are removed through coalescence with droplets of all sizes is given by

$$\Omega(m,t) = \int_0^{L-m} F(m:\mu)d\mu \tag{15}$$

The limit of integration in Eq. (15) is necessary to maintain consistency, since a finite upper size limit, L, has been placed on the droplet-size distribution.

Valentas and Amundson (1966) have obtained a mathematical expression for a basic coalescence frequency by considering the coalescence process in some respect analogous to a second-order chemical reaction. The coalescence frequency can be deduced by alluding to this mechanism as follows;

$$F(m:\mu) = \lambda(m,\mu)\; h(m)\; h(\mu)\; N(t)\; A(m)\; N(t)\; A(\mu) \tag{16}$$

where $h(m)$ and $h(\mu)$ are the characteristic collision frequencies and $\lambda(m,\mu)$ is the coalescence efficiency. Substitution of Eq. (16) into Eqs. (14) and (15) gives, respectively,

the number input in the mass range by coalescence
$= \Lambda(m,t)$
$$= \int_0^{m/2} [\lambda(m - \mu,\mu)\; h(m - \mu)\; h(\mu)\; N(t)\; A(m - \mu)\; N(t)\; A(\mu)]d\mu \tag{17a}$$

and

the number output from the mass range by coalescence
$= \Omega(m,t)$
$$= h(m)\; N(t)\; A(m) \int_0^{L-m} \lambda(m,\mu)\; h(\mu)\; N(t)\; A(\mu)\; d\mu \tag{17b}$$

Finally the term $\Gamma(m,t)$, which is the rate of loss of the droplets of mass in the range of m to $m + dm$ that are lost through breakage, coalescence, and escape, is given by combining Eqs. (11b), (12b), and (17b). Thus,

$$\Gamma(m,t) = N(t) A(m,t) [f(m) + g(m) + h(m) \omega(m)] \tag{18}$$

where

$$\omega(m,t) = \int_0^{L-m} \lambda(m,\mu) h(\mu) N(t) A(\mu) d\mu \tag{19}$$

Substitution of Eqs. (11a), (12a), (17a), and (18) into Eq. (10) gives an integrodifferential equation as the balance over an arbitrary mass range in the vessel:

$$\frac{d}{dt} [N(t) A(m,t)]$$

accumulation in mass range

$$= \underbrace{n_a\, a(m,t)}_{\text{flow into mass range}} + \underbrace{\int_m^L \nu(\mu) g(\mu) \beta(\mu:m) N(t) A(\mu,t) d\mu}_{\text{breakage into mass range}}$$

$$+ \underbrace{\int_0^{m/2} \lambda(m-\mu,\mu) h(m-\mu) N(t) A(m-\mu,t) h(\mu) N(t) A(\mu,t) d\mu}_{\text{coalescence into mass range}}$$

$$- [\underbrace{f(m)}_{\substack{\text{flow out of}\\\text{mass range}}} + \underbrace{g(m)}_{\substack{\text{breakage out of}\\\text{mass range}}} + \underbrace{h(m) \omega(m,t)}_{\substack{\text{coalescence out of}\\\text{mass range}}}] N(t) A(m,t) \tag{20}$$

Under steady state conditions in the absence of coalescence, Eq. (20) reduces to

$$NA(m) = \frac{n_a\, a(m)}{f(m) + g(m)} + \frac{1}{f(m) + g(m)} \int_m^L [\nu(\mu) g(\mu) \beta(\mu:m) NA(\mu)] d\mu \tag{21}$$

which is a Volterra integral equation for $NA(m)$. In the above development the reaction within the droplets is not considered and numerical solution is required since the analytical solution of Eq. (20) is not known.

If Eq. (20) is used to determine the effect of various choices of mixing functions on the drop-size distribution in the vessel, the functions h, λ, g, β, f and ν would have to be measured experimentally. Measurement of each function in a separate specialized experiment would be a difficult task. The number of these mixing functions, however, could be reduced by combining some of them to form more complicated functions. In particular $h(m)$, $h(\mu)$ and $\lambda(m,\mu)$ could be written as $\pi(m,\mu)$, and $\nu(\mu)$ and $\beta(m:\mu)$ could be written as $\theta(m,\mu)$. The function $f(m)$ can be considered a constant by assuming perfect mixing in the vessel. This would then reduce the number of functions to three, namely ν and θ for breakage, and π for coalescence.

Verhoff (1969) has shown that it is possible to simplify the expression of a number balance equation [Eq. (14)] further by considering other spaces, e.g., volume, area, or diameter distributions using volume, area, or diameter as the independent variables. Verhoff (1969) has developed a volume balance using drop volume as the independent variable. The use of a volume balance should simplify the kernel description of breakage because in a number description of breakage the number of drops is not conserved whereas in a volume description of breakage the volume is conserved. This may allow elimination of $\nu(m)$ in Eq. (14). However, coalescence is basically a number process and its description in a volume balance may become more complicated. Let

V = fluid volume of a continuous stirred tank
Q = input volumetric flow rate
ϕ_1 = dispersed phase ratio in the influent
ϕ_2 = dispersed phase ratio in the vessel
$p_o(v)dv$ = fraction of drops with volume between v and $v + dv$ in the influent
$p(v)dv$ = fraction of drops with volume between v and $v + dv$ in the vessel

It will be assumed that the effluent volume distribution is the same as the volume distribution [assumption of "perfect mixing"]; this is the same as the assumption that $f(m)$ is a constant in the Valentas model and is true in many real systems. Furthermore, it is assumed that the vessel is homogeneous, that is, that $p(v)$ applies approximately everywhere in the vessel and that the mixing kernels also are approximately the same everywhere. It is also assumed that the system is operating under steady state condition.

The volume balance is formed for an arbitrary drop volume range of v to $v + dv$ in the vessel. There are three ways (flow, breakage, and coalescence) whereby volume is transferred into and out of this range. Flow carries volume into and out of the vessel and thus into and out of a volume range in the vessel, whereas breakage and coalescence transfer volume between volume ranges in the vessel. Breakage shifts volume from larger volume ranges to smaller volume ranges, and coalescence transfers volume from smaller volumes to larger volumes. Figure 2 shows a schematic diagram of the system and also the processes involved in a balance in a volume range dv.

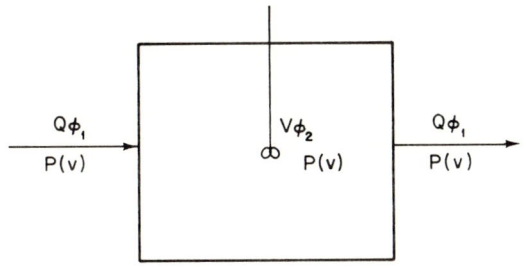

a. Schematic drawing of system

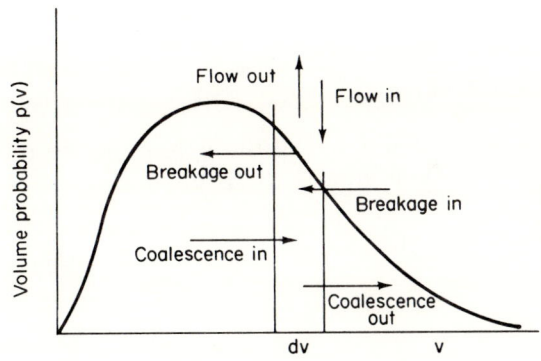

b. Processes involved in mixing

FIG. 2. Schematic diagram of system and also the schematic diagram for a differential volume dv.

First of all consider the flow terms. It can be seen that

the volume input into the volume range by flow $= Q\, \phi_1\, p_0(v)\, dv$ \quad (22a)

Since the system is operating at steady state and since there is no reaction in the vessel, the system volume and dispersed phase volume cannot change with time. Hence,

the volume output from the volume range by flow $= Q\, \phi_1 p(v)dv$ (22b)

Next consider the breakage phenomenon. Let $K_B(v',v)dv$ be the breakage function which can be defined as

$K_B(v',v)dv =$ the volume flux from the drop of volume range v' to $v' + dv'$ into the drop volume range v to $v + dv$ per unit time per unit volume of drop volume v'

By definition, $K_B(v',v) = 0$ whenever $v > v'$ because it is impossible for breakage of drop volume v' to cause a flux into a volume range larger than v'. Hence, the flux into a volume range v to $v + dv$ from a volume range v' to $v' + dv'$ by breakage is given by

$$V\, \phi_2\, K_B(v',v)\, p(v')dv'dv$$

When all the fluxes of volume range v' to $v' + dv'$ which are greater than v are summed, one yields

the flux input in the volume range by breakage $= V\, \phi_2 \int_{v'=v}^{v'=\infty} K_B(v',v)\, p(v)dv'dv$ (23a)

Similarly, the volume flux out of the volume range v to $v + dv$ and into the volume range v'' to $v'' + dv''$ is given by

the flux output from the volume range by breakage $= V\, \phi_2 \int_{v''=0}^{v''=v} K_B(v,v'')\, p(v)dv''dv$ (23b)

Finally one must discuss the coalescence phenomenon by assuming that only binary coalescence occurs in the vessel. Number probability is used in the derivation of coalescence terms because coalescence is basically a number process, that is, it occurs between two drops and not between the fractional volume of drops. Hence, the number probability function will be related to the volume probability function by

$$f_n(v)dv = \frac{\dfrac{p(v)dv}{v}}{\displaystyle\int_0^\infty \dfrac{p(v)dv}{v}} = \frac{p(v)dv}{vc_L} \quad (24)$$

where

$$c_L = \int_0^\infty \frac{p(v)dv}{v} \qquad (25)$$

is used to normalize the number probability function. The total number of drops in the size range v to $v + dv$ is then given by

$$n(v) = N\, f_n(v)dv \qquad (26)$$

where

$$N = \frac{\text{total number of}}{\text{drops in the vessel}} = \frac{V\,\phi_2}{\bar{v}} = V\,\phi_2\,c_L \qquad (27)$$

and

$$\bar{v} = \frac{\int_0^\infty v\, f_n(v)dv}{\int_0^\infty f_n(v)dv} = \frac{1}{c_L}$$

Substitution of Eqs. (24) and (27) into Eq. (26) gives

$$n(v) = V\,\phi_2\, \frac{p(v)dv}{v} \qquad (28)$$

Next defining $\gamma(v',v-v')$ to be the velocity coalescence cross-section product (the coalescence efficiency times the relative velocity collision cross-section product as defined in gas kinetic theory), it is possible to write the transfer of volume when two drops of sizes v' and $v-v'$ coalesce and give a drop of size v. The volume transferred into the size range v to $v + dv$ where $v = v' + (v - v')$ by this coalescence is given by

$$\frac{v\,\gamma(v',v-v')\,n(v')\,n(v-v')}{V} \qquad (29)$$

Substitution of Eq. (28) into Eq. (29) gives

$$V\,\phi_2\,v\,K_c(v',v-v')\,p(v')\,p(v-v')dv'\,d(v-v') \qquad (30)$$

where

$$K_c(v',v-v') = \text{coalescence kernel}$$

$$= \frac{\phi_2\,\gamma(v',v-v')}{v'(v-v')} \qquad (31)$$

Note that the coalescence kernel is symmetric, that is,

$$K_c(v',v-v') = K_c(v-v',v') \qquad (32)$$

This important property will be used later. The total flux into a volume range v to $v + dv$ is the integral overall v' and $v - v'$ such that $v = v' + (v - v')$. To evaluate this integral it will be assumed that $d(v - v')$ is approximately equal to dv. Hence,

the flux input into the volume range by coalescence $= V \phi_2 \int_{v'=0}^{v'=\frac{v}{2}} v \, K_c(v',v-v')p(v')p(v-v')dv'dv \qquad (33a)$

This integral is taken from 0 to $v/2$ instead of to v in order not to count each coalescence twice. The flux out of the volume range v to $v + dv$ by coalescence is the flux caused by the coalescence of drops in the interval with drops outside the interval. Hence, the total flux out from a volume range is given by

$$V \phi_2 \int_{v'=0}^{v'=\infty} v'' \, K_c(v,v')p(v)p(v')dv'dv$$

where

$$v'' = v + v'$$

Hence,

the flux output from the volume range by coalescence $= V \phi_2 \int_{v'=0}^{v'=\infty} v \, K_c(v,v')p(v)p(v')dv'dv$

$$+ V \phi_2 \int_{v'=0}^{v'=\infty} v' \, K_c(v,v')p(v)p(v')dv'dv \qquad (33b)$$

Since the system is operated at steady state, the sum of fluxes in must be equal to the sum of fluxes out. Hence, one can obtain, by means of Eqs. (22a), (22b), (23a), (23b), and (33b) the balance over an arbitrary volume range in the vessel as

$$W_r[p(v) - p_0(v)] = \int_v^\infty K_B(v',v)p(v')dv'$$

$$-\int_0^v K_B(v,v')p(v)dv'$$

$$+\int_0^{v/2} v\, K_c(v',v-v')p(v')p(v-v')dv'$$

$$-\int_0^\infty v\, K_c(v,v')p(v)p(v')dv'$$

$$-\int_0^\infty v'\, K_c(v,v')p(v)p(v')dv'$$

where

$$W_r = \frac{Q}{V}\frac{\phi_1}{\phi_2} \tag{34}$$

Since $K_B(v,v') = 0$ whenever $v' > v$ and also since $K_c(v,v') = K_c(v',v)$ [see Eq. (32)], the mixing kernel can be defined as follows:

$$K(v,v') = \begin{cases} K_B(v,v') & \text{when } v > v' \\ K_c(v,v') & \text{when } v \leq v' \end{cases} \tag{35}$$

This kernel is substituted into Eq. (34) to give rise to a final equation with one unknown two-dimensional kernel. Thus,

$$W_r[p(v) - p_0(v)] = \int_v^\infty K(v',v)p(v')dv'$$

$$-\int_0^v K(v,v')p(v)dv'$$

$$+\int_0^{v/2} v\, K(v',v-v')p(v-v')p(v')dv'$$

$$-\int_0^v (v+v')\, K(v',v)p(v)p(v')dv'$$

$$-\int_v^\infty (v+v')\, K(v,v')p(v')p(v)dv' \tag{36}$$

Verhoff (1969) showed that if by experiment the input and output distributions were known, it still would not be possible to determine the mixing kernel. Thus to determine this mixing kernel at least an interactory process must be added to the equation to give a bivariate analysis. This gives rise to a set of two-dimensional equations. Hence, a two-dimensional experiment should be performed in order to develop an expression for the mixing kernel. Moreover, the added quantity in the two-dimensional experiment must interact in the mixing process, e.g., reaction in the dispersed phase, mass transfer from the dispersed phase, or dye dispersion in the dispersed phase; the last is the easiest experimentally. Verhoff (1969) has developed several space balances for the two-dimensional experiment to determine the mixing kernel from the experimental data.

Considering a volume balance in an area $dvdc$ (that is, volume distributed over volume and concentration), in the volume-concentration space and assuming instantaneous mixing of the dye upon coalescence in a steady state continuous stirred tank system, Verhoff has obtained

$$W_r[p(v,c) - p_0(v,c)] = \int_v^L K(x,v)p(x,c)dx$$

$$- \int_0^v K(v,x)p(v,c)dx$$

$$+ \int_0^{v/2} \int_0^c \frac{v^2}{v-x} K(x,v-x)p(x,y)p\left(v-x, \frac{cv-xy}{v-x}\right)dydx$$

$$- \int_0^v \int_0^c (v+v')K(x,v)p(x,y)p(v,c)dydx$$

$$- \int_0^\infty \int_0^c (v+v')K(v,x)p(x,y)p(v,c)dydx \qquad (37)$$

Figure 3 shows how the coalescence and breakage processes redistribute volume in the volume-concentration space. Coalescence transfers volume to some average points in the concentration dimension and to higher values in the volume dimension. Breakage causes volume to be transferred only in the volume dimension and only to lower values.

Verhoff (1969) applied three techniques [least-square estimation, pseudomaximum likelihood estimation, and polynomial fit method] to estimate the mixing kernel for several experiments. The mixing kernels obtained by using the balance equation are still in question because the uniqueness question is unresolved and the resulting kernel values seem unreasonable physically.

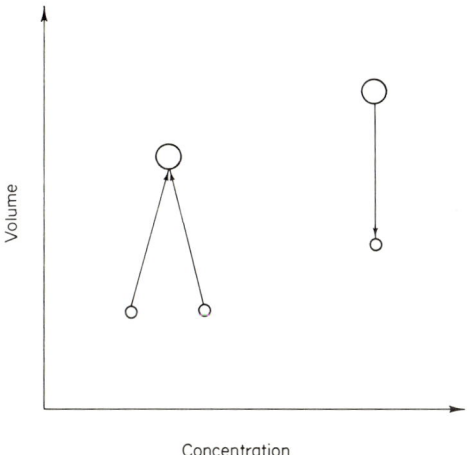

FIG. 3. Volume transfer in volume-concentration space.

D. MEASUREMENTS OF DISPERSION-COALESCENCE FREQUENCY AND DROP-SIZE DISTRIBUTION

1. Measurements of Dispersion-Coalescence Frequency

So far no definitive work has been published concerning the relationship between the dispersion-coalescence frequency, ω_i, and the various operational conditions. Madden and Damerell (1962) reported that the dispersion-coalescence frequency, ω_i, increases with the mixing velocity and with the holdup of the dispersed phase, ϕ, as shown in Fig. 4. According to the reports of different investigators the frequency, ω_i, is proportional to powers of N_I ranging from 1.5 to 3.3. Results reported include the 1.5 to 3.3 power (R. S. Miller et al., 1963), or the 1.9 to 2.25 power (Howarth, 1967), or the 2.4 power (Madden and Damerell, 1962), or the 2.5 power (Komasawa et al., 1969), or the 1.8 power (Komasawa, 1970) of N_I. The frequency ω_i is also said to be proportional to the 0.5 power (Madden and Damerell, 1962), or the 0.6 power (Howarth, 1967), or the 0.7 power (Komasawa et al., 1969), or the 1.1 power (R. S. Miller et al., 1963) of ϕ. Komasawa (1970) has derived similar relationships between ω_i and ϕ and between ω_i and N_I as

$$\omega_i = 84k\, P\, \phi^{0.87}\, N_I^{1.8} \quad \text{when} \quad 0.01 < \phi \leq 0.1$$
$$\omega_i = 61k\, P\, \phi^{0.73}\, N_I^{1.8} \quad \text{when} \quad 0.1 \leq \phi < 0.5$$
(38)

where

$$k = \frac{(D)^{11/5}}{(T^2H)^{1/3}(\gamma/\rho)^{2/5}} \qquad (39)$$

In the above expression D is the diameter of the impeller, T is the diameter of the stirred tank, H is the depth of liquid, γ is the surface tension, ρ is the coalescence probability function and ϕ is the volume fraction of dispersed phase.

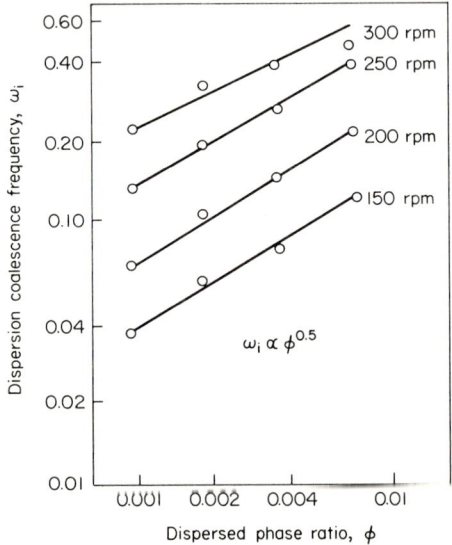

FIG. 4. The effect of holdup of the dispersed phase, ϕ, on the dispersion-coalescence frequency, ω_i (Madden and Damerell, 1962).

Experiments to measure the dispersion-coalescence frequency, ω_i, in a dispersed phase system have been carried out by Kramers and co-workers as reported by Rietema (1964) and also by many other investigators such as Shinnar (1961), Matsuzawa and Miyauchi (1961), Kintner et al. (1961), Madden and Damerell (1962), R. S. Miller et al. (1963), Hillestad (1964), Groothuis and Zuiderweg (1964), Matsuda (1966), Howarth (1967), Otake and Komasawa (1968), and Komasawa et al. (1969). In all these studies the homogeneous mixing model is used to interpret the results and care is taken to prevent wetting of the wall by the dispersed phase in order to avoid dead corners.

The experimental methods which have been used so far are summarized by Komasawa (1970) and are described, in brief, here.

a. A small amount of the continuous phase containing insoluble tracer dispersed phase is poured into the stirred tank which is already filled with large quantities of the continuous phase and the untracer dispersed phase. How fast the tracer disperses from the tracer dispersed phase to the continuous phase and to the untracer dispersed phase can then be observed. Some light absorbent material can be used as an insoluble tracer, so the variation of transmitivity can be observed with respect to time. Therefore this method is called transmitivity measurement method (see R. S. Miller *et al.*, 1963; Matsuda, 1966; Komasawa *et al.*, 1969).

b. Two different dispersed phases with the two different densities (the specific gravity of one is a little higher than that of the continuous phase and the specific gravity of the other is a little lighter than that of the continuous phase) are separately supplied to the stirred tank. The relationship between the ratio of the amount of the two dispersed phases supplied and the ratio of the amount of the two dispersed phases in an outlet can be observed (see Groothuis and Zuiderweg, 1964).

c. Mixing velocity is suddenly changed and then the variation in the diameter of liquid drops can be observed (see Howarth, 1967).

d. Instead of the insoluble tracer material described above in method a, some reacting material can be used and this reacting material will react with the unreacted dispersed phase which is already in the stirred tank and the reaction rate can then be observed [see Kramers' and co-workers' work reported by Rietema (1964), Madden and Damerell (1962), and Hillestad (1964)].

e. In a dispersed phase a reaction other than that of the first order is forced to react in a continuous stirred tank operation and reaction rate can then be observed (see Matsuzawa and Miyauchi, 1961; Otake and Komasawa, 1968).

The works of the different investigators on the experimental measurement of the dispersion-coalescence frequency, ω_i, in a dispersed phase system are summarized in Table I. Furthermore, Rietema (1964) and Komasawa (1970) have correlated various data according to the relationship between the frequency, ω_i, and the rate of dispersion energy of mixing, $p_v (= \epsilon \, g_c/\rho)$, as shown in Fig. 5. Komasawa (1970) has pointed out that the order of correlation differs greatly. The difference in diameter of liquid drops due to differences in physical properties, flow condition, etc., and also difference of geometrical shape alone could not explain the large differences in the measured values of ω_i. The variation in coalescence probability, P, due to the irregularity of the condition of liquid-liquid interface may be the main factor for the differences in measured ω_i. However, Rietema (1964)

TABLE I

SUMMARY OF THE MEASUREMENTS OF DISPERSION-COALESCENCE FREQUENCY, ω_i

Authors [reference] (year)	Experiment				Procedure	Remarks
	Measuring method[a]	Liquid of continuous phase	Liquid of dispersed phase	Type of impeller		
Shinnar (1961)	Direct microscopic method	Water and polyvinyl alcohol added	Molten microcrystalline wax	Paddle	Sampling was carried out by siphoning about 20 ml of liquid through a 10-mm glass tube into 400 ml of a cold aqueous solution of the colloid. The mean diameter of the droplets and the drop-size distribution were determined by counting and measuring 500 to 1000 particles under a microscope.	1. Negligible mixing for certain values of impeller speed and for some other speeds mixing rates were much higher. 2. Using the theory of isotropic turbulence, he explained the above phenomena and postulated the prevention of coalescence to be caused by the turbulent fluctuations. 3. Detail discussion is presented in Section II,A.
Matsuzawa and Miyauchi (1961)	e	Benzene	Water		Try to measure the frequency, ω_i, by means of a 2.5th order chemical reaction using $FeCl_3$ and $SnCl_2$ which takes place in the dispersed phase.	1. Qualitative conclusion was drawn that there is a negligible mixing and a reactor was very close to complete segregation condition ($\omega_i = 0$).

Madden and Damerell (1962)	d	Toluene and iodine added	Water	Turbine	A pure water dispersed into the iodine toluene solution. Then a small quantity of dispersed phase containing a high concentration of sodium thiosulfate was injected into the vessel. The frequency, ω_i, was determined by measuring the rate of disappearance of iodine in the continuous phase as it reacted with sodium thiosulfate in the dispersed phase.	The two components are mixed before entering in the vessel, so ω_i can only be measured indirectly from interpretation of the conversion-mean residence time curve 1. The mass transfer from the continuous phase to the dispersed phase or vice versa may induce Marangoni effects which may increase or decrease the frequency ω_i. 2. Good for small frequency range only. 3. ω_i is measured only for the aqueous dispersed phase because of the requirement that the chemical reaction be nearly instantaneous. 4. Interfacial tension may not be the same for all drops and hence the drop size may depend on the amount and type of reactants which the drop contains.

(Continued)

TABLE I (Continued)

Authors [reference] (year)	Measuring method[a]	Experiment		Type of impeller	Procedure	Remarks
		Liquid of continuous phase	Liquid of dispersed phase			
R. S. Miller et al. (1963)	a	Water	Oil	Turbine	Measurements of the dispersed phase mixing rate were made by a light transmission technique. A small quantity of highly colored drops was added to a stirred vessel containing uncolored or weakly colored drops dispersed in the continuous phase.	1. The plot of light transmission versus time obtained from the experiment was interpreted using homogeneous mixing model. 2. This method can be used with the aqueous phase or the organic phase dispersed.
Kramers and co-workers (see Rietema, 1964)	d	Toluene + carbon tetrachloride	Water	Paddle	Two types of tracer reactions were considered. In the first type dispersed phase contained sulfuric acid and potassium iodate while in the second type dispersed phase contained ferrocyanide with few drops containing ferric chloride. The course of reaction was followed by measuring the light absorption.	1. Remarks 1, 2, and 3 of Madden and Damerell also apply in this case. 2. This work was reported by Rietema (1964). 3. The results are plotted in Fig. 5.

Groothuis and Zuiderweg (1964)	b	Water	Toluene + carbon tetrachloride	Turbine	Two streams of dispersed phase of different densities fed to the vessel. The density of lighter stream was adjusted such that if a drop from this stream coalesced with any drop of density heavier than the water the resultant drop would be heavier than water. The output stream was separated into two fractions of dispersed phase, one heavier and the other lighter than water. The mixing rate was deduced from the change in dispersed phase fraction heavier than water as it passes through the vessel.	1. This method can in principle be used also with both phases dispersed, although their results are obtained with an organic phase dispersed.
Hillestad (1964)	d	Organic material	Water	Turbine	Measurements were made by light-transmission technique. Experimental technique was similar to that of R. S. Miller et al. (1963). Instead of following the spread of dye on the vessel, he followed the course of the ferric chloride–potassium ferricyanide reaction in aqueous dispersed drops.	

(Continued)

TABLE I (*Continued*)

Authors [reference] (year)	Experiment				Remarks
	Measuring method*	Liquid of continuous phase	Liquid of dispersed phase	Type of impeller	Procedure	

Authors [reference] (year)	Measuring method*	Liquid of continuous phase	Liquid of dispersed phase	Type of impeller	Procedure	Remarks
Matsuda (1966)	a	Nitrotoluene	Concentrated H_2SO_4	Paddle	—	1. His results reported by Komasawa (1970). 2. His results are plotted in Fig. 5.
Howarth (1967)	c	Water	Toluene + carbon tetrachloride	Turbine	Dispersion-coalescence frequency measured by measuring the change of interfacial area when the stirring speed has been changed instantaneously from one value to a lower value.	1. It is assumed that the turbulence is isotropic and the decay of turbulence in the vessel is instantaneous. 2. Homogeneous mixing model is used to develop a relationship between ω_i as the dependent variable and average drop size and rate of change in interfacial area per unit time as two independent variables.

Komasawa et al. (1969)	a	Water + salt added	Benzene + carbon tetrachloride	Turbine	—	3. He also found that the addition of small quantities of simple electrolytes to the continuous phase reduced frequency ω_i remarkably.
						1. His results were reported by Komasawa (1970).
						2. The results are plotted in Fig. 5.
Komasawa et al. (1969)	a	Oil (i-C_8H_{18} + CCl_4)	Potassium sulfate solution	Turbine	Transmission technique was used for the measurements.	1. Results are plotted in Fig. 5.

*Letters in this column, a–e, refer to the methods described in Section II,D,1.

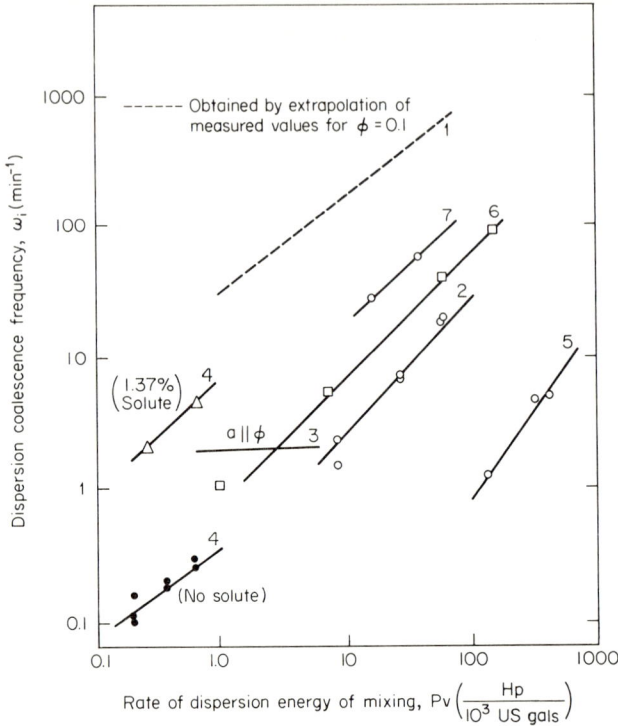

Fig. 5. The dispersion-coalescence frequency versus the rate of dispersion energy of mixing for $\phi = 0.1$ (Rietema, 1964; Komasawa, 1970). 1. Madden and Damerell (1962). 2. R. S. Miller et al. (1963). 3. Rietema (1964). 4. Groothuis and Zuiderweg (1964). 5. Matsuda (1966). 6. Komasawa et al. (1969). 7. Komasawa et al. (1972).

has pointed out that even though the results obtained by R. S. Miller et al. (1963) and those of Groothuis and Zuiderweg (1964) for the case of no solute seem to fit remarkably well on one correlating curve, the total picture is very complicated. Much remains to be done before the frequency, ω_i, can be predicted with reasonable accuracy for a given phase system at given operating conditions. The following general conclusions have been drawn by Rietema (1964) with some caution:

(i) ω_i increases with the mixing velocity (that is, with power input) and with the dispersed phase ratio ϕ.

(ii) Mass transfer from the drops to the continuous phase seems to increase ω_i strongly.

(iii) When the dispersed phase is an organic liquid, ω_i is much lower than when the dispersed phase is an aqueous liquid.

(iv) Both Madden and Damerell (1962) and R. S. Miller *et al.* (1963) found that addition of inorganic salts to the aqueous phase gives more reproducible results.

(v) For commercial processes ω_i is neither infinity nor zero and so ω_i is a factor which is of high importance in reactor design to obtain both a high degree of conversion and a good selectivity.

2. *Measurements of Drop-Size Distributions*

Many engineering operations depend on effective mass transport across liquid-liquid interfaces. This transport is directly proportional to the interfacial area which is strongly affected by the distribution of drop sizes in the vessel. Both breakup and coalescence play an important role in determining the observed drop-size distributions. Hence, the drop-size distribution is one of the most difficult properties of a dispersion to predict theoretically or to measure experimentally.

The measurements of drop-size distributions by microscopic inspection, by Coulter counters, by light transmitting or light scattering, or by photographic techniques are the primary physical measurements made on the dispersed system by many investigators. But no one method has been found which allows the rapid accurate determination of drop-size distributions containing a wide range of diameters.

An excellent review on the measurement of drop-size distributions is given by Collins (1967). In this section his review (in brief), in addition to other works, will be summarized.

A typical example of the determination of drop-size distributions by direct microscopic observation is given by Shinnar (1961). His experiment involved mixing Shellwax in a heated vessel. A sample of the dispersion is removed from the vessel and cooled quickly. The solidified wax drops were then measured under a microscope. For each distribution presented, he counted approximately 1000 drops. Other methods, such as coating the outside of drops with a polymer, have been suggested by Madden and McCoy (1964) for obtaining drop-size distributions by microscopic inspection.

A Coulter counter was used by Sprow (1967b) to measure drop-size distributions of an emulsion of methyl isobutyl ketone in salt water. This device measures drop sizes by measuring the change in electrical resistance through a small aperture as the nonconducting drop passes through it. The change in resistance is related to drop size by calibration.

Langloise *et al.* (1954) used a light transmittance technique to measure the sauter mean diameter, D_{32}, of a coarse emulsion generated in a stirred tank. Rodger *et al.* (1956) designed a light transmittance probe and measured D_{32} within 10% of the value obtained from photographs of a dispersion in a stirred tank containing drops with diameters ranging from 50 to 2000 microns. Scott et al. (1958) measured D_{32} downstream of an orifice mixer by the same method. Cengle et al. (1961) studied the effects of flow rate, mixing time, and dispersed phase concentration on light transmittance. Lloyd (1959) determined average diameters of colored emulsions by reflectance.

Light scattering was used by Lindsey et al. (1964) to measure drop-size distribution in a batch stirred vessel. The drop-size distribution is related to a plot of the product of light intensity times scattering angle squared versus the scattering angle. To obtain the drop-size distribution itself, a differential analysis of the intensity versus scattering angle curve must be performed. For results to be valid the drops must range in size from 1 to 50 microns.

Direct photography of a flowing liquid-liquid dispersion has been employed by many investigators. Kintner et al. (1961) presented a review of many photographic techniques which had been adapted to bubble and drop research. R. A. S. Brown and Govier (1961) used a high speed motion picture camera to study the motion of large (diameter 6,000 to 12,000 microns) oil drops in water flowing in a tube at low flow rates. Sleicher (1962) took both still pictures and high speed motion pictures of drops (diameter 2,000 to 8,000 microns) at very low concentrations. Scott et al. (1958) photographed water in kerosene dispersion (diameter 20 to 250 microns). Ward (1964) developed a method which allowed him to photograph drops (diameter 1 to 800 microns) in a dispersed phase concentration up to 50% by volume while they were flowing at velocities up to 16 feet per second. H. T. Chen and Middleman (1967) photographed drops in the tank with the camera focused on a plane midway between the tank wall and its axis. The drops were measured directly from the negatives using a 32-power traveling microscope. The system was checked by measuring a known distribution of glass beads in the vessel. Collins (1967) also used the photographic technique to measure the drop-size distribution which resulted from the action of the turbulent field produced by pipe flow of the liquids. Recently, Yoshida and Yamada (1971) determined the average size of oil droplets in connection with fermentation of hydrocarbons (dispersion of kerosene in water) by means of microscopic photography. The droplets ranged from 10 to 500 microns.

All of the above methods have basic limitations. They all work best at low phase fraction. The light transmitting, light scattering, and Coulter counter methods operate best with very small drops. The photographic and microscopic inspection techniques are very time consuming. Finally none of the above methods allows another variable such as concentration of dye in the drop to be measured.

Adler *et al.* (1954) have tried to eliminate the long tedious effort needed to obtain drop-size distributions from photographs by using the sweep of a narrow light beam and a photo cell to measure drop-size distribution from photographic negatives. Recently Verhoff (1969) has developed a method for the measurement of the bivariate drop-volume-dye-concentration distribution in a dispersed phase mixing vessel. The method involves extracting a sample of the dispersed phase from the vessel and immediately coating it with an anticoalescence agent. The drops are then forced at a constant flow rate through a small capillary. A logarithmic photometer focused on the capillary through a microscope produces an output of rectangular pulses. The length of each pulse is a measure of the volume of the drop and the height of each pulse is a measure of dye concentration in the drop. The analysis of the photometer output is performed easily and quickly on the computer. Verhoff (1969) has also compared the drop-size distributions in the vessel for the experiments with each other and with the log normal distribution.

The application of the traditional determination methods (Becher, 1965) in characterizing the emulsions, that is, measurement of absorption and light dispersion, electronic particle counting and sizing, sedimentation analysis is greatly hindered by the exceptional conditions existing during the fermentation process. The fineness and stability of the emulsion, number of cells and air bubbles, viscosity, etc., are strongly influenced by the stage of culture development. However, Katinger *et al.* (1970) have pointed out that the microscopic method enables the direct determination of the drop-size distribution in the presence of these disturbing factors. In their modified method the sample of agitated culture medium is drained directly into a gelatine solution at 30°C and cooled at once in ice water until the gelatine has hardened. The high viscosity and the stabilizing effect of the gelatine hinders the emulsified drops from coalescing. Then the pieces are cut from the gel at different places and transferred onto microscopic slides. Particle sizes are determined microphotographically to obtain particle-size distribution, interfacial area, and other parameters of the emulsions. Prokop *et al.* (1971) have used this modified method of Katinger *et al.* (1970) to estimate the drop-size distribution of oil drops for the system which is composed of *n*-hexadecane dissolved in dewaxed gas oil. They used the size distribution to calculate the interfacial area between the dispersed and continuous phases.

E. Coalescence Phenomena of Liquid Drops

In order to understand the effect of coalescence phenomena, it is obviously necessary to first have some understanding of the fundamental mechanism involved. Much of the considerable research effort in this field has been concentrated on the coalescence of drops at an interface and not with coalescence between adjacent drops. Cockbain and McRoberts (1953) were the first to study the coalescence of a drop with a flat interface. The work of many other investigators such as Gillespie and Rideal (1956), Elton and Picknett (1957), Linton and Sutherland (1958), Charles and Mason (1960a,b), MacKay and Mason (1961), and Lang (1962) and many more followed. The general picture is that when a drop approaches a flat liquid interface some deformation of the drop and the interface occurs. Deformation depends on the force with which the two are pressed together. As long as no coalescence has occurred then the two are separated by a film of the other liquid and this film is continuously thinning with time. After some time the thinning film breaks at some place and the content of the drop flows into the continuous phase. Charles and Mason (1960a,b) made an extensive study of partial coalescence and produced some quite remarkable high speed photographs of the phenomena.

Physical conditions have a considerable influence on droplet stability. Gillespie and Rideal (1956), Linton and Sutherland (1958), Nielsen *et al.* (1958), and Charles and Mason (1960b) have pointed out that the stability decreases with increasing temperature and with decreasing viscosity ratio of the component phases. One effect of temperature is to decrease the viscosity of the film separating the drop from the interface and thus accelerate drainage. Increase in density differences between the phases will increase the gravitational force exerted by the drop at the interface causing film drainage. Cockbain and McRoberts (1953) have pointed out that the time necessary for coalescence will be affected by temperature, drop size, surface active agents, impurities that collect at the interface, etc.

The materials which are absorbed even in small amounts (impurities) at the interface affect the liquid-liquid phase phenomena greatly and the existence of these materials inhibits coalescence. There are two kinds of inhibitors; the first inhibits by repulsion of electric double layers around the liquid drops which prevent their direct contact and the second inhibits the breakage of the liquid film by forming an absorption layer and a hydration layer around the liquid drops. Known ionized surface activators, high polymers such as polyvinyl alcohol (PVA) and methylcellulose, large molecule electrolytes such as carboxylmethylcellulose (CMC), strong electrolytes such as sodium chloride are the typical inhibitors.

The effect of electrolyte on dispersion-coalescence frequency was studied by Howarth (1967). They showed that a small amount of electrolyte stabilizes the liquid drops remarkably. The rate of stabilization decreases with increasing electrolyte. The effects of large molecules and surface activators on the dispersion coalescence have not been directly measured yet. However, Church and Shinnar (1961), Shinnar (1961), and Shinnar and Church (1960) reported that the addition of 0.1% PVA to the continuous phase prevents the coalescence completely and Sleicher (1962) also reported that the addition of 0.01% of CMC prevents the coalescence of the liquid drops in the turbulent flow. R. S. Miller et al. (1963) and also Komasawa et al. (1969) reported that in the measurement of frequency, ω_i, by the transmitivity method the addition of 0.05 gm mole/liter of sodium chloride to the continuous phase improved considerably the reproducibility of the measurements. This improvement may be the result of the artificial stabilization of liquid drops by addition of the electrolyte and hence canceling the effect of contamination which may differ from measurement to measurement.

Groothuis and Zuiderweg (1964) showed that when mass transfer takes place from the dispersed phase to the continuous phase an increased coalescence rate is observed. The separating film between the drops more rapidly approaches equilibrium than the remaining part of the continuous phase. Hence when mass transfer takes place from the dispersed phase to the continuous phase, this separating film will have a higher concentration of the transferring component. Since the interfacial tension between the two phases in equilibrium generally decreases at a higher concentration of the mutually soluble transferring component, the interfacial tension in the separating film will be smaller than in the remaining part of the drop interface. The gradient of interfacial tension which thus arises induces a liquid flow in the direction of the gradient and liquid is drawn away from the separating film. The separating film becomes thinner more rapidly, which ultimately results in an increased coalescence rate. On the other hand, when the solution transfers from the continuous phase to the liquid drops, the coalescence of liquid drops is hindered because the drainage of the film component in the continuous phase is retarded.

The theory of Groothuis and Zuiderweg (1964) has been confirmed for drops coalescing on a flat interface (MacKay and Mason, 1961), and for pairs of drops rising in an extraction column (Smith et al., 1963). Jeffreys and Lawson (1965) carried out a study of the benzene-water-acetone system. Their results are in good general agreement with the finding of Groothuis and Zuiderweg (1964). Finally, from Fig. 5 it can be seen that the dispersion-coalescence frequency, ω_i, of Madden's and Damerell's (1962) curve is much higher than that of Komasawa's curve (1972). This is true since in the former case the mass transfer was considered in the measurement.

III. Mathematical Modeling

Biological growth processes are complex processes which involve substrate transfer through the liquid medium and cell membrane and also involve a series of enzymatic reactions. The models described in the previous section are only for the dispersion-coalescence phenomena of drops and no reaction terms are included. These dispersion-coalescence models described in the previous section must be modified when biochemical reactions are present.

A model of biological growth with two liquid phases present should include the kinetics of the reaction, the transport of organisms between the phases, the transport of substrate from one phase to another, heat transfer, the distribution of microorganisms and substrate between the phases, the size distribution of dispersed drops, and the dispersion-coalescence phenomena of dispersed drops. It is very difficult to include all of the above phenomena in a model. It is inevitable that some simplifying assumptions be made if one wishes to obtain a useful or working model.

Recently Johnson (1964), Humphrey (1967), Munk et al. (1966), Dostalek et al. (1968), and Erdtsieck and Rietema (1969) have considered the adsorption of oil drops and also the distribution of microorganisms and substrate between the phases in their work. They have concluded that these factors definitely affect the growth of microorganisms. However, recent evidence by Yoshida et al. (1971) indicates that substrate transport through the continuous phase plays an important role. Laine et al. (1967) have suggested that one should increase the surface area of oil drops by increasing agitation, since growth rate may be limited by the ratio of surface area of oil drops to cell mass. Aiba et al. (1969a,b) reported the effect of rotation speed of impeller on specific growth rate for the cultivation of *Candida guilliermondii* (Y-7) in batch fermentation. They have assumed that with very small accommodated oil drops, cell growth follows the Monod equation. Recently, Aiba and Haung (1970) indicated that the saturation constant, K_s, is very important in the design and operation of fermentors with two liquid phases. Results presented by Mimura (1970) show the effect of the ratio of substrate concentration to cell concentration, (S_0/X_0), on the macroscopically observed specific growth rate.

Some simple models to predict growth behavior in batch processes when two liquid phases exist were studied by Erickson *et al.* (1969, 1970; Erickson and Humphrey, 1969a,b) and by Fan *et al.* (1969). Erickson *et al.* (1969) have developed mathematical models which can be used to describe batch growth in fermentations with two liquid phases present in which the growth limiting substrate is dissolved in the dispersed phase. They have considered the possibilities of growth occurring at the surface of the dispersed phase and also in the continuous phase. They have considered three cases: the first case assumes that all growth occurs at the surface of the dispersed phase, and the second and third cases assume that growth occurs both at the interface and in the continuous phase. The second case also assumes that substrate equilibrium is continuously established between the two phases while the third case assumes that the substrate concentration in the continuous phase is limited by the rate of transport of substrate to that phase. The relatively long region of linear growth assumed with these models is quite different from the relatively long region of exponential growth which characterizes most other batch fermentations. This region of linear growth is probably due to surface area limited growth since both the growth at the drop surface and the transport of substrate to the continuous phase depend on the surface area. If the surface area is relatively constant, the batch growth rate is constant and proportional to the rate with which the substrate is utilized by the cells.

Erickson and Humphrey (1969a) have extended their models to systems in which the dispersed phase is pure substrate and there is a decrease in the interfacial area due to substrate consumption. Once again the same three cases have been examined by them. According to the experimental results considered by them, the concept of a growth rate proportional to the surface area of the dispersed liquid phase may be useful in explaining the rates of growth on *n*-alkanes. The surface area may limit growth either by allowing only a fraction of the total cell population to be at a drop surface or by limiting the rate of substrate transport to the continuous phase or both.

Erickson *et al.* (1969; Erickson and Humphrey, 1969a) have used these simple models to examine the effects of the interfacial area, substrate phase equilibrium, saturation constant in the kinetic equations, and mass transfer coefficient on the growth behavior in batch fermentation processes after making many simplifying assumptions. Recently, however, a more realistic and more adequate model has been developed by Erickson *et al.* (1970) to represent batch growth in fermentations with two liquid phases present in which the growth-limiting substrate is dissolved in the dispersed phase. This new model takes into account the drop-size distribution, the rate of adsorption of cells on the drop surface, rate of desorption of cells from the drop surface, substrate transport between phases, phase equilibrium, and growth kinetics. It also considers the effect of coalescence and redispersion of oil drops in the system. They have used a discrete uniform distribution and a discrete normal distribution which has been obtained from an experimental distribution curve as drop-size distributions. They have examined limiting cases and also investigated the relative importance of each parameter encountered in the model. They have reported that the adsorption parameter can greatly affect growth in fermentors. A much greater lag period is obtained for a small value of the adsorption parameter as compared to a larger value of this parameter because of the slower rate of adsorption of cells onto the drop surface where growth occurs. A lag period in batch growth studies with gas oil fermentations has been reported by Munk et al. (1966) and by Dostalek et al. (1968), and it may be that the lag period is at least partially caused by the time required for the cells to be adsorbed onto the drop surface. In this work it has also been found that there is an effect of drop-size distribution on batch growth. When all drops are of the same size, growth continues on all drops until all of the substrate is exhausted. However, when a drop-size distribution is present and there is little or no coalescence and redispersion, the larger drops require more time for their substrate to be consumed than the smaller drops and thus the time required to complete the fermentation is increased. It was also found that the coalescence and redispersion parameter has a greater effect on the growth for uniform drop-size distribution than for normal drop-size distribution for a given set of other parameters. This occurs because the variance of the size distribution for uniform distribution is greater than that for normal distribution, and it is the size distribution which causes the concentration distribution. It should be noted that once the mixing kernel from the previous section is found, it can be used to estimate the coalescence and dispersion-frequency coefficient in the model by Erickson et al. (1970).

Dunn (1968) has reported an interfacial kinetics model for hydrocarbon oxidation. His communication is directed toward distinguishing between growth at the oil-water interface and that supported by trace quantities of hydrocarbon dissolved in water. He has pointed out the way by which this problem can be experimentally investigated; however, no experimental results have been presented.

Recently Prokop et al. (1971) have investigated experimentally the effects of inoculum size, dispersed phase volume, and substrate concentration on the batch growth of *Candida lipolytica* in a model system composed of n-hexadecane dissolved in a dewaxed gas oil. They have performed sixteen different experiments and found that all of the batch growth curves exhibited a linear growth region with the length of the region ranging from 1.5 to 9.5 hours. Furthermore, they have showed that the rate of linear growth varied both with the change in the dispersed phase volume and with the initial dispersed phase substrate concentration. Although they have presented only a qualitative analysis of their results, a quantitative effort to use the results of their experiments to estimate the parameters in the models presented previously by Erickson et al. (1970) is in progress. They have pointed out that the simple model of all growth occurring only at drop surfaces (see Erickson et al., 1969) with uniform dispersed phase substrate concentration is not adequate. The increase in linear growth rate with increasing values of initial substrate concentration are not predicted by this simple model. They have also pointed out that either continuous phase growth or growth on small segregated drops or both contribute significantly to the linear growth rate especially when the dispersed phase substrate concentration is large.

Recent experimental results presented by Fiechter (1971) for growth on n-alkanes show the importance of mixing in hydrocarbon fermentations. By improving mixing and drop breakage he was able to improve substrate transport to the cells and reduce the time required to complete a batch fermentation. The importance of mixing and substrate transport has also been discussed recently by Humphrey and Erickson (1971).

Very little experimental data on continuous fermentation processes with two liquid phases present have appeared in the literature (Munk et al., 1969; Dostalek et al., 1968). However, it is important to examine the behavior of continuous systems because of the widespread interest in the fermentation of hydrocarbons. Erickson and Humphrey (1969b) investigated the implication of the kinetic models for a continuous cultivation in a chemostat by using models introduced by them previously (Erickson et al., 1969; Erickson and Humphrey, 1969a) along with a complete mixing flow model. They have considered two types of dispersed systems. The first type assumes that the composition of the dispersed phase is such that increased substrate utilization results in a decreased substrate concentration with no change in the interfacial area. In the second type of system, the dispersed phase is assumed to be pure substrate;

therefore, the substrate concentration in the dispersed phase remains constant but the interfacial area is affected by changes in dilution rate. Once again the three special cases were examined for each type of system and the effects of the interfacial area, phase equilibrium constant, and mass transfer coefficient on system performance were studied. Their results indicate that the interfacial area is an important parameter which must be considered in design.

Aiba and Huang (1971) have investigated the growth of yeast cells on n-alkanes in a continuous fermentor at several agitation rates. They analyzed their results using the general mixing model of Valentas and Amundson (1966), the drop-size distribution work of H. T. Chen and Middleman (1967), and the accommodation concept used previously by Aiba *et al.* (1969a,b). They assumed the net production rate of accommodated drops to be proportional to the interfacial area to the δ power and then found values of the proportionality constant and δ which fit their experimental data.

In continuous flow chemical reactors the conversion and selectivity are influenced by mixing, and therefore the effect of mixing on this process must be studied. Hence, in recent years the significance of mixing in designing fermentors, biological waste treatment systems, and other process systems in which biological growth occurs has begun to be appreciated. The mixing within continuous flow systems, in general, can be described in terms of two components, macromixing and micromixing (see Danckwerts, 1953, 1958; Kramers, 1958; Zwietering, 1959; Douglas, 1964; Ng, 1965; Weinstein and Adler, 1967; Rippin, 1967). The macromixing component specifies the variation in the residence time experienced by molecules flowing through a system and is completely given by the residence time distribution. The micromixing component specifies the details of mixing, that is, the variation of environment experienced by the molecules during their passage through the system.

Many types of continuous reactor systems that can be used for fermentation have been summarized by Herbert (1961). Reusser (1961) has discussed the use of a series of completely mixed reactors and the use of a plug flow reactor for the production of novobiocin. Grieves *et al.* (1964) have adopted the mixing model of Cholette and Cloutier (1959) for modeling the activated sludge process. Bischoff (1966) has considered optimal design of the continuous fermentation reactor and concluded that a system in which a continuous stirred tank reactor is followed in series by a plug flow reactor is the best. Hansford and Humphrey (1966) have shown that segregation effects may affect the yield in yeast fermentation.

Fan et al. (1970a, 1971a, 1972), Erickson et al. (1968), and G. K. C. Chen (1971) have investigated a number of problems related to the modeling and optimal design of biochemical reactor systems with longitudinal mixing. The flow model used by them is the axial dispersion model which is characterized by the Peclet number. A computational approach, which involves collocation points, is introduced by Fan et al. (1971a) to obtain an approximate solution of an arbitrary degree for a highly nonlinear unsymmetrical boundary value problem. Furthermore, using the diffusion (axial dispersion) model in conjunction with the Monod growth kinetics, they have shown that to maintain partial (incomplete) fluid mixing is often desirable for biochemical processes. Under sterile feed conditions, the optimal degree of fluid mixing, that is, the optimal Peclet number, appears to range between 0 and 2 with the exact value depending on the values of the kinetic constants and initial substrate concentration. The axial dispersion (longitudinal mixing) which occurs in the aeration chamber of the activated sludge system is investigated to determine the effects of axial dispersion on system performance. An executive type computer-aided program has been developed by Fan et al. (1972) for the simulation and optimal design of biological waste treatment processes. The diffusion model, cost functions which can be easily updated, and a pattern search technique for determining the optimal design conditions are features of this program.

Tsai et al. (1969) and Tsai (1970) have shown the importance of micromixing in growth processes. They have considered the effect of micromixing on microbial growth processes in an isothermal continuous stirred tank reactor and also in a plug flow reactor. It was concluded by them that the effect of micromixing on a growth process is appreciable and that segregation is unfavorable to the growth process. Furthermore, Fan et al. (1971b) have extensively studied the simultaneous effect of macromixing and micromixing on growth processes in flow reactors. In this work they have considered two general types of flow systems which represent a wide variety of macromixing states. These flow systems are the system of n continuous stirred tank reactors in series and the series combination of a continuous stirred tank reactor and a plug flow reactor. In the same work they have considered two additional states of micromixing besides the four micromixing states treated by Zwietering (1959) by considering a system consisting of two continuous stirred tank reactors in series as an empirical model of a flow system. Similarly they have also considered an additional micromixing state besides the three micromixing states proposed by Kramers (1958) by considering a continuous stirred tank reactor in series followed by a plug flow reactor as an empirical model of a flow system.

Tsai (1970) has successfully applied the two-environment model of micromixing proposed by Ng and Rippin (1964), by Ng (1965), and by Rippin (1967) to the growth processes which take place in a chemostat. For this model, the critical washout dilution rate does not appear to be affected by the degree of micromixing. Fan et al. (1970b) have also considered separately the effect of mixing on washout and the steady state performance of continuous culture. Finally but not least this group (Tsai et al., 1971) have applied a reversed two-environment model of micromixing to growth processes by assuming the feed enters a maximum mixedness environment and then passes to a segregated environment.

The above ideas of mixing should be extended to biological growth processes with two liquid phases. The results reported by Erickson and Humphrey (1969b) assumed that there was complete mixing of substrate in the fermentor; that is, it was assumed that all drops in the dispersed phase have the same substrate concentration. If the hydrocarbon drops are partially segregated, this may not be true since the smaller drops will tend to have their substrate consumed before the larger drops. Hence, additional work is needed in order to understand the effect of segregation on fermentations with two liquid phases. However, the terms macromixing and micromixing which have been applied to one liquid phase systems now have to be rephrased so as to account for the heterogeneity of the two liquid phase systems.

As mentioned before systems with two liquid phases generally have a continuous phase and a dispersed phase. The cells may be primarily in one of the two liquid phases or they may be concentrated at the interface between the two liquid phases. In this type of process, the intermixing of the contents of the dispersed phase (the exchange of matter from drop to drop) may be large or in other cases it may be very small. Often substrate is contained in the dispersed phase and if each drop has its own segregated and isolated life during its stay in the reactor, there may be differences in substrate concentration from drop to drop. The size distribution of the dispersed drops may also be important. If growth occurs at the surface or if the substrate leaves the drop due to diffusion, the supply of substrate in the smaller drops will be exhausted before that in the larger drops. The distribution of cells in the two phases also may be very important. Hence, micromixing and macromixing will be complicated concepts for heterogeneous systems. There are many ways of defining the segregation in a liquid-liquid system (see Rietema, 1958, 1964; Kobota and Omi, 1967).

Rietema (1964) has considered the influence of segregation on the conversion rate of chemical reactions involving two liquid phases. He has shown that the segregation can greatly increase the residence time required to obtain a desired conversion.

Acknowledgments

This project received financial support, which is appreciatively acknowledged, from the Kansas State University Agricultural Experiment Station, the Kansas Water Resources Research Institute, and the Offices of Water Resources Research, U.S. Department of Interior (KWRRI A-029), the Water Quality Office, Environmental Protection Agency (Project WP-01141-02 and 17090 ELL), and the Public Health Service (Research Career Development Award No. 1K4 GM35397-01). The authors also wish to express their sincere appreciation to Mr. Yuji Horie for helping out in translating the Japanese literature into English.

Nomenclature

$a(m,t)$	=	number probability function in the influent
$A(m,t)$	=	number probability function in the vessel
$A(h_0)$	=	necessary energy required to separate two drops of unit radius from an initial minimum distance h_0 to infinity
c_L	=	constant defined by Eq. (25)
d_p	=	drop diameter
$(d_p)_{max}$	=	maximum stable size of drop
$(d_p)_{min}$	=	minimum stable size of drop
D	=	impeller diameter
D_{32}	=	Sauter mean diameter
$f(m)$	=	escape frequency
$f\left(\dfrac{\mu_d}{\mu_c}\right)$	=	function of μ_c and μ_d
$f_n(v)$	=	number probability density of drop volume
$F(m;\mu)$	=	basic coalescence frequency
$g(m)$	=	breakage frequency
$h(m)$	=	collision frequency
H	=	depth of liquid
k,K,K_1,K'_1,K_2,K'_2	=	constants
$K(v,v')$	=	mixing kernel defined by Eq. (35)
$K_B(v,v')$	=	breakage kernel defined in text
$K_c(v,v')$	=	coalescence kernel defined in text
L	=	maximum drop size in the vessel
m	=	droplet mass
m_A	=	mass feed rate
m_B	=	mass effluent rate
n_A	=	number feed rate of droplets
n_c	=	number of drops involved in one coalescence
$n(v)$	=	number of drops in vessel in size range v to $v + dv$
N	=	number of droplets in vessel

N_I	=	impeller speed, rpm
N_{We}	=	Weber number
p	=	number of drops which coalesce per unit time
$p(v)$	=	volume probability density function in vessel of drop volume
$P_0(v)$	=	input volume probability density function of drop volume
$p(v,c)$	=	volume probability density function in vessel of drop volume and drop concentration
$P_0(v,c)$	=	input volume probability density function of drop volume and drop concentration
P	=	coalescence probability function
P_v	=	dispersion energy of mixing
Q	=	input volumetric flow rate
T	=	diameter of stirred tank
v, v', v''	=	volume of drop
V	=	fluid volume in a continuous stirred tank
W_r	=	ratio of $Q\Phi_1$ to $Q\Phi_2$
x,y	=	integration variables
$\beta(m{:}\mu)$	=	breakage density defined in text
$\Gamma(m,t)$	=	number output from the mass range by breakage, coalescence, and flow
ϵ	=	local rate of energy dissipation per unit mass of the fluid
$\bar{\epsilon}$	=	average local rate of energy dissipation per unit mass of the fluid
η	=	Kolmogoroff length, that is, microscale of turbulance
η_d	=	maximum drop size in the feed as used by Valentas and Amundson
$\lambda(m,\mu)$	=	coalescence efficiency
$\Lambda(m,t)$	=	number input in the mass range by coalescence
μ	=	drop size
μ_c	=	viscosity of continuous phase
μ_d	=	viscosity of dispersed phase
$\nu(v',v'')$	=	velocity coalescence cross-section product
ν	=	kinematic viscosity
$\nu(m)$	=	average number of drops formed per breakage
ρ_c	=	density of continuous phase
ρ_d	=	density of dispersed phase
σ	=	interfacial tension
$\omega(m,t)$	=	function defined by Eq. (19)
ω_i	=	dispersion coalescence frequency
$\Omega(m,t)$	=	function defined by Eq. (15)
ϕ	=	volume fraction of dispersed phase
ϕ_1	=	dispersed phase ratio in the influent
ϕ_2	=	dispersed phase ratio in the vessel
$\phi(m,t)$	=	number input in the mass range by breakage

References

Adler, C. R., Mark, A. M., Marshall, W. R., Jr., and Parent, J. R. (1954). *Chem. Eng. Progr.* **50**, 14.

Aiba, S., and Haung, K. L. (1970). *Kagaku Kogaku* **34**, 686.

Aiba, S., and Haung, K. L. (1971). *Kagaku Kogaku* **35**, 202.

Aiba, S., Moritz, V., Someya, J., and Haung, K. L. (1969a). *J. Ferment. Technol.* **47**, 203.

Aiba, S., Haung, K. L., Moritz, V., and Someya, J. (1969b). *J. Ferment. Technol.* **47**, 211.
Becher, P. (1965). "Emulsions: Theory and Practice." Van Nostrand-Rheinhold, Princeton, New Jersey.
Belk, T. E. (1965). *Chem. Eng. Progr.* **61**, 72.
Bischoff, K. B. (1966). *Can. J. Chem. Eng.* **44**, 281.
Brodkey, R. S. (1967). "The Phenomena of Fluid Motions." Addison-Wesley, Reading, Massachusetts.
Brown, A. H., and Hanson, C. (1966). *Brit. Chem. Eng.* **11**, 695.
Brown, R. A. S., and Govier, G. W. (1961). *Can. J. Chem. Eng.* **39**, 159.
Cengle, J. A., Knudson, J. G., Landsberg, A., and Farngui, A. A. (1961). *Can. J. Chem. Eng.* **39**, 189.
Charles, G. E., and Mason, S. G. (1960a). *J. Colloid Sci.* **15**, 105.
Charles, G. E., and Mason, S. G. (1960b). *J. Colloid Sci.* **15**, 235.
Chen, G. K. C. (1971). Ph.D. Dissertation, Kansas State University, Manhattan.
Chen, H. T., and Middleman, S. (1967). *AIChE J.* **13**, 989 (1967).
Cholette, A., and Cloutier, L. (1959). *Can. J. Chem. Eng.* **37**, 105.
Church, J. M., and Shinnar, R. (1961). *Ind. Eng. Chem.* **53**, 479.
Cockbain, E. G., and McRoberts, T. S. (1953). *J. Colloid Sci.* **8**, 440.
Collins, S. B. (1967). Ph.D. Dissertation, Oregon State University, Corvallis.
Curl, R. L. (1963). *AIChE J.* **9**, 175.
Danckwerts, P. V. (1953). *Chem. Eng. Sci.* **2**, 1.
Danckwerts, P. V. (1958). *Chem. Eng. Sci.* **8**, 93.
Dostalek, M., Munk, V., Volfova, O., and Fencl, Z. (1968). *Biotechnol. Bioeng.* **10**, 865.
Douglas, J. M. (1964). *Chem. Eng. Progr., Symp. Ser.* **60**, 1.
Dunn, I. J. (1968). *Biotechnol Bioeng.* **10**, 891.
Elton, G. A. H., and Picknett, R. G. (1957). *Proc. Int. Congr. Surface Activ., 2nd, 1957* Vol. 1, p. 288.
Erdtsieck, B., and Rietema, K. (1969). *Antonie van Leeuwenhoek; J. Microbiol. Serol.* **35**, F19.
Erickson, L. E., and Humphrey, A. E. (1969a). *Biotechnol. Bioeng.* **11**, 467.
Erickson, L. E., and Humphrey, A. E. (1969b). *Biotechnol. Bioeng.* **11**, 489.
Erickson, L. E., Chen, G. K. C., and Fan, L. T. (1968). *Chem. Eng. Progr., Symp. Ser.* **64**, 97.
Erickson, L. E., Humphrey, A. E., and Prokop, A. (1969). *Biotechnol. Bioeng.* **11**, 449.
Erickson, L. E., Fan, L. T., Shah, P. S., and Chen, M. S. K. (1970). *Biotechnol. Bioeng.* **12**, 713.
Fan, L. T., Shah, P. S., and Erickson, L. E. (1969). *Proc. Conf. Appl. Continuous Simulation Language, 1969* p. 63.
Fan, L. T., Chen, G. K. C., Erickson, L. E., and Naito, M. (1970a). *Water Res.* **4**, 274.
Fan, L. T., Erickson, L. E., Shah, P. S., and Tsai, B. I. (1970b). *Biotechnol. Bioeng.* **12**, 1019.
Fan, L. T., Chen, G. K. C., and Erickson, L. E. (1971a). *Chem. Eng. Sci.* **26**, 379.
Fan, L. T., Tsai, B. I., and Erickson, L. E. (1971b). *AIChE J.* **17**, 689.
Fan, L. T., Erickson, L. E., and Chen, G. K. C. (1972). *Chem. Eng. Progr., Symp. Ser.* (in press).
Fiechter, A. (1971). Seminar on Biochemical Engineering, Kansas State University, Manhattan, April 16.
Gal-Or, B., Klinzing, G. E., and Tavlarides, L. L. (1969). *Ind. Eng. Chem.* **61**, 21.
Gillespie, T., and Rideal, E. K. (1956). *Trans. Faraday Soc.* **52**, 173.
Grieves, R. B., Milbury, W. F., and Pipes, W. O. (1964). *J. Water Pollut. Contr. Fed.* **36**, 619.

Groothuis, H., and Zuiderweg, F. I. (1964). *Chem. Eng. Sci.* **19**, 63.
Hansford, G. S., and Humphrey, A. E. (1966). *Biotechnol. Bioeng.* **8**, 85.
Harada, M., Arima, K., Eguchi, W., and Nagata, S. (1962). *Mem. Fac. Eng., Kyoto Univ.* **24**, 431.
Herbert, D. (1961).*Sci (Soc. Chem. Ind., London) Monogr.* **12**, 21.
Hillestad, J. G. (1964). Ph.D. Dissertation, Purdue University, Lafayette, Indiana.
Hinze, J. O. (1955). *AIChE J.* **1**, 289.
Hinze, J. O. (1959). "Turbulence." McGraw-Hill, New York.
Howarth, W. J. (1967). *AIChE J.* **13**, 1007.
Humphrey, A. E. (1967). *Biotechnol. Bioeng.* **9**, 3.
Humphrey, A. E. (1968). *Proc. West. Hemisphere Nutr. Congr., 2nd 1968.*
Humphrey, A. E. (1969). *Chem. Eng. Progr., Symp. Ser.* **65**, 60.
Humphrey, A. E., and Erickson, L. E. (1971). *Int. Symp. Continuous Cult. Microorganisms, 5th; 1972, J. Appl. Chem. Biotech.* **22**, 125.
Imoto, T., and Lee, S. (1967). *Chem. Eng. Progr., Ser. Japan* **1**, 88.
Imoto, T., and Lee, S. (1970). "Polymer Reaction Engineering." Nikkan Kogyo, Japan.
Jeffreys, G. V., and Hawksley, J. L. (1965a). *AIChE J.* **11**, 413.
Jeffreys, G. V., and Hawksley, J. L. (1965b). *AIChE J.* **11**, 418.
Jeffreys, G. V., and Lawson, G. B. (1965). *Trans. Inst. Chem. Eng.* **43**, T294.
Johnson, M. J. (1964). *Chem. Ind. (London)* p. 1532.
Jordan, G. V. (1965). *Chem. Eng. Progr.* **61**, 64.
Katinger, H., Nobis, A., and Meyrath, J. (1970). *Experientia* **26**, 565.
Kintner, R. C., Horton, T. J., Graumann, R. E., and Amberkar, S. (1961). *Can. J. Chem. Eng.* **39**, 235.
Kobota, H., and Omi, S. (1967). *Chem. Eng. Progr., Ser. Japan* **1**, 111.
Kolmogoroff, A. N. (1949). *Dokl. Akad. Nauk USSR* **66**, 825.
Komasawa, S. (1970). *Jap. Soc. Chem. Eng.* **22**, 1.
Komasawa, S., Morioka, S., and Otake, T. (1969). *J. Chem. Eng. Jap.* **2**, 208.
Komasawa, S., Morioka, S., and Otake, T. (1972). *Kagaku Kogaku* (in press).
Kramers, H. (1958). *Chem. Eng. Sci.* **8**, 45.
Laine, D., Vernet, C. H., and Evans, G. H. (1967). *World Petrol. Congr., 7th, 1967* p. 197.
Lang, S. B. (1962). "Hydrodynamic Mechanism for the Coalescence of Liquid Drops." Lawrence Radiat. Lab. Rep. UCRL 10097. University of California, Berkeley.
Langloise, E. E., Gullberg, J. E., and Vermeulen, T. (1954). *Rev. Sci. Instrum.* **25**, 360.
Lindsey, E. E., Chappelear, D. C., Sullivan, D. M., and Augstkalas, V. A. (1964). *J. Polym. Sci., Part C* **5**, 55.
Linton, M., and Sutherland, K. L. (1958). *J. Colloid Sci.* **13**, 441.
Lloyd, N. E. (1959). *J. Colloid Sci.* **14**, 441.
Luss, D. (1966). Ph.D. Dissertation, University of Minnesota, Minneapolis.
Luss, D., and Amundson, N. R. (1967). *Chem. Eng. Sci.* **22**, 267.
MacKay, G. D. M., and Mason, S. G. (1961). *Nature (London)* **191**, 488.
Madden, A. J., and Damerell, G. L. (1962). *AIChE J.* **8**, 233.
Madden, A. J., and McCoy, B. J. (1964). *Chem. Eng. Sci.* **19**, 506.
Malek, I., and Ricica, J. (1970). *Folia Microbiol. (Prague)* **15**, 377.
Mateles, R. I., and Tannenbaum, S. (1968). "Single Cell Protein." MIT Press, Cambridge, Massachusetts.
Matsuda, H. (1966). Master's Thesis, Osaka University, Osaka, Japan.
Matsuzawa, H., and Miyauchi, T. (1961). *Chem. Eng. Sci.* **25**, 282.
Miller, A. H., Selden, C. E., and Atkinson, W. R. (1965). *Phys. Fluids* **8**, 1921.
Miller, R. S., Ralph, J. H., Curl, R. L., and Towell, G. D. (1963). *AIChE J.* **9**, 196.

Mimura, A. (1970). *J. Ferment. Technol.* **48**, 449.
Mimura, A., Sugeno, M., Ooka, T., and Takeda, I. (1971a). *J. Ferment. Technol.* **49**, 245.
Mimura, A., Watanabe, S., and Takeda, I. (1971b). *J. Ferment. Technol.* **49**, 255.
Munk. V., Dostalek, M., Fencl, Z., Vernerova, J., and Silinger, V. (1966). *Proc. Int. Congr. Microbiol., 9th, 1966* p. 1.
Munk, V., Dostalek, M., and Volfova, O. (1969). *Biotechnol. Bioeng.* **11**, 383.
Nielsen, L. E., Wall, R., and Adams, G. J. (1958). *J. Colloid Sci.* **13**, 44.
Ng, D. Y. C. (1965). Ph.D. Dissertation, Imperial College, University of London, London.
Ng, D. Y. C., and Rippin, D. W. T. (1964). *Proc. Eur. Symp. Chem. Eng., Third 1965* p. 161.
Onda, K., and Takeuchi, H. (1970). *Jap. Soc. Chem. Eng.* **22**, 26.
Otake, T., and Komasawa, S. (1968). *Kagaku Kogaku* **32**, 475.
Prokop, A., Erickson, L. E., and Paredes-Lopez, O. (1971). *Biotechnol. Bioeng.* **13**, 241.
Resnick, W., and Gal-Or, B. (1968). *Advan. Chem. Eng.* **7**, 296.
Reusser, F. (1961). *Appl. Microbiol.* **9**, 361.
Rietema, K. (1958). *Chem. Eng. Sci.* **8**, 103.
Rietema, K. (1964). *Advan. Chem. Eng.* **5**, 237.
Rippin, D. W. T. (1967). *Chem. Eng. Sci.* **22**, 247.
Rodger, W. S., Trice, V. G., Jr., and Rushton, J. H. (1956). *Chem. Eng. Progr.* **52**, 515.
Rushton, J. H., Costich, E. W., and Everett, H. J. (1950). *Chem. Eng. Progr.* **46**, 395.
Schmitz, R. A., and Amundson, N. R. (1963a). *Chem. Eng. Sci.* **18**, 265.
Schmitz, R. A., and Amundson, N. R. (1963b). *Chem. Eng. Sci.* **18**, 391.
Schmitz, R. A., and Amundson, N. R. (1963c). *Chem. Eng. Sci.* **18**, 447.
Schmitz, R. A., and Amundson, N. R. (1963d). *Chem. Eng. Sci.* **18**, 915.
Scott, L. S., Hayes, W. B., and Holland, C. D. (1958). *AIChE J.* **4**, 346.
Shinnar, R. (1961). *J. Fluid Mech.* **10**, 259.
Shinnar, R., and Church, J. M. (1960). *Ind. Eng. Chem.* **52**, 253.
Sleicher, C. A., Jr. (1962). *AIChE J.* **8**, 471.
Smith, A. R., Caswell, J. E., Larson, P. P., and Cavers, S. D. (1963). *Can. J. Chem. Eng.* **41**, 150.
Snell, I. (1943). *Ind. Eng. Chem.* **35**, 107.
Soo, S. L. (1967). "Fluid Dynamics of Multiphase Systems." Ginn (Blaisdell), Boston, Massachusetts.
Spielman, L. A., and Levenspiel, O. (1965). *Chem. Eng. Sci.* **20**, 247.
Sprow, F. B. (1967a). *Chem. Eng. Sci.* **22**, 435.
Sprow, F. B. (1967b). *AIChE J.* **13**, 995.
Tavlarides, L. L., Coulaloglou, C. A., Zeitlin, M. A., Klinzing, G. E., and Gal-Or, B. (1970). *Ind. Eng. Chem.* **62**, 6.
Tsai, B. I. (1970). Ph.D. Dissertation, Kansas State University, Manhattan.
Tsai, B. I., Erickson, L. E., and Fan, L. T. (1969). *Biotechnol. Bioeng.* **11**, 181.
Tsai, B. I., Fan, L. T., Erickson, L. E., and Chen, M. S. K. (1971). *J. Appl. Chem. Biotechnol.* **21**, 307.
Valentas, K. J., and Amundson, N. R. (1966). *Ind. Eng. Chem., Fundam.* **5**, 533.
Valentas, K. J., Bilous, O., and Amundson, N. R. (1966). *Ind. Eng. Chem., Fundam.* **5**, 271.
Vanderveen, J. H. (1962). Lawrence Radiat. Lab Rep. UCRL 8733. University of California, Berkeley.
Veltkamp, G. W. (1964). Letter to the Editor, *Chem. Eng. Sci.* (as quoted in Rietema, 1964).

Verhoff, F. H. (1969). Ph.D. Dissertation, University of Michigan, Ann Arbor.
Vermeulen, T., Williams, G. M., and Langlois, F. G. (1955). *Chem. Eng. Progr.* **51,** 85.
Ward, J. P. (1964). Ph.D. Dissertation, Oregon State University, Corvallis.
Waterman, L. C. (1965). *Chem. Eng. Progr.* **61,** 51.
Weinstein, H., and Adler, R. J. (1967). *Chem. Eng. Sci.* **22,** 65.
Weiss, L. H., Fick, J. L., Houston, R. H., and Vermeulen, T. (1962). "Dispersed Phase Distribution Patterns in Liquid Liquid Agitation." Lawrence Radiat. Lab Rep. UCRL 9787. University of California, Berkeley.
Yamazaki, H., and Ichikawa, A. (1970). *Kagaku Kogaku* **34,** 219.
Yoshida, F., and Yamada, T. (1971). *J. Ferment. Technol.* **49,** 235.
Yoshida, F., Yamane, T., and Yagi, H. (1971). *Biotechnol. Bioeng.* **13,** 215.
Zwietering, T. N. (1959). *Chem. Eng. Sci.* **11,** 1.

Microbiology and Fermentations in the Prairie Regional Laboratory of the National Research Council of Canada 1946–1971

R. H. HASKINS

*National Research Council of Canada,
Prairie Regional Laboratory,
Saskatoon, Saskatchewan, Canada*

I.	Historical	415
II.	Microbial Studies	419
	A. Culture Collections	419
	B. Fundamental Studies	420
	C. Analytical Technique Development	420
	D. Fermentation Technology	421
	E. Metabolic Products	423
	F. Metabolic Pathways	429
	G. Cell Biology and Physiology	430
	H. Degradation of Pollutants	432
	I. Biological Nitrogen Fixation	432
	J. Yeasts and Yeastlike Fungi	432
III.	Higher Plant Studies and Plant Cell Culture	433
IV.	In Retrospect—and the Future	434
	References	435

I. Historical

Plans were considered for a regional laboratory in the Canadian west early in World War II, but did not materialize until 1943. Delays were numerous, but on February 22, 1945, the Hon. C. D. Howe, as Chairman of the Privy Council Committee for Scientific and Industrial Research, announced that Dr. R. K. Larmour, Professor of Chemistry, University of Saskatchewan, had been appointed Director of the National Research Council's Prairie Regional Laboratory which was to be built in Saskatoon, Saskatchewan. By the end of June, tenders had been called for the construction of the research laboratory on the campus of the University of Saskatchewan in that city.

This new laboratory was to be concerned primarily with investigations into the utilization of agricultural crops, especially those produced in surplus, and those materials which largely go to waste. The laboratory would be equipped to undertake all phases of laboratory and pilot plant investigations in this field.

Before the end of 1945, early elements of the staff were already at work devoting time to straw utilization, while an N.R.C. Oil Seeds Laboratory staff housed in University buildings was studying the possibility of making linseed oil into an edible product. Early in 1946, Dr. J. B. Marshall of N.R.C. Ottawa staff and a former graduate of the University of Saskatchewan, was appointed assistant administrator. By May 1947, the four main lines of investigation being considered were: vegetable oils, starch and cellulose, microbiological fermentations, and the development of basic tools for rapid, accurate measurements.

In October of 1947, Dr. Larmour left to join the Maple Leaf Milling Co., Ltd., and Dr. G. A. Ledingham of the Division of Applied Biology, N.R.C., Ottawa, and a former graduate of the University of Saskatchewan, was appointed to direct the Laboratory.

The Official Opening, June 8, 1948, of the nearly finished building was marked by various ceremonies and a Special Convocation at the University. The program planned for the laboratory was described then as *"utilization of agricultural products, waste materials and by-products of established processes"* and *"the furthering of pure research which creates the reservoir from which the stream of applied science may be fed for solving problems of the future."*

By the end of 1948, the staff of 10 scientific personnel and 23 technicians and service personnel had been organized into four sections: Administrative—Dr. Ledingham and Dr. Marshall; Fermentations—Dr. A. C. Neish, Dr. R. H. Haskins, Dr. (Mrs.) W. M. Dion, Mr. A. C. Blackwood, and Miss M. Lambert; Utilization—Dr. H. R. Sallans and Mr. C. G. Youngs; and Residues and Engineering—Mr. A. L. Babb. During the following year, Dr. J. A. Thorn, Dr. Ping Shu, Dr. J. M. Roxburgh, and Mr. R. Tink joined the Microbiology and Fermentations Section, and Dr. W. B. McConnell, Dr. R. U. Lemieux, Dr. J. E. Stone, Mr. K. L. Phillips, and Miss H. C. J. Brice joined the Crop Utilization Section. The Crop Utilization Section was rounded out during 1950 with the addition of Dr. B. M. Craig and Dr. L. R. Wetter to the scientific staff. Later, an Engineering and Process Development Section was organized to include Dr. Sallans, Dr. Shu, Dr. Roxburgh, and Mr. Phillips, Mr. Tink, and Mr. J. D. Salloum.

By the end of 1959, the staff totaled 31 (13) professionals, 12 (6) postdoctorate, 38 (13) technical and 14 service personnel (the numbers in parentheses indicate those involved with microbiology and fermentation studies). On September 30, 1971, the comparable figures were: 24 (10), 4 (3), 36 (11), and 17, a decrease from the peak year 1959.

The staff, for the work set out for it to do, seemed nicely balanced with "experts" in mycology, bacteriology, microbial biochemistry, fats and oils, proteins, carbohydrates, crop residues, and engineering. Adequate stores, wood-, metal- and glass-working shops and a growing library completed the picture. The organization was small enough that all were able to work together, yet large enough to provide a reasonable range of skills, specialties, and, eventually, equipment.

Some changes affecting microbiological studies have occurred over the years as the laboratory and its programs developed. A new section concerned with the physiology and biochemistry of higher plants was organized under Dr. Neish, and moved into a new annex in 1958 (Fig. 1). Dr. Haskins became head of the Physiology and Biochemistry of Fungi Section in 1956, and Dr. F. J. Simpson of the parallel Bacteriology Section in 1961 when Dr. Blackwood left for MacDonald College faculty. These two sections were combined in 1970 under Dr. Haskins.

FIG. 1. The Prairie Regional Laboratory on the campus of the University of Saskatchewan in Saskatoon, Saskatchewan. The annex, with greenhouse attached, was opened in 1958, the Division of Building Research addition in 1968.

Microbiological work was also developed in the Engineering and Process Development Section with Dr. J. F. T. Spencer working with yeasts and Dr. P. S. S. Dawson developing continuous fermentation techniques. An extension of "fermentation beyond laboratory scale" to include "laboratory scale studies" created some overlap of research areas. The wisdom of such overlap was questioned by some but the closeness of cooperation among the staff made the question academic.

The Laboratory was deeply grieved and suffered great loss with the death of its Director, Dr. Ledingham, in 1962 (Haskins, 1963). Dr. Sallans became director in 1963, and on his retirement in 1970 was replaced by Dr. Craig; both were former graduates of the local university.

In the early years the activities and program of the laboratory were reviewed periodically by Review Committees appointed by the N.R.C. These Committees, by their recommendations and suggestions, on occasion influenced the direction of the work undertaken, as evidenced during the 1960's when they suggested that more emphasis be placed on cellular biology and consequently less on metabolic products. In more recent years with retrenchment of available finances, staff reductions and such, and with the increasing demands that scientific effort should be more toward the solution of current problems (i.e., applied research with early returns) the Review Committees have been replaced by Advisory Boards and a greater control of the programs undertaken.

An outstanding feature of laboratory life has been N.R.C.'s postdoctorate fellowship scheme, which from 1950 on has brought 126 postdoctoral fellows (more than 56 of these became involved with microbiological problems) from many countries throughout the world to work in the laboratory for 1 to 2 years. This scheme provided a welcome influx of new ideas and personalities, keeping the permanent staff alert, refreshed, and aware of scientific advances and problems throughout the world.

This paper appearing as it does in this microbiological journal is restricted to the laboratory's microbiological studies and achievements, with only a mention of the other activities, which even a microbiologist would concede have been important!

The laboratory is now well equipped for research in microbiology, the chemistry of natural products, biochemistry, enzymology, and plant physiology. An intermediate range electron microscope facility with the latest freeze-etch equipment is in the Microbial Physiology and Biochemistry Section. Techniques of infrared, ultraviolet, proton magnetic resonance, optical rotatory dispersion, and mass spectrometry are being used, and a wide variety of GLC (gas-liquid chromatography) units and other sophisticated instruments are available for use of all sections. Time-lapse cine-photomicroscopes, and an analytical ultracentrifuge are in use.

The laboratory maintains a microanalytical service unit, a tracer and counting laboratory, an air-conditioned greenhouse, cold rooms, pilot plant facilities (up to 200 gal), mechanical, electrical, and glass-blowing shops, and a well-equipped photographic and drafting group. Facilities for batch, continuous, and synchronous cultivation of microorganisms and plant cells, and an amino acid analyzer round out the picture.

II. Microbial Studies

By February, 1949, the fermentation section was well into its task of studying the effects of molds, yeasts, actinomycetes, and bacteria on various agricultural products, with a view to conversion of these to useful chemicals, antibiotics, pharmaceuticals, feedstuffs, vitamins, etc. Fungi were grown on a wide range of materials on a carbon balance basis to discover those able to produce metabolites quickly and in quantity. For bacteria the fastest utilizers of glucose were examined for production of glycerol, lactic acid, 2,3-butanediol, etc., in the early studies.

A. Culture Collections

An adequate culture collection to a microbiologist is as a shelf of chemicals to a chemist. Starting with small collections of fungi used in citric acid and other studies in Ottawa, the P.R.L. Collection has been built up so that a wide range of isolates is on hand in addition to named cultures from other institutions. At first, special attention was paid to those organisms not generally used in industrial fermentations. The working collections now contain some 2750 fungi, 1200 yeast strains, and about 600 bacterial cultures. These are available to other laboratories on a personal favor basis, but many have not yet been identified. Each of the fungus cultures is conserved under sterile mineral oil, though many are also lyophilized or stored in soil. The yeasts are either lyophilized or stored under mineral oil. The bacteria have been lyophilized.

Lack of staff over the years has prevented the full potential of the collections from being reached, and many of the isolates have not yet been thoroughly examined. However, time was found for a consideration of a few fundamental problems. Studies of factors affecting the survival of lyophilized fungal spores showed that the better survival obtained with the centrifuge freeze-drying method was due, among other things, to the less severe temperatures reached and the shorter freezing times involved. Degassing of spore suspensions prior to freezing and conditioning in moist atmosphere after opening the lyophil tubes enhanced survival in many cases.

B. Fundamental Studies

Inevitably, when one collects microorganisms from nature, new or unfamiliar forms are found. This may be especially so on the flat and dusty prairies where holding a plate out of the window on a normal (i.e., windy) day brings inocula from afar!

Some of the more interesting organisms studies at P.R.L. include *Volucrispora aurantiaca* Haskins and related forms, a heterothallic inoperculate discomycete not yet identified, and *Trichosporonoides oedocephalis* Haskins and Spencer. Each of these was studied because of the metabolic substance(s) it formed. *Volucrispora aurantiaca* produced the terpenyl quinone, volucrisporin. The discomycete produced large yields of the perylene hydroxyquinone, mycochrysone; snail digestive juice proved useful in the orderly isolation of ascospores of this species. *Trichosporonoides oedocephalis* among other things produced erythritol. While very similar to the genus *Trichosporon* it featured aerial oedocephalum-like fruiting structures and occasionally exhibited backward-directed hyphal branching.

Isaria orctacea was used by Dr. W. A. Taber to study the nutritional factors affecting morphogenesis of the synemma. The Saskatoon area abounds in alkali lakes and sloughs. Evidence was presented for the existence of acid-sensitive actinomycetes in the soil, and alkali-dependent species of streptomycetes were characterized.

The influence of the inoculum on variability in comparative nutritional experiments with fungi and the effect of sellenite and tellurite on cellular division of yeast-like fungi were considered.

Recently, *Botryodiplodia theobromae,* responsible for a soft rot of yams in storage in tropical countries came under study and the relation of light to the induction of pycnidia formation was demonstrated and control methods were studied.

C. Analytical Technique Development

In a newly organized laboratory, new apparatus has to be set up and calibrated, and new techniques devised to cope with the problems encountered in the new fields of research being opened up. A considerable number of methods were assembled, perfected, or developed for the estimation of metabolites in fermentation solutions. These received wide informal distribution in the several revisions of Dr. Neish's Report 46-8-3 "Analytical methods for bacterial fermentations" (now out of print). Studies of others along this line included: estimation of reducing sugars in starch hydrolyzates by paper chromatography, a micro-method for determination of sugars, and determination of substances such as ethyl alcohol or free amino acids by microdiffusion techniques.

Various compounds necessary for use in projected studies had to be synthesized by methods suitable for the introduction of ^{14}C. This work was facilitated by the inclusion of a tracer laboratory in the annex, which was opened in October 1958.

A new apparatus was designed for rapid carbon determination by wet combustion. Partition chromatography proved useful for the analyses of mixtures of simple aliphatic alcohols. An agar diffusion assay method was developed for the antifungal polyene, candidin. Methods for the determination of catechols, pyrazole, pyrimidine, hydrazine, and various carboxylic acids, and for the accurate determination of specific rotation of metabolites, were developed and extensive use was made of gas chromatographic techniques. Spectrophotometric methods were widely used for pyridazine carbolic acids with O-nitrobenzaldehyde, and methionine with pentacyanoaminoferrate.

D. Fermentation Technology

The laboratory had been provided with a fermentations pilot plant of 15-, 200-, and 1500-gal fermentors, various recovery equipment, distilling columns, etc. The "bugs" had to be taken out of the equipment and various test runs conducted. The plant was essentially an alcohol plant with facilities for recovery of products dissolved in the fermentation broth. Additional equipment was acquired that could be used where the products had to be recovered from the culture solids.

While this work was under way, studies were made on the extraction of polyhydroxy compounds from dilute aqueous solutions by cyclic acetal formation, with special attention to the batch and continuous extraction of glycerol. In the laboratory, studies of the dissimilation of glucose by yeast at poised hydrogen ion concentrations were undertaken using an apparatus developed in Ottawa in connection with the production of glycerol. Control of pH was very precise. Maximum utilization of substrate was shown in a number of cases to require a very narrow range of pH control. The highest yield of glycerol obtained in one test was 29% by weight of the sugar fermented. The apparatus proved to be more useful for anaerobic than for aerobic fermentations because of contamination problems, which were overcome later.

Various difficulties with equipment were met. Series of 5-liter and other fermentors were developed, complete with all controls, and these proved useful until such equipment became available commercially. Control of foam in the larger fermentors was improved with a new type of foam breaker. Carbon oxidation-reduction balances were determined and estimations made of fermentation efficiencies of aerobic fermentations in the various equipment.

Dr. Shu, in studying oxygen uptake in shake flask fermentation, developed ingenious techniques for automatic translation of aeration data between different fermentors and continued to a study of control of oxygen uptake in deep tank fermentors. Studies on fermentation aeration started by Dr. Roxburgh, were continued mainly by Dr. Phillips with well-designed experiments from the effect of antifoam agents on oxygen transfer in deep tank fermentations, to horizontal fermentors and reactor systems for processes with extreme gas-liquid transfer requirements. He is continuing to study the fundamentals of gas-liquid mass transfer, with emphasis on oxygen transfer in all types of aerobic fermentations; the results to be applied to increased yields of fermentation products and to more efficient treatment of sewage.

Fermentations on a large scale had long since become commonplace. Certain problems with batch fermentations had to be met. Continuous fermentations were promising. Extensive studies in this field were initiated by Dr. Dawson. While developing his apparatus for continuous culture, he was able to undertake studies which not only tested his equipment but also provided information required by others in the laboratory. A continuous flow culture apparatus with a cyclone column unit was developed for filamentous fungi and has proved useful for the other types of microorganisms.

The intracellular amino acid pool of *Candida utilis* during growth in batch and in continuous culture was studied. Changes in the cell cycle of this organism revealed the operational flexibility of the phased culture technique. The use of the apparatus, and continuous phased culture as a technique for growing, analyzing, and using microbial cells, has been extended in P.R.L. to higher plant cell culture. A versatile dome fermentor suitable for batch, continuous, and synchronous cultivation of microorganisms or tissue culture cells, has been developed. In a comparative study, *in vivo*, of enzyme activities in batch, continuous, and phased culture of a pseudomonad grown on phenylacetic acid, some of the disadvantages of batch cultivation and the relative advantages of chemostat and continuous phased cultures were revealed.

E. METABOLIC PRODUCTS

As isolates were added to the culture collections they were screened for their ability to produce new or useful products. This work was organized on the basis of carbon balance studies and on growth on a series of media containing various agricultural materials ranging from glucose and starch to wheat straw and oilseed cake. New and potentially useful products were obtained from these surveys, some of which are mentioned more fully below. On the other hand, when, for various reasons, certain specific products were required, the isolates in the collections were surveyed for ability to produce such materials.

In either case, once the product had been detected, factors affecting its production were systematically investigated, yields were increased by cultural modification and selection. Finally methods of extraction and purification were developed. The more promising procedures and products were scaled up to pilot plant level.

1. Glycerol

At first the bacteriologists aimed at the conversion of waste or surplus carbohydrates to chemicals such as sugar alcohols, glycerol, lactic acid, and 2,3-butanediol. They concentrated chiefly on glycerol production — it was a useful chemical required in large quantities at a good price. Wheat starch and sugar beet molasses were cheap sources of carbohydrate in western Canada.

Attempts were made to establish conditions that were necessary to give rapid fermentation with high consistent yields of glycerol from a starch substrate using *Bacillus subtilis* (Ford's strain). A strain of *Gliocladium roseum* proved useful for the production of saccharifying agents for this fermentation. Such agents were necessary since the Ford strain produced amylases only in aerobic culture. In a survey, no isolates were found which gave better yields than did the *B. subtilis*. The mechanisms of the production of 2,3-butanediol and glycerol during dissimilation of glucose by *B. subtilis* were studied.

Tracer techniques were applied to the study of carbohydrate metabolism by Ford's type in the production of glycerol and later extended to other studies and biosynthetic pathways. The production of glycerol by honey yeasts was investigated early in Dr. Spencer's studies with yeasts which are described more fully below (Section II, E).

2. Ustilagic Acid

Carbon balance studies by Haskins showed that a locally collected strain of *Ustilago maydis* rapidly produced considerable quantities of an unknown product. Factors affecting the production of the material were studied, and selection procedures were used to boost the yields as high as 28 gm/liter in 36–48 hr. The material, ustilagic acid, was isolated and purified and was shown by Lemieux and others to be a cellobiolipid, a complex of cellobiose with dihydroxyhexadecanoic, caprylic, and caproic acids.

Because of the high antibiotic properties of the organism, the fermentation was an easy one to handle. The ustilagic acid could be extracted readily from the culture solids by the addition of methanol and recovered by dilution of the methanol solution with water. Tests showed that the antibiotic activity of the organism resided in the ustilagic acid. *In vitro* studies showed it to be active against many fungi, actinomycetes, and gram-positive bacteria. However further studies (Reed and Holder, 1953) showed the material to be less than satisfactory for conventional therapeutic use. A preliminary investigation of the pharmacological properties of ustilagic acid (V. Chivers-Wilson) showed the acid and its sodium salt to be remarkable in their lack of such properties! A biological analysis of the metabolic products using crossbred chicks (E. Walter) indicated the presence of either a gonadotrophic or an androgenic hormone in the culture mixture. Treatment of a case of lumpy jaw in a calf resulted in rapid but short-lived remission. Postmortem examination showed the infection had been too advanced for hope of success. Unfortunately, the acid is rapidly degraded into its less effective components at alkaline pH's, and is too insoluble in water to be of much value.

In the meantime, scale-up studies resulted in good production and recovery. Patents were obtained on the production of ustilagic and related acids. In continuing structural studies, Lemieux showed that ustilagic acid could be used as a relatively cheap starting material for the synthesis of macrocyclic musks, especially "Exaltolide," of value to the perfume industry.

Other isolates of *Ustilago* examined during these studies were found to produce good yields of various mannose-erythritol compounds.

3. Amino Acids

When work on the ustilagic acid fermentation was tapered off because of lack of interest in some of the patents by possible developers, new surveys were undertaken. Lysine is the most important amino acid lacking from wheat, the main crop of the Canadian prairies, therefore organisms in the collections, mainly those capable of growing in yeastlike fashion, were screened for ability to produce lysine.

One of the best yielders was a *Ustilago* sp. Factors affecting the production were studied, and the yields were raised to a level possibly competitive with the processes used at the time for fermentative lysine production. Patents were taken out for the production of lysine, arginine, and glutamic acid. However the success of the Japanese mutants of *Micrococcus* (=*Brevibacterium*) *glutamicus* put a damper on further work with the smut fungi at that time as effort had to be directed elsewhere.

Interest in feed supplementation did not wane. A survey of microorganisms by Dr. F. Reusser led to some work involving the culture of certain mushrooms in submerged culture. An organism thought to be *Tricholoma nudum* showed promise in cooperative studies with the University as a source of amino acids, vitamins, and lipids as determined from mice feeding trials. All the necessary B-vitamins for normal growth were present, and methionine and phenylalanine were indicated to be the first and probably the only limiting amino acids. The product was relatively rich in tryptophan. The production of fructose as a residue of sucrose fermentation by *"Tricholoma nudum"* was demonstrated.

4. Ergot-Type Alkaloids

The laboratory became interested in *Claviceps* and ergot alkaloids in 1950. However adequate amounts of water-soluble alkaloids were not found either in submerged culture or in locally produced ergot bodies. The studies were discontinued until the arrival of Dr. Taber and Dr. L. C. Vining on the staff. Much fundamental work was required before production could be considered. Nutritional studies of strains of *Claviceps purpurea* were followed by studies of the influence of certain factors on the *in vitro* production of ergot alkaloids. Techniques for estimation of the alkaloids present were developed. Tryptophan was shown to be a precursor. The phosphate metabolism was considered and the relationships between nucleic acid synthesis and ergot alkaloid production worked out (with Dr. C. deWaart). Similar alkaloids were identified in several higher plants. Extensive studies on the biosynthesis of the alkaloids and on the physiology of production were undertaken by Taber and Vining but staff changes in 1964 prevented furtherance of these studies.

5. Pigments

Growing a wide range of fungi on a variety of media reemphasized the ability of fungi to produce compounds of many colors. A closer look at some of these pigments was considered a legitimate part of the survey of fungi for production of new and useful products. The organism producing the metabolite was studied, then the factors affecting pigment production were investigated. Methods of extraction and purification were developed and biogenesis and structure clarified. Some of the new pigments studied and characterized were:

Volucrisporin (from *Volucrispora aurantiaca* Haskins) — a meta-hydroxylated terphenyl quinone, produced in addition to β-carotene and other carotenoids in response to nutritional imbalance (I).

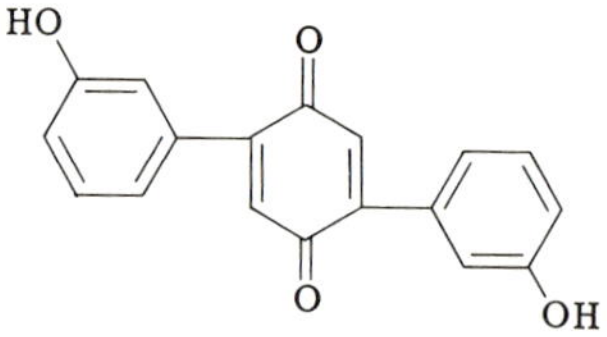

(I)

Mycochrysone (from a heterothallic inoperculate discomycete) — a complex naphthoquinone with an epoxide group (II).

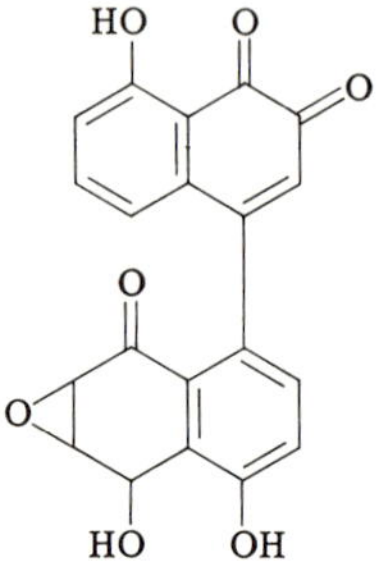

(II)

Cephalochromin (from *Cephalosporium* sp.) a complex dinaphtho-dihydro pyranone (III).

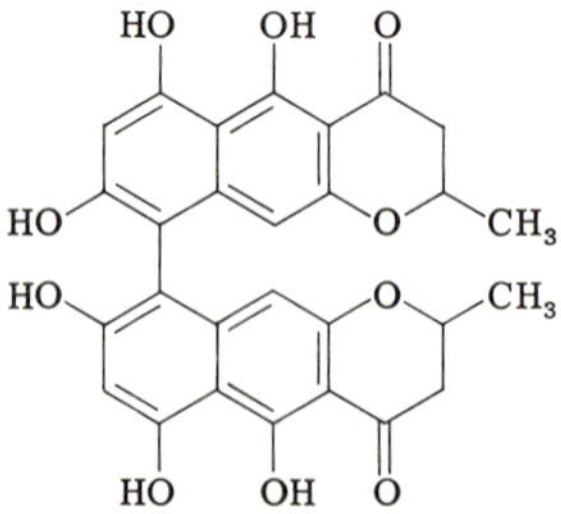

(III)

Monochaetin (from *Monochaetia compta*) a complex 1*H*-2-benzopyranone (IV).

(IV)

Pigments from *Trichoderma viride* (pachybasin, chrysophanol, and emodin); from *Acremonium* sp. and carpophores of *Amanita muscaria* (oosporein); from *Pseudomonas indigofera* (pyocyanine); from *Penicillium funiculosum;* and from *P. herquei*. Pulcherrimin and related compounds (from *Candida pulcherrima, Bacillus cereus,* and *Micrococcus violagabriellae*) were studied intensively by MacDonald.

6. Antibiotics

In 1950 Blackwood screened many bacteria for the production of antibiotics of possible effectiveness against plant diseases. Loose smuts of wheat and barley were selected for the initial testing. Thiolutin was shown in field trials to be capable of controlling loose smut in barley grown from heavily infected seed. However, germination was reduced by the seed treatment.

Many substances with antibiotic properties turned up in various cultures. Investigated were the actinomycins, candidin, endomycins, and various other polyene antibiotics (Taber and Vining). A caerulomycin was described and antibiotic factors from various organisms were studied. These included ustilagic acid, atrovenetin, and amidomycin. Mycological chemists (Dr. J. C. MacDonald and Dr. G. P. Slater) were interested in the structures and the biosynthesis of compounds related to aspergillic acids. Techniques were developed for the production of valinomycin, a highly toxic antibiotic useful in studies of K^+ transport in living systems. For a while their laboratory was the only source of the material until they were able to turn their methods of production and problems of supply to Calbiochem, thus freeing themselves to "get back to work"!

7. Enzymes

All biological processes depend on the action of enzymes, therefore both fundamental and practical work in this area was necessary.

Degradation and biosynthetic pathway studies involved isolation and purification of special enzymes and systems and these are described elsewhere (Sections II,A,6, II,B,1 and 3). The enzyme work considered in this section is where the enzymes were required products for which the collections were surveyed.

A large number of bacilli and molds were found to produce enzymes that attack pentosan gums obtained from wheat. These proved useful for reducing the troublesome "squeegee starch" encountered in the recovery of starch and gluten from wheat. Yields of prime starch recovered were raised from 60% to 85–93% of that present in the flours.

After working for a time with fungal amylases so that saccharifying agents would be ready when needed for fermentation studies, Dion surveyed fungi for the ability to produce proteases. Factors affecting production were studied, characteristics of extracellular proteases examined, and certain of the enzymes purified. Two proteases isolated from *Mortierella renispora* merited closer attention by Wetter. Dion was able to work with polyphenol oxidases for a short while before leaving the staff. Shu continued the studies on the production of amylolytic enzymes by submerged culture of *Aspergillus niger*. Simpson surveyed microorganisms for the production of pentosanases that attack the pentosans of wheat flour. This was part of the problem in which other sections in the laboratory were attempting the production of vital gluten. Several bacterial pentosanases showed promise in the recovery of starch from "squeegee starch." Eveleigh began a study of the function and mode of action of polysaccharase with the view to the use of such enzyme systems for removal of cell walls to produce protoplasts, following his successful work with *Pythium* protoplasts.

8. Miscellaneous Products

Some work was completed by Neish and others on the production and properties of 2,3-butanediol, a wartime project initiated in the Ottawa laboratories. The mechanism of citric acid formation from glucose by *Aspergillus niger* received further study. Mannans produced by various fungi were studied in joint projects with the Carbohydrate Laboratory. Sepedonin, a tropolone metabolite from *Sepedonium chrysospermum,* thelephoric acid from lichens and higher fungi, aspartic acid from *Lactobacillus arabinosus,* and glutaconic acid came in for attention, while good yields of itaconic acid were indicated in *Ustilago* fermentations. The fermentation of glutamic acid by *Pepotococcus aerogenes* was found to yield glutaconic acid.

The attention of Child and Haskins was attracted by the strong phenolic odor of cultures of *Valsa friesii*. Tests were made and factors affecting production were studied. It was shown that the organism could produce *m*-cresol in yields as high as 1 gm/liter on a chemically defined medium. Knowledge of the ability of *Valsa friesii* to produce *m*-cresol may be helpful to biosynthetic studies, such as those on the antibiotic "patulin," and may provide a means of obtaining radio-labeled intermediates for those studies.

F. METABOLIC PATHWAYS

1. *Biosynthetic Pathways*

Studies of metabolic pathways are essential to an understanding of the means by which organisms build up metabolites and are able to degrade compounds. Such understanding contributes to clarification of the molecular structure of the compounds concerned. Biosynthetic pathways of aromatic compounds were investigated by Neish, of terphenylquinones by Dr. G. Read and Vining, of diketopiperazines, and of a new galactobiosyl glucose formed from lactose by *Chaetomium globosum* (Gorin). Studies were made on the pathway of methanol assimilation by a bacterium (Kaneda and Roxburgh).

The utilization of tryptophan in the biosynthesis of echinulin was clarified and the configuration of the compound determined. Aspergillic and related acids were investigated biosynthetically by MacDonald in the hope of obtaining antibiotics with more favorable potential than aspergillic acid itself. The biosyntheses of aspergillic acid, and hydroxy-, neo-, and neohydroxy aspergillic acids were worked out.

Pyocyanine, an antibiotic pigment produced by certain strains of *Pseudomonas aeruginosa*, was investigated biosynthetically because of the interesting versatile biosynthetic ability of the organism.

2. *Degradative Pathways*

Work done in the early days of the laboratory on the dissimilation of many types of sugars by certain organisms especially *Aerobacter aerogenes* and *Leuconostoc mesenteroides* resulted in publication of 22 papers. Several enzymes were characterized and degradation steps were followed by means of tracer studies.

Other studies were concerned with the microbial metabolism of cinnamic acid, the metabolism of aromatic compounds with different side chains, and the changes produced in hop extract components by yeasts. This latter had a connection of interest with the brewing industry and their use of hop extracts in their art.

Although the chemistry of flavones and related compounds had been the subject of extensive chemical investigations, there was little information on the mechanism of biodegradation. Carbon monoxide was now shown to be produced during the enzymatic degradation of rutin (Dr. D. W. S. Westlake and Dr. F. J. Simpson). Factors affecting the production of the enzyme system by which *Aspergillus niger* degrades rutin were examined and the enzymes rutinase and quercetinase purified and characterized. The carbon monoxide producing system was worked out. Various rumen bacteria capable of anaerobic rutin degradation were isolated and identified. Identification was made of the products produced in the anaerobic degradation of rutin and related flavonoids by a *Butyrivibrio* sp.

G. Cell Biology and Physiology

1. Plant Rusts and Tissue Culture Studies

Although the responsibilities of a Director were never light, Dr. G. A. Ledingham spent considerable time at the bench. Returning to his earlier interest in the rust fungi, he began a course of studies with with his postdoctoral fellows and staff members in the hope of getting wheat rusts into artificial culture as did Cutter (1959) for certain other rusts. Perhaps when grown in this manner they would produce metabolites of value.

Various studies on the respiration of resting and germinating uredospores of wheat stem rust were conducted. Carbohydrate metabolism of the spore (Shu), influence of various compounds on germination (F. L. M. Turel), cell wall splitting enzymes and various spore components (C. F. van Sumere, Vining, F. W. Hougen, A. P. Tulloch, *et al.*), polyphenol-polyphenol oxidase systems in uredospores, and the relation of self-inhibition to oxidative metabolism (G. L. Farkas), cytochrome oxidase and oxidation pathway in uredospores (G. A. White), metabolism of labeled valerate, propionic, and other short-chain fatty acids (H. Reisener, McConnell), and metabolism of labeled pelargonate (S. Suryanaryanan) are indicative of the wide range of sophisticated studies conducted.

With Dr. Turel, safflower leaf tissue cultures were developed in connection with investigations of the possibility of getting flax rust into artificial culture. The production of aerial mycelium and uredospores by *Melampsora lini* on flax leaves in tissue culture was achieved, enabling investigation of the utilization of labeled substrates by isolated mycelium and uredospores. After leaving P.R.L.

Dr. Turel (1969) continued this work and successfully obtained flax rust in synthetic culture.

With Dr. P. G. Williams, mitochondrial fractions from stem rust uredospores were prepared and their properties examined. This was followed by fine-structure studies on the uredospores. Dr. Williams continued in this field after leaving P.R.L. and was successful in growing an Australian wheat rust in artificial culture (Williams et al., 1966).

2. Physiology of Root-Rot Control

The discovery by Haskins in 1963 that sexual reproduction of a *Pythium* sp. on potato dextrose agar required the presence of a substance such as β-sitosterol or cholesterol (or one with similar structure and which similarly satisfied certain specific structural requirements) opened up a whole new field for those interested in oospore production in the Pythiaceae. This field had lain dormant for 26 years.

The important thing was that *Pythium* and other Pythiaceous fungi were shown to require an exogenous source of sterol to reproduce sexually on certain commonly used media. Such sterols were shown by Dr. J. H. Sietsma, Dr. J. J. Child, and others to be taken up, for the most part unchanged, by the cell and incorporated mainly into the protoplasmic membrane with the remainder in the mitochondrial and endoplasmic reticulum fractions.

Many aspects of metabolism are affected by the presence of the sterol. Growth is stimulated. Greater extremes, both high and low, of temperature are tolerated. Leakage of constituents from the mycelium is altered. Cellulase and laminaranase activities are decreased with consequent effect on morphology. The mycelium becomes susceptible to polyene antibiotics.

During this work, production of *Pythium* protoplasts was achieved with the pure enzymes used revealing the composition of the cell wall material. The effects of both cholesterol and polyene antibiotics on the permeability of the protoplasmic membrane (Dr. G. Defago, J. J. Child) and on sexual reproduction itself are under study. This is to gain a better knowledge of the control of the sexual process in these organisms, which as ubiquitous plant pathogens cause considerable losses in agricultural crops.

That the work of Sietsma in this field has been recognized by the award of the Netherlands' Kluyver Prize is a matter of satisfaction to this laboratory.

Fundamental studies are continuing on the metabolism, cell membrane integrity, and reproduction of *Pythium* root rots and related fungi with the aim of developing more effective controls and consequently increased crop yields.

3. Enzymes in Metabolic Pathways

Included in the research on enzymes in metabolic pathways were studies of a stereoisomerase in L-arabinose fermentation, preparation of various phosphate intermediates, hexokinase activities of yeasts, pyruvic dehydrogenase system of *Clostridium pasteurianum*, xylulokinase and other kinases of *Aerobacter aerogenes*, isomerases, etc. (Simpson was particularly active in these investigations.) Tryptophanase from *Bacillus alvei*, glucanases, and cellulases were studied by Dr. D. E. Eveleigh.

H. Degradation of Pollutants

Early work, concerned with alkyl benzene sulfonate complex, was discontinued with the advent of biodegradable detergents. Currently Dr. E. R. Blakley is investigating the mechanisms by which aromatic, alicyclic halogenated compounds and related pesticides are degraded by microorganisms. The enzymes involved in the conversions are isolated and studied with the view of studying the factors influencing the biodegradability of related pesticides.

Dr. Slater is undertaking a study of waste products in spent bleach from pulp mills to identify the chlorophenols and chloro-compounds of lower molecular weight than lignin formed by conversion of the residual lignin. Material from pulp mill lagoons is being examined for microorganisms which have the ability to degrade the lignin and for the lignin degradation products present in the spent bleach liquor.

I. Biological Nitrogen Fixation

In the study of biological nitrogen fixation by Dr. T. A. LaRue and Dr. W. G. W. Kurz there is the possibility of increasing plant protein production by increasing the biological conversion of molecular nitrogen to ammonia. This involves fundamental studies on the nitrogen-fixing enzyme nitrogenase. Current studies are on nitrogenase production by bacteria, control of nitrogenase activity, nitrogenase in blue-green algae, and batch and synchronous culture of *Azotobacter*.

J. Yeasts and Yeast-like Fungi

For a long time, Dr. Spencer and Dr. P. A. J. Gorin have been active in the fields of ecology, metabolism, identification, and biochemical taxonomy (based on their polysaccharides) of various yeasts and related fungi. Early work was concerned with the production of polyhydric alcohols by osmophilic yeasts. This involved tracer studies with labeled glucose, recovery of the alcohols from yeast fermentation mixtures, examination of factors influencing production, and studies on biosynthesis of erythritol and glycerol.

The fermentation of long-chain compounds by *Torulopsis* was concerned with the structure of the hydroxy fatty acids obtained by fermentation of fatty acids and hydrocarbons. The formation of D-arabitol by osmophilic yeast was studied. During these studies several new species of yeasts were described and named. Extracellular glycolipids and mannans from *Rhodotorula* were described.

In connection with the flavonoid studies mentioned elsewhere (Section II,E,1), the utilization of flavonoid compounds by yeasts and yeast-like fungi was investigated. The extracellular polysaccharides and lipids of yeast were studied as were a number of new metabolites and the utilization of various compounds.

These studies led to an investigation of the proton magnetic resonance (p.m.r.) spectra of mannans in a wide variety of yeasts. These spectra were shown to be useful as an aid in the identification and classification of yeasts. The systematics of the various yeast genera were reviewed from this standpoint using the p.m.r. spectra of the mannans and the mannose-containing polysaccharides present in the cells. A comparison was made of the p.m.r. spectra of cell-wall mannans and galactomannans of selected yeasts with their chemical structure. The results showed that polysaccharides with similar p.m.r. spectra have related chemical structures and may therefore arise from phylogenetically related yeasts. Some 60 papers have resulted from these studies with Spencer and Gorin the most frequent authors.

Current work is designed to investigate the use of biochemical taxonomy as an improvement in the identification and classification of yeasts and fungi, especially those important in commercial processing as pathogens or introduced into the environment as a result of human activities; and to identify, isolate, and produce new industrial chemicals from metabolic byproducts through fermentation.

III. Higher Plant Studies and Plant Cell Culture

Other activities of the Prairie Regional Laboratory have developed over the years. At present there are groups specifically concerned with: proteins; high protein seed crops (field peas); carbohydrates; vegetable fats and oils; oil seed crops (rapeseed); forest products (fibrous product utilization, chemotaxonomy of conifers by GLC analysis of terpenes); composition and analysis of plant leaf waxes and beeswax; and plant cell cultures.

This last group merits the attention of microbiologists for this is an area in which microbiological techniques have been adopted for studies of cultures of higher plant cells. The plant cell group is studying (1) conditions required for the production of plant cells by continuous systems as compared to the batch culture techniques in use at present (Kurz); (2) methods for the control of stage of maturity of such plant cells to produce homogeneous cultures as an aid to physiological and biochemical studies (Kurz); (3) analytical methods for determination of hormones in plant cell cultures for application in the studies on cell growth, morphogenesis, and differentiation (Gorin); (4) the conditions required for the hybridization of somatic cells to provide the genetic material for the development of new plant species (Dr. R. A. Miller); and (5) the production of complete plants from single plant cells by control of the environment, including nutrients, with the view of possibly producing new plant species or hybrids not obtainable by other methods (Dr. O. L. Gamborg).

IV. In Retrospect — and the Future

The Laboratory in its first 25 years, has been, for the most part, a happy and productive place in which to work. Ample freedom has been granted to its investigators within the broad confines of the purposes for which the laboratory was built. The equipment has been excellent and up to date. Cooperation among the staff has been of the "instant" variety! What an aid it has been to the microbiologist to be able to get his complete amino acid or fatty acid analyses done quickly by an expert; to the chemist to have someone knowledgeable grow the mysterious organisms and produce the metabolites he needs; or to the mycologist to know that the sterols he is using are indeed pure, having been checked by the most sophisticated methods! Through it all, many opportunities were taken to aid in the solution of regional problems of diverse kinds. Much has been added to scientific knowledge.

At the start of its second quarter-century, the laboratory has entered a new era. A series of investigations has been undertaken, the successful conclusion of which, it is hoped, will have earlier and more readily apparent benefits for the Canadian taxpayer! To each of the projects, each staff member contributes as much as his skills and training make feasible, determined that the next 25 years shall be at least as happy and productive as the first 25 have been!

References

More than 980 scientific papers have been published from the laboratory since its inception; 479 of these have been concerned with microbiological topics. These cannot all be cited here. The author declines to list those he considers to be of greater importance than others and apologizes to those whose names do not appear! A limited number of copies of a complete listing is available from the Director's Office, National Research Council of Canada, Prairie Regional Laboratory, Saskatoon, Saskatchewan, Canada.

Cutter, V. M. (1959). *Mycologia* **51**, 248–295.
Haskins, R. H. (1963). *Mycologia* **55**, 365–370.
Reed, R. W., and Holder, M. A. (1953). *Can. J. Med. Sci.* **31**, 505–511.
Turel, F. L. M. (1969). *Can. J. Bot.* **47**, 821–823.
Williams, P. G., Scott, K. J., and Kuhl, J. L. (1966). *Phytopathology* **56**, 1418–1419.

Author Index

Numbers in italics refer to the pages on which the complete references are listed.

A

Aaronson, H. 147, *156*
Aaronson, S., 147, *156*
Abdel-Fattah, A. F., 54, 56, *70*
Abdel-Samie, M., 54, 56, *70*
Abderhalden, R., 10, *11*
Ablondi, F. B., 54, 56, *72*
Adams, G. J., 400, *413*
Adams, J. R., 193, 195, 196, *210*
Adams, S. L., 105, 111, 115, 120, *142*
Adesnik, M., 308, *331*
Adler, C. R., 399, *410*
Adler, R. J., 406, *414*
Adler, W. H., 217, *228*
Äyräpää, T., 106, 134, *142*
Agui, N., 180, 204, *209*
Ahiko, K., 62, *69*
Ahrenst-Larsen, B., 77, 78, 80, 83, 86, 88, 89, 90, 91, 95, 97, 103, 108, *142*
Aiba, S., 367, 402, 406, *410*, *411*
Ainsworth, G. C., 54, *66*
Aizawa, K., 163, 170, 190, 192, 197, *209*
Akiba, T., 338, 342, *363*
Alais, C., 42, 46, 60, 61, *66*, *70*, *72*
Albertsson, J. G., 361, *363*
Albu-Weissenberg, M., 16, 22, *37*
Aldridge, W. N., 44, *66*
Alford, E. D., 140, *143*
Algranati, I. D., 301, *332*
Allen, E. H., 301, *331*
Allende, J. E., 303, *334*
Allison, J. L., 299, *331*
Allnoch, H., 241, 242, 244, 245, 248, *295*
Almashi, K. K., 80, 83, 86, 88, *142*
Al'-Nuri, M. A., 324, *331*
Alvarez, W. C., 7, *11*
Amargier, A., 192, 193, 196, *212*, *214*
Amberkar, S., 388, 398, *412*
Ambike, S. H., *227*
Ambrose, C. T., 301, *331*
Amerine, M. A., 76, 139, *142*
Amos, H., 301, *331*
Amundson, N. R., 368, 373, 375, 378, 406, *412*, *413*
Anagnostakis, S. L., 60, *70*

Ananthakrishnan, C. P., 50, 52, 54, *71*
Anantharamaiah, S. N., 50, 52, 54, *71*
Anderson, H. W., 298, 299, *332*
Andreasen, A. A., 105, 111, 115, 120, *142*
Andreen, B. H., 359, *365*
Andrews, J. F., 351, *363*
Anfinsen, C. B., 16, *36*
Annibaldi, S., 50, 51, *66*
Anraku, N., 311, *331*
Anson, M. L., 43, *66*
Anthony, C., 341, 345, 346, *363*
Antila, V., 63, *69*
Apirion, D., 302, *332*
Arai, S., 45, *68*
Archetti, I., 169, 172, *210*
Arditti, J., 224, *229*
Arima, K., 43, 49, 50, 51, 56, 57, 58, 61, 62, 66, 69, 73, 375, *412*
Armentrout, S., 301, *336*
Arnold, C. G., 313, *334*
Aronson, A. I., 308, *331*
Arrambide, E., 219, *228*
Arvy, L., 174, *209*
Asmundson, C. M., 136, *144*
Asthana, H., 339, 344, 358, 361, *363*
Atkinson, W. R., 373, *412*
Atlas, D., 15, 21, *36*
Augstkalns, V. A., 398, *412*
Augustyn, O. P. H., 93, 141, *142*
Aunstrup, K., 57, 63, *66*
Aveyard, M., 224, *230*
Avrameas, S., 19, 24, *35*, *38*
Axén, R., 16, 22, 35, *37*
Azoulay, E., 138, *143*

B

Babbar, I. J., 41, 42, 45, 49, 50, 51, 52, 54, 66, *71*, *72*
Babel, F. J., 43, 47, 51, 62, 67, 69, *71*
Bach, F. H., 217, *229*
Bachler, M. J., 15, 20, *35*
Baddiley, J., 311, *335*
Bailey, B. K., 221, *230*
Bailey, J. A., 224, *227*
Bailey, J. M., 222, *228*
Bain, B., 301, *335*

Chattopadhyay, S. K., 17, 20, 23, 26, *36*
Chaudhary, S. S., 77, 78, 84, 87, 88, 89, 90, 92, 94, 95, 96, 97, 98, 99, 100, 102, 103, 104, 106, 107, 108, 109, 110, 113, 116, 118, 120, 121, 123, 124, 125, 126, 127, 128, 129, *142*
Chazov, E. I., 4, *11*
Chebotarev, A. I., 55, 56, 67
Chen, G. K. C., 407, *411*
Chen, H. T., 398, 406, *411*
Chen, J. H., 48, 63, 69
Chen, J. S., 175, 178, 179, *209*
Chen, L. L., 43, 69
Chen, M., 223, *228*
Chen, M. S. K., 403, 404, 405, 408, *411, 413*
Chen, W.-P., 49, 72
Chesnut, V. K., 101, *145*
Chiba, Y., 45, *71*
Chibata, I., 15, 16, 20, 22, 27, 28, *35*
Chisholm, D. R., 246, 247, 248, 292, *296*
Chisholm, J. A., 359, *365*
Chistyakova, A. V., 195, *209*
Chiu, R.-J., 170, 173, 183, *201, 209*
Cholette, A., 406, *411*
Choteau, J., 138, *143*
Choudhery, A. K., 52, 53, 67
Christensen, P. A., 51, 67
Church, J. M., 369, 371, 372, 376, 401, *411, 413*
Church, R. L., 218, *228*
Ciak, J., 299, 300, *331, 333*
Claridge, C. A., 240, 252, *294*
Clarke, B. J., 93, 131, 135, *143*
Cleeland, R., 244, 245, 246, 247, *294, 295*
Clements, A. N., 167, *209*
Cloutier, L., 406, *411*
Cockbain, E. G., 400, *411*
Coffman, J. R., 82, 83, 85, 90, *145*
Cohen, G. N., 327, *331*
Cole, A. A., 207, *209*
Cole, E. W., 79, 80, 87, 94, 111, *143*
Cole, M. B., 207, *209*
Cole, R. D., 351, *363*
Cole, R. J., 151, *156*
Collier, B., 42, 43, 67
Collins, E., 79, 82, 87, 94, *143*
Collins, S. B., 372, 397, 398, *411*
Colmore, J. P., 249, *294*
Colombo, G., 180, *212*
Conn, E. E., 137, *143*

Conn, H. W., 39, 48, 49, 51, 67
Conner, H. A., 77, 78, 79, 80, 81, 83, 84, 86, 88, 89, 90, 91, 92, 96, 97, 98, 99, 102, 103, 104, 105, 106, 107, 109, 110, 111, 112, 113, 114, 116, 117, 118, 120, 123, 124, 127, 128, 129, 130, 134, 138, *143*
Connolly, J. D., 223, *228*
Conti, G., 318, *331*
Conti, S. F., 341, *364*
Controulis, J., 298, *331, 334*
Converse, J. L., 198, *209*
Conway, T. W., 303, *334*
Coon, M. J., 138, *144*
Cooney, C. L., 339, 344, 355, 356, 360, 361, *364*
Coons, A. H., 301, *331*
Cooperbund, S. R., 217, *228*
Corcoran, J. W., 301, *334*
Coriell, L. L., 189, *210, 211*
Cory, J., 159, 181, 198, 201, 202, 203, *214*
Costich, E. W., 370, *413*
Cotton, N. T., 361, *365*
Coty, V. F., 338, 358, *364*
Coulaloglou, C. A., 368, *413*
Counce, S. J., 172, *209*
Coutsogeorgopoulos, C., 302, 304, *331*
Cresswell, P., 29, *36*
Creveling, R. K., 134, *143*
Cron, M. J., 234, 235, 246, 253, 254, 256, *295*
Crock, E. M., 21, 28, 29, *36, 38*
Crooks, H. M., Jr., 298, *331, 334*
Cross, J. H., 164, 174, 178, 183, 188, *210*
Crothers, W. C., 181, *212*
Cruickshank, R., 10, *11*
Crutchfield, G., 21, *36, 66, 68*
Cunningham, I., 202, *209*
Curl, R. L., 373, 387, 388, 389, 392, 393, 396, 397, 401, *411, 412*
Cutter, V. M., 430, *435*
Czulak, J., 44, 45, 67

D

Dagley, S., 303, *332*
Dale, E. C., 14, 21, *38*
Damerell, G. L., 387, 388, 389, 391, 396, 397, 402, *412*
Damodaran, M., 50, 67
Danckwerts, P. V., 406, *411*
Danner, G. A., 357, *364*
Das, H. K., 302, *332*

AUTHOR INDEX

Das, J., 17, 21, 26, 36
Dashek, W. V., 219, 228
David, J. R., 217, 230
Davies, J., 301, 336
Davies, S. L., 350, 364
Davis, B. D., 300, 332
Davis, F. C., 307, 332
Davis, J. G., 42, 47, 67
Davis, R. H., 342, 364
Davis Szekeres, A., 72
Dawson, R. F., 221, 228
Day, M. F., 207, 209
Deal, C., 222, 230
De Feo, J. J., 223, 229
De La Fuente, G., 135, 145
De Lamater, E. D., 300, 332
Delfs, E., 222, 229
Dellweg, H., 77, 78, 79, 85, 95, 98, 120, 131, 142, 143
de Madrid, A. T., 198, 212
Demain, A. L., 299, 328, 332
Demal, J., 163, 209
deMan, J. M., 42, 67
De Moss, J. A., 301, 302, 305, 332, 335, 336
De Pizzol, J., 140, 145
Determann, H., 26, 38
de Urse, C. A., 219, 228
Deutsch, V., 181, 209
De Vault, R. L., 256, 296
Devine, L. F., 243, 248, 249, 295
DeVries, H., 309, 332
DeVries, M. J., 117, 146
Dewane, R. A., 45, 46, 49, 51, 57, 67
Dhar, M. M., 299, 332
Dienes, L., 300, 324, 332
Dimitrov, D., 50, 52, 55, 68
Dimmick, P. S., 140, 144
Dioguardi, N., 10, 11
Divatia, M. A., 42, 71
Diwan, B. N., 311, 334
Djordjevic, B., 301, 332
Dobberstein, R. H., 221, 228
Dobrovol's'ka, H. M., 195, 209
Dolan, T., 200, 213
Dolfini, S., 181, 183, 209
Dorfman, A., 219, 228
Dostalek, M., 343, 359, 360, 363, 402, 404, 405, 411, 413
Dougherty, K., 167, 214
Douglas, J. M., 404, 411

Douglas, P. E., 141, 144
Drane, W., 207, 212
Drawert, F., 77, 79, 80, 81, 84, 85, 86, 91, 92, 93, 98, 99, 100, 102, 104, 106, 108, 110, 114, 117, 119, 124, 134, 143
Drum, D. E., 2, 11
Dryden, F. D., 352, 365
Duchâteau-Bosson, G., 161, 211
Dudani, A. T., 41, 42, 45, 49, 50, 51, 52, 54, 66, 71, 72
Dulaney, E. L., 324, 332
Dulbecco, R., 166, 209
Duma, R. J., 243, 295
Dunitz, J. D., 303, 332
Dunn, I. J., 404, 411
Dunnick, J. K., 325, 332
Dunnill, P., 16, 18, 20, 22, 25, 31, 32, 37
Dunstan, P., 346, 363
Durand, G., 18, 37
Durova, Zh. I., 55, 56, 67
Durr, A., 223, 228
Dutheil, H., 46, 66
Dutta, S. M., 50, 52, 54, 71
Dworkin, M., 337, 338, 340, 345, 364
Dyachenko, P. F., 41, 56, 68
Dyr, J., 77, 79, 81, 83, 93, 95, 102, 143

E

Ebata, M., 306, 332
Eberhardt, H., 313, 315, 326, 333
Echalier, G., 169, 172, 181, 183, 194, 198, 209, 210, 212
Edelsten, D., 63, 64, 65, 68, 69
Edwards, B. A., 21, 38
Edwards, J. L., 45, 68
Egami, F., 306, 332
Egorov, N. S., 324, 331
Eguchi, W., 375, 412
Ehrhardt, J. D., 220, 230
Ehrlich, F., 106, 115, 134, 143
Ehrlich, J. E., 298, 299, 319, 332, 334
Eide, P. E., 172, 175, 209
Eillers, N. J., 147, 156
Elizarova, T. N., 340, 364
El Kousy, L., 63, 68
Elliott, J. A., 44, 69
Ellis, B. J., 161, 167, 173, 182, 189, 210
El-Negoumy, A. M., 41, 68
Elsden, S. R., 354, 363
Elson, D., 301, 336
Elsworth, R., 361, 364

Elton, G. A. H., 400, *411*
Emanuiloff, I., 50, 52, *68*
Emery, A. N., 18, *35*
Emmons, D. B., 47, *68*
English, J., 352, *365*
Epton, R., 14, 20, *35*
Erdelyi, A., 324, *334*
Erdtsieck, B., 402, *411*
Erickson, L. E., 368, 399, 403, 404, 405, 407, 408, *411, 412, 413*
Ernback, S., 16, 22, *35*, 37
Ernstrom, C. A., 41, 42, 43, 48, *68*, 70, *71*
Esau, P., 138, 139, *145*
Evans, D. A., 310, *333*
Evans, G. H., 402, *412*
Evans, J. E., 50, 52, 53, 54, *71*
Everett, H. J., 370, *413*
Everson, T. C., 42, *68*
Ewers, J., 14, *37*
Ezekiel, D. H., 308, *332*

F

Fagerson, I. S., 134, 140, *144*
Fahey, J. L., 217, *228*
Falb, R. D., 14, 21, *36*
Fan, L. T., 368, 403, 404, 405, 407, 408, *411, 413*
Fardig, O. B., 254, 256, *295*
Farese, R. V., 301, *332*
Farngui, A. A., 398, *411*
Farrell, J., 140, *143*
Fassin, W., 299, *332*
Faulkner, P., 191, 192, 196, *210, 214*
Fawwal, I., 60, *68*
Fedorko, J., 241, 242, 244, 245, 248, *295*
Fedoroff, S., 158, *210*
Feil, M. F., 78, 79, 80, 81, 82, 93, 99, *144*
Feingold, D. S., 300, *332*
Feldman, L. I., 63, *68*
Fencl, Z., 402, 404, 405, *411, 413*
Ferrier, L. K., 66, *68*
Fick, J. L., 372, *414*
Fiechter, A., 405, *411*
Filippusson, H., 17, 23, *36*
Filshie, B. K., 201, *210*
Finegold, I., 217, *228*
Finn, R. K., 339, 343, *365*
Fioriti, J. A., 140, *143*
Firkin, F. C., 309, *332*
Fischer, E. H., 301, *331*

Fischer, R. G., 200, *213*
Fish, N. L., 43, 63, *70*
Fitzgerald, R. J., 3, *11*
Flath, R. A., 85, 102, 104, 110, 111, *145*
Fleischman, A. I., 5, *11*
Florkin, M., 161, *211*
Folkers, K., 225, *229*
Folkes, J. P., 301, *332*
Foltmann, B., 41, 47, 48, *68*
Foltz, E. L., 287, *296*
Forget, B. G., 308, *333*
Forss, D. A., 134, 140, *143, 146*
Foster, J. W., 337, 338, 340, 342, 345, 362, *364*
Foulkes, P., 14, 20, *35*
Fox, A. S., 163, 172, 183, *210*
Fox, P. F., 43, 45, 47, 48, 56, *68*
Freeman, K. B., 304, 310, *332*
Freeman, N. K., 137, *145*
Frentz, R., 42, *72*
Friend, I., 224, *230*
Fritig, B., 220, *230*
Fujimaki, M., 45, *68*
Fujisawa, K., 231, 232, 237, 239, 241, 242, 243, 244, 245, 246, 247, 248, *295*
Fukada, T., 195, *210*
Fukaya, M., 180, 204, 205, *209, 214*
Fukimara, T., 338, 342, *363*
Fukumoto, J., 53, 55, 57, 58, *68*, 72
Fulco, L., 204, *212*
Furlenmeier, A., 231, 232, 233, 239, 252, *294, 295*
Furuya, T., 221, 223, 225, *228*
Fuse, N., 15, 16, 20, 22, 27, 28, *37*
Fusillo, M. H., 301, *332*

G

Gabe, M., 174, *209*
Gajzago, I., 42, 63, *72*
Gale, E. F., 300, 301, *332*
Galetto, W. G., 77, 83, 84, 86, 89, 90, 91, 94, 96, 97, 98, 102, 103, 106, 107, 108, 109, 110, 111, 113, 114, 116, 118, 121, 123, 124, 126, 127, 139, *143, 146*
Gallicchio, V., 322, *332*
Gal-Or, B., 368, *411, 413*
Gamborg, O. L., 220, 221, *229, 230*
Gancedo, C., 135, *145*
Gandhi, N. R., 42, *71*
Garcia-Patrone, M., 301, *332*
Garnier, J., 41, *68*

Garzon, S., 192, *211*
Gaulden, M. E., 207, *210, 212*
Gavrilova, L. P., 313, *335*
Gaw, Z.-Y., 173, 182, 190, 192, 196, *210*
Gehrig, R. F., 134, *143*
Germanova, K. I., 303, 304, *335*
Gertzman, D. P., 63, 64, 67
Gey, G. O., 219, 222, 228, 229
Gey, M. K., 222, *228*
Ghose, T. K., 32, *36*
Giam, C. S., 319, *331*
Giauffret, A., 173, *210*
Gibson, F., 312, 332, 333
Gillan, R. H., 47, *68*
Gilles, J., 45, 62, 69, *71*
Gillespie, T., 400, *411*
Glassmeyer, C. K., 19, 28, *36*
Glick, J. L., 216, *228*
Godchaux, W., 310, *332*
Godfrey, J. C., 234, 235, 246, 247, 248, 254, 256, 257, 258, *295, 296*
Godman, G., 219, *228*
Goldberg, I. H., 300, *332*
Goldlust, M. B., 218, *230*
Goldman, R., 19, 26, 28, *36*
Goldschmidt, R., 160, *210*
Goldstein, A., 302, *332*
Goldstein, J. L., 225, *228*
Goldstein, L., 15, 16, 21, 23, 28, 29, *36*
Gonzalez, N. S., 301, *332*
Goodwin, R. H., 193, 195, 196, *210*
Goosdev, V. A., 183, *211*
Gordon, D. F., Jr., 44, *68*
Gorini, C., 39, 48, 50, 51, *68*
Gorini, L., 43, *68*
Gorman, M., 220, *228*
Gottlieb, D., *296*, 298, 299, 319, 322, 323, 324, 326, 328, 329, *332, 333*
Gourevitch, A., 240, 246, 247, 248, 258, *294, 295, 296*, 326, *332*
Govier, G. W., 398, *411*
Govindarajan, V. S., 50, *67*
Goyena, H., 219, *228*
Grace, J., Jr., 217, *229*
Grace, T. D. C., 157, 159, 161, 162, 167, 173, 174, 180, 181, 182, 188, 189, 190, 193, 195, 196, 197, 201, 205, 206, *209, 210, 212, 214*
Grajevskaja, R., 302, *331*
Granger, H., 217, *228*

Graumann, R. E., 388, 398, *412*
Graves, B. S., 287, *296*
Graves, J. M. H., 223, *228*
Gray, E., 312, *334*
Gray, W. J. H., 308, *334*
Green, G. M., 18, 24, *37*
Green, J. A., 217, *228*
Green, M. L., 21, *36*, 66, *68*
Greenberg, B., 169, 172, *210*
Greenberg, D. M., 45, *68*
Greene, A. E., 188, 189, *210*
Grieves, R. B., 406, *411*
Grimberg, M. V., 44, *68*
Grimmer, G., 137, *143*
Grobstein, C., 177, *210*
Groger, D., 220, *230*
Groothuis, H., 372, 388, 389, 393, 396, 401, 402, *412*
Gros, F., 301, 308, *333, 334*
Gross, W., 310, *332*
Grossfeld, H., 219, *228*
Grove, M. J., 25, *36*
Grubhofer, N., 17, 23, *36*
Grunberg, E., 232, 241, 242, 244, 245, 246, 247, *294, 295*
Gudkov, A. V., 45, 49, 63, *73*
Guenther, K. R., 355, *364*
Guertin, M., 308, *334*
Guilbault, G. G., 17, 21, 26, *36*
Gullberg, J. E., 397, *412*
Gunderson, M. F., 50, *70*
Gurgo, C., 302, *332*
Gust, H.-P., 42, *69*
Gutfriend, H., 1, *11*
Gutman, M., 22, *36*
Guymon, J. F., 134, *143*

H

Habeeb, A. F. S. A., 19, 21, *36*
Häggström, L., 339, 342, 358, 359, *364*
Haffen, K., 219, *228*
Hageman, R. H., 306, *335*
Hagemeyer, K., 60, *68*
Hagen, K. W., 200, *210*
Hagerman, C. R., 243, 248, 249, *295*
Hagihara, B., 43, *68*
Hahn, F. E., 298, 299, 300, 301, 303, 312, *331, 332, 333*
Haider, K., 218, *228*
Halder, D., 310, *332*

Hall, N. L., 181, *212*
Halpern, M., 217, *228*
Hamdy, A., 63, 64, 65, *68*, *69*
Hamer, G., 341, 358, 361, *364*
Hammer, B. W., 43, 50, *69*
Hancock, I. C., 311, *335*
Hancock, R., 301, 310, *333*
Haňka, L. J., 147, 153, 154, 155, *156*
Hannoun, C., 198, *210*
Hanoch, A., 244, *296*
Hansen, H. L., 77, 78, 80, 83, 86, 88, 89, 90, 91, 95, 97, 103, 108, *142*
Hansford, G. S., 406, *412*
Hanson, C., 372, *411*
Hanson, J. B., 309, 310, *333*, *335*
Hara, O., 316, *333*
Harada, M., 375, *412*
Harada, T., 94, 141, *143*
Harborne, J. B., 224, *229*
Hardwick, W. A., 78, 79, *144*
Hardy, P. J., 119, 121, 123, 126, 128, *143*
Harms, H., 218, *228*
Harold, F. V., 93, 131, 135, *143*
Harrington, A. A., 340, 344, 345, 346, *364*
Harris, D. A., 324, *336*
Harrison, D. E. F., 358, *364*
Harrison, G. A. F., 79, 82, 87, 94, *143*
Harrison, J. E., 338, 348, *365*
Hartman, R. E., 299, *331*
Hartman, R. S., 299, *331*
Hartzman, R. J., 217, *229*
Harwalkar, V. R., 44, *69*
Harwood, R. F., 200, *210*
Hashimoto, T., 324, *334*
Hashimoto, Y., 193, 195, *210*
Haskins, R. H., 418, 431, *435*
Hata, T., 44, *70*
Hattingh, W. H. J., 350, 351, 352, *365*
Hattori, A., 55, *72*
Haung, K. L., 367, 402, 406, *410*, *411*
Havewala, N. B., 30, 31, *38*
Havlova, J., 49, *70*
Hawke, J. C., 135, *143*
Hawkins, N. G., 77, 78, 80, 82, 83, 89, 96, 102, 109, 113, 118, 123, 127, 128, *143*
Hawksley, J. L., 373, *412*
Hayashi, R., 44, *70*
Hayashi, T., 218, *230*
Hayes, J. E., 298, 303, *333*
Hayes, W. B., 398, *413*

Haynes, B., 310, *333*
Hazem, A., 313, *334*
Heble, M. R., 223, *229*
Hedley, B., 357, *364*
Hegedus, V. E., 42, 63, *72*
Heinz, D. E., 134, *143*
Heinz, R., 301, *335*
Heitor, F., 192, *210*
Helmke, V., 79, 80, 87, 94, 111, *143*
Hengel, R., 299, *332*
Hennig, K., 80, 105, 110, *143*
Herbert, D., 361, *364*, 406, *412*
Herbert, E., 310, *332*
Hersh, L. S., 18, 25, *38*
Hesseltine, C. W., 58, *72*
Hicks, G. P., 17, 26, 31, *36*, *37*
Hildebrand, R. P., 93, 131, 135, *143*
Hildebrandt, A. C., 216, *229*
Hill, I. D., 362, *364*
Hill, R. D., 41, 43, 67, *69*
Hill, R. J., 41, *69*
Hiller, G., 357, *364*
Hillestad, J. G., 388, 389, 393, *412*
Himeno, M., 195, *210*, *212*
Hink, W. F., 157, 161, 163, 167, 168, 169, 173, 178, 182, 188, 189, 193, 195, 197, 200, *210*, *211*
Hinman, J. W., 253, 254, *295*
Hinuma, Y., 217, *229*
Hinze, J. O., 369, 376, *412*
Hippel, P. H., 41, *72*
Hirabayashi, T., 94, 141, *143*
Hirose, Y., 90, 91, 97, 99, 100, 103, 104, 105, 108, 109, 111, 112, 113, 114, 115, 116, 117, 118, 119, 120, 121, 123, 125, 126, 127, 128, 134, *143*
Hirotani, M., 223, *228*
Hirsch, P., 341, *364*
Hirth, L., 223, *228*
Hirumi, H., 169, 173, 202, *210*
Ho, P. L., 219, *228*
Ho, R., 312, *333*
Hoare, D. S., 343, 351, 352, *365*
Hodges, A. B., 324, *333*
Hodges, T. K., 309, 310, *333*, *335*
Hoeksema, H., 253, 254, *295*
Hoeprich, P. D., 243, 248, 249, *295*
Holder, M. A., 424, *435*
Holland, C. D., 398, *413*
Holtzer, H., 219, *229*

Homas, A. L., 224, *230*
Hooper, I. R., 234, 235, 246, 254, 256, 257, 258, *295*
Hoover, S. R., 41, *67*
Hopps, H. E., 298, 301, 303, 312, *333*
Horikawa, M., 163, 172, 183, *210*
Hornby, W. E., 17, 21, 23, 26, 29, 32, *36*, *37*
Horton, T. J., 388, 398, *412*
Hosokawa, K., 307, *333*
Houck, J. C., 218, *229*
Houston, R. H., 372, *414*
Howarth, W. J., 387, 388, 389, 394, 401, *412*
Hrdlicka, J., 77, 79, 81, 83, 93, 95, 102, *143*
Hsu, C. J., 16, 23, *37*
Hsu, S. H., 164, 174, 178, 181, 183, 188, *210*
Huang, C., 61, *69*
Huang, W. Y., 222, *229*
Hudson, J. O., 362, *365*
Huffee, M., 244, *230*
Hukuhara, T., 193, 195, *210*
Hume, D., 4, *11*
Humphrey, A. E., 34, *36*, 339, 344, 358, 361, *363*, *364*, 367, 402, 403, 405, 406, 408, *411*, *412*
Hunt, D. E., 155, *156*
Hunt, G. A., 258, *295*
Hunter, I. R., 77, 78, 80, 82, 83, 89, 96, 102, 109, 113, 118, 123, 127, 128, *143*
Hunter, K., 135, 136, *143*
Hunter, M. E., 300, *332*
Hunter, R. J., 140, *144*
Hurwitz, C., 302, 304, *333*
Hussang, R. V., 50, *69*
Hutner, S. H., 147, *156*
Hutton, W. E., 337, *364*

I

Ibrahim, R. K., 224, *229*
Ichikawa, A., 368, *414*
Ignoffo, C. M., 169, 173, 178, 182, 189, 193, 195, 197, *210*, *211*
Ikai, M., 45, *71*
Ikeda, R. M., 78, 81, 83, 84, 90, 91, 97, 105, 108, 118, 121, 123, 126, 127, 128, 129, *143*, *146*
Ilany-Feigenbaum, J., 43, 46, *67*, *69*
Imai, N., 50, *69*
Imai, S., 105, 111, *144*

Imai, T., 50, 51, *69*
Imoto, T., 369, 372, *412*
Ingall, D., 318, *333*
Ingraham, J. L., 134, 136, *144*, *145*
Irie, S., 50, *69*
Irie, Y., 50, 51, *69*
Ishihara, R., 202, *211*
Ishiyama, S., 248, 249, *295*
Itoh, M., 59, *70*
Itoh, T., 307, *334*
Ittycheriah, P. I., 174, *211*
Ivanova, T. V., 52, *70*
Iwasaki, S., 43, 49, 50, 51, 56, 57, 58, 61, 62, *66*, *69*
Iya, K. K., 49, 50, 52, 54, *71*, *72*
Iyengar, M. K. K. 49, *72*

J

Jack, E. L., 137, *145*
Jackson, S. F., 218, *230*
Jackson, W. G., 254, *295*
Jacob, J., 137, *143*
Jago, G. R., 44, 45, *69*, *72*
Jamison, V. W., 362, *365*
Jansen, E. F., 19, *36*
Jansen, R. J., 309, *333*
Jardetsky, O., 303, 304, *333*
Jarvin, B., 6, 10, *11*
Jeffreys, G. V., 373, 401, *412*
Jely, E., 77, 79, 81, 83, 93, 95, 102, *143*
Jennings, W. G., 134, *143*
Jensen, D. D., 173, 183, *213*
Jensen, J. S., 64, *68*
Jeuniaux, C., 161, *211*
Johnson, D. P., 248, 249, *295*
Johnson, H., 200, *213*
Johnson, I. S., 219, 220, *228*
Johnson, M. J., 342, 357, 359, 360, *365*, 402, *412*
Johnson, P. A., 341, 345, 346, *364*
Johnson, R. R., 140,*143*
Jolles, P., 164, 165, 167, *211*
Jollow, D., 136, *143*
Jonak, J., 306, *331*
Jones, M. J., 312, *332*
Jordan, B. R., 308, *333*
Jordan, G. V., 373, *412*
Joshi, P. N., 311, *334*
Joslyn, D. A., 298, 299, *332*
Judy, K. J., 205, 206, *211*

Julian, G. R., 302, 303, *333*
Jurriens, G., 140, *143*

K

Kahan, B. D., 217, *230*
Kahn, A. W., 299, *332*
Kahn, J. H., 76, 77, 78, 79, 80, 81, 83, 84, 86, 88, 89, 90, 91, 92, 96, 97, 98, 99, 102, 103, 104, 105, 106, 107, 109, 110, 111, 112, 113, 114, 116, 117, 118, 120, 123, 124, 127, 128, 129, 130, 134, 138, *143*
Kakpakov, V. T., 183, *211*
Kallio, R. E., 138, *144*, 340, 344, 345, 346, *364*
Kamen, M. D., 351, *363*
Kametani, R., 299, 319, *335*
Kamibayashi, A., 87, *146*
Kanari, H., 299, 319, *335*
Kanazawa, Y., 50, 51, *69*
Kanda, T., 192, *214*
Kaneda, T., 339, 340, 345, 347, *364*
Kaneshiro, W. M., 151, *156*
Kanner, L. C., 302, *332*
Kaplan, S. A., 248, *295*
Karapetian, M. G., 303, 304, *335*
Karr, A. E., 231, 232, 233, 239, *294*
Kassell, B., 47, *69*
Katchalski, E., 16, 19, 22, 23, 26, 28, 29, *35, 36, 37*
Katoo, M., 105, *143*
Katinger, H., 399, *412*
Kato, H., 45, *68*
Kato, J., 16, *36*
Katz, E., 312, 324, *333, 336*
Katz, I., 134, *144*
Katz, J., 326, *333*
Katz, S., 241, 242, 244, 245, 248, *295*
Kaul, B., 223, *229*
Kawade, Y., 195, *210*
Kawaguchi, H., 231, 232, 237, 239, 241, 242, 243, 244, 245, 246, 247, 248, 249, *295*
Kawaguchi, K., 223, *228*
Kawahara, O., 299, 319, *335*
Kawai, M., 56, 57, 58, 59, *69, 70*
Kawase, S., 195, *211*
Kay, G., 14, 20, 29, 31, 32, 34, *36, 37, 38*
Kayahara, K., 116, *143*
Keay, L., 43, *69*
Kedam, O., 19, 26, 28, *36*

Keency, M., 134, *144*
Keighley, G., 301, *331*
Keil, J. G., 234, 235, 246, 251, 254, 255, 256, 257, 258, *295*
Kellehev, G., 16, 23, *37*
Keller, O., 231, 232, 233, 239, 252, *294, 295*
Kellerman, G. M., 136, *143*
Kemp, M. B., 346, 347, 349, *364*
Kenard, R. J., 357, *364*
Kendall, M. S., 50, 53, 54, *70*
Kennel, S. J., 17, 22, *37*
Kenny, C. P., 224, *229*
Kepner, R. E., 77, 78, 80, 81, 82, 83, 84, 85, 86, 87, 88, 89, 90, 91, 92, 93, 94, 95, 96, 97, 98, 99, 100, 101, 102, 103, 104, 105, 106, 107, 108, 109, 110, 111, 112, 113, 114, 115, 116, 117, 118, 119, 120, 121, 122, 123, 124, 125, 126, 127, 128, 129, 130, 134, 135, 137, 139, 140, 141, *142, 143, 144, 145, 146*
Keränen, A. J. A., 135, *145*
Keshavamurthy, N., 50, 52, 54, *71*
Keyes, P. H., 3, *11*
Khanna, P., 221, 224, *229, 230*
Kikuchi, T., 52, 61, 62, *69*
King, K., 310, *332*
Kinsella, J. E., 140, *144*
Kintner, R. C., 388, 398, *412*
Kinzer, G. W., 140, *143*
Kira, H., 53, *72*
Kirilov, S., 302, *331*
Kirsch, E. L., 350, 352, *364*
Kiss, E., 63, *69*
Kitamura, S., 163, 169, 173, 178, 181, 183, 188, *211*
Kitazawa, T., 192, *214*
Kjeldgaard, N. O., 308, 320, *333*
Klein, U. P., 299, *332*
Klinzing, G. E., 368, *411, 413*
Kluyver, A. J., 351, *364*
Knight, D. J., 41, *72*
Knight, S. G., 55, 56, 57, 69, 134, *143*
Knudson, J. G., 398, *411*
Kobota, H., 408, *412*
Koehler, P. E., 139, *144*
Koenig, W. A., 79, 80, 81, 83, 86, 87, 88, 89, 91, 92, 93, 96, 100, 101, 102, 103, 104, 105, 106, 107, 109, 110, 112, 113, 114, 116, 117, 118, 119, 120, 121, 122,

123, 124, 125, 126, 127, 128, 129, 130, 134, 138, 139, *144*
Kojima, H., 221, 225, *228*
Kojima, M., 326, *333*
Kok, I. P., 195, *209*
Kolar, J. R., 147, *156*
Kolmogoroff, A. N., 369, 375, *412*
Kolosov, M. N., 303, 304, *335*
Komano, T., 195, *212*
Komasawa, S., 369, 387, 388, 389, 394, 395, 396, 401, 402, *412, 413*
Komoda, H., 85, 134, *146*
Kondo, E., 205, *214*
Kong, Y. L., 240, *296*
Kono, M., 316, *333*
Kordan, H. A., 225, *229*
Kornelli, B. M., 77, 81, 84, 87, 88, 89, 90, 91, 94, 95, 96, 97, 98, 102, 103, 104, 106, 109, 110, 113, 116, 118, 119, 123, 124, 126, 127, *145*
Kosaki, M., 134, *146*
Kosegarten, D., 223, *229*
Koshiyama, H., 231, 232, 237, 239, 241, 242, 243, 244, 245, 246, 247, 248, *295*
Kostick, J. A., 32, *36*
Kowal, J., 219, 222, *229*
Kowalska, W., 55, 56, *71*
Kraemer, P. M., 219, *229*
Kramers, H., 406, 407, *412*
Krampl, V., 140, *143*
Krejci, J., 217, *229*
Krenkel, H., 42, *69*
Krikorian, A. D., 216, 218, 220, 223, *229*
Krikova, N. N., 339, 342, *364*
Kritchevsky, D., 222, *230*
Kroon, A. M., 309, 332, *333*
Kruger, W., 42, *69*
Krupka, G., 54, 56, *72*
Krywienczyk, J., 191, 192, 196, *211*
Kuccan, Z., 302, *333*
Kuentzel, S. L., *156*
Kuhl, J. L., 431, *435*
Kuhns, D. M., 301, *332*
Kuila, R. K., 50, 52, 54, *71*
Kunitz, M., 41, 43, *69*
Kuno, G., 168, *211*
Kurland, C. G., 308, *333*
Kuroda, Y., 178, *211*
Kurstak, E., 192, *211*
Kurtti, T. J., 174, 175, 178, 180, 202, *211*

Kusuda, Y., 90, 91, 97, 99, 100, 103, 104, 105, 108, 109, 111, 112, 113, 114, 115, 116, 117, 118, 119, 120, 121, 123, 125, 126, 127, 128, 134, *143*
Kwaan, H. C., 218, *228*
Kwong, A., 18, 25, *36*
Kyla-Siurola, A. L., 63, *69*

L

Lacks, S., 301, *333*
Laine, B., 402, *412*
Lallier, R., 302, *333*
Lambeth, D., 1, *11*
Lambin, S., 311, *335*
Landman, O. E., 311, *331*
Landsberg, A., 398, *411*
Landureau, J.-C., 164, 165, 167, 168, 173, 181, 183, *209, 211*
Lang, S. B., 400, *412*
Langloise, G. E., 371, 397, *412, 414*
Lanzavecchia, G., 43, *68*
Lardy, H. A., 1, *11*
Large, P. J., 347, *364*
Lark, C., 305, *333*
Lark, K. G., 305, *333*
LaRoe, E. G., 77, 78, 79, 80, 81, 83, 84, 86, 88, 89, 90, 91, 92, 96, 97, 98, 99, 102, 103, 104, 105, 106, 107, 109, 110, 111, 112, 113, 114, 116, 117, 118, 120, 123, 124, 125, 127, 128, 129, 130, 134, 138, 139, *143, 144*
Larsen, W. P., 177, 207, *211*
Larson, M. K., 60, *69*
Larson, P. P., 401, *413*
Larsson, P. O., 26, *34, 37*
Larue, T. A. G., 220, *229*
Laskowski, M., 46, *69*
Laurence, H. S., 217, *229*
Laursen, P., 221, *230*
Laverdure, A.-M., 174, *211*
Lawrence, A. J., 346, 347, 350, *364*
Lawrence, R. C., 45, *69*
Lawrence, W. C., 141, *144*
Lawson, G. B., 373, 401, *412*
Laxminarayana, H., 56, 57, *71*
Layman, D. L., 218, *229*
Leadbetter, E. R., 337, 362, *364*
Leake, C. D., 301, *334*
Lebek, G., 317, *333*
Ledford, R. A., 48, 63, *69*

Lee, A., 145
Lee, S., 369, 372, 412
Legator, M., 323, 324, 328, 329, 333
Leibowitz, J., 46, 67
Lein, J., 240, 258, 294, 295, 326, 332
Leitner, F., 292, 296
Leloup, A. M., 163, 180, 209, 211
Lender, T., 174, 211
Lengyl, Z. L., 324, 334
Lenhoff, H. M., 19, 36
Lennarz, W. J., 138, 144
Leoni, L., 318, 331
Leopold, R. A., 205, 211
Lesseps, R. J., 170, 211
Levenspiel, O., 374, 413
Levi-Montalcini, R., 175, 178, 179, 209, 213
Levin, Y., 15, 16, 21, 23, 36
Levine, A. J., 305, 333
Levine, D. W., 339, 344, 361, 364
Levine, L., 217, 230
Levinthal, C., 308, 331
Levitov, M. M., 303, 304, 335
Levy, E. J., 142, 144
Lewandowski, L. J., 307, 333
Lewis, M. J., 141, 144
Lewis, R. F., 352, 363
Libiková, H., 200, 211
Lieberman, I., 326, 333
Liebich, H. M., 79, 80, 81, 83, 86, 87, 88, 89, 91, 92, 93, 96, 100, 101, 102, 103, 104, 105, 106, 107, 109, 110, 112, 113, 114, 116, 117, 118, 119, 120, 121, 122, 123, 124, 125, 126, 127, 128, 129, 130, 134, 138, 139, 144
Light, R. J., 312, 333
Lilly, M. D., 14, 16, 18, 20, 21, 22, 25, 29, 30, 31, 32, 34, 36, 37, 38
Lin, F. M., 140, 144
Lindqvist, B., 41, 69
Lindsay, R. C., 112, 145
Lindsey, E. E., 398, 412
Line, W. F., 18, 25, 36
Ling, L.-N., 163, 172, 183, 210
Lingens, F., 313, 315, 326, 333
Linker, A., 219, 228
Linnane, A. W., 136, 143, 309, 332
Linton, M., 400, 412
Lipmann, F., 302, 303, 326, 333, 334
Lite, S. W., 169, 172, 177, 214
Liu, H. H., 175, 181, 210, 213

Liu, H.-Y., 201, 209
Liu, N. T., 173, 182, 190, 192, 196, 210
Lloyd, D., 300, 309, 310, 333, 334, 335
Lloyd, N. E., 398, 412
Loeb, A., 50, 69
Loughheed, T. C., 44, 67, 161, 214
Louloudes, S. J., 167, 193, 195, 196, 210, 214
Lowe, D. A:, 329, 333
Lowenstein, L., 301, 335
Lubnow, R. E., 44, 62, 71
Luciani, J., 192, 193, 212, 214
Lund, E. D., 144
Luss, D., 368, 412
Luttinger, J. R., 258, 295
Lynen, F., 135, 144
Lynn, J., 14, 21, 36
Lyons, J. M., 136, 144

M

Maaløe, O., 308, 320, 333
Maarse, H., 83, 87, 91, 104, 110, 113, 114, 119, 121, 122, 124, 125, 134, 144
Maas, W. K., 326, 333
Mabrouk, S. S., 54, 56, 70
Maccacaro, G. A., 325, 331
McCarthy, D., 201, 212
McCarty, P. L., 351, 352, 365
McClesky, C. S., 341, 345, 364, 365
McComb, J., 53, 73
McCoy, B. J., 397, 412
McCullough, J. L., 8, 11
McDaniel, L. E., 324, 333
McDermott, G. N., 352, 365
McDonald, C. E., 43, 69
MacDonald, F. J., 42, 67
McDongall, B., 312, 333
McDonnell, C. D., 256, 296
McDougall, B., 312, 332
McFadden, W. H., 134, 145
McGarrity, G. J., 189, 211
McGoodwin, E. B., 218, 229
McGrath, R., 323, 335
McHale, J. S., 178, 182, 183, 213
MacKay, G. D. M., 400, 401, 412
McKenna, E. J., 138, 144
McKenzie, I. J., 61, 70
MacKinlay, A. G., 41, 70
Mackler, B., 310, 333
MacLennan, D. G., 361, 365
MacLeod, R., 201, 209

Macpherson, I., 222, *230*
McRoberts, T. S., 400, *411*
McWhirter, J. R., 361, *363*
Madden, A. J., 387, 388, 389, 391, 396, 397, 402, *412*
Madoff, S., 300, 324, *332*
Maeda, H., 15, 20, 21, *36, 37*
Magasanik, B., 305, *333*
Maggiora, L., 81, 82, 83, 85, 86, 88, 89, 90, 93, 94, 96, 97, 98, 99, 100, 101, 102, 103, 104, 105, 106, 107, 108, 109, 111, 112, 113, 116, 117, 118, 120, 121, 124, 126, 134, 135, 137, 141, *144, 146*
Majerska, M., 203, *213*
Malachouris, N., 63, 64, *67*
Malek, I., 367, *412*
Malek, Y. A., 313, *334*
Malik, V. S., 298, 299, 304, 319, 321, 322, 323, 324, 325, 326, 328, 329, 330, *334, 336*
Málková, D., 199, *211*
Mancy, D., 235, 236, 237, *295*
Mandelstam, J., 301, 310, *334*
Mandin, J., 203, *212*
Mandl, I., 2, 9, 10, *11*
Mandron, P., 178, *213*
Manecke, G., 17, 23, *36*
Mann, M., 9, *11*
Manoilov, S. E., 20, 21, *37*
Mao, W. H., 164, 174, 178, 183, 188, *210*
Maramorosch, K., 159, 169, 170, 173, 175, 177, *210, 212*
Marchant, R., 308, 310, *334, 335*
Mardashev, S. R., 15, 20, *37*
Margalith, P., 45, *70*, 133, *144*
Margoliash, E., 1, *11*
Marhoul, Z., 199, *211*
Maricz, J., 231, 232, 233, 239, *294*
Marinelli, L., 78, 79, 80, 81, 82, 93, 99, *144*
Mark, A. M., 399, *410*
Marks, E. P., 179, 180, 205, *211*
Maroti, M., 223, *230*
Maroudas, N. G., 318, *336*
Marr, A. G., 136, *144*
Marschner, F., 357, *364*
Marshall, J. J., 32, *36*
Marshall, W. R., Jr., 399, *410*
Martignoni, M. E., 166, 170, 174, 194, 196, *211*
Martin, B., 180, *211*
Martin, G. R., 218, *229*

Martin, S. M., 218, *230*
Martirosova, L. A., 55, *68*
Masek, J., 49, *70*
Mason, M. E., 139, *144*
Mason, R. D., 24, *38*
Mason, R. T., 5, *11*
Mason, S. G., 400, 401, *411, 412*
Masuda, M., 86, 87, 90, 91, 92, 93, 100, 101, 104, 105, 106, 107, 110, 111, 112, 115, 138, 139, 141, *144*
Mateles, R. I., 339, 343, 355, 356, 358, 360, *364, 365*, 367, *412*
Mathes, M. C., 223, *229*
Matoba, T., 44, *70*
Matsubara, H., 43, *68*
Matsuda, H., 388, 389, 394, 396, *412*
Matsudo, C., 50, 51, *69*
Matsumura, F., 207, *214*
Matsuoka, Y., 217, *229*
Matsuzawa, H., 388, 389, 390, *412*
Matthijsen, R., 66, *70*
Mattiasson, B., 22, 24, 26, 34, 35, *36, 37*
Mattingly, R. F., 222, *229*
Maubois, J. L., 41, 61, *68, 70*
Maul, S., 362, *365*
Mazzone, H. M., 174, 191, 193, *211*
Medora, R., 223, *229*
Medvedeva, N. B., 191, 192, 193, *211*
Mehltretter, C. L., 15, *37*
Mehta, D. D., 356, *365*
Meitner, P. A., 47, *69*
Meito Sangyo, K. K., *70*
Melachouris, N. P., 47, 50, 52, *70*
Melrose, G. J. H., 19, *37*
Mercer, E. H., 193, 195, 196, *209*
Merkel, J. R., 315, *334*
Messing, R. A., 18, 19, 25, *37*
Mettler, N. E., 199, *212*
Metzger, J. F., 301, *332*
Meunier, J., 180, *213*
Meyer, K., 219, *228*
Meyers, B. R., 241, 242, 243, 244, 245, 248, *295, 296*
Meyrath, J., 399, *412*
Michaeli, D., 241, 242, 243, 244, 245, 248, *295, 296*
Micheel, F., 14, *37*
Miciarelli, A., 180, *212*
Mickelsen, R., 43, 48, 63, *70*
Mickiene, N., 42, *67*
Middleman, S., 398, 406, *411*

Midgley, J. E. M., 308, *334*
Miglio, G., 77, 78, 79, 85, 95, 98, 120, 131, 142, *143*
Mika, E. S., 221, *230*
Mikolajcik, E. M., 52, 53, 67
Milbury, W. F., 406, *411*
Miller, A. H., 373, *412*
Miller, A. L., 326, *334*
Miller, R. S., 387, 388, 389, 392, 393, 396, 397, 401, *412*
Mills, S. E., 218, *229*
Milner, A. J., 222, *229*
Miloserdova, V. D., 191, 192, *212*
Mimura, A., 367, 368, 402, *413*
Mirelman, D., 244, *296*
Mise, K., 315, 317, *334*
Misiek, M., 246, 247, 248, 292, *296*
Mitsuhashi, J., 159, 164, 170, 174, 175, 177, 181, 182, 183, 188, 193, 194, 201, 206, *210, 212*
Mitsuhashi, S., 316, *333*
Mitz, M. A., 14, 15, 20, 21, *37*
Miura, G. A., 218, *229*
Miyachi, N., 116, *143*
Miyajima, S., 195, *211*
Miyaki, T., 231, 232, 237, 239, 241, 242, 243, 244, 245, 246, 247, 248, *295*
Miyamura, S., 315, *334*
Miyauchi, T., 388, 389, 390, *412*
Mizuno, D., 318, *334*
Mizusaki, S., 225, *229*
Mizuuchi, T., 80, *145*
Mocquot, G., 41, 61, 68, *70*
Moelker, H. C. T., 66, *70*
Moews, P. C., 47, 60, 67, *70*
Mohan, S., 224, *229*
Molavi, A., 244, *296*
Molho, D., 313, *334*
Molho-Lacroix, L., 313, *334*
Money, C., 21, *36*
Monib, M., 313, *334*
Monier, R., 308, *333*
Monod, J., 327, *331*
Monro, R. E., 303, *331*
Monson, P., 18, *37*
Montaldi, E., 221, *230*
Moore, E. B., 159, *213*
Moore, G. E., 217, *229*
Moore, S. F., 351, 352, *365*
Morgan, C., 300, *334*

Morgenstern, L., 225, *229*
Mori, S., 116, *143*
Mori, T., 15, 16, 20, 22, 27, 28, 32, *37*
Morieson, A. B., 93, 131, 135, *143*
Morio, H., 207, *209*
Morioka, S., 387, 388, 389, 395, 396, 401, 402, *412*
Moritz, V., 339, 344, 358, 361, *363*, 367, 402, 406, *410, 411*
Morpurgo, G., 318, *331*
Morris, T. A., 61, *70*
Morse, D. E., 308, *334*
Mosbach, K., 17, 22, 24, 26, 34, 35, *36*, *37*
Mosbach, R., 17, 26, *37*
Moshonas, M. G., *144*
Mosichev, M. S., 50, 52, 54, 63, *72*
Mosna, G., 183, *209*
Mothes, K., 220, *230*
Mowat, J. H., 54, 56, *72*
Mrowetz, G., 49, 62, 64, *71*, *72*
Muck, G. A., 134, 140, *144*
Mudd, J. B., 134, *144*
Mueller, G. A., 207, *212*
Mukai, N., 56, 57, 58, 59, *69, 70*
Mulherkar, L., 311, *334*
Muller, C. J., 77, 83, 85, 90, 91, 93, 94, 96, 97, 98, 102, 103, 106, 108, 109, 113, 123, 124, 134, 140, 141, *142, 144*
Munk, V., 402, 404, 405, *411, 413*
Murray, E. D., 50, 53, 54, *70*
Murray, E. G. D., 49, 67
Murrell, C. B., 147, *156*

N

Nag, T. N., 224, *229*
Nagata, S., 375, *412*
Nagawa, K., 316, *333*
Nagle, S. C., 168, 181, 198, *209, 212*
Naguib, M., 338, *365*
Naha, P. M., 305, *334*
Naito, M., 407, *411*
Naito, T., 232, 239, 241, *295*
Nakada, D., 307, *336*
Nakai, H., 195, *210*
Nakai, M., 43, *68*
Nakamoto, T., 303, *334*
Nakanishi, T., 59, *70*
Nakase, T., 362, *365*
Nakaya, R., 317, *334*
Nameroff, M., 219, *229*

AUTHOR INDEX

Narayanaswami, S., 223, *229*
Nasu, S., 201, *212*
Nath, K. R., 48, 63, *69*
Nathan, H. A., 147, *156*
Nathans, D., 303, *335*
Naudts, I. M., 49, 54, 63, *70*
Nawar, W. W., 134, 140, *144*
Negoro, H., 19, 28, *37*
Neidhardt, F. C., 308, *334*
Neil, G. L., *156*
Nelson, G. D., 78, 79, *144*
Nelson, J. H., 44, 62, *71*
Netter, E., 10, *11*
Nettleton, D. E., Jr., 234, 235, 246, 253, 254, 256, *295*, *296*
Netzer, A., 43, 46, *69*
Neudoerffer, T. S., 134, *144*
Neurath, A. R., 25, *37*
Newell, J. A., 139, *144*
Newmark, H. L., 247, 248, *296*
Newton, B. A., 300, *334*
Ng, D. Y. C., 406, 408, *413*
Nicholl, W. G., 309, *334*
Nichols, W. W., 188, 189, *210*, *212*
Nickell, L. G., 224, *229*
Nicoli, J., 202, *212*
Niefind, H.-J., 77, 78, 79, 85, 95, 98, 120, 131, 142, *143*
Nielsen, L. E., 400, *413*
Nielsen, T. K., 49, 64, *70*
Nikolaev, A'. Ya., 15, 20, *37*
Nilsson, I., 225, *229*
Nilsson, J. L. G., 225, *229*
Nimo, N. M., 357, *364*
Ninet, L., 235, 236, 237, *295*
Niri, L., 324, *334*
Nishiitsutsuji-Uwo, J., 180, *211*
Nishikawa, H., 338, 339, 344, 358, 359, 360, 361, *365*
Nishimura, K., 86, 87, 90, 91, 92, 93, 100, 101, 104, 105, 106, 107, 110, 111, 112, 115, 138, 139, 141, *144*
Nobis, A., 399, *412*
Noguchi, M., 225, *229*
Nomura, M., 307, *333*
Norris, J. R., 341, 358, *364*
Northcote, D. H., 221, *229*
Novais, J. M., 18, *35*
Novak, G., 60, 61, *66*
Novelli, G. D., 301, 326, *332*, *333*

Novick, R. P., 318, *334*
Novikova, M. A., 304, *335*
Nunez, J., 219, *229*
Nusbaum, M., 5, *11*
Nuzhina, A. M., 6, *11*
Nykänen, L., 87, 89, 90, 91, 98, 100, 103, 105, 106, 112, 115, 117, 118, 129, 140, *144*, *145*
Nyong, L. V., 50, 52, 54, *72*

O

Obata, Y., 90, *144*
Oberlander, H., 204, *212*
O'Brien, P., 8, *11*
Ochoa, A., 356, *365*
Oele, J. M., 140, *143*
Ofner, P., 218, *230*
Ogata, K., 19, *37*, 338, 339, 344, 358, 359, 360, 361, *365*
Ogawa, M., 90, 91, 97, 99, 100, 103, 104, 105, 108, 109, 111, 112, 113, 114, 115, 116, 117, 118, 119, 120, 121, 123, 125, 126, 127, 128, 134, *143*
Ogle, J. D., 19, 28, *36*
Ogutuga, D. B. A., 221, *229*
Ohanessian, A., 169, 172, 181, 183, 194, 209, *212*
O'Herron, F. A., 254, 256, *295*
Ohmori, T., 231, 232, 237, 239, 241, 242, 243, 244, 245, 246, 247, 248, *295*
Ohsugi, M., 338, 339, 344, 358, 359, 360, 361, *365*
Ohta, S., 360, *365*
Oi, S., 44, *70*
Okami, Y., 324, *334*
Okamoto, S., 317, 318, *334*
Okanishi, M., 231, 232, 238, 239, 241, 242, 243, 244, 245, 246, 247, 248, *295*
Okumura, S., 362, *365*
Okun, M. R., 225, *229*
Okunuki, K., 43, *68*
Oldham, S., 6, 10, *11*
O'Leary, W. M., 325, *332*
Oleinick, N. L., 301, *334*
Oleniacz, W. S., 5, *11*
Olson, A. C., 19, *36*
Olson, N. F., 14, 20, 35, 66, *68*
Oltmanns, O., 313, 315, 326, *333*
Omi, S., 408, *412*
Onda, K., 368, *413*

O'Neill, S. P., 16, 20, 22, 31, 32, 37
Ono, H., 87, *146*
Onodera, K., 195, *210, 212*
Ooka, T., 367, 368, *413*
Oosthuizen, J. C., 42, 50, 52, *70, 71*
Organon, N. V., 66, *70*
Oringer, K., 54, 56, *70*
Ormsbee, R. A., 202, *214*
Orosin, B., 52, *70*
Oruntaeva, K. B., 48, *70*
Osati, T., 299, 313, *335*
Osawa, S., 307, *334*
Osman, H. G., 54, 56, *70*
Oster, I. I., 207, *209*
Otaka, E., 307, *334*
Otake, T., 387, 388, 389, 395, 396, 401, 402, *412, 413*
Otsuka, K., 105, 111, *144*
Ottensen, M., 63, *70*
Otteson, M., 19, *37*
Ough, C. S., 88, 139, *144*
Ourisson, G., 223, *228*
Ousby, S. C., 361, *365*
Overbeck, J., 338, *365*
Owen, C. A., 218, *230*
Oyaas, J. E., 319, *334*
Ozawa, Y., 15, 20, *37*

P

Page, I. H., 7, *11*, 218, *230*
Palamand, S. R., 78, 79, *144*
Paleva, N. S., 56, *70*
Pan, W. W., 356, *365*
Papermaster, B. W., 217, *230*
Pardee, A. B., 305, *334*
Paredes-Lopez, O., 399, 405, *413*
Parent, J. R., 399, *410*
Park, J. T., 301, 310, *333*
Parkhurst, J. D., 352, *365*
Parks, O. W., 134, *144*
Parliament, T. H., 140, *144*
Pastan, I., 308, *335*
Patel, A. B., 17, 20, 23, 26, *36*
Patterson, B. D., 221, *229*
Pattillo, R. A., 219, 222, *229*
Patton, S., 134, 140, *144, 146*
Paul, F. J., 198, 203, *213*
Paul, S. D., 197, *212, 213*
Paulus, H., 312, *334*
Pavlovic-Hournac, M., 219, *229*

Peacock, M. G., 202, *214*
Pearson, E. A., 351, *363*
Pecherer, B., 231, 232, 233, 239, 252, *294, 295*
Pecht, M., 15, 16, 21, 23, *36*
Peckinpaugh, R. O., 248, 249, *295*
Pedersen, A. H., 60, 62, 65, *70*
Peel, D., 339, 340, 347, *364, 365*
Peleg, J., 173, 182, 188, 199, 200, *212*
Pellegrino, M., 217, *230*
Pence, J. W., 77, 78, 79, 80, 82, 83, 87, 89, 94, 96, 102, 109, 111, 113, 118, 123, 127, 128, *143*
Pennington, S. N., 17, 23, *36*
Perazzolo, C. A., 301, *332*
Perlman, D., 188, *212*
Perlman, R. L., 308, *335*
Permanand, B., 224, *229*
Pestka, S., 302, *334*
Peters, D., 173, 183, 201, *212*
Petersen, W. E., 39, *70*
Peterson, A. C., 50, *70*
Petina, T. A., 50, 52, 55, 56, 67, 68, *72*
Petrasovits, A., 47, *68*
Pfeiffer, S. E., 218, *228*
Phillips, K. C., 338, 340, 343, *365*
Pick, E., 217, *229*
Picknett, R. G., 400, *411*
Pierce, G. B., Jr., 219, *229*
Pierce, W. E., 248, 249, *295*
Pihan, J.-C., 180, *212*
Pine, M. J., 353, *365*
Pinsky, A., *72*
Pipes, W. O., 406, *411*
Pisano, M. S., 5, *11*
Pisarnitskii, A. F., 117, 125, 128, *144*
Pittillo, R. F., 155, *156*
Platova, T. P., 183, *211*
Polishhook, R. D., 5, *11*
Pollack, O. J., 222, *229*
Pollock, M. R., 317, 326, *334*
Polukarova, L. G., 183, *211*
Pomerat, C. M., 301, *334*
Poole, R. K., 309, *334*
Popova, N. V., 56, *70*
Porath, J., 16, 22, 35, *37*
Porterfield, J. D., 198, *212*
Potter, V. R., 309, *334*
Poulson, D. F., 201, *210*
Pourquier, M., 203, *212*

AUTHOR INDEX

Poutier, F., 173, *210*
Power, F. B., 101, *145*
Powers, W., 357, *364*
Poznanski, S., 55, 56, *71*
Poznansky, M. J., 32, *36*
Pozsar Hajnal, K., 42, 63, *72*
Praprotnik, V., 63, *70*
Preobrazhenskii, A. A., 77, 81, 84, 87, 88, 89, 90, 91, 94, 95, 96, 97, 98, 102, 103, 104, 106, 109, 110, 113, 116, 118, 119, 123, 124, 126, 127, *145*
Prescott, J. H., 357, *365*
Pressman, D., 217, *229*
Prestidge, L. S., 305, *334*
Preud'Homme, J., 235, 236, 237, *295*
Price, K. E., 234, 235, 241, 242, 244, 246, 247, 248, 254, 256, 292, *295, 296*
Pridham, T. G., *296*
Priest, R., 218, *230*
Prince, M. P., 50, 53, 54, *70*
Prins, J., 49, 64, *70*
Prokop, A., 399, 403, 405, *411, 413*
Pudney, M., 182, 183, 188, 198, 201, *212, 214*
Puhalla, J. E., 60, *70*
Puhan, Z., 50, 53, 61, *70*
Pulvertaft, R. J. V., 300, 324, *334*
Puputti, E., 77, 79, 81, 85, 86, 88, 91, 92, 98, 106, 107, 108, 115, 129, 139, *144, 145*
Pursiano, T. A., 326, *332*
Pye, E. K., 34, *36*

Q

Quayle, J. R., 339, 340, 341, 345, 346, 347, 349, *364, 365*
Quiocho, F. A., 19, *37*
Quiot, J.-M., 173, 183, 192, 193, 196, *210, 212, 214*

R

Ragan, E. A., 254, 256, *295*
Rahman, S. B., 188, *212*
Raison, J. K., 136, *144*
Ralph, J. H., 387, 388, 389, 392, 393, 396, 397, 401, *412*
Ramaranova, O. Kh., 50, 52, *67*
Ramet, J. P., 61, *70*
Ramsey, H. H., 306, *334*

Ramshaw, E. H., 119, 121, 123, 126, 128, *143*
Rannenberg, H., 313, *334*
Rapp, A., 80, 81, 91, 93, 98, 99, 104, 110, 114, 119, 121, 124, *143*
Rappaport, L., 219, *229*
Ray, D. S., 305, *334*
Raymond, R. L., 362, *365*
Rebstock, M. C., 298, 322, *331, 334, 335*
Reddy, D. V. R., 170, 201, *209*
Reed, G., 41, 42, 44, 46, 47, 48, *71*
Reed, R. W., 424, *435*
Řeháček, J., 200, 201, 203, *210, 212, 213*
Řeháček, Z., 200, *213*
Reif, R. E., 309, *334*
Reilly, H. C., 324, *336*
Reinbold, G. W., 39, *72*
Reinecke, J. P., 179, 206, *213*
Reisfeld, R. A., 217, *230*
Rendi, R., 301, *334*
Reps, A., 55, 56, *71*
Resmini, P., 61, *71*
Resnick, W., 368, *413*
Reusser, F., 406, *413*
Rhode, S. L., III, 248, 249, *295*
Ribadeau-Dumas, B., 41, *68*
Ribbons, D. W., 338, 348, *365*
Ribéreau-Gayon, P., 84, 89, 94, 96, 97, 98, 103, 106, 107, 108, 109, 113, 116, 118, 121, 123, 126, 127, *146*
Richard-Molard, C., 183, *213*
Richards, F. M., 19, *37*
Richardson, B. L., 161, 167, *210*
Richardson, G. H., 42, 44, 62, *71*
Richardson, J., 173, 183, *213*
Richardson, T., 14, 20, *35*, 66, *68*
Richardson, U. I., 217, 218, *230*
Richter, I., 220, *230*
Ricica, J., 367, *412*
Rickert, W., 63, *70, 71*
Rideal, E. K., 400, *411*
Riesel, E., 23, *37*
Rietema, K., 373, 374, 388, 389, 392, 396, 402, 408, 409, *411, 413*
Rimon, A., 22, 23, *36, 37*
Rimon, S., 23, *37*
Rinaldini, L. M., 166, 170, *213*
Rippin, D. W. T., 406, 408, *413*
Ristich, S. S., 188, *212*
Roach, G. I., 309, *334*

Robb, J. A., 166, *213*
Robbins, J. D., 206, *213*
Robbins, P. W., 222, *230*, 319, 323, 326, *332*
Robertson, A. L., 218, *230*
Robertson, N. F., 224, *230*
Robertson, P. S., 62, *71*
Robinson, E. K., 361, *363*
Robinson, P. J., 18, 25, *37*
Rocklin, R. E., 217, *230*
Rodger, W. S., 371, 397, *413*
Rogers, H. J., 301, 310, *334*
Rommel, F. A., 218, *230*
Ronkainen, P., 83, *145*
Root, M. A., 219, *228*
Rose, A. H., 135, 136, *142*, *143*
Rose, G., 159, 205, *213*
Rose, H. M., 300, *334*
Rosenkranz, H. S., 300, *334*
Rossomano, V. Z., 240, *294*
Rothblatt, G. H., 222, *230*
Rothfus, J. A., 17, 22, *37*
Rotini, O. T., 60, *71*
Rousche, M. A., 254, 256, *295*
Roxburgh, J. M., 339, 340, 345, 347, *364*
Roy, K. L., 201, *213*
Royer, G. P., 18, 24, *37*
Ruben, S., 351, *363*
Rubin, H., 217, *228*
Rushton, J. H., 370, 371, 397, *413*
Rutter, D. A., 7, 8, *11*
Ruttle, D., 58, *72*
Rychlik, I., 306, *331*
Ryle, A. P., 47, *71*

S

Saadeh, M., 359, *365*
Saeki, Y., 56, 59, *71*
St. Amant, P., 351, 352, *365*
Sairam, T. V., 221, *230*
Sakai, F., 195, *210*, *212*
Sakakihara, Y., 5, *11*
Sala, F., 323, *335*
Salton, M. R., 306, *335*
Sameshima, H., 6, *11*
Saminski, E., 302, *331*
Sampathkumar, B., 49, *66*
Samra, Z., 317, 318, *335*
Samuels, H. H., 218, *230*
Sanders, G. P., 39, *71*
Sanderson, A. R., 29, *36*
Sandler, S., 134, *144*
Sandquist, A., 218, *230*
Sands, L., 316, *335*
Sang, J. H., 175, *213*
Sannabadthi, S. S., 56, 57, *71*
Saracchi, S., 61, *71*
Sardinas, J. L., 40, 45, 49, 52, 56, 60, *71*
Sargent, J. A., 221, *230*
Saska, J., 168, *214*
Sato, F., 163, *209*
Sato, G. H., 219, *230*
Sato, R., 306, *332*
Sato, S., 218, *230*
Sato, T. R., 17, 23, *37*
Sato, Y., 45, *71*
Satomura, Y., 44, *70*
Savage, G. M., 147, *156*
Sawada, A., 44, *70*
Sbrenna, G., 180, *212*
Scallion, R. J., 166, 174, 194, 196, *211*
Scannell, J., 240, *296*
Schaechter, M., 300, *333*
Schaefer, J., 87, 99, 104, 105, 110, 111, 113, 114, 115, 117, 138, *145*
Schaeffer, P., 299, *335*
Schait, A., 78, 79, 80, 81, 82, 93, 99, *144*
Schaller, F., 180, *213*
Schejter, A., 19, *37*
Schenk, D. K., 161, 167, *210*
Schenk, T. E., 199, *213*
Scheuerbrandt, G., 135, *145*
Schilt, P., 61, *67*
Schleich, H., 66, *71*
Schleif, R. F., 307, *335*
Schleith, L., 17, 23, *36*
Schlesinger, R. W., 199, *213*
Schlessinger, D., 302, *332*
Schluter, A. S., 61, *70*
Schmanke, E., 64, *72*
Schmid, J., 245, *296*
Schmitz, H., 234, 235, 246, 254, 256, *295*, *296*
Schmitz, R. A., 368, *413*
Schneider, H. A. W., 224, *230*
Schneider, I., 159, 169, 170, 178, 181, 182, 183, 188, 202, *213*
Schnellen, G. T. P., 351, *364*
Schocher, A. J., 231, 232, 233, 239, 252, *294*, *295*

Scholler, J., 225, *229*
Scholtan, W., 245, *296*
Schormuller, J., 45, *71*
Schrader, L. E., 306, *335*
Schreiber, R. H., 254, 256, 257, 258, *295*
Schroeder, E. D., 351, 352, *365*
Schulz, M. E., 49, 62, *71*
Schwarberg, R. L., 44, *71*
Schwartz, D. P., 134, *144*
Schwartz, Y., 45, *70*, 133, *144*
Schweet, R., 301, *331*, *335*
Scott, K. J., 431, *435*
Scott, L. S., 398, *413*
Scott, R. B., 7, *11*
Scott Blair, G. W., 42, *71*
Seecof, R. L., 172, 175, 179, 194, *213*
Seitov, Z. S., 48, *70*
Seitz, E. W., 44, *69*
Seki, M., 338, 342, *363*
Sekiguchi, Y., 45, *71*
Selden, C. E., 373, *412*
Selegny, E., 19, *35*
Self, D. A., 32, *37*
Sell, H., 49, 62, *71*
Sells, B. H., 307, *332*
Senez, J., 138, *143*
Sengel, P., 178, *213*
Sen Gupta, K., 161, 192, 202, *213*
SentheShanmuganathan, S., 134, *145*
Sequi, P., 60, *71*
Sergeeva, E. G., 49, 63, *73*
Seshan, K. R., 179, *213*
Shadduck, J. A., 200, *210*
Shah, P. S., 403, 404, 405, 408, *411*
Shapiro, M., 193, 195, 197, *211*
Sharma, V. K., 218, *229*
Sharp, A. K., 20, 30, *36*
Shaw, P. D., 323, 328, *332*
Shaw, P. E., 140, *145*
Shaw, W. V., 315, 316, 317, 326, *335*
Sheehan, B. T., 356, 359, *365*
Sheffield, C. W., 352, *365*
Shemyakin, M. M., 303, 304, *335*
Sher, S., 147, *156*
Sherman, J. D., 318, *333*
Sherry, S., 4, 5, *11*
Shichiji, S., 56, 59, *71*
Shields, G., 175, *213*
Shifrine, M., 147, *156*
Shimmin, P. D., 45, *67*

Shimwell, J. L., 50, 52, 53, 54, *71*
Shinnar, R., 369, 370, 371, 372, 376, 388, 390, 397, 401, *411*, *413*
Shipley, P. A., 77, 78, 79, 80, 81, 83, 84, 86, 88, 89, 90, 91, 92, 96, 97, 98, 99, 102, 103, 104, 105, 106, 107, 109, 110, 111, 112, 113, 114, 116, 117, 118, 120, 123, 124, 125, 127, 128, 129, 130, 134, 138, 139, *143*, *144*
Shitara, J., 134, *146*
Shockman, G. D., 310, *335*
Shortridge, K. F., 201, *212*
Shovers, J., 61, *71*
Shvetsov, Iu. B., 303, 304, *335*
Shyluk, J. P., 221, *230*
Sicher, S., 312, *331*
Siddiqueullah, M., 323, *335*
Siewert, R., 64, *67*
Sihto, E., 89, 90, 91, 98, 103, 105, 118, *145*
Silinger, V., 402, 404, *413*
Silman, I. H., 16, 22, *37*
Silman, J. H., 19, 26, 28, *36*
Siminoff, P., 149, *156*
Simpson, P. J., 220, *228*
Sims, R. J., 140, *143*
Singh, A., 50, 52, 54, *71*
Singh, K., 57, *71*
Singh, K. R. P., 170, 172, 173, 175, 177, 182, 188, 197, *209*, *212*, *213*
Singleton, V. L., 138, 139, *145*
Sinsheimer, R. L., 305, *333*
Sinskey, A. J., 32, *36*, 360, *365*
Sizer, I. W., 2, *11*
Skelton, G. S., 43, *71*
Skoog, F., 221, *230*
Slaughter, J. C., 84, *145*
Slavyanova, V. V., 41, 56, *68*
Sleicher, C. A., Jr., 398, 401, *413*
Smadel, J. E., 298, 303, 312, *333*
Smeby, R. R., 218, *230*
Smiley, K. L., 15, 20, 25, 26, 27, 28, 33, *35*, *36*, *37*
Smirnova, I. A., 195, *209*
Smirnova, T. A., 52, *70*
Smith, A. R., 401, *413*
Smith, C., 181, *213*
Smith, C. G., 147, 153, 154, 155, *156*, 299, 321, *335*
Smith, D. E., 82, 83, 85, 90, *145*
Smith, D. G., 306, 308, 313, *334*, *335*

Smith, G. N., 305, 306, 313, 314, 326, 335
Smith, L. M., 137, 145
Smith, N. D., 134, 144
Smith, N. R., 49, 67
Smith, R. H., 218, 230
Smith, R. M., 298, 299, 319, 332, 334
Smith, R. T., 217, 228
Smith, W. K., 223, 228
Snell, I., 372, 413
Snow, N. S., 41, 67
Soda, K., 326, 335
Söhngren, N. L., 338, 365
Sohi, S. S., 169, 173, 174, 177, 181, 182, 191, 192, 196, 202, 211, 213
Sokolski, W. T., 147, 151, 156
Sols, A., 135, 145
Somers, P. J., 14, 20, 35
Someya, J., 367, 402, 406, 410, 411
Somkuti, A. C., 43, 62, 71
Somkuti, G. A., 43, 51, 62, 67, 71
Sompolinsky, D., 317, 318, 335
Soo, S. L., 368, 413
Soxhlet, F., 41, 71
Spaansen, C. H., 173, 183, 212
Sparkes, B. G., 224, 229
Speck, M. L., 44, 68
Sperl, G. T., 343, 351, 352, 365
Spickett, R. G. W., 43, 71
Spiegelberg, H., 231, 232, 233, 239, 252, 294, 295
Spiegelman, S., 308, 331
Spielman, L. A., 374, 413
Spinell, D. M., 3, 11
Spirin, A. S., 313, 335
Sprow, F. B., 370, 371, 374, 397, 413
Spyrides, G. J., 303, 334
Srinivasan, R. A., 41, 42, 45, 49, 50, 51, 52, 54, 56, 57, 66, 71, 72
Staba, E. J., 221, 223, 224, 228, 229, 230
Stadhouders, J., 45, 72
Stadtman, T. C., 350, 351, 352, 353, 354, 363, 365
Stahl, W. A., 142, 144
Stanley, M. S. M., 173, 181, 192, 195, 213, 214
Stapert, E. M., 151, 156
Stark, W., 140, 143
Starkey, B. J., 311, 335
Steck, W., 221, 230
Steers, E., 287, 296, 315, 334

Steffen, C., 61, 70
Steinbuch, M., 168, 211
Stephanos, S., 174, 211
Stephens, R. E., 207, 209
Stephenson, M., 350, 351, 352, 365
Stern, I. J., 9, 11
Sternberg, M. Z., 64, 72
Stevens, K. L., 85, 102, 104, 110, 111, 134, 145
Stevens, T. M., 199, 213
Steward, F. C., 216, 218, 220, 223, 229
Stickland, L. H., 350, 351, 352, 365
Stobaugh, R. B., 357, 364
Stocks, P. K., 341, 365
Stohs, S. J., 221, 223, 228, 229, 230
Stokoe, W. N., 134, 145
Stokes, J. L., 136, 145
Stollar, B. D., 199, 213
Stollar, V., 199, 213
Stolle, K., 220, 230
Stoner, C. D., 310, 335
Stoudt, T. H., 3, 11
Stow, M., 311, 335
Strandberg, G. W., 15, 20, 25, 26, 27, 28, 35, 36, 37
Strating, J., 80, 86, 137, 143, 145
Stratton, C. D., 322, 335
Strauss, N., 306, 335
Strawinski, R. J., 311, 315, 363, 364
Street, H. E., 221, 230
Stumpf, P. K, 134, 135, 137, 143, 145
Stupp, Y., 23, 37
Su, Y., 49, 72
Suarez, G., 303, 335
Subramanian, S. S., 50, 67
Sugeno, M., 367, 368, 413
Suhadolnik, R. J., 324, 326, 331
Suitor, E. C., 175, 181, 198, 203, 210, 213
Sukegawa, K., 61, 62, 69
Sullivan, D. M., 398, 412
Sullivan, J. J., 44, 45, 72
Sulser, H., 140, 145
Summaria, L. J., 14, 20, 21, 37
Sundaram, P. V., 20, 21, 37
Suomalainen, H., 77, 79, 81, 83, 85, 86, 87, 88, 89, 90, 91, 92, 98, 100, 103, 105, 106, 107, 108, 112, 115, 117, 118, 129, 133, 135, 139, 140, 144, 145
Surinov, B. P., 20, 21, 37
Sussman, B. J., 9, 11

AUTHOR INDEX

Sutherland, K. L., 400, *412*
Suzuki, H., 15, 20, 21, *36*, *37*
Suzuki, M., 324, *334*
Suzuki, Y., 62, 69, 315, 317, *334*
Svendsen, J., 19, *37*
Swain, T., 225, *228*
Swanson, A. L., 305, *335*
Swanson, C. L., 256, 257, 258, *295*
Sweet, B. H., 178, 182, 183, *213*
Sykes, J. A., 159, *213*
Sykes, R. M., 350, 352, *364*
Syono, K., 221, *228*
Sypherd, P. S., 305, 306, *335*
Szybalski, W., 300, 301, *332*

T

Taber, W. A., 319, *331*
Taguchi, T., 116, *143*
Taira, T., 104, 117, *145*
Takafuji, S., 61, 62, *69*
Takahashi, M., 217, *229*
Takami, H. S., 192, *214*
Takamine, J., 48, 56, *72*
Takamine, J., Jr., 48, 56, *72*
Takayama, T., 80, *145*
Takeda, I., 367, 338, *413*
Takeda, K., 299, 319, *335*
Takeuchi, H., 368, *413*
Tamaki, E., 225, *229*
Tamura, G., 49, 50, 51, 56, 57, 58, 61, 62, *66*, *69*, *73*
Tanabe, Y., 225, *229*
Tanaka, H., 90, *144*, 326, *335*
Tang, J., 46, 47, *72*
Tannahill, A., 339, 343, *365*
Tannenbaum, S., 367, *412*
Tannenbaum, S. R., 360, *365*
Tanzer, M. L., 218, *228*
Tarodo de la Fuente, B., 42, *72*
Tashjian, A. H., 217, 218, 219, *228*, *230*
Tatum, J. H., 140, *145*
Tavlarides, L. L., 368, *411*, *413*
Taylor, R. J., 140, *142*
Telesnina, G. N., 304, *335*
Telling, R. C., 361, *364*
Teltscher, H., 312, *332*
Tendler, M. D., 49, 51, *72*
Ten Noever de Brauw, M. C., 83, 87, 91, 104, 110, 113, 114, 119, 121, 122, 124, 125, 134, *144*

Tepley, M., 49, *70*
Teplitz, R. L., 179, *213*
Teranishi, R., *145*
Teraoka, H., 301, *335*
Terry, T. D., 90, *145*
Tetenyi, P., 223, *230*
Thakur, M. L., 224, *229*
Thomas, D., 19, *35*
Thomas, E., 221, *230*
Thomasow, J., 64, *72*
Thompson, J. A., 189, *214*
Thompson, K., 200, *213*
Thoukis, G., 134, *145*
Tilles, J. G., 217, *228*
Timms, R., 87, 99, 104, 105, 110, 111, 113, 114, 115, 117, 138, *145*
Tipograf, D. Ya., 50, 52, 54, 56, 63, *68*, *72*
Titsworth, E., 245, 246, 247, *295*
Titus, J. L., 218, *230*
Tjhio, K. H., 225, *228*
Tobias, J., 134, 140, *144*
Tochikura, T., 344, 359, 360, *365*
Toerien, D. F., 350, 351, 352, *365*
Tokita, F., 45, *72*
Tolkachev, A. N., 45, *73*
Tomkins, G. M., 218, *230*
Tong, W., 219, *230*
Toolens, H. P., 45, *72*
Tosa, T., 15, 16, 20, 22, 27, 28, 32, *37*
Towell, G. D., 387, 388, 389, 392, 393, 396, 397, 401, *412*
Toyoda, S., 52, 61, 62, *69*
Trager, W., 190, 192, 197, 202, *214*
Trakhtenberg, D. M., 253, *294*
Treffers, H. P., 306, *335*
Tressl, R., 81, 100, 102, 104, 108, 110, 114, 117, 119, 124, 134, *143*
Trice, V. G., Jr., 371, 397, *413*
Tripconey, D., 192, *214*
Trop, M., *72*
Trucco, R. E., 46, *72*
Truhaut, R., 311, *335*
Tsai, B. I., 368, 407, 408, *411*, *413*
Tsao, D. P. N., 223, *229*
Tsugawa, R., 362, *365*
Tsugo, T., 50, 56, *72*
Tsuji, H., 55, *72*
Tsukiura, H., 231, 232, 237, 239, 241, 242, 243, 244, 245, 246, 247, 248, *295*
Tsuru, D., 53, 55, 57, *68*, *72*

Tuckey, S. L., 47, 50, 52, 70
Tucknott, O. G., 95, 96, 105, 111, 112, 146
Turel, F. L. M., 431, 435
Turk, J. L., 217, 229
Turner, G., 300, 309, 310, 334, 335
Tuszynski, W. B., 41, 72

U

Ueda, H., 248, 249, 295
Umezawa, H., 326, 333
Umezawa, H. T., 299, 319, 335
Unanue, R. L., 172, 175, 213
Underkofler, L. A., 3, 6, 11
Unowsky, J., 317, 335
Updike, S. J., 17, 26, 31, 36, 37
Urbach, G., 140, 143
Urvölgyi, J., 203, 213

V

Vaccaro, S. E., 5, 11
Vago, C., 162, 163, 170, 173, 180, 182, 183, 190, 191, 192, 193, 196, 203, 209, 210, 212, 214
Vagujfalui, D., 223, 230
Vahldieck, N. P., 361, 363
Valentas, K. J., 373, 375, 378, 406, 413
Vallee, B. L., 2, 11
Valulis, B., 308, 332
Vambutas, V. K., 300, 335
Vamos Vigyazo, L., 42, 63, 72
VanderSlice, R. W., 336
Vanderveen, J. H., 372, 374, 413
Van der Ven, B., 140, 145
Van Straten, S., 76, 146
Van Wyk, C. J., 77, 80, 82, 84, 85, 88, 89, 90, 91, 93, 94, 95, 96, 97, 98, 103, 106, 107, 108, 109, 113, 115, 116, 117, 118, 122, 123, 124, 134, 137, 141, 142, 145
Varma, M. G. R., 182, 183, 188, 198, 199, 212, 214
Varmus, H. E., 308, 335
Vary, P. S., 342, 357, 359, 360, 365
Vas, M. R., 301, 335
Vasey, R. B., 361, 365
Vaterlaus, B. P., 231, 232, 233, 239, 252, 294, 295
Vattier, G., 202, 212
Vaughn, J. L., 159, 167, 173, 181, 182, 189, 191, 192, 193, 195, 196, 210, 213, 214
Vavra, R. W., 200, 210

Vazquez, D., 301, 303, 317, 331, 335
Veliky, I., 218, 230
Veltkamp, G. W., 374, 413
Venables, S. A., 310, 333
Venema, A., 86, 137, 145
Verhoff, F. H., 373, 380, 385, 386, 399, 414
Veringa, H. A., 45, 49, 51, 72
Vermeulen, T., 371, 372, 397, 412, 414
Vernerova, J., 402, 404, 413
Vernet, C. H., 402, 412
Verney, E. L., 219, 229
Veselov, A. I., 50, 52, 54, 72
Veselov, I. Ya., 50, 52, 55, 56, 63, 68, 70, 72
Veyama, H., 338, 342, 363
Vezina, C., 57, 71
Villee, D. B., 222, 229
Villforth, F., 80, 105, 110, 143
Vining, L. C., 298, 299, 319, 321, 323, 324, 326, 330, 334, 335, 336
Vinter, V., 310, 336
Vishniae, W., 353, 365
Visser, M. K., 112, 145
Vogel, T., 301, 336
Vogel, Z., 301, 336
Vogt, M., 166, 209
Volfova, O., 402, 404, 405, 411, 413
Volonterio, G., 61, 71
Voss, F., 49, 62, 71

W

Wade, H. E., 7, 8, 11
Wadzinski, A. M., 338, 348, 365
Wagener, G. W. W., 95, 145
Wagener, W. W. D., 95, 145
Wagner, D. E., 5, 11
Wagner, R. P., 170, 211
Wagner, T., 16, 23, 37
Wahlin, J. G., 48, 50, 51, 72
Wake, R. G., 41, 69, 70
Waksman, S. A., 324, 336
Walker, J. B., 326, 334
Wall, R., 400, 413
Walley, B. F., 45, 68
Wallis, R. C., 169, 172, 177, 214
Walter, M. M., 62, 71
Wan, J., 17, 26, 29, 35
Wang, C. M., 207, 214
Wang, D. I. C., 32, 36, 339, 344, 355, 356, 361, 364, 365

Wang, H. L., 58, 72
Ward, J. P., 398, *414*
Warner, J. F., 243, *295*
Wat, C., 330, *336*
Watanabe, S., 367, *413*
Watanabe, T., 318, *336*
Waterhouse, D. F., 206, *212*
Waterman, L. C., 373, *414*
Watson, D. M., 26, 28, *35*
Waugh, D. F., 41, 72
Weakley, F. B., 15, *37*
Webb, A. D., 76, 77, 78, 80, 81, 82, 83, 84, 85, 86, 87, 88, 89, 90, 91, 92, 93, 94, 95, 96, 97, 98, 99, 100, 101, 102, 103, 104, 105, 106, 107, 108, 109, 110, 111, 112, 113, 114, 115, 116, 117, 118, 120, 121, 122, 123, 124, 125, 126, 127, 128, 129, 130, 134, 135, 137, 139, 140, 141, *142*, *143*, *144*, *145*, *146*
Weber, F., 61, *70*
Weber, M. J., 302, *336*
Weetall, H. H., 2, 8, *11*, 17, 18, 24, 25, 27, 30, 31, 32, 34, 35, *36*, *37*, *38*
Weibel, M. K., 35, *38*
Weinberger, H. J., 300, 324, *332*
Weinstein, H., 406, *414*
Weinstein, L., 241, 242, 243, 244, 245, 248, *295*, *296*
Weisberger, A. S., 301, *336*
Weisblum, B., 301, *336*
Weislogel, P. O., 318, *336*
Weiss, E., 197, *214*
Weiss, L. H., 372, *414*
Weissbach, H., 312, 324, *333*, *336*
Weliky, N., 14, 21, 31, *38*
Wellner, D., 22, *37*
Weniger, J. P., 220, *230*
Wenzel, K. W. O., 117, *146*
West, F. R., Jr., 221, *230*
West, R. A., Jr., 147, *156*
Westlake, D. W. S., 298, 321, 323, 329, *333*, *335*, *336*
Weston, P. D., 24, *38*
Weurman, C., 76, 80, *143*, *146*
Wharton, C. W., 21, 28, 29, *36*, *38*
Wheeler, K. P., 21, *38*
Wheelock, J. V., 41, 72
Whelan, W. J., 32, *36*
Whitaker, J. R., 43, 60, 68, 69, 72
Whitaker, W. D., 252, *296*

White, A. E., 303, *332*
White, J. F., 181, *214*
Whitehill, A. R., 54, 56, 72
Whitney, R. McL., 134, 140, *144*
Whittam, R., 21, *38*
Whittenbury, R., 338, 340, 343, 350, *364*, *365*
Wieland, T., 26, *38*
Wild, D. G., 303, *332*
Wildenradt, H. L., 141, *144*
Wilhelm, J. M., 301, *334*
Wilkens, W. F., 140, *144*
Wilkerson, R., 249, *294*
Wilkie, D., 318, *336*
Wilkinson, J. F., 338, 340, 343, *365*
Williams, A. A., 95, 96, 105, 111, 112, *146*
Williams, G. M., 371, *414*
Williams, P. G., 431, *435*
Williamson, D. H., 318, *336*
Willmer, E. N., 175, *214*
Wilson, H. L., 39, 72
Wilson, R. J. H., 20, 29, 31, 34, *36*, *38*
Winder, W. C., 42, 68
Winshell, E., 316, *335*
Wise, M., 312, *333*
Wisseman, C. L., Jr., 298, 301, 303, 312, *333*
Wolf, S., 46, 72
Wolfe, A. D., 299, *331*
Wolfe, R. S., 350, *364*
Wolfe, S., 301, *336*
Wolin, E. A., 350, *364*
Wolin, M. J., 350, *364*
Wolnak, B., 359, *365*
Wong, N. P., 134, *146*
Woodruff, H. B., 324, *336*
Woolley, D. W., 311, *336*
Worrell, C. S., 305, 306, 313, 314, 326, *335*
Wyatt, G. R., 161, 204, *214*
Wyatt, S. S., 161, 173, 204, *214*
Wykes, J. R., 16, 22, 32, *37*

Y

Yagi, H., 402, *414*
Yagi, S., 180, 204, 205, *209*, *214*
Yagi, Y., 217, *229*
Yamada, M., 85, 134, *146*
Yamada, T., *414*
Yamada, Y., 326, *333*
Yamaguchi, M., 87, *146*

Yamamoto, T., 53, 55, 57, 68, 72, 326, 335
Yamane, T., 402, *414*
Yamanoi, A., 362, 365
Yamaoka, T., 218, *230*
Yamashita, M., 45, *68*
Yamauchi, K., 50, 56, 72
Yamazaki, H., 368, *414*
Yamazaki, O., 44, *70*
Yasumura, Y., 219, *230*
Yasunobu, K. T., 53, *73*
Yeigan, C. D., *336*
Yogawa, H., 134, *146*
Yoneyama, H., 85, *146*
Yoshida, F., 402, *414*
Yoshida, T., 324, *336*
Yoshizawa, K., 85, 134, *146*
Young, R. M., 307, *336*
Yu, J., 43, 49, 50, 61, 62, *66*, 69, *73*
Yulius, A. A., 50, 52, *67*

Yunkers, C. E., 159, 181, 197, 198, 201, 202, 203, *214*

Z

Zacharski, L. R., 218, *230*
Zahid, N. D., *227*
Zamir, A., 301, *336*
Zatman, L. J., 341, 345, 346, *363*
Zeitlin, M. A., 368, *413*
Zhdanov, G. D., 304, *335*
Zia, T. U., 173, 182, 190, 192, 196, *210*
Zielinska, Z. M., 168, *214*
Ziteer, E. M., 170, *211*
ZoBell, C. E., 337, *364*
Zubeckis, E., 134, *144*
Zuiderweg, F. I., 372, 388, 389, 393, 396, 401, 402, *412*
Zupančičová, M., 203, *213*
Zvyagintsev, V. I., 45, 49, 63, *73*
Zwietering, T. N., 406, 407, *414*

SUBJECT INDEX

A

Animal cell metabolites
  hormones, 219–220
  lipids, 221–223
  nucleic acids, 216
  proteins, 216–218
  steroids, sterols, 221–223

B

Beers, volatile constituents, 75–146

C

Chloramphenicol
  biosynthesis, 321–331
  differential effects on esterases, 305
  differential effects on membrane-bound enzyme, 306
  differential effects on nitrate reductase, 306
  differential effects on proteins and protein synthesis, 305
  differential effects on regulatory proteins, 306
  effect on antibiotic synthesis, 312
  effect on biosynthesis of aromatic compounds, 311–312
  effect on cell wall biosynthesis, 310–311
  effect on energy-producing system, 308–310
  effect on morphology, 300
  effect on RNA metabolism, 308
  growth-inhibitory effect, 299
  mode of action, 301–303
  mutagenic effect, 318
  structure-activity relationships, 303–304
  therapeutic use, 318–319
  resistance mechanisms, 312–318
Correlative microbial assays, organisms used, 148–151
Coumermycin, 231
  antimicrobial spectrum, 234
  relationship to novobiocin, 234–235
  relationship to RP 18,631, 235–236
  Streptomyces albocinerescens DS 21,647 (NRRL 3419)
  Streptomyces hygroscopicus DS 9,751 (NRRL 3418)
  Streptomyces rishiriensis, 231, 236, 238
  Streptomyces roseochromogenes var. *oscitans* DS 12,976 (NRRL 3504) 237–238
  Streptomyces spinichromogenes, 237–238
  Streptomyces spinicoumarensis, 237–238
  Streptomyces X-7763 (*S. hazeliensis*), 231, 237–238
  sugordomycin, 231
Coumermycin A_1
  analytical, 240–241
  biogenesis, 240
  chemistry, 232–233
  in vitro activity, 241–244
  in vivo activity, 245–248
  production processes, 239–241
  resistance to, 244
  serum binding, 245
  side effects, 248–249
Coumermycins, semi-synthetic
  BL-C43, 285–294
  3′-esters, 253
  3′-nitrogen derivatives, 253–277
  2′-hydroxy, 252
  4-hydroxyl modification, 251
  pharmacology, 278–285

F

Fungal rennet
  Basidiomycetes, 58–59
  Byssochlamys fulva, 55
  commercial products, 59
  Endothia parasitica, 60–61
  Mucor miehei, 63–65
  Mucor pusillus, 61–63
  Rhizopus chinensis, 55–58
  Rhizopus niveus, 58
  Rhizopus oligosporus, 58

I

Insect tissue culture
  characteristics of cell lines, 185–189

ecdysones, effect on, 203–206
equipment/supplies, 158–160
establishment of cell lines, 180–185
insecticides, 206–207
media formulation, 160–166
nutrition and metabolism, 166–168
organs and tissues, 172–174
primary cultures, 169–172
protozoa studies, 202
rickettsial studies, 212–213
virus studies, 190–202

M

Medical enzymes
  blood clot dissolution, 4–5
  cancer treatment, 6–8
  connective tissue dissolution, 8–10
  digestive aids, 2–3
  infection treatment, 5–6
Methanol, microbial utilization
  aerobic, 338–339
    biochemistry, 339–346
    cell production, 346–350
    microbiology, 338–339
  anaerobic, 350–355
    biochemistry, 352–354
    energetics, 354–355
    microbiology, 350–352
  single-cell protein, 355
    amino acids in, 362
    cell composition, 358–360
    economics, 356–358
    engineering, 360–361
Microbial rennet
  assays
    milk clotting, 41–43
    lipase, 43
    protease, 43
  bacterial rennet, 49–54
    Bacillus cereus, 52–53
    Bacillus mesentericus, 52
    Bacillus polymyxa, 51–52
    Bacillus subtilis, 53–54
    Serratia marcescens, 51
  coagulation process, 40–41
  fungal rennet, 56–61
    Entomophthorales, 54–55

P

Plant cell culture
  alkaloids from, 220–221
  antibiotics from, 223–224
  glycosides from, 220–221
  pigments from, 224–225
Prairie Regional Laboratory of the National Research Council, Canada
  amino acid research at, 424–425
  antibiotic research at, 427
  enzyme research at, 427–428
  ergot-type alkaloid research at, 425
  glycerol fermentation research at, 423
  historical aspects, 415–419
  microbiological research at, 419
  plant cell culture research at, 433–434
  ustilagic acid research at, 424

S

Sherry, volatile constituents from, 75–146

W

Whiskeys, volatile constituents from, 75–146
Wines, volatile constituents from, 75–146

CONTENTS OF PREVIOUS VOLUMES

Volume 1

Protected Fermentation
  Miloš Herold and Jan Nečásek

The Mechanism of Penicillin Biosynthesis
  Arnold L. Demain

Preservation of Foods and Drugs by Ionizing Radiations
  W. Dexter Bellamy

The State of Antibiotics in Plant Disease Control
  David Pramer

Microbial Synthesis of Cobamides
  D. Perlman

Factors Affecting the Antimicrobial Activity of Phenols
  E. O. Bennett

Germfree Animal Techniques and Their Applications
  Arthur W. Phillips and James E. Smith

Insect Microbiology
  S. R. Dutky

The Production of Amino Acids by Fermentation Processes
  Shukuo Kinoshita

Continuous Industrial Fermentations
  Philip Gerhardt and M. C. Bartlett

The Large-Scale Growth of Higher Fungi
  Radcliffe F. Robinson and R. S. Davidson

AUTHOR INDEX — SUBJECT INDEX

Volume 2

Newer Aspects of Waste Treatment
  Nandor Porges

Aerosol Samplers
  Harold W. Batchelor

A Commentary on Microbiological Assaying
  F. Kavanagh

Application of Membrane Filters
  Richard Ehrlich

Microbial Control Methods in the Brewery
  Gerhard J. Hass

Newer Development in Vinegar Manufactures
  Rudolph J. Allgeier and Frank M. Hildebrandt

The Microbiological Transformation of Steroids
  T. H. Stoudt

Biological Transformation of Solar Energy
  William J. Oswald and Clarence G. Golueke

SYMPOSIUM ON ENGINEERING ADVANCES IN FERMENTATION PRACTICE

Rheological Properties of Fermentation Broths
  Fred H. Deindoerfer and John M. West

Fluid Mixing in Fermentation Processes
  J. Y. Oldshue

Scale-up of Submerged Fermentations
  W. H. Bartholemew

Air Sterilization
  Arthur E. Humphrey

Sterilization of Media for Biochemical Processes
  Lloyd L. Kempe

Fermentation Kinetics and Model Processes
  Fred H. Deindoerfer

Continuous Fermentation
W. D. Maxon

Control Applications in Fermentation
George J. Fuld

AUTHOR INDEX — SUBJECT INDEX

Volume 3

Preservation of Bacteria by Lyophilization
Robert J. Heckly

Sphaerotilus, Its Nature and Economic Significance
Norman C. Dondero

Large-Scale Use of Animal Cell Cultures
Donald J. Merchant and C. Richard Eidam

Protection Against Infection in the Microbiological Laboratory: Devices and Procedures
Mark A. Chatigny

Oxidation of Aromatic Compounds by Bacteria
Martin H. Rogoff

Screening for and Biological Characterizations of Antitumor Agents Using Microorganisms
Frank M. Schabel, Jr., and Robert F. Pittillo

The Classification of Actinomycetes in Relation to Their Antibiotic Activity
Elio Baldacci

The Metabolism of Cardiac Lactones by Microorganisms
Elwood Titus

Intermediary Metabolism and Antibiotic Synthesis
J. D. Bu'Lock

Methods for the Determination of Organic Acids
A. C. Hulme

AUTHOR INDEX — SUBJECT INDEX

Volume 4

Induced Mutagenesis in the Selection of Microorganisms
S. I. Alikhanian

The Importance of Bacterial Viruses in Industrial Processes, Especially in the Dairy Industry
F. J. Babel

Applied Microbiology in Animal Nutrition
Harlow H. Hall

Biological Aspects of Continuous Cultivation of Microorganisms
T. Holme

Maintenance and Loss in Tissue Culture of Specific Cell Characteristics
Charles C. Morris

Submerged Growth of Plant Cells
L. G. Nickell

AUTHOR INDEX — SUBJECT INDEX

Volume 5

Correlations between Microbiological Morphology and the Chemistry of Biocides
Adrien Albert

Generation of Electricity by Microbial Action
J. B. Davis

Microorganisms and the Molecular Biology of Cancer
G. F. Gause

Rapid Microbiological Determinations with Radioisotopes
Gilbert V. Levin

The Present Status of the 2,3-Butylene Glycol Fermentation
Sterling K. Long and Roger Patrick

Aeration in the Laboratory
W. R. Lockhart and R. W. Squires

Stability and Degeneration of Microbial Cultures on Repeated Transfer
Fritz Reusser

Microbiology of Paint Films
Richard T. Ross

The Actinomycetes and Their Antibiotics
Selman A. Waksman

Fusel Oil
A. Dinsmoor Webb and John L. Ingraham

AUTHOR INDEX — SUBJECT INDEX

Volume 6

Global Impacts of Applied Microbiology: An Appraisal
Carl-Göran Hedén and Mortimer P. Starr

Microbial Processes for Preparation of Radioactive Compounds
D. Perlman, Aris P. Bayan, and Nancy A. Giuffre

Secondary Factors in Fermentation Processes
P. Margalith

Nonmedical Uses of Antibiotics
Herbert S. Goldberg

Microbial Aspects of Water Pollution Control
K. Wuhrmann

Microbial Formation and Degradation of Minerals
Melvin P. Silverman and Henry L. Ehrlich

Enzymes and Their Applications
Irwin W. Sizer

A Discussion of the Training of Applied Microbiologists
B. W. Koft and Wayne W. Umbreit

AUTHOR INDEX — SUBJECT INDEX

Volume 7

Microbial Carotenogenesis
Alex Ciegler

Biodegradation: Problems of Molecular Recalcitrance and Microbial Fallibility
M. Alexander

Cold Sterilization Techniques
John B. Opfell and Curtis E. Miller

Microbial Production of Metal–Organic Compounds and Complexes
D. Perlman

Development of Coding Schemes for Microbial Taxonomy
S. T. Cowan

Effects of Microbes on Germfree Animals
Thomas D. Luckey

Uses and Products of Yeasts and Yeast-like Fungi
Walter J. Nickerson and Robert G. Brown

Microbial Amylases
Walter W. Windish and Nagesh S. Mhatre

The Microbiology of Freeze-Dried Foods
Gerald J. Silverman and Samuel A. Goldblith

Low-Temperature Microbiology
Judith Farrell and A. H. Rose

AUTHOR INDEX — SUBJECT INDEX

Volume 8

Industrial Fermentations and Their Relations to Regulatory Mechanisms
Arnold L. Demain

Genetics in Applied Microbiology
S. G. Bradley

Micotoxins
Alex Ciegler and Eivind B. Lillehoj

Microbial Ecology and Applied Microbiology
Thomas D. Brock

The Ecological Approach to the Study of Activated Sludge
Wesley O. Pipes

Control of Bacteria in Nondomestic Water Supplies
Cecil W. Chambers and Norman A. Clarke

The Presence of Human Enteric Viruses in Sewage and Their Removal by Conventional Sewage Treatment Methods
Stephen Alan Kollins

Oral Microbiology
Heiner Hoffman

Media and Methods for Isolation and Enumeration of the Enterococci
Paul A. Hartman, George W. Reinbold, and Devi S. Saraswat

Crystal-Forming Bacteria as Insect Pathogens
Martin H. Rogoff

Mycotoxins in Feeds and Foods
Emanuel Borker, Nino F. Insalata, Colette P. Levi, and John S. Witzeman

AUTHOR INDEX — SUBJECT INDEX

Volume 9

The Inclusion of Antimicrobial Agents in Pharmaceutical Products
A. D. Russell, June Jenkins, and I. H. Harrison

Antiserum Production in Experimental Animals
Richard M. Hyde

Microbial Models of Tumor Metabolism
G. F. Gause

Cellulose and Cellulolysis
Brigitta Norkrans

Microbiological Aspects of the Formation and Degradation of Cellulosic Fibers
L. Jurášek, J. Ross Colvin, and D. R. Whitaker

The Biotransformation of Lignin to Humus — Facts and Postulates
R. T. Oglesby, R. F. Christman, and C. H. Driver

Bulking of Activated Sludge
Wesley O. Pipes

Malo-lactic Fermentation
Ralph E. Kunkee

AUTHOR INDEX — SUBJECT INDEX

Volume 10

Detection of Life in Soil on Earth and Other Planets. Introductory Remarks
Robert L. Starkey

For What Shall We Search?
Allan H. Brown

Relevance of Soil Microbiology to Search for Life on Other Planets
G. Stotzky

Experiments and Instrumentation for Extraterrestrial Life Detection
Gilbert V. Levin

Halophilic Bacteria
D. J. Kushner

Applied Significance of Polyvalent Bacteriophages
S. G. Bradley

Proteins and Enzymes as Taxonomic Tools
Edward D. Garber and John W. Rippon

Transformation of Organic Compounds by Fungal Spores
Claude Vézina, S. N. Sehgal, and Kartar Singh

Microbial Interactions in Continuous Culture
Henry R. Bungay, III and Mary Lou Bungay

Chemical Sterilizers (Chemosterilizers)
Paul M. Borick

Antibiotics in the Control of Plant Pathogens
M. J. Thirumalachar

AUTHOR INDEX—SUBJECT INDEX

CUMULATIVE AUTHOR INDEX—CUMULATIVE TITLE INDEX

Volume 11

Successes and Failures in the Search for Antibiotics
Selman A. Waksman

Structure–Activity Relationships of Semisynthetic Penicillins
K. E. Price

Resistance to Antimicrobial Agents
J. S. Kiser, G. O. Gale, and G. A. Kemp

Micromonospora Taxonomy
George Luedemann

Dental Caries and Periodontal Disease Considered as Infectious Diseases
William Gold

The Recovery and Purification of Biochemicals
Victor H. Edwards

Ergot Alkaloid Fermentations
William J. Kelleher

The Microbiology of the Hen's Egg
R. G. Board

Training for the Biochemical Industries
I. L. Hepner

AUTHOR INDEX—SUBJECT INDEX

Volume 12

History of the Development of a School of Biochemistry in the Faculty of Technology, University of Manchester
Thomas Kennedy Walker

Fermentation Processes Employed in Vitamin C Synthesis
Miloš Kulhánek

Flavor and Microorganisms
P. Margalith and Y. Schwartz

Mechanisms of Thermal Injury in Nonsporulating Bacteria
M. C. Allwood and A. D. Russell

Collection of Microbial Cells
Daniel I. C. Wang and Anthony J. Sinskey

Fermentor Design
R. Steel and T. L. Miller

The Occurrence, Chemistry, and Toxicology of the Microbial Peptide-Lactones
A. Taylor

Microbial Metabolites as Potentially Useful Pharmacologically Active Agents
D. Perlman and G. P. Peruzzotti

AUTHOR INDEX—SUBJECT INDEX

Volume 13

Chemotaxonomic Relationships Among the Basidiomycetes
Robert G. Benedict

Proton Magnetic Resonance Spectroscopy — An Aid in Identification and Chemotaxonomy of Yeasts
P. A. J. Gorin and J. F. T. Spencer

Large-Scale Cultivation of Mammalian Cells
R. C. Telling and P. J. Radlett

Large-Scale Bacteriophage Production
K. Sargeant

Microorganisms as Potential Sources of Food
Jnanendra K. Bhattacharjee

Structure–Activity Relationships Among Semisynthetic Cephalosporins
M. L. Sassiver and Arthur Lewis

Structure–Activity Relationships in the Tetracycline Series
Robert K. Blackwood and Arthur R. English

Microbial Production of Phenazines
J. M. Ingram and A. C. Blackwood

The Gibberellin Fermentation
E. G. Jefferys

Metabolism of Acylanilide Herbicides
Richard Bartha and David Pramer

Therapeutic Dentifrices
J. K. Peterson

Some Contributions of the U.S. Department of Agriculture to the Fermentation Industry
George E. Ward

Microbiological Patents in International Litigation
John V. Whittenburg

Industrial Applications of Continuous Culture: Pharmaceutical Products and Other Products and Processes
R. C. Righelato and R. Elsworth

Mathematical Models for Fermentation Processes
A. G. Fredrickson, R. D. Megee, III, and H. M. Tsuchiya

AUTHOR INDEX – SUBJECT INDEX

Volume 14

Development of the Fermentation Industries in Great Britain
John J. H. Hastings

Chemical Composition as a Criterion in the Classification of Actinomycetes
H. A. Lechevalier, Mary P. Lechevalier, and Nancy N. Gerber

Prevalence and Distribution of Antibiotic-Producing Actinomycetes
John N. Porter

Biochemical Activities of Nocardia
R. L. Raymond and V. W. Jamison

Microbial Transformations of Antibiotics
Oldrich K. Sebek and D. Perlman

In Vivo Evaluation of Antibacterial Chemotherapeutic Substances
A. Kathrine Miller

Modification of Lincomycin
Barney J. Magerlein

Fermentation Equipment
G. L. Solomons

The Extracellular Accumulation of Metabolic Products by Hydrocarbon-Degrading Microorganisms
Bernard J. Abbott and William E. Gledhill

AUTHOR INDEX – SUBJECT INDEX